Continuous Time Active Analog Filters

Filter circuits play a vital role in diverse applications such as audio/video signal processing, biomedical devices, instrumentation and control. The replacement of many analog circuits by digital filters and modules gave the impetus for further advancements in continuous-time as well as discrete-time analog filters. Today there are important applications such as analog simulation for retransmission, power quality modulation, and prevention of signal-noise-ration problems in multistage amplifiers where analog filters are either essential or preferred over digital filters.

This book is conceived as base material for a first course on active network synthesis at the advanced undergraduate level. It presumes that the reader has already studied basic network theory and analog electronics. After studying this book the student should be able to select an appropriate design technique for a specific application and also be equipped for advanced studies on continuous-time, switched capacitor or current mode filters.

Focusing mainly on continuous-time domain techniques, this text discusses a wide spectrum of topics. It discusses active filter circuits and their analysis in both frequency and time domains. It includes practical application-based examples and case studies on topics including medical instrumentation, audio/video signal processing, anti-aliasing and signal conditioning modules. More than 130 solved examples and 370 design problems are interspersed throughout the book. Graphs are plotted throughout the text using the computer simulation tool PSpice.

Muzaffer Ahmad Siddiqi retired as Professor and Chairman from the Department of Electronics Engineering at Aligarh Muslim University, India. His research includes linear analog integrated circuits, digital integrated circuits, VLSI design, active network synthesis, microelectronics, and analog signal processing. He has over 43 years of teaching experience and is the author of the book *Dynamic RAM: Technology Advancements*.

Continuous Time Active
Analog Filters

Muzaffer Ahmad Siddiqi

CAMBRIDGE
UNIVERSITY PRESS

CAMBRIDGE
UNIVERSITY PRESS

University Printing House, Cambridge CB2 8BS, United Kingdom

One Liberty Plaza, 20th Floor, New York, NY 10006, USA

477 Williamstown Road, Port Melbourne, VIC 3207, Australia

314–321, 3rd Floor, Plot 3, Splendor Forum, Jasola District Centre, New Delhi – 110025, India

79 Anson Road, #06–04/06, Singapore 079906

Cambridge University Press is part of the University of Cambridge.

It furthers the University's mission by disseminating knowledge in the pursuit of education, learning and research at the highest international levels of excellence.

www.cambridge.org
Information on this title: www.cambridge.org/9781108486835

First published in 2020

Printed in India by Nutech Print Services, New Delhi 110020

A catalogue record for this publication is available from the British Library

ISBN 978-1-108-48683-5 Hardback

Additional resources for this publication at www.cambridge.org/9781108486835

Cambridge University Press has no responsibility for the persistence or accuracy of URLs for external or third-party internet websites referred to in this publication, and does not guarantee that any content on such websites is, or will remain, accurate or appropriate.

To the memory of my parents

Contents

Preface

Recent advances in solid-state technology and the well-known advantages of integrated circuits (ICs) have made microminiaturized filters and systems an important field of study. For a particular application, the main object of the designer is to select/design a circuit, which is optimized to maximize performance and minimize cost through economic integration. At present, several methods are available to adapt the design of linear networks. Among these, networks (filters) employing resistance, capacitance, and linear active devices, i.e., active RC filters, have become very popular because of their functional versatility, ease of IC implementation and cost reduction. Now, good quality active RC filters are being produced using integrated operational amplifiers (OAs), mainly because of the commercial availability of high quality OAs at low cost.

It is for this reason that the majority of the, presently, available books on linear active filter design and the curriculum at almost all universities include OA-RC based circuit and systems design either as a full course or part of a course on linear integrated circuit design. However, it is only part of a story, which began with passive electric filter design. Approximation was used to develop filter structures, predominantly in the form of either singly or doubly terminated lossless ladders. Great strides were made in developing ladder structures and related information for different approximations, and related information like their transfer function in terms of ratio of polynomials or in terms of poles and zeros, and required values of elements used in the ladders. Not only are passive filter realizations still used today, but the large literature on it also finds application in active network synthesis. Hence, a discussion on passive filter design and properties of ladder structure finds place in books on network analysis/synthesis.

Development of filter circuits from passive to active did not stop at the OA-RC case. The simultaneous development of digital filters and systems, and the replacement of analog sub-systems with digital sub-systems, motivated further advancements in continuous-time as well as discrete-time analog filters. Employing the design procedures used with OA-RC filter synthesis, high speed sampling circuits such as the switched capacitor (SC) provided good alternatives to digital filters in many applications. As a result, SC filter design became an entity which needed a separate (and specialized) study and followed procedures parallel to that of the active RC case; it had its own techniques as well. However, for a large number of applications, continuous-time techniques were preferred, because they did not need anti-aliasing or smoothing filters; moreover, no clock and associated issues, such as clock feed-through, fold-over noise, etc., are present. Though, in some cases, continuous-time filter design needed specially designed OAs or devices other than OAs, such as operational transconductance amplifiers (OTAs), which are basically current mode (CM) devices. Further boost to CM filtering came with the use of current conveyors (CCs) and related devices; the application of CM devices such as CCs was researched extensively in the 1970s. Recently, other CM devices like current-feedback operational amplifiers (CFOAs) and current differencing (CDTA) have become available, creating further interest in CM filters. These devices have created another specialized field of active filter, for studies.

Hence, the subject of electric filter design has become very wide including passive filter design and its impact on active filter design, active RC filter design which is mainly OA-RC based, SC filter design and CM filter design with a number of new active devices. Of course, each area of study is not entirely separate and has many overlapping regions. It can be safely said that techniques employed with active RC filters using OA are fundamental for further studies towards the in-depth investigation of OA-RC filter design and/or SC and CM filters. Courses need to be developed as advances are made in techniques and curricula must evolve for fundamental and advanced courses.

It is obvious that such a wide area cannot be covered even in two courses of 3/4 credits. However, practically speaking, with the ever expanding areas of important studies, there is pressure to find even one 3/4 credit slot for specialized subjects like analog active filter design in a four-year undergraduate engineering curriculum. Based on the aforementioned constraint, the major aim of this book is to give enough material on continuous-time active filters using OAs, so that it provides a good understanding of the core material of active filter design in a one-semester course. Only an introduction to SC filters and some material on CM filters is included to provide connectivity with the active RC filter synthesis.

The book is written for use in a first course on active network synthesis at the advanced undergraduate level. Hence, it is assumed that the reader has already studied basic network theory and analog electronics. For example, basic filter classification, OA's characteristics, and so on, have been only briefly reviewed in Chapter 1. It needs mentioning that the material given in these 17 chapters is a bit more than can be covered in a one-semester course and the individual instructor has the choice to leave some chapters partly or fully.

A significant contribution in the book is in the form of practical applications of the continuous-time analog filter. It is expected that this component will be well received by the students and will act as a motivator for taking up studies in the area of analog filters. In addition to two chapters on practical applications, more examples are also included in other chapters.

Chapters 1 and 2 provide an introduction to the subject and related aspects of the active device (OA) and components commonly used in the realization of active filters. The chapters also review the classification of filters on the basis of signal frequency separation and the basics of different approaches to active filter synthesis like the *direct form* and the *cascade form*. Description of the characteristics of the first-order and second-order low pass (LP), band pass (BP), high pass (HP), band elimination (BE) and all pass (AP) filtering, their basic parameters, and effect of the frequency dependent gain of the OA on filter circuits is discussed for second-order sections as well.

Study of Butterworth-maximally flat, equal ripple and elliptic approximations have been undertaken in Chapter 3. Finding locations of the transfer function poles and normalized element values for different approaches have been discussed. Also included is an introduction to finite zeros with a maximally flat pass band. Since all the approximations are initiated in LP form, transformations are required to apply these approximations to other type of responses like BP, HP and BE responses. Chapter 4 deals with this issue; impedance scaling is also included to complement the topic of frequency scaling. A simpler alternative to network transformation is also discussed for converting an LP passive prototype to other responses.

A significant issue in the form of designing delay filters is studied in Chapter 5, which was not given distinct coverage while magnitude approximation was discussed. Bessel approximation filters, which are more suited for controlling delays, are studied. Delay equalization using first- and second-order AP filters have also been studied in this chapter.

Before taking up different synthesis methods, sensitivity studies of active networks are discussed in Chapter 6. Beginning with some simpler aspects of sensitivities, like incremental sensitivity, the discussion covers transfer function sensitivity and sensitivity of second- and higher-order functions.

In Chapters 7 and 8, important topics related to the realization of second-order sections is taken up employing single and multi amplifiers, respectively. Single amplifier sections, employing single-feedback and multiple-feedback while using single input as well as differential input are discussed. General active RC feedback approach leading to some well-known biquadratics is explained. In Chapter 8, the state variable multi amplifier biquad and active compensation of integrators is studied. Development of the general biquadratic section using the conventional summation method and modified summation method is included. Also included are generalized impedance converter (GIC) based second-order sections and development of the general biquadratic section from it.

After laying the foundation and discussing the issues important for active RC filter design, other important approaches in the direct form of synthesis, like element substitution and operational simulation, are studied in Chapter 9. Since element substitution is mainly concerned with the inductance fabrication problem in IC form, it is taken up in detail, followed by frequency dependent negative resistance (FDNR) simulation and their utilization in filter realization. Instead of applying direct substitution of inductance (or FDNR), an alternate approach is also taken up which is known as the *operational simulation* scheme. It is based on the modeling of circuit equations and current–voltage relations in the circuit. The technique is illustrated with the simulation of an LP ladder, a BP ladder and a general ladder.

The basics of an important higher filter design approach, namely the cascade design, is discussed in Chapter 10. Optimization of the process, aiming at increasing the filter dynamic range, is discussed. Also included is an introduction to the tuning of second-order filter sections, which are the basic building blocks (BBBs) employed in the cascade approach.

Chapters 11 and 12 are devoted to practical application examples employing continuous-time active analog filters and some case studies. Only some of the more common areas of application of analog filters are included. Chapter 11 includes some simple signal processing biomedical applications. An introduction to instrumentation amplifiers is also given as it is an essential component in such applications. Different aspects of audio signal processing and the essential requirements with widespread usage of analog filters in such applications are discussed in

Chapter 12. The latter part of the chapter shows the importance of analog low pass filters as anti-aliasing and reconstruction filters in otherwise digital signal processing.

Another useful direct form of filter realization known as follow the leader feedback (FLF) has been described in Chapter 13. Its basic idea, structure and its different approaches are discussed. Introduction of finite transmission zeros through the use of feed-forward path is discussed in the *shifted companion* form. Also included is the general FLF structure study leading to the *primary resonator block* technique.

Switched capacitor (SC) circuit(s) forms an important study in itself and some good books are available exclusively on it or along with continuous-time filters. As these are only approximated to continuous-time filters to a certain degree, its study is suitable in sampled-data form; hence, only an introduction is given here in Chapter 14.

In Chapter 15, filter realization using OTAs and capacitors is discussed. BBBs, like simulation of resistance, integrators, voltage amplification and current and voltage addition are illustrated with examples. The conversion process from single-ended to differential output is studied and applied to OTA based filters. First- and second-order sections using two-integrator loops are included. Simulation of inductors is used for the element substitution process, and operation simulation process is also studied to realize OTA-C based higher-order filters.

Recently, CM filter realization has gained momentum and needs to be studied in a specialized form. Though OTA is also a current source device and CM filters have been obtained using OTAs, many other CM devices are being used; CC and its variations being one of the most commonly used devices. Chapter 16 begins with a brief depiction of CCII and some BBBs using CCs. First-order CM filters and second-order filter realizations with different approaches are discussed. Biquadratic section developments using one, two and more CCs are shown. Inductance and FDNR simulation is explained for application to direct form synthesis. CDTA, another CM device, has also been shown to realize CM elements and filter circuits.

In Chapter 17, the frequency dependent model of OAs is used to obtain active R and active C circuits, using OAs and resistances, and OAs and capacitances, respectively. Their advantages lie in their ability to extend the operating frequency range and obtain filter parameters in terms of ratio of resistances/capacitances. In the direct form of element substitution approach, the grounded and floating form of all the passive elements are simulated and then employed for filter realization. First-order sections and second-order filter sections are studied and then used in the cascade design in active R and active C form. Since critical frequency of these filters depends on the value of gain bandwidth product of OAs, which has large tolerance, available methods to minimize errors on this count are also briefly mentioned.

I have tried to follow a middle path between depth and width of the subject. It became necessary to limit the number of topics so that adequate detail for understanding a topic could be provided, and the undergraduate student remains interested in the subject. It is hoped that after completing a course based on the text, the student is able to design active filters and become qualified enough for further advanced studies on continuous-time or SC filters or CM filters.

The solutions manual for practice problems in this book can be accessed at *www.cambridge.org/9781108486835*. Your suggestion for further improvements are welcome. Kindly write to me at *siddiqima48@gmail.com*.

Acknowledgments

I am thankful to my former colleagues at the Electronics Engineering Department of Aligarh Muslim University, India, who provided a lot of encouragement and assistance to me. I am not mentioning everyone by name, but my gratitude is nonetheless sincere. The book would not have been completed without their helpful suggestions. I would also like to acknowledge the contribution of my students who provided feedback over several years when the subject was taught.

Contents of Chapter 17 are mostly from two PhD theses: one is my own and the other is of my former colleague Professor I. A. Khan, who graciously permitted use of the material. Drawing of a large number of figures was a very important part of the project and special thanks are due to Mr Mohammad Sulaiman, technical staff of the department, who spent many evenings on the job.

I am thankful to a number of companies and corporates who permitted the use of figures and data from their data sheets and for furnishing valuable information that was required in the context of providing practical examples of active analog filters.

It is important to express my gratitude for the support and encouragement to the editorial staff, and specially to Mr Gauravjeet Reen at the initial stage of the project and then Ms Taranpreet Kaur at the latter stage, at Cambridge University Press.

I would be failing if I do not acknowledge the support of my wife, Shagufta, and children Zeba, Subuhi, Soofia, Bushra, and Belal.

Chapter 1

Analog Filter: Concepts

1.1 Introduction

The study of electric network theory has two related but distinct spheres, namely *network analysis* and *network synthesis*. The two terms are easily identified in terms of the *network*, the *excitation* or input, and the *response* or output as shown in Figure 1.1. Here *network* means some combination of passive elements, such as the resistor, the capacitor and the inductor, and (not necessarily) dependent energy sources. The *excitation* or input is an electrical energy source connected to the network. The *response* or output is observed in different forms such as voltage across certain element(s), current through certain element(s), or energy dissipated in a resistor. This response is observed either in the *time domain* or in the *frequency domain*. In the time domain, the output voltage/current variation is observed/measured with respect to time, whereas in the frequency domain, the same observation is taken in terms of frequency.

Figure 1.1   Network: a physical entity comprising passive (and active) components.

When the response of a given network to a certain excitation is observed/evaluated, it is known as *network analysis*; whereas, the process of finding a network for a given excitation and response is known as *network synthesis*. A significant difference between analysis and synthesis is that while analyzing the network, for a certain input or excitation, the response solution is unique; however, during the network synthesis process, it may be possible to obtain a number of solutions and even no solution for a given combination of excitation and response. It is this possibility of obtaining many solutions which has led to the development of a large number of synthesis procedures. Classification of the various synthesis methods shall be discussed in Section 1.5. Sensitivity concept, being an important issue in the practical utilization of filters, is introduced in Section 1.6 and discussed in more detail in a separate chapter, Chapter 6

The basic principles behind the classification of various types of network functions that filter signals on the basis of their frequency, and the terminologies used, are explained in Sections 1.2 and 1.3. An important stage of conversion/evolution from passive filters (networks which use only passive elements) to active filtering (using at least one active device) is discussed in Section 1.4.

One very important reason for changing from passive filters to active filters in the integrated circuit (IC) form was the inability of realizing practically feasible inductors in passive filters. For a long time, active RC (resistance–capacitance) filter structures using resistance, capacitance and operational amplifiers (OAs) were synonymous to active filters. Though the usage of OA is still prevalent, other active devices are also being used now in a big way.

It is assumed that the reader is familiar with OAs; hence, only a brief discussion about these amplifiers is included in Section 1.7. The application of OA as an integrator is reviewed in Section 1.8. A brief discussion about the characteristics of resistors and capacitors fabricated in the IC form is also included in Section 1.9 with the assumption that a detailed study of the realization of these passive components is undertaken in the separate subject 'microelectronics' or an allied subject.

1.2   Network Functions

External connections from a network to another network or excitation and response observations are made through terminals or ports: terminals are commonly paired to form a *port*. For example, in Figure 1.2, there are four terminals in the network forming two ports.

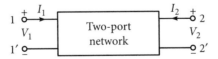

Figure 1.2   A two-port network showing conventional direction of current and voltage references.

For a single-port network, only one voltage and one current are identified and only one network function (and its reciprocal) is defined, which is given as:

$$Z(s) = V(s)/I(s) \qquad (1.1)$$

This function is known as the *driving-point impedance function* because the terminals are connected to the driving force or energy source. Generally, it is simply known as the impedance of a network. Reciprocal of the impedance function is known as the *driving-point admittance function* or simply admittance, $Y(s)$.

For the two-port network in Figure 1.2, two currents and two voltages must be defined as shown. The four quantities $V_1(s)$, $I_1(s)$, $V_2(s)$ and $I_2(s)$ can be related to each other in six different forms (and their six reciprocals), when only two quantities are taken at a time. For example, in Figure 1.2, the driving point impedance at input terminal port 1 for a passive termination at port 2 is given as:

$$Z_{11}(s) = V_1(s) / I_1(s) \tag{1.2}$$

Similarly, the driving point impedance at port 2 for a passive termination at port 1 is given as:

$$Z_{22}(s) = V_2(s) / I_2(s) \tag{1.3}$$

Example 1.1: Find the driving point impedance function for the network shown in Figure 1.3.

Solution: As input voltage V_1 is connected across the two impedances Z_1 and Z_2 in parallel, total impedance shall be the parallel combination of these two impedances.

$$Z_1 = \frac{s}{2} + \frac{3}{2s} = \frac{s^2 + 3}{2s} \text{ and } Z_2 = \frac{s}{8} + \frac{5}{8s} = \frac{s^2 + 5}{8s}$$

$$\frac{1}{Z} = \frac{1}{z_1} + \frac{1}{Z_2} \rightarrow Z(s) = \frac{s^4 + 8s^2 + 15}{10s^3 + 34s} \tag{1.4}$$

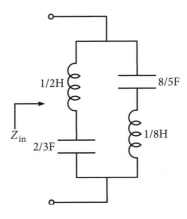

Figure 1.3   A simple LC circuit for illustrating definition of driving point function.

Driving point impedance is a ratio of polynomials expressed in the complex frequency variable s. The degree of the numerator is one more that of the denominator as shown as an example in equation (1.4).

1.2.1 Transfer function

Consider Figure 1.2. Four out of the possible six network functions, which are defined for the two-port network, are such that they relate a voltage or a current of one port to the voltage or current of the other port. These functions are defined for a given driving force at the input port and a terminating condition at the output port; they are known as *transfer functions*. For the symbolic representation of the transfer function, a convenient convention is for the function to be subscripted first by the input and then by the output; for example, the voltage ratio transfer function is defined by the following relation:

$$V_2(s) \,/\, V_1(s) = G_{12}(s) \tag{1.5}$$

The current ratio transfer function is given by:

$$-I_2(s) \,/\, I_1(s) = \alpha_{12}(s) \tag{1.6}$$

The transfer impedance function and the transfer admittance function are, respectively, as follows:

$$V_2(s) \,/I_1(s) = Z_{12}(s) \text{ and } -I_2(s)/V_1(s) = Y_{12}(s) \tag{1.7}$$

In equations (1.6) and (1.7), $I_2(s)$ has a negative sign because of the conventionally chosen direction of the second-port current going into the port.

Since each transfer functions is a ratio of the Laplace transform of current/voltage, these functions are quotients of rational polynomials in the complex frequency variable s. For example, looking at Figure 1.4, the network's description given in the time domain is stated as follows:

$$L\left(di \,/\, dt\right) + iR + \left(1\,/\,C\right)\int i\,dt = v_1\left(t\right) \tag{1.8}$$

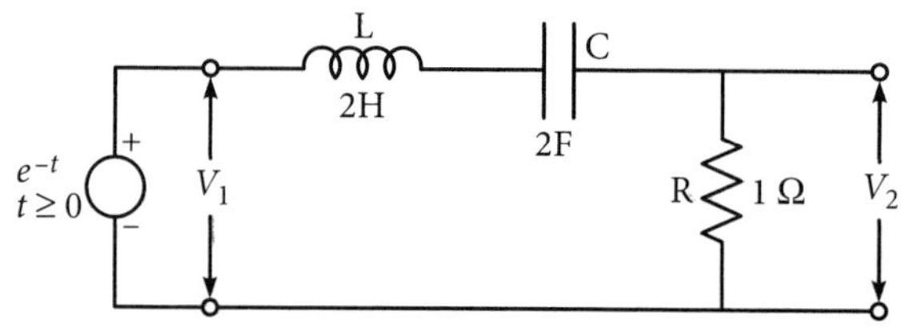

Figure 1.4 A simple RLC circuit for illustrating voltage-ratio transfer function.

Taking the Laplace transform of equation (1.8) for the given element values of the circuit, we get:

$$\left(2s+1+\frac{1}{2s}\right)I(s)=\frac{1}{(s+1)}$$
(1.9)

As $V_2(s) = I(s) \times 1$, and $V_1(s) = 1/(s + 1)$, we get the following expression for the voltage ratio transfer function:

$$V_2(s) / V_1(s) = (2s)/(4s^2 + 2s + 1)$$
(1.10)

It is important to note that in equation (1.10), no component of the driving input or excitation contributes to the expression of the transfer function; in fact, the transfer function depends only on the element values and their interconnections. A general form of the transfer function may be expressed as a ratio of polynomials in the complex frequency s.

$$H(s)=\frac{N(s)}{D(s)}=\frac{a_m s^m + a_{m-1}s^{m-1}+\ldots\ldots\ldots+a_1 s+a_0}{b_n s^n + b_{n-1}+s^{n-1}+\ldots\ldots\ldots\ldots+b_1 s+b_0}$$
(1.11)

In equation (1.11), with $i = 0$ to m and $j = 0$ to n, the coefficients a_i and b_j are real constants, m is the degree of the numerator and n is the degree of the denominator. Factorization of both the polynomials $N(s)$ and $D(s)$ gives the following important form of the transfer function.

$$\frac{N(s)}{D(s)}=\frac{a_m}{b_n}\frac{(s-z_1)(s-z_2)\ldots\ldots\ldots(s-z_{m)}}{(s-p_1)(s-p_2)\ldots\ldots..(s-p_n)}$$
(1.12)

Here, the roots of the numerator, $z_1, z_2, \ldots, z_m$ are called *zeros*, and the roots of the denominator $p_1, p_2, \ldots, p_n$ are called *poles* of the transfer function, respectively. Consideration of zeros and poles of a network function is extremely important as they describe the behavior of the network function in the frequency domain.

1.3   Basic Filtering Action

As mentioned earlier, network synthesis is the process of finding circuit elements, such that their values and interconnections conforms to the given relation between excitation and response. Since excitation and response are both conveniently represented by the transfer functions and y or z functions, the first step in the design of a *filter* is to convert the specified magnitude, phase or related entity as a function of frequency to a form of the transfer function. Electronic filters are circuits that perform signal processing in such a way that it separates signals on the basis of frequency. Ideally, the filter suppresses signals in a certain band of frequency, known as *stop band,* completely and passes signals in other bands of frequency, known as *pass bands,* without any attenuation or gain. Practically, however, it is not possible to suppress a signal completely in the stop band, though the suppression is not less than a permissible level, and signals may not pass without some attenuation in the pass band (unless intentional gain is added).

1.3.1   Types of filters

There are four commonly used types of filters: they are classified based on the terminology of pass band and stop band mentioned in the previous section. The types as shown in Figure 1.5(a–d) and are defined as follows:

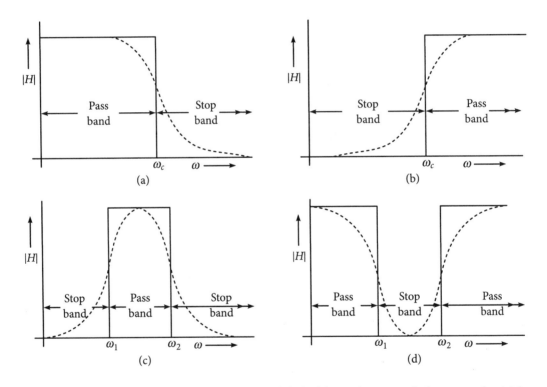

Figure 1.5   Firm lines show ideal response and dashed lines shows practical response for (a) low pass, (b) high pass, (c) band pass, and (d) band stop filters.

Low Pass Filters: When the frequency range of the pass band is from $\omega = 0$ to $\omega = \omega_c$ and the stop band extends from $\omega = \omega_c$ to ∞, the filter is known as a *low pass filter* (LPF) as shown in Figure 1.5(a). Here, ω_c is called the *cut-off frequency.*

LPFs are widely used in many diverse applications. For example, in acoustics, LPFs are used to filter out high frequency signals from the transmitting sound that would otherwise cause echo at higher sound frequencies. In audio speakers, LPF reduces the high frequency hiss sound produced in the system and inputs the clearer sound to the sub-woofers. LPFs are also used in equalizers and audio amplifiers. An important application of LPF is as an anti-aliasing filter placed before the analog to digital convertor (ADC) and as a reconstruction filter placed after the ADC. In a variety of applications, including those in medical instrumentation, LPFs are used to eliminate high frequency noise.

High Pass Filter: The complement of LPF is the *high pass filter* (HPF), where the stop band ranges from $\omega = 0$ to ω_c and the pass band extends from $\omega = \omega_c$ to ∞, as shown in Figure 1.5(b).

HPFs can be used wherever noise at low frequencies is to be eliminated, such as in medical instrumentation and audio systems. They find applications in loudspeakers to reduce the low-frequency noise. HPFs are also employed in those applications where high frequency signals are to be amplified. For example, a popular application is the *treble boost*. In some applications, combinations of HPFs with LPFs yield band pass filters or band stop filters. HPFs are used to pass high frequency signals to a tweeter and to block interfering and potentially damaging signals to loudspeakers. They are also used extensively in crossover of audio signals.

Band Pass Filters: In a *band pass filter* (BPF), the pass band ranges from ω_1 to ω_2; signals in the rest of the frequency range, which is known as the stop band, are stopped as shown in Figure 1.5(c).

BPFs are extensively used in all types of instruments, such as seismology and medical instruments like electroencephalograms and electrocardiograms. There is widespread use of BPFs in audio signal processing, where signals in a certain frequency range are to be passed (maybe with amplification) and signals in the remaining band of frequencies are to be rejected. In communication systems, the transmitter as well as the receiver employs BPFs to avoid interference from unwanted signals.

Band Stop Filter: A *band stop filter* (BSF) (or a notch filter) complements BPF with the stop band ranging between ω_1 and ω_2 and rest of the frequency band is the pass band as shown in Figure 1.5(d).

These filters are mainly used in public address systems and speaker systems to ensure rejection of power supply frequency interference. Similarly, BSFs are crucial for line noise reduction in telephonic signal transmission. These filters are also employed in many electronic communication devices to eliminate/reduce interference from harmonics. BPFs are used in the medical instruments, such as electrocardiogram machines, to reject unwanted signals.

In Figure 1.5, straight lines show the ranges of pass and stop bands for ideal filters. It has been proved mathematically in literature that such ideal characteristics are not realizable exactly with finite number of elements [1.1]. The approximated responses for the filter, which are practically realizable, are shown as dotted lines in Figure 1.5. The transfer function for such practically realizable filter networks are described by the *real rational function* given in equation (1.11). It is important to note that the transfer function expressed as the ratio of polynomials in the complex frequency s must satisfy certain conditions for the filter to be realizable practically. For example, the coefficients in the two polynomials in equation (1.11), $N(s)$ and $D(s)$, a_i, $i = 0$ to m and b_j, $j = 0$ to n are should be real numbers. In addition, the degree of the denominator n must be larger than or at least equal to the degree of the numerator, m, that is, $n \geq m$. Magnitude of the transfer function in equation (1.11) when evaluated at the $j\omega$-axis, that is, $|H(j\omega)|$, is a continuous function of frequency ω as shown by the dotted lines in Figure 1.5(a–d), rather than the characteristics given by the solid lines. It is to be noted that the sharpness of the curves in Figure 1.5 depends on the value of the coefficients in the denominator $D(s)$; hence, sharpness can be controlled by the designer.

It is a common practice in filter circuit design to represent the characteristics of the circuit in terms of linear output signal magnitude, or gain characteristic $|H(j\omega)|$ or in terms of the logarithmic attenuation characteristic $\alpha(\omega)$. Attenuation characteristics of the main four types of filters, corresponding to Figure 1.5(a–d), are shown in Figure 1.6(a–d). Here, logarithmic attenuation $\alpha(\omega)$ is related to the gain magnitude as follows:

$$\alpha(\omega) = 20 \log |H(j\omega)| \text{ dB} \tag{1.13}$$

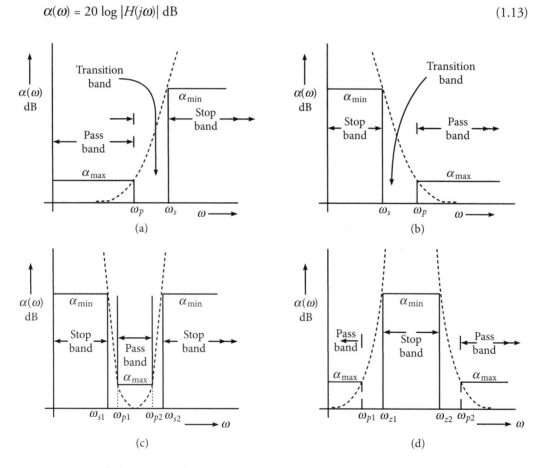

Figure 1.6   Ideal and practical characteristics in term of logarithmic attenuation for the (a) low pass, (b) high pass, (c) band pass, and (d) band stop filters.

It is to be noted that in the normal course, α will be positive for $|H(j\omega)| > 1$ and negative for $|H(j\omega)| < 1$. However, in filter design, the negative sign for α is normally not used even when

$|H(j\omega)| < 1$. It is the context which informs the analyzer whether positive α implies gain during filtering or attenuation during loss in signal magnitude. The negative sign for α is used only when it becomes necessary to avoid any confusion.

Filter characteristics of Figure 1.5, which are drawn again in Figure 1.6, in terms of attenuation, show clearly that for a practical filter, the boundary of the pass and the stop band is separated by a definite region, which is called the *transition band*. This is not so in the case of the ideal filter response, where transition from the pass band to the stop band is instant and no transition region exists. It is the width of this transition region which decides the sharpness of the filter characteristic, an important parameter in filter design.

Characteristics of important four types of filter shown in Figure 1.6 are shown as solid lines and continuous dotted lines. The required practical characteristics are shown by the continuous dotted lines; however, the characteristics are not fixed, in the sense that they are governed by two parameters α_{max} and α_{min} shown on the solid lines. Here, α_{max} is the maximum attenuation and implies that the attenuation should always be less than this value in the pass band; α_{min} is the minimum attenuation and implies that the attenuation should always be more than this value in the stop band. Hence, the most common method of providing specifications for a filter is to mention α_{min}, α_{max} and the edge frequencies of the pass band and stop band ω_1 and ω_2, respectively. The mathematical process of finding the appropriate transfer function $H(s)$ in the form of equation (1.11) is known as *approximation*. There are different methods of approximations: classical forms, which are discussed in Chapter 3, and other methods based on the approximating phase or delay requirements of a filter, which are discussed in Chapter 4.

Another significant point to note is that the level of attenuation $\alpha(\omega)$ has been shown to reach the level of zero dB. In reality, the level may be at zero dB, below or above zero dB depending on the kind of circuitry used for filter realization and whether a gain (active) device has been used in the circuit or not. However, this level of attenuation does not violate the specifications α_{min}, α_{max} or pass band or stop band edge frequencies. Therefore, absolute value of the attenuation in the pass band is not a worrying factor for the designer as shall be evident later in the examples.

1.4   From Passive to Active Filtering

Over the years, passive elements have been used extensively for building *passive filters*. The filter could use all the three elements, resistance (R), inductance (L), and capacitance (C), or only two elements like RC, LC, or RL. Excellent literature is available on the conditions of realization, and methods that enable the realization of such passive filters [1.1]. There are quite a few advantages that such filters offer.

i.   Passive filters do not require any power supply for operation.

ii.   Two-port networks employing only LC components, commonly known as *lossless* networks as they did not dissipate any power, were found to be very useful as they could provide sharp transition between stop band and pass band.

iii. One of the most common forms of passive realization that was, and is still used, the *ladder structure*. LC ladder structures have been studied extensively and data, such as structural combination, element values, poles, and zero location for their transfer function, are available for different types of approximated filters [1.2]. One of the main advantages of the LC ladders is their low *element sensitivity* in the pass band. This means that realized filter parameters are comparatively much less affected due to changes (unavoidable or intentional) in the values of the elements used.

iv.   Passive filters can and are still used at some higher frequencies where use of inductance is not very problematic and other types of filters face limitations.

v.   Study of passive filters, especially the ladder structure, is important from the point of view of *active filter* realizations as well, as many active realizations use passive ladder structures as the starting point of synthesis because of their excellent sensitivity properties.

One of the major limitations of passive filters lies in its use of inductors, which are coils of wires on some kind of core material. Not only are the inductors bulky, radiate electromagnetic energy, and can result in parasitic mutual inductance, they are also practically not realizable in integrated circuit (IC) form. With the increase in the use of electronic filters in a big way and advances in technology, the use of inductors in filters has become a big limitation. The advent of active devices, especially the availability of operational amplifiers (OAs) in IC form, has reduced the need of inductors. These active filters use active device(s), mainly OAs (and later operational transconductance amplifiers, current conveyors and other active devices as well), capacitors and/or resistors. Such filter circuits can be implemented in a very small space in IC form, cutting down the cost of ICs heavily [1.3].

In passive filters, output power was always less than the input power; whereas, in active filters, use of active devices provides *gain*, which is very helpful especially when the input signal magnitude is small.

When OA (or some other active device) is used, and the output is taken at the output terminal of the OA, the output impedance of the active filter is (almost all the time) low. Such a condition is suitable for connecting these filter circuits in a chain, better known as *cascade*, even if the input impedance of a proceeding cascaded filter circuit is not very high. Cascades are commonly used in active filters for enhancing the filters' characteristics, resulting in *higher-order* filters (where n of equation (1.11) is large).

As mentioned earlier, passive filters are used at high frequencies; this needs a bit more explanation. Electronic filters can also be classified in terms of the reactive elements employed; these elements control the useful frequency range of operation of the filter. For example, resonant cavities are found suitable at microwave frequencies; whereas, excellent selectivity is provided by the piezoelectric crystal unit in the frequency range 5 to 150 MHz. The intermediate radio frequency reception range between 100 kHz and 10 MHz utilizes mainly ceramic filters. Even mechanical filters can provide medium to high selectivity with good tunability in the audio band down to about 0.1 Hz. Another important frequency spectrum lies between 20 kHz and 100 kHz, where lumped inductor and capacitors are used. However, as mentioned in Sections 1.4 and 1.5, active elements were introduced, particularly in the lower and sub-audio bands (though not limited to this frequency range). This effectively amounted to replacing passive inductors, thereby introducing the subject of this book: namely, active RC filters. As active

RC filters became widely used, the subject matter gained considerable momentum. However, with advancements in technology, digital and sampled data posed great challenges to active RC filters. Communication channels were progressively digitized and the filtering operation was done directly on digital signals. Without going into a discussion on the limitations (or comparison) of digital and sampled data filters, it can safely be said that analog filters are still in great demand. There are a number of applications where analog filters become essential or are preferred over digital or sampled data filters.

Active analog filters manufactured in IC form are in great demand in the field of communication, instrumentation, medical science and many more areas (some of the application areas were mentioned in Section 1.3.1), and their design is a specialized subject. However, this book provides an introduction to this topic in order to enable one to design filters with less demanding specifications in the domain of continuous-time filters and make the reader ready to undertake the specialized study.

Before proceeding further, let us keep in mind some of the following points while using or designing active analog filters.

(a) Active analog filters need dc power supply for active devices; the power is dissipated and it produces heat which needs to be removed without increasing the working temperature.

(b) The range of the operating frequency is limited by the type of active device used, its network topology and the magnitude of the input signal. Though the whole range of operating frequency may extend from around a fraction of Hz to a few GHz, the actual operating frequency range needs to be evaluated for each specific type of active filter with specific active device(s).

(c) Depending on the type of technology used for ICs, the tolerance of the values of passive elements and parameters of the active devices are different. Tolerance affects parameters and characteristics of the realized filter. It is for this reason that it is important to study the *sensitivity* effect on filter parameters due to the tolerance of the elements used for their practical utilization.

(d) As active analog filters provide gain, small magnitude signals can also be processed. However, active devices do generate noise and care has to be taken so that the signal does not get polluted by noise. In addition, rise in the level of signal due to gain should not go above the saturation level of the active device at any stage (final or intermediate) of the filtering; the devices operate in linear range.

1.5 Active RC Filters

To decrease the size of electronic filters in accordance with modern technology, replacement of inductors with appropriate elements becomes essential. Utilization of active devices, along with resistance and capacitance elements, which simulated inductance, led to active RC filters being considered as an alternative to passive filters. In the beginning, transistors were used as active devices; however, integration of OAs in 1970s gave active RC circuits a big boost. Parameters of OAs being close to ideal, active RC synthesis remained synonymous with OA–

RC network synthesis for a long time. Inspite of the frequency-dependent gain of OAs, and the advent of new active devices, OAs are still used in most voltage-mode filtering circuits. Depending on the application, either general purpose but cheaper OAs or specially designed but costlier OAs are used.

The new devices, which are also being increasingly used, are basically current-mode devices. They include operational transconductance amplifiers (OTAs), current conveyors (CCs) and its variants, and other current mode devices like current differencing transconductance amplifiers (CDTAs).

In linear active networks, the signal processing properties are largely dependent on the time constants, that is, RC products. Therefore, in IC filters, generally high-quality R and C elements are realized; effort is also made to obtain filter parameters in terms of element ratios as resistance or capacitance fabricated in ratio forms are more precise[1.3].

Whether we use OA-based design (OA–RC) or OTA based design (OTA–RC) or some other variations (these will be discussed at a later stage), the basic synthesis processes are similar. The processes are as follows:

i.   Cascade form synthesis.

ii. Direct form synthesis using element substitution or through *operational simulation*.

In the cascade form active synthesis, the nth order transfer function, which was given in equation (1.11), is decomposed into second-order functions (if n is even, and one first/third-order sections if n is odd). These second-order functions are then individually realized and the resulting non-interactive blocks are cascaded to obtain the overall transfer function $H(s)$. Each second-order building block can be expressed as follows:

$$H_i(s) = \frac{N(s)}{b_2 s^2 + b_1 s + b_0} \tag{1.14}$$

where $N(s)$, the numerator polynomial of second-order functions, is chosen according to the required nature of the response of the section; whereas, the denominator polynomial defines the two important filter parameters, viz., the pole frequency ω_o and pole Q (Q). These are related to the denominator coefficient as follows:

$$\omega_o = (b_0 / b_2)^{1/2} \text{ and } Q = (b_0 b_2)^{1/2} / b_1 \tag{1.15}$$

There are some specific advantages while realizing second-order sections, hence great emphasis has been given to the optimum design of second-order sections. In addition, different forms of feedback and utilization of second-order (and first-order) sections have resulted in the *follow the leader feedback* type of filters as well.

In the element substitution approach of the direct form of synthesis technique, the network is directly realized from the transfer function $H(s)$ of equation (1.11). A number of direct form techniques are available; the following two are the most common.

(a) Inductance simulation (IS) approach.

(b) Frequency dependent negative resistance (FDNR) approach.

These techniques and their implementation shall be taken up in later chapters.

Operational Simulation of Ladders: Using passive RLC ladders as the starting point, active RC networks can also be realized in a slightly different direct form in what is known as *operational simulation*. In this approach, instead of simulating the elements, the process simulates the operation of the ladder by appropriately modeling the circuit equations and voltage–current relations of the elements used. Such circuit equations are represented by block diagrams in a signal flow graph.

1.5.1 Active R and active C filters

Conventionally, the design of active RC circuits assumes OA to be ideal with a large frequency-independent voltage gain ($A \rightarrow \infty$). However, OAs exhibit a frequency-dependent gain of low pass nature. This kind of non-ideality restricts the use of OA–RC circuits to mostly audio frequency range while using economically cheaper off-the-shelf OAs. Suitable compensating schemes have been developed to design filters with lesser dependence on the amplifier gain characteristics. In an alternate approach, the frequency-dependent gain of the OAs was directly used in the design; this helped in increasing the useful frequency range of operation to the MHz range while using commercially available OAs. The non-ideality of OAs has been used to develop active R and active C circuits, which employ OAs and resistors and OAs and capacitors only, respectively. However, because of certain limitations, these types of filter structures have not been manufactured in bulk. The limitations included the large tolerance of the bandwidth of OAs and their dependence on temperature and biasing voltage. Suggested methods to overcome these limitations in active R and active C filters are also available.

1.6 Sensitivity Concepts

Due to the availability of a number of choices for the synthesis of a particular transfer function, it is possible to use a number of active circuits for a single specification. Hence, the designer/user has to decide on the *best possible* choice amongst the available designs. To take this decision, certain criterion is chosen, such as number of active and passive elements used, the useful frequency range of operation and ease of design. However, while realizing the circuit in practice, one significant problem is that element values differ from their nominal values due to the fabrication process tolerance. In addition, the elements' values change with working temperature and variation in voltage and environment. Inaccurate modeling of the passive as well as active elements and parasitic elements also contribute towards increasing the complexities of the circuit. Since all the coefficients of the transfer function, hence its poles and zeros, depend on the passive and active component values, it is obvious that the gain and characteristics of the transfer function also deviate from the specified form. It is important to evaluate the deviations in different performance parameters caused by the possible change in circuit element values. The amount of introduced error depends on the amount of component tolerance and the *sensitivity* of the circuit's performance parameter to these tolerances. As a result, while comparing different available active circuits to choose the best possible alternative, one of the important considerations is the study of *sensitivity*.

1.7 Operational Amplifier

OA is the most prominent active device used in active filters. It is expected that the reader is familiar with the basics of OAs. Therefore, here, the discussion will be restricted to only those aspects of OAs that are directly related to their usage in analog filters.

Figure 1.7 shows the pin connection diagram of the most commonly used OA, type 741. It needs dual power supply and has two terminals, one for inverting and non-inverting inputs and one for the output. Dual OA and quad OA ICs are also available with matching characteristics. OA is basically a high-gain differential amplifier that can be modeled in the simplest form as shown in Figure 1.8. Output voltage of the OA is the difference of the two input voltages multiplied by the high gain factor A; it can be expressed as follows:

$$V_o = A(V_+ - V_-) \tag{1.16}$$

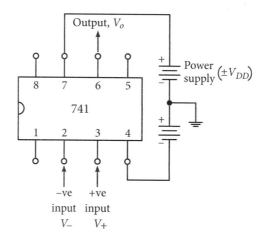

Figure 1.7   Signal input/output and power supply connections for an operational amplifier.

The differential gain A is frequency dependent in a practical OA. Therefore, it is represented by the single-pole roll-off model given here:

$$A(s) = \frac{A_o \omega_a}{s + \omega_a} = \frac{B}{s + \omega_a} \simeq \frac{B}{s} \tag{1.17}$$

Here, A_o, the ratio of the single-ended output voltage to the differential input voltage, is called *open-loop gain*, or dc gain (i.e., gain at $\omega = 0$), ω_a is the open-loop bandwidth and $B = A_o\omega_a$ is the gain–bandwidth product. For a practical OA, $A_o \approx 10^5$, $\omega_a \approx 2\pi \times 10$ rad/s, that is, $B = 2\pi \times 10^6$ rad/s and the approximated expression in equation (1.17) is valid for all $\omega \gg \omega_a$; input resistance at each input (R_i) is normally more than 1 MΩ and the output resistance (R_o) is less than 100 Ω. For the purpose of analysis, initially, OA is assumed to be ideal with the following parameters: $A_o = \infty$, $\omega_a = 0$, $B = \infty$, $R_i = \infty$ and $R_o = 0$; appropriate corrections are made at a later stage. However, during simulation, an appropriate model has to be used, which would involve a large number of parameters depending on the type of model used for the OA.

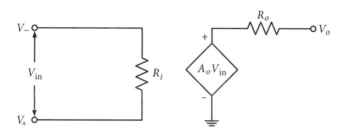

Figure 1.8   Simple model of a practical operational amplifier.

Open-loop Gain and Bandwidth: It is generally preferred to have the value of the open-loop gain A as high as practically feasible. Obviously, during analysis, assuming the value of gain as infinite will create less error if its absolute value is higher. Figure 1.9 shows the typical variation of the open-loop gain with frequency; this is a replica of the frequency response of a general-purpose OA like type 741. Roll-off at the rate of 20 dBs/decade is due to the presence of a compensating capacitor in the OA; the roll-off begins at a small frequency ω_a. Figure 1.9 also depicts the idea of the gain bandwidth of the OA, that is, the frequency where the gain falls to unity or 0 dB. As the gain bandwidth product is a constant, use of OA in the close-loop form increases its 3 dBs, or half-power bandwidth, while decreasing its close-loop gain.

Slew Rate: Signal magnitude needs to be controlled so that the level of output remains below the supply voltage ($\pm V_{DD}$), otherwise, the signal will be clipped and distorted. Another important reason of distortion in OAs is due to its slew rate (SR) limitation. Because of the current driving capability of the transistors used in OAs, output voltage cannot change at a faster rate than specified. The maximum rate of change of the output voltage in terms of SR is as follows:

$$SR = |dv_o(t)/dt|_{max} \tag{1.18}$$

Therefore, for undistorted output signal, the bandwidth becomes limited. Its relation with SR is given as:

$$Bandwidth = SR/(\pi V_{pp}) \tag{1.19}$$

Here V_{pp} is the peak-to-peak output voltage. For a typical value of slew rate for the 741 type OA, 0.5V/μs, if maximum useful frequency is fixed at 100 kHz, V_{pp} shall be limited to (0.5 × $10^6/\pi10^5$) = 1.59 volts.

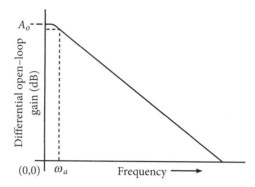

Figure 1.9   Typical variation of open-loop gain of operational amplifier with frequency.

Input Currents: OAs have a differential stage and transistors at this stage require bias current to flow so that the transistors can operate in the saturation region. For bias current to flow, a dc path must be made available for both the inputs to either ground or to the output of the OA. Normally, the mentioned path remains available; but in circuits involving only capacitors/switches, precaution is needed, as shall be shown later.

1.7.1   Analysis of circuits using operational amplifiers

For accurate analysis, simulation of filters using OAs needs better and appropriate models. However, at the first instance, manual analysis assumes ideal OA with infinite open-loop gain and input resistance and zero output resistance. This means that both input terminals are forced to be at the same potential and no current flows in them. Hence, when the non-inverting terminal is grounded, which happens in a majority of circuits, the inverting terminal is forced to be at the ground potential; the inverting terminal is now said to be at *virtual ground*. This concept of *virtual grounds* has been found to be very useful during analysis. With the suggested conditions of the OA, it is preferable to employ Kirchhoff's current law (KCL) at the input/output terminals of the OAs, making analysis simpler.

The analysis procedure mentioned here gives fairly accurate results at low frequencies (up to a few kHz range). Deviations from the expected results start taking place when the working frequency is increased. Hence, the next step is to consider finite values of the open-loop gain A, resulting in the appearance of a non-zero differential voltage between the inverting and non-inverting input terminals. The gain A can be substituted by an appropriate numerical value or by its frequency-dependent expression in equation (1.17). In specific cases where the effect of finite R_i and R_o needs to be analyzed, there values are taken into consideration; otherwise, these are taken as ideal. Considering A to be infinite is very helpful in manual analysis when conventional circuit analysis methods are used.

In the next section, two-integrator circuits are analyzed using the aforementioned method, with $A\rightarrow\infty$ as well as A being represented by the frequency-dependent model.

1.8   Integrators Using Operational Amplifiers

A large number of simple and not so simple applications of OAs are in practice. However, the integrator is probably the most common OA application in active filtering. The integrating function has been employed for simulating inductors (say in ladder structure of any order) or for the realization of second-order filter sections. Assuming an ideal OA with infinite open-loop gain, for the circuit shown in Figure 1.10, with no signal current going into the OA, application of KCL at the inverting terminal gives:

$$\frac{\left(V_{in}-V_x\right)}{R}=\frac{\left(V_x-V_{out}\right)}{\left(1/sC\right)}$$

(1.20)

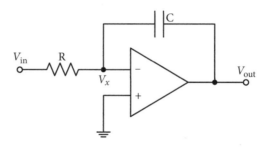

Figure 1.10 Operational amplifier as inverting integrator.

Since $V_x = 0$, the transfer function of the circuit, working as an inverting integrator, is obtained as:

$$\frac{V_{out}}{V_{in}}=-\frac{1}{sCR}$$

(1.21)

However, with the single-pole roll-off model of equation (1.17) having frequency-dependent finite OA gain, the inverting terminal potential in Figure 1.10 shall be $(-V_o/A)$ instead of virtual zero. Simple analysis gives the closed-loop transfer function:

$$\frac{V_{out}}{V_{in}}=-\frac{1}{sCR}\frac{1}{1+\frac{1}{A}\left(1+\frac{1}{sCR}\right)}$$

(1.22)

Expression in equation (1.22) will obviously reduce to equation (1.21) for $A \to \infty$. Hence, a larger value of A is always preferred as it means lesser deviation from the ideal condition.

Another commonly used variation of the integrator is shown in Figure 1.11(a). For a non-ideal OA, with A being finite, the transfer function is obtained as follows:

$$\frac{V_o}{V_{in}} = -\frac{R_2}{R_1\left(1+sCR_2\right)} \frac{1}{1+\dfrac{1}{A}\left\{1+\dfrac{R_2}{R_1(1+sCR_2)}\right\}} \tag{1.23}$$

$$\simeq -\frac{1}{R_1 C\left(s+1/CR_2\right)} \text{ for } A \to \infty \tag{1.24}$$

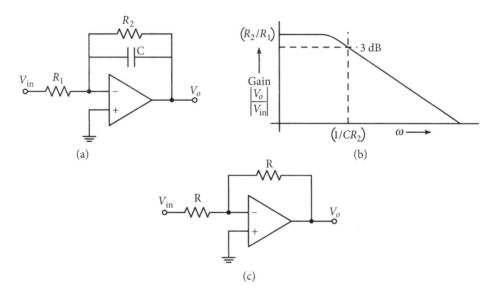

(a)

(b)

(c)

Figure 1.11 (a) Operational amplifier as lossy inverting integrator, (b) its frequency response, and (c) OA as inverter.

Obviously, equation (1.23) will reduce to equations (1.22) and (1.21), respectively, for $R_2 \to \infty$ and further with $A \to \infty$. At high A, the response of equation (1.24) is like a first-order low pass filter (LPF) shown in Figure 1.11(b) with 3 dB frequency (cutoff frequency) and the dc gain as given here:

$$\omega_{LP} = 1/CR_2 \tag{1.25}$$

$$h_{dc} = (R_2/R_1) \tag{1.26}$$

A non-inverting integrator can be obtained by cascading the inverting integrator with an OA based inverter. For equal value resistors in the inverter shown in Figure 1.11(c), gain is (−1) when $A \to \infty$. However, with A represented by the approximated frequency-dependent model of equation (1.17), the circuit's gain is given as:

$$\text{Gain} = -\frac{1}{1 + 2/A(s)} \to -\frac{1}{1 + 2s/B} \tag{1.27}$$

This gain expression implies that when an inverter is added to obtain a non-inverting integrator, it adds a bit more non-ideality to the inverting integrator. Compensation methods are used for both inverting and non-inverting integrators to minimize the amount of non-ideality. This will be studied in more detail in Chapter 7.

1.9 Passive Components in Monolithic Filters

In the early stages of the development of active filters, OAs and discrete resistors and capacitors were mounted on a printed circuit board (PCB). The next stage of miniaturization was called *hybrid technology*. In hybrid technology, RC components were fabricated using the (i) thick film, or (ii) thin film technique and the OAs and RC components were bonded in a single package reducing the overall size considerably and improving the reliability of the interconnections between OAs and RC components. In the thick film technique, resistors are fabricated by depositing a composite paste of conducting material and glass on an insulating surface. Depending on the conducting material used, sheet resistance (resistance per square of sheet or strip forming the resistance) of the paste is in the range of 1 Ω/square and 100 MegΩ/square. Hence, resistors in the range of 0.5 Ω to 1 GΩ could be realized with tolerance in the range of 20% to 50%. Reduction in tolerance became possible in the thin film technique through deposition of metal films on insulating material. Fabricated resistance using thin film was much smaller than that realized using thick film as its sheet resistance was in the range of 10 Ω to 10 MegΩ. Capacitances were also fabricated in hybrid technology using thin and thick films.

With advancements in fabrication technology employing bipolar, MOS and CMOS technologies leading to system on chip (SOC) expertise, efforts were taken in the direction of the fabrication of monolithic filters in which passive elements could also be integrated simultaneously with active devices. At present, further advancements are taking place in the fabrication of passive elements in integrated form using new materials and improved physical configurations.

Resistors: Technologies may differ, but the most common method of fabricating resistance is through layering rectangular sheets of semiconducting material. The value of the realized resistance is equal to the number of squares multiplied by the sheet resistance. Obviously, a larger value resistance requires a larger chip area.

While fabricating a transistor, a diffusion layer has also been used to realize resistors without adding any separate processing step. Usually, the base diffusion resistor is the most commonly used resistor in a bipolar process. For such resistors, a matching tolerance of +0.2% between resistors is possible. If the base diffusion area is pinched by diffusing an *n+* diffusion layer over a *p*-type base diffusion, sheet resistance is increased to 2–10 kΩ/square. Resistances so obtained are called *pinched resistors*; they realize larger resistance, but tolerance becomes worse, ranging up to 50%.

An advanced technique of resistance fabrication is through ion implantation. A very thin layer of implant (0.1–0.8 μm) leads to a very high value of sheet resistance (100–1000 Ω/square). Matching tolerance is also good, being approximately 2%.

In a different technique, resistances are fabricated through active devices. Observing the current–voltage relation of a BJT (bipolar junction transistor) or a MOS (metal oxide semiconductor) transistor, it is easy to see that for a certain range of operation, these devices behave as quite a stable resistance. In MOS or CMOS (complementary MOS) technology, transistors operating in the linear region are used to realize *active resistors*. Resistance at the fabrication/design stage is controlled by the width to length (*W/L*) ratio of transistors, and at a later stage through gate voltage V_G.

Important advantages of active resistors compared to passive resistors are as follows: (i) required chip area in active resistors is very small for the same value of passive resistance, and (ii) value of the resistance is easily controllable.

Capacitors: Quite a few techniques are available for the fabrication of capacitors in monolithic IC technology: for example, using pn junctions, MOSFETs (metal oxide semiconductor field effect transistors) and polysilicon capacitors. In the BJT process, capacitors are formed between semiconductor junctions. However, the obtained value of the capacitor is very small: 0.05–0.5 pF/mil² and large chip areas are needed for even small value capacitors.

It is well known that reverse bias semiconductor junctions create a depletion region. This region acts as an insulator sandwiched between doped silicon on two sides, resulting in a capacitor depending on the width of the depletion region. Along with some parasitic capacitors, a depletion region capacitance of 0.001 pF/μm² can be realized.

In the MOS technology process, an MOS capacitor is formed between the *n+* diffusion regions, while forming the channel region, a polysilicon layer, and a thin layer of silicon dioxide or silicon nitride between them.

Polysilicon capacitors are the most commonly used ones, as the fabrication process suits MOS technology. In this technique, basically, the gate of the transistor is made of polysilicon. Thin oxide is deposited on top of a polysilicon layer which acts as the insulating layer over another bottom polysilicon plate. This type of capacitor too has parasitic capacitance and shall be considered later while discussing switched capacitor circuits in Chapter 15.

References

[1.1]   Van Valkenburg, M. E. 1976. *Introduction to Modern Network Synthesis*. New York: Wiley Eastern Limited.

[1.2]   Zverev, A. I. 1967. *Handbook of Filter Synthesis*. New York: Wiley.

[1.3]   Moschytz, G. S. 1974. *Linear Integrated Circuit Fundamentals* Part I. New York: Van Nostrand Reinhold.

Practice Problems

1-1   Find the driving point impedance function $Z(s)$ for the network shown in Figure P1.1 and find its poles and zeroes.

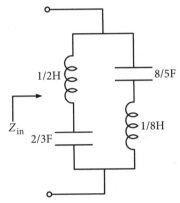

Figure P1.1

1-2   Plot the magnitude and phase of the following transfer functions for $s = j\omega$, $0 \leq \omega \leq 10$.

(a) $\dfrac{1}{(s+2)(s+4)}$

(b) $\dfrac{s}{(s+3)(s+6)}$

(c) $5\dfrac{(s-2)}{(s+2)}$

(d) $\dfrac{s^2}{(s^2+s+1)}$

1-3   Find the transfer function for the network shown in Figure P1.2.

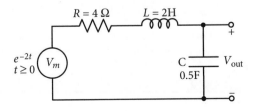

Figure P1.2

1-4   Sketch the following ideal responses in one figure and classify them.

(i)   pass band from 0 to 10 krad/s; rest is stop band.

(ii) pass band from 15 krad/s to infinity; rest is stop band.

1-5   Repeat problem 1.4 for the following:

(i)   pass band from 10 to 12 krad/s; rest is stop band.

(ii) stop band from 18 to 20 krad/s; rest is pass band.

1-6   Sketch the magnitude responses for the following values of attenuations and classify them.

(i) $\alpha_{max} = 2$ dBs, from 0 to 10 krad/s and $\alpha_{min} = 40$ dBs from 20 krad/s to infinity.

(ii) $\alpha_{max} = 1$ dBs, from 40 krad/s to infinity and $\alpha_{min} = 50$ dBs from 0 to 20 krad/s.

(iii) $\alpha_{max} = 0.5$ dBs, from 14 to 18 krad/s and $\alpha_{min} = 40$ dBs from 0 to 8 krad/s and from 24 krad/s to infinity.

(iv) $\alpha_{max} = 1$ dB, from 0 to 6 krad/s and from 14 krad/s to infinity, $\alpha_{min} = 50$ dBs from 8 to 12 krad/s

1-7   Apply $(1/s)$ transformation to the network shown in Figure P1.2. Determine the new element values and fine its voltage-ratio transfer function again. Does it remain same as before in problem 1-3 or not?

1-8   Determine the input impedance of the network of Figure P1.3, (a) with OA as ideal and (b) when OA is represented by its first-pole roll-off model. Note: Unless specified, approximated model of OA, $A \approx B/s$ shall be used.

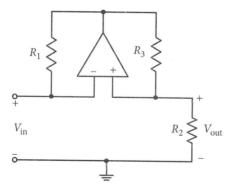

Figure P1.3

1-9   A 741 type OA employing power supply of $\pm$ 15 volt was used to construct an inverting amplifier with voltage gain of (-5). Sinusoidal signal input frequency was 50 kHz. Find the largest input signal before it gets distorted due to the slew rate limitation; assume standard value of SR.

1-10   (a) Find the transfer function for the amplifier shown in Figure P1.4 ($Z_1 = Z_2 = 5$ kΩ) and ($Z_1 = 5$ kΩ and $Z_2 = 50$ kΩ). The amplifier is non-ideal and is modeled with single-pole roll-off model with $A_o = 10^5$ and the first pole $\omega_a = 2$ rad/s. Determine the 3-dB bandwidth of the amplifiers.

(b) Repeat problem in part (a) for Figure P1.5.

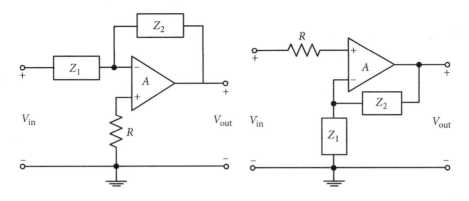

Figure P1.4   **Figure P1.5**

1-11   Derive the transfer function of the integrator circuit shown in Figure P1.6. Find its critical frequency when OA used is near ideal and non-ideal with unity gain bandwidth of 10^5 rad/s and $R_1 = R_2 = R_3 = 10$ k Ohm, $R_4 = 5$ k Ohm and $C = 0.1$ nF.

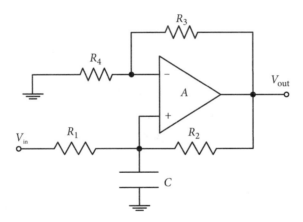

Figure P1.6

1-12   Determine the transfer functions of the circuits shown in figures P1.7 and P1.8, considering OA as (a) ideal and (b) non-ideal with unity gain bandwidth of 10^5 rad/s. In Figures P 1.7 and P 1.8 capacitors $C_1 = C_2 = 1$ pF, $R_1 = 10$ k Ohm, $R_2 = 5$ k Ohm, $R = 1$ k Ohm and $C = 1$ nF. Find critical frequencies when OA is considered ideal and non-ideal.

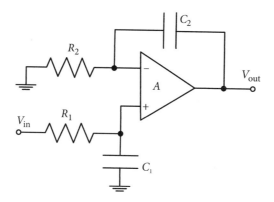

Figure P1.7

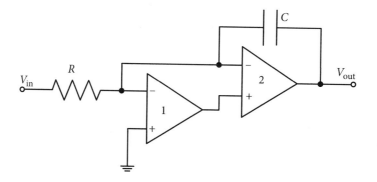

Figure P1.8

Chapter 2

First- and Second-order Filters

2.1   Introduction

Circuit designers/users evaluate filters and the order of the filters needed is based on what are the given specifications. The filter order n can be small or large. There are techniques by which filters of order $n \geq 2$ can be realized directly. Filters with order one or two can be used as such depending on the requirement; they can also be combined to provide filters of higher-order. Therefore, it is necessary to study the basic principles underlying the behavior of first- and second-order filter sections and the important terms used for their parameters before studying realization of higher order filters.

A first-order section can easily be realized using RC components only; but such sections suffer from certain limitations as shall be shown in Section 2.2. Hence, it is advisable to use first-order active filters with inverting or non-inverting amplifiers. A comparative study of active first-order filters, along with a discussion on the non-ideal effect of operational amplifiers (OAs) on their frequency response is given in Section 2.3 and 2.4. Terminologies used for second-order active sections and characteristics associated with low pass (LP), high pass (HP), band pass (BP), band reject (BR), and all pass (AP) responses are included in Sections 2.5 to 2.11. Constraints of the finite bandwidth of the OA on second-order filters are briefly discussed in Section 2.12. Three application examples in Sections 2.3.2, 2.3.3, and 2.7.1 are included to show the utility of these simple filter structures.

2.2   First-order Filter Sections

The transfer function of a physically realizable filter using a finite number of elements has to be a real rational function [2.1]. The rational function is a ratio of polynomials in the complex frequency s. It is repeated here from Chapter 1.

$$H(s) = \frac{N(s)}{D(s)} = \frac{a_m s^m + a_{m-1} s^{m-1} + \ldots + a_1 s + a_0}{b_n s^n + b_{n-1} s^{n-1} + \ldots + b_1 s + b_0} \tag{2.1}$$

In this equation, the order of the transfer function is n, with $m \le n$.

For $m = n = 1$, the transfer function of equation (2.1) reduces to the following.

$$H(s) = \frac{N(s)}{D(s)} = \frac{(a_1 s + a_0)}{(b_1 s + b_0)} \tag{2.2}$$

As $H(s)$ in equation (2.2) is a ratio of two polynomials representing a straight line, it is also called a *bilinear function*. The transfer function $H(s)$ of equation (2.2) can be modified into a desirable form in terms of pole p_1 and zero z_1 zero as follows:

$$H(s) = \frac{a_1}{b_1} \frac{(s + a_0/a_1)}{(s + b_0/b_1)} = k \frac{s + z_1}{s + p_1} \tag{2.3}$$

For the transfer function to be physically realizable in stable form, its pole must not be in the right-hand side of the s plane [2.1]. Hence, b_0 and b_1 will have to be positive and finite (or both negative and finite), though a_0 and a_1 can be positive, negative, or even zero (one of the two, either a_0 or a_1). The zero z_1 can be located anywhere on the real axis. To realize the transfer function $H(s)$ with passive elements, a simple arrangement as shown in Figure 2.1 can be used. Different variations are possible depending on the choice of impedances Z_1 and Z_2. Some of the combinations are as follows:

For $Z_1 = R_1$ and $Z_2 = 1/sC_1$, as shown in Figure 2.2(a), the realized transfer function is $H(s) = (1/R_1 C_1)/(s + 1/R_1 C_1)$

For $Z_1 = 1/sC_2$ and $Z_2 = R_2$, as shown in Figure 2.2(b), $H(s) = s/(s + 1/R_2 C_2)$

For $Z_1 = (R_1 + 1/sC_1)$ and $Z_2 = (R_2 + 1/sC_2)$ as shown in Figure 2.2(c), the transfer function becomes:

$$H(s) = \frac{R_2}{R_1 + R_2} \frac{(s + 1/R_2 C_2)}{\left[s + (C_1 + C_2)/C_1 C_2 (R_1 + R_2) \right]} \tag{2.4}$$

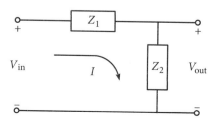

Figure 2.1   First-order bilinear transfer function realization using passive elements.

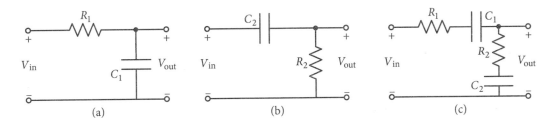

Figure 2.2   Few first-order transfer function realizations using resistors and capacitors only.

The bilinear function of equation (2.4) has a pole p_1 and a zero z_1; their expressions are as follows:

$$p_1 = \frac{(C_1 + C_2)}{C_1 C_2 (R_1 + R_2)} \text{ and } z_1 = \frac{1}{R_2 C_2} \tag{2.5a}$$

The gain of the transfer function of equation (2.4) is as follows:

$$\text{DC gain} = \frac{C_1}{(C_1 + C_2)} \text{ and gain at high frequencies} = \frac{R_2}{R_1 + R_2} \tag{2.5b}$$

It is important to note that in this passive circuit, the voltage gain at any frequency will never be more than unity.

2.3   Active First-order Filters

In the previous section, it was shown that first-order transfer functions can easily be realized using passive elements. However, there are quite a few limitations in such networks. For example, all the values of pole (s) and zero (s) are not realizable. For example, an ideal integrator is not realizable. An important limitation is a resultant disturbance in the transfer function when the network gets loaded. Significant changes in the response take place because of the loading effect. Hence, it is preferable to realize the elements of the circuit in active form, which will additionally provide gain as well, which is generally needed.

The following sections describe a few simple applications of OAs being used as first-order active filters.

2.3.1   Use of inverting amplifiers

If the elements used in Figure 1.11(a) (lossy inverting integrator) are replaced by impedances Z_1, Z_2 (or admittances Y_1, Y_2), the transfer function of the circuit gets modified as shown in equation (2.6).

$$H(s) = -(Z_2/Z_1) \text{ or } -(Y_1/Y_2)$$

(2.6)

The impedances Z_1, Z_2 (or Y_1, Y_2) can be any series/parallel combination of R_1 and C_1; R_2 and C_2 are as shown in Figure 2.3; the transfer function of the circuit is obtained as follows:

$$H(s) = -(Z_2/Z_1) = -\frac{R_2}{R_1}\frac{s+(1/R_2C_2)}{s+(1/R_1C_1)} = -k\frac{s+z_1}{s+p_1}$$

(2.7)

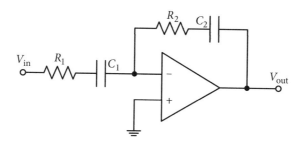

Figure 2.3   Realization of active first-order section using an inverting amplifier.

Comparing this equation with equation (2.3), we can realize a first-order section with the following parameters:

$$k = \frac{R_2}{R_1}; \text{ zero } z_1 = \frac{1}{R_2C_2} \text{ and pole } p_1 = \frac{1}{R_1C_1}$$

(2.8)

If the designer is aware of the location (value) of the pole and zero and the gain at $s = 0$ (or $s = \infty$), element values can be easily calculated using equation (2.8). It is to be noted that equation (2.7) has three parameters, whereas element values to be found are four. Hence, one element value (or an element ratio) has to be assumed.

Example 2.1: Design a first-order active bilinear function having its zero at 1000 rad/s, pole at 2000 rad/s and gain of 20 dBs at very low frequencies.

Solution: Corresponding to 20 dBs, gain on linear scale is $k = 10^{20/20} = 10$ at very low frequencies (or $s = 0$). Using equation (2.7):

$$H(0) = -(C_1/C_2) = -10$$

If the OA is used in inverting mode, the gain will have a negative value of -10. Corresponding to the given pole and zero values, $1/R_2C_2 = 1000$ and $1/R_1C_1 = 2000$. Selecting $C_1 = 1\ \mu F$, we get $C_2 = 0.1\ \mu F$, $R_2 = 10\ k\Omega$ and $R_1 = 0.5\ k\Omega$. The desired circuit with element values is shown in Figure 2.4(b). Figure 2.4(c) shows the circuit's PSpice (Simulation Program for

Integrated Circuits Emphasis: a simulator program used to verify circuit designs and predict their behavior) simulation with an input voltage of 0.1 volt. At low frequencies, the output voltage is approximately 1.0 V with a maximum of 2.0 V or gain of (−20).

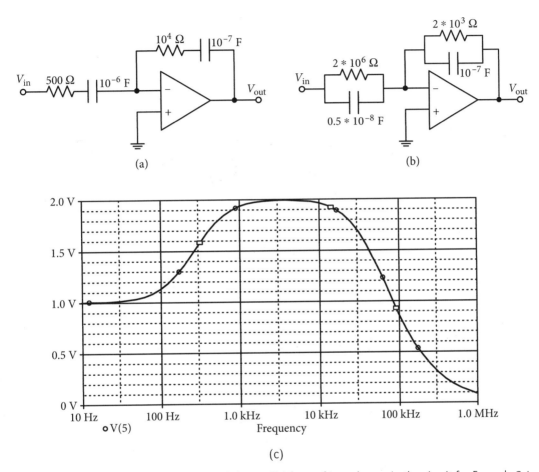

(a)   (b)

(c)

Figure 2.4 (a) Using series form and (b) parallel form of impedances in the circuit for Example 2.1. (c) Magnitude response of the active bilinear circuit for the circuit in Example 2.1.

2.3.2 Bass cut/boost and treble cut/boost filters – application example

In audio systems, low frequencies, which are typically in the range of a few Hz to 100 Hz, are called *bass* notes. Mid-range frequency signals, typically ranging between 100 and 1000 Hz, are called *middle* notes. High frequencies are called *treble* notes; they are typically above 1000 Hz. Low frequencies are responsible for the deep sound of bass guitars and drums. Most instruments create sounds in the mid-range frequency; these include guitars, brass or string instruments, and even the human voice. High frequencies are responsible for the *sparkle* sound of cymbals and clarity of voices. Sound gets muffled if the treble note is missing or weak. All the three types of notes are enhanced or *boosted* else weakened or *cut* to improve the quality. If no boost or cut is applied, the response is said to be flat. The following example is a simple practical illustration employing an inverting amplifier for audio systems.

In equation (2.7), $H(s) = -Z_2/Z_1$. Hence, if $Z_1 = R_1$ and Z_2 is a parallel combination of R_2, with resistance R_3 and capacitance C_1 in series, as shown in Figure 2.5(a), we obtain the following relation:

$$H(s) = -\frac{R_2}{R_1} \frac{R_3}{R_2 + R_3} \left\{ \frac{s + (1/C_1 R_3)}{s + 1/C_1 (R_2 + R_3)} \right\} \qquad (2.9)$$

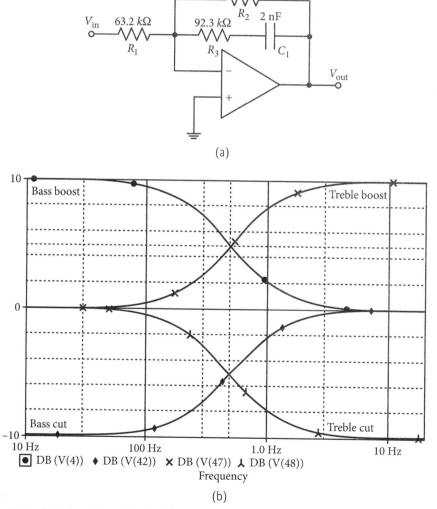

(a)

(b)

Figure 2.5   (a) A bass boost (and treble cut) circuit. (b) Response of the bass boost/cut and treble boost /cut by 10 dBs.

The circuit in Figure 2.5(a) realizes a function which performs a bass boost action with the following expressions for gain at high and low frequencies:

$$\text{Gain at dc} = R_2/R_1 \qquad (2.10a)$$

$$\text{Gain at high frequencies} = \frac{R_2}{R_1}\frac{R_3}{R_2+R_3} \qquad (2.10b)$$

For getting a boost of 10 dBs, if R_2 is selected as 200 kΩ, we need an R_1 of 63.2 kΩ, and for a high frequency gain of unity, equation (2.10b) gives R_3 = 92.3 kΩ. Pole frequency f_c of the bass boost circuit is decided by the choice of the capacitor C_1, with its expression as follows:

$$f_c = \frac{1}{2\pi C_1 (R_2+R_3)} \qquad (2.11)$$

With C_1 = 2 nF, f_c will be 272 Hz for the selected resistances.

The circuit in Figure 2.5(a) can also be used to function as a *treble cut*. To get a –10 dB treble cut and 0 dB gain at dc with a pole frequency of 272 Hz, equation (2.10) and (2.11) gives the required values of elements as: R_1 = 63.2 kΩ, R_2 = 200 kΩ, R_3 = 92.3 kΩ and C_1 = 2nF. Figure 2.5(b) shows the simulated responses of the designed bass boost and treble cut filter section.

If Z_1 is a series combination of resistance R_4, with resistance R_5 and capacitance C_2 in parallel, and Z_2 = R_6 as shown in Figure 2.6, its transfer function will be as follows:

$$H(s) = \frac{R_6(1+sC_2R_5)}{(R_4+R_5)+sC_2R_4R_5} \qquad (2.12)$$

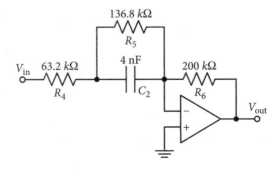

Figure 2.6　A treble boost (and bass cut) circuit.

While realizing a treble boost circuit, expressions of gain at higher frequency and at dc are as follows:

Treble boost gain = R_6/R_4 (2.13a)

dc gain = $R_6/(R_4 + R_5)$ (2.13b)

For a treble boost of 10 dB, selecting R_6 = 200 kΩ, we require an R_4 = 63.2 kΩ, for unity dc gain, R_5 = 136.8 kΩ from equation (2.13b). The expression of the pole frequency is:

$$f_c = \frac{(R_4 + R_5)}{2\pi C_2 (R_4 R_5)}$$ (2.14)

Hence, with C_2 = 4 nF, f_c will be 837 Hz.

The circuit in Figure 2.6 can also be used for bass cut function. To get a −10 dB bass cut, 0 dB gain at higher frequencies and pole frequency of 837 Hz, equations (2.13) and (2.14) gives element values R_4 = 63.2 kΩ, R_5 = 136.8 kΩ, R_6 = 63.2 kΩ and C_2 = 4 nF. Figure 2.5(b) also shows the simulated responses of the treble boost and bass cut circuits, verifying the design.

2.3.3 Fluorescence spectroscopy: application example

Frequency domain fluorescence measurements in atomic and molecular physics can be modeled in terms of first-order low pass filters (LPFs). Hence, as fluorescence can be mathematically equated to analog filters, a unified treatment of the entire fluorescence chain is possible by cascading their transfer functions [2.2].

Without going into the theoretical background of the representation of fluorescence, let us see the utility of simple LPFs as a useful practical application. It is observed that fluorescence from a three-level system (Figure 2.7(a)) can be represented by a Laplace transform equation [2.2]. This Laplace representation has been realized using two first-order cascaded LPFs. The cascaded filter is simulated for (i) a very fast relaxation from level 2 to 3 in Figure 2.7(b) and (ii) for a slower relaxation from level 2 to 3. The life time of the first-stage LPF for both the cases was set at 1 second. For the second-stage LPF, values of the components were selected by the inspection of the transfer function of a near-practical fluorescence measurement case with a life time of 10^{11} s and 10^3 s. Values of the elements for both the first filter and the two cases of the second filter are obtained from the equation of life time = $(1/2\pi RC)$:

R_1 = 159 kΩ, C_1 = 1 μF for stage 1 and (a) R_2 = 1.59 kΩ, C_2 = 100 nF for case (i) and
(b) R_2 = 0.159Ω, C_2 = 10 pF for case (ii) (2.15)

Simulated response of the two cases is shown in Figure 2.7(c), where curve '-Vp(51)' represents the decay rate of 10^{11} and curve '-Vp(5)' represents the decay rate of 10^3.

The equivalent circuit for the transfer function with two life time components is shown in Figure 2.8(a). The frequencies used for the slow and fast transition rates in the circuit were 100 Hz and 1 MHz. For these frequencies, the design values of the components are as follows:

R_1 = 15.9 kΩ, C_1 = 1 μF R_2 = 15.9 kΩ, C_2 = 100 pF, $R_4 = R_5$ = 10 kΩ, R_3
varies from 10 kΩ to 11.11 kΩ, 12.5 kΩ, 16.66 kΩ, 25 kΩ, 50 kΩ and 100 kΩ (2.16)

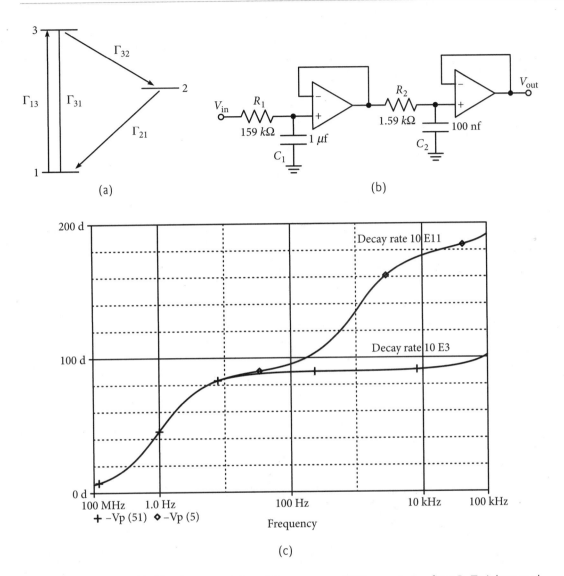

Figure 2.7   (a) Three-level representation of fluorescence. {With permission from R. Trainham et al. [2.2]} (b) Low pass filter realization of the transfer function for the three-level fluorescence. {With permission from R. Trainham et al. [2.2]} (c) Phase shifts corresponding to the two cases of fast and slow RC time constants for the cascaded low pass filter in Figure 2.7(b).

R_3 was varied to change the weight age of the transition rate of the slow component. The signals from the two LPFs were added and the final response is shown for two different ratios of the intensity of the slow component to the intensity of the fast component in Figure 2.8(b). Responses given by the filters of Figures 2.7(b) and 2.8(a) match very well with calculated theoretical responses [2.2].

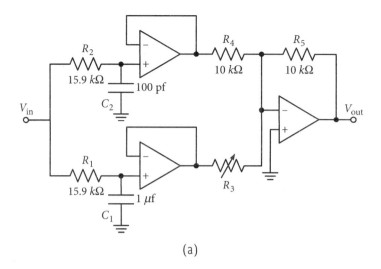

(a)

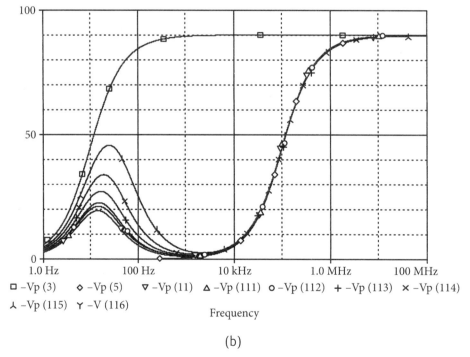

(b)

Figure 2.8   (a) Realization of the transfer function for the two life time fluorescence. {With permission from R. Trainham et al. [2.2]} (b) The family of curves of phase shifts for different mixtures of two life times separated by four decades.

2.3.4   Use of non-inverting amplifiers

A non-inverting amplifier using OA can also be used to realize a first-order bilinear function by replacing its resistors with general impedances Z_1, Z_2 (or Y_1, Y_2) as shown in Figure 2.9.

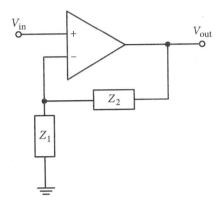

Figure 2.9   Non-inverting amplifier for the realization of a first-order active bilinear transfer function.

The transfer function is as follows:

$$H(s) = 1 + (Z_2/Z_1) = 1 + (Y_1/Y_2) \tag{2.17}$$

Comparison of equation (2.17) with equation (2.3) gives the following relation.

$$(Z_2/Z_1) = k\frac{(s + z_1)}{(s + p_1)} - 1 = \frac{s(k-1)(kz_1 - p_1)}{(s + p_1)} \tag{2.18}$$

As impedances Z_1, and Z_2 are positive entities, the following constraints are to be met to keep the numerator positive:

$$k \geq 1 \text{ and } (kz_1 - p_1) \geq 0 \tag{2.19}$$

The designer needs to be careful about the strict constraint out of the two in equation (2.19). The absolute value is not important as it only adds to the gain which can also be controlled with a cascaded amplifier/ attenuator. Element values depend on the way the impedances Z_1 and Z_2 are realized. For example, if Z_1 is a series combination of R_1 and C_2 and Z_2 is a parallel combination of R_2 and C_2 or vice versa, the circuit will realize a second-order transfer function. With Z_1 and Z_2 both having resistors and capacitors in series, the flow of biasing current in the inverting node of the OA is blocked (this needs to be overcome by connecting a high value resistor in parallel with the input capacitor C_1).

Example 2.2: Design a first-order bilinear transfer function using a non-inverting amplifier having the pole and zero of Example 2.1 with gain $k = 5$.

Solution: As, $k > 1$, the first constraint in equation (2.19) is valid for selecting Z_1 as a parallel combination of R_1 and C_1 and Z_2 as a parallel combination of R_2 and C_2 in the non-inverting amplifier of Figure 2.9. The obtained transfer function is as follows:

$$H(s) = \frac{V_{out}}{V_{in}} = \frac{(C_1 + C_2)}{C_2} \frac{s + (R_1 + R_2) / R_1 R_2 (C_1 + C_2)}{\left(s + (1 / R_2 C_2)\right)} \tag{2.20}$$

Hence, $H(0) = 1 + (R_2/R_1)$, and at high frequencies, say

$$H(10\text{kHz}) = 1 + (C_1/C_2) \tag{2.21}$$

Expression of its pole and zero are obtained as follows:

$$p_1 = \frac{1}{R_2 C_2} \text{ and } z_1 = \frac{(R_1 + R_2)}{R_1 R_2 (C_1 + C_2)} \tag{2.22}$$

With $k = H(10\text{kHz}) = 5$, equation (2.21) gives $(C_1/C_2) = 4$; hence, selecting $C_2 = 0.1\ \mu\text{F}$, we get $C_1 = 0.4\ \mu\text{F}$. For the pole at 2000 rad/s and the zero at 1000 rad/s, use of equation (2.22) gives the values of R_1 and R_2 as 5 kΩ and 1.25 kΩ, respectively.

All the values of the elements used are shown in Figure 2.10(a) and the PSpice simulated response is shown in Figure 2.10(b). For an input voltage of 1.0 V, the circuit has a minimum output voltage of 1.25 V at low frequencies and a maximum of 5.044 V at around 10 kHz. A gain of 5 can be verified at around 10 kHz, the circuit's useful frequency range. The voltage peaks at nearly 81 kHz is due to the effect of the frequency-dependent gain of the OA; the peak voltage is controlled by the supply voltages of the OA. Additionally, if the input voltage is increased beyond nearly 1.0 volts, the output gets distorted due to the effect of the slew rate constraint.

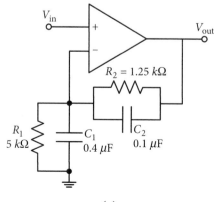

(a)

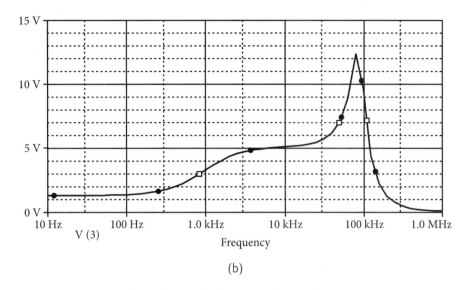

(b)

Figure 2.10   (a) Circuit realizing the transfer function of Example 2.2 using a non-inverting amplifier.
(b) Response of the first-order bilinear circuit of Figure 2.10(a).

2.4   Effect of Operational Amplifier's Pole on Integrators

Analysis of the circuits done in the chapter so far assumed OAs as ideal. Finite values of the circuit's input and output resistors do come in the picture but the most significant effect of the resistors is that of the finite and frequency-dependent open-loop gain. This non-idealness is mainly responsible for the use of OA (with the commonly used type, like 741) based circuits being restricted to low frequencies (in the audio frequency range) as was shown in Example 2.2.

This section will look into the performance variation of the first-order filters discussed in Section 2.3 when the OA is represented by its first-pole roll-off model given in equation (1.17). Effect of finite values of R_i and R_o is not considered here for two reasons. First, their values are not far from ideal; hence their effect is minimal and their introduction will only increase the complexity unnecessarily. Second, in critical cases, the effect of the finite values of the resistors can be absorbed in the components employed in the filter realizations.

For the realization of the inverting amplifier of the first-order section shown in Figure 2.3, the ratio of the output to the input voltage is as follows:

$$\frac{V_{\text{out}}}{V_{\text{in}}} = -\frac{Z_2}{Z_1} \frac{1}{1 + (1 + Z_2/Z_1)/A(s)} \tag{2.23}$$

With $-\dfrac{Z_2}{Z_1} = -k\dfrac{s+z_1}{s+p_1}$ from equation (2.7) and $k = \dfrac{R_2}{R_1}$ from equation (2.10a), limitation of the use of OA can be observed by using the approximated model given by equation (1.17) while neglecting its first-pole frequency of ω_a. The integrator model gives sufficiently accurate results at high frequencies; though at very low frequencies, ω_a cannot be neglected for correct results. Hence, equation (2.23) can be modified to the following:

$$\frac{V_{out}}{V_{in}} = -k\frac{(s+z_1)}{(s+p_1)+(s/B)[(s+p_1)+k(s+z_1)]} = -k\frac{(s+z_1)}{s^2\left(\dfrac{1+k}{B}\right)+s\left(1+\dfrac{p_1+kz_1}{B}\right)+p_1} \quad (2.24)$$

For an ideal OA, B being infinite, equation (2.24) will reduce to equation (2.7), and when B is finite, larger B or smaller values of z_1 and p_1 will lessen the parasitic effect. Frequency dependence of the OA gain has increased the denominator order by one, resulting in two poles. One of the poles will be near the original pole p_1 as $(p_1 + \Delta p_1)$ and a second pole, p_2 will be far from p_1, but the distance between the two (or the effect of non-ideality) will depend on the value of B (unfortunately, this is not the same for all OAs) and the value of the parameters p_1, z_1, and k.

Example 2.3: Design a first-order circuit using an inverting amplifier which will have a pole at 2×10^5 rad/s, zero at 10^5 rad/s and a gain of 2 at low frequencies. Find the effect of the frequency-dependent open-loop gain of the OA with B as (i) 10^6 rad/s, (ii) 0.5×10^6 rad/s and (iii) 10^5 rad/s. Compare the results with the simulated responses of the circuit.

Solution: In a similar way as in Example 2.1, the structure of the circuit is similar to that in Figure 2.11(a) or 2.11(b) with the following values if OAs were taken as ideal during analysis.

$$R_1 = 250\Omega, R_2 = 1000\Omega, C_1 = 2 \times 10^{-8} F \text{ and } C_2 = 10^{-8} F$$

$$R_1' = 500\Omega, R_2' = 1000\Omega, C_1' = 2 \times 10^{-8} F \text{ and } C_2' = 0.5 \times 10^{-8} F$$

With frequency-dependent gain, $A(s) \cong B/s$, the expression of gain is obtained as follows:

$$\frac{V_{out}}{V_{in}} = -2\frac{(s+10^5)}{s^2\left(\dfrac{1+2}{B}\right)+s\left(1+\dfrac{2\times10^5+2\times10^5}{B}\right)+2\times10^5} \quad (2.25a)$$

For further analysis and in order to determine the effect of the frequency dependence of the gain of the OA, three cases are taken with different values of the gain bandwidth product B.

i.   For $B = 10^6$ rad/s, equation (2.25a) can be modified as follows:

$$\frac{V_{out}}{V_{in}} = -2\frac{(s+10^5)}{3\times10^{-6}s^2+1.4s+2\times10^5} \quad (2.25b)$$

When $B = 10^6$ rad/s, we get conjugate poles $p_{1,2} = (-2.333 \pm j\,1.1055) \times 10^5$ rad/s.

In this case, the poles are not too far from the design pole value, and the characteristics show small deviation with peak gain having small reduction. The peak value is 3.9735 (against a theoretical value of 4), that too only at a higher frequency of 250.55 kHz, nearly one-fourth of B.

ii. For $B = 0.5 \times 10^6$ rad/s, we get conjugate poles $p_{1,2} = (-1.5 \pm j1.0408) \times 10^5$ rad/s.

In this case, the characteristics gets deviated a bit more, with peak gain going down to 3.95 at 159.3 kHz, still at a reasonably high frequency.

iii. For $B = 10^5$ rad/s, we get two real poles at -1.0×10^5 rad/s and -0.666×10^5 rad/s, and the characteristics get highly deviated with the gain doing down to 3.73, a deviation of 6.75%, at 78.7 kHz, because of the second real pole positions.

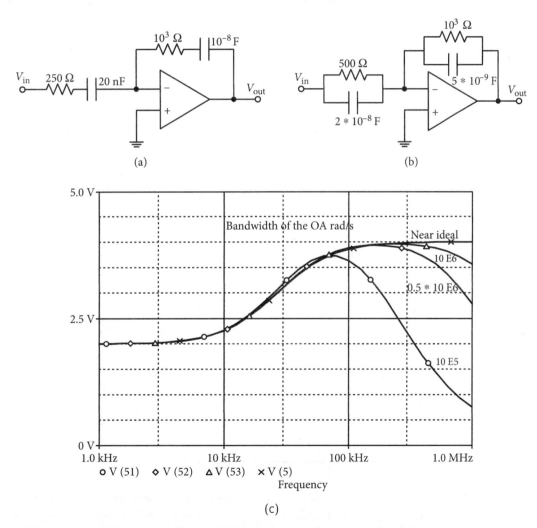

Figure 2.11 (a) Series, and (b) parallel forms of circuits for Example 2.3. (c) Magnitude response of the circuit of Figure 2.11(a) with ideal OA and OA with bandwidth = 10^6, 0.5×10^6 and 10^5 rad/s.

It is clear from these observations and the response, when OA is ideal in Figure 2.11(c), that the higher the ratio of gain bandwidth product to the working pole frequency, the more the non-ideality effect gets reduced. It may be noted that the value of the dc gain k also plays an important role in the amount of non-ideal effect; the larger the value of k, the more deviation in pole location. In the example provided here, the pole value was intentionally selected high enough and close to B to highlight the non-ideality effect of OAs.

It is observed that when the first-order section is realized using a non-inverting amplifier, the non-ideality of the operational amplifier affects the filter in a similar way. Order of the filter is increased by one for each OA; the original pole deviates and the amount of deviation depends on the ratio of B with the pole value and on the gain value k.

Example 2.4: Design a first-order section using an ideal non-inverting amplifier with the same specifications as in Example 2.3. Evaluate the effect of the non-ideality of the OA through simulation results.

Solution: Using the non-inverting circuit of Figure 2.10 and taking OA as ideal, the obtained expression of the gain is as follows:

$$\frac{V_{out}}{V_{in}} = \frac{(C_1 + C_2)}{C_2} \cdot \frac{s + \dfrac{(R_1 + R_2)}{R_1 R_2 (C_1 + C_2)}}{s + (1/C_2 R_2)} \tag{2.26}$$

Values of the elements for the given specifications $p_1 = 2 \times 10^5$ rad /s, $z_1 = 10^5$ rad/s and dc gain of 2 from equation (2.26) are obtained as follows:

$$R_1 = R_2 = 0.5 \text{ k}\Omega, \ C_1 = 0.03 \ \mu\text{F and } C_2 = 0.01 \ \mu\text{F}$$

The circuit with the aforementioned element values is shown in Figure 2.12(a) and the simulated response of the ideal case is shown in Figure 2.12(b) having a dc gain of 2 and a high frequency gain of 4.

For the non-inverting amplifier circuit of Figure 2.5, expression of the ratio of the output to input voltage with $A(s) \cong B/s$ is given as follows:

$$\frac{V_{out}}{V_{in}} = \left(1 + \frac{Z_2}{Z_1}\right) \frac{1}{1 + \dfrac{1}{A(s)}\left(1 + \dfrac{Z_2}{Z_1}\right)} = k \frac{s + z_1}{\dfrac{k}{B}s^2 + \left(1 + \dfrac{k}{B}z_1\right)s + p_1} \tag{2.27}$$

Equation (2.27) results in two poles. Value of the poles will again depend on the value of B, its ratio with p_1 and the value of the gain k.

For $B = 10^6$ rad/s, $p_{1,2} = (-3 \pm j1) \times 10^5$ and peak gain of 3.8 occurs at 126.36 kHz.

For $B = 0.5 \times 10^6$ rad/s, $p_{1,2} = (-1.75 \pm j1.39) \times 10^5$ and response drops with peak gain, dropping to 3.7 at a frequency of 100 kHz.

For $B = 10^5$ rad/s, $p_{1,2} = (-0.75 \pm j0.661) \times 10^5$, the response further worsens, having a peak gain of only 3.26, and an error of 18.5% at a much lower frequency, 62.28 kHz.

All the three responses, along with the case when OA is almost ideal, are shown in Figure 2.12(b). It is to be noted that distortion in the non-inverting amplifier case is comparatively much larger than the inverting amplifier case.

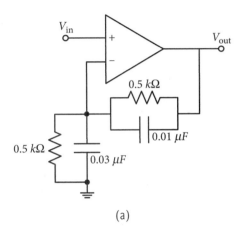

(a)

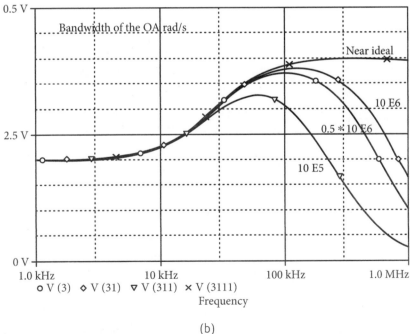

(b)

Figure 2.12 (a) Non-inverting amplifier circuit for Example 2.4. (b) Magnitude response of the circuit of Figure 2.12(a) with ideal OA and OA with bandwidth = 10^6, 0.5×10^6 and 10^5 rps.

2.5   Biquadratic Section: Parameters ω_o and Q_o

Moving on from first-order sections to second-order ones, transfer functions of second-order sections can be easily obtained by cascading two first-order sections. The overall transfer function is simply the product of the individual transfer function of the first-order sections, provided they satisfy the condition of cascading, that is, they have very high input impedance and very low output impedance. However, such second-order sections realize poles only on the negative real axis (with OAs considered ideal), which can be realized even with only passive elements. This is where the following second-order section, given by equation (2.28), comes in. The equation gives a basic module that is used in different ways to construct various higher-order filter sections, for example, in cascade or multiple feedback form.

$$H(s) = \frac{N(s)}{D(s)} = \frac{a_2 s^2 + a_1 s + a_0}{b_2 s^2 + b_2 s + b_0} \tag{2.28}$$

It is, therefore, useful to concentrate first on such sections expressed in terms of poles and zeroes as follows:

$$H(s) = \frac{a_2}{b_2} \frac{(s + z_1)(s + z_2)}{(s + p_1)(s + p_2)} \tag{2.29}$$

The importance of the aforementioned section, which is commonly known as a *biquadratic section* comes from the condition that the poles are complex conjugate; the zeroes may or may not be complex conjugate. With the condition that the poles are in conjugate pair form, the transfer function can be expressed in terms of real and imaginary parts of the zeros $R_e(z_1)$ and $I_m(z_1)$, and the real and imaginary parts of the poles, $R_e(p_1)$ and $I_m(p_1)$ as follows:

$$H(s) = k \frac{s^2 + \left[2R_e(z_1)\right]s + R_e^2(z_1) + I_m^2(z_1)}{s^2 + \left[2R_e(p_1)\right]s + R_e^2(p_1) + I_m^2(p_1)} = k \frac{s^2 + (\omega_z / Q_z)s + \omega_z^2}{s^2 + (\omega_o / Q_o)s + \omega_o^2} \tag{2.30}$$

Here ω_o is the pole frequency, which is given in terms of its real and imaginary parts as follows:

$$\omega_o^2 = R_e^2(p_1) + I_m^2(p_1) \tag{2.31a}$$

At frequency ω_o, the gain function becomes approximately the highest. At the zero frequency ω_z, the gain function become approximately the least and its relation is given as follows:

$$\omega_z^2 = R_e^2(z_1) + I_m^2(z_1) \tag{2.31b}$$

Conjugate zeros and poles, and their real and imaginary components are shown in Figure 2.13.

For the biquadratic function of equation (2.28) or equation (2.30), dc gain is given as follows:

$$20\log_{10}|H(j0)| = 20\log_{10}\left(k\,\omega_z^2/\omega_o^2\right) \tag{2.32}$$

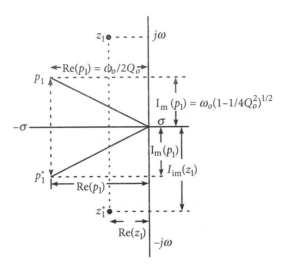

Figure 2.13 Conjugate zeroes and poles of a second-order section, located on the s plane and their relation with Q_o.

And the asymptotic gain for ω reaching infinity is as follows:

$$20\log_{10}|H(j\infty)| = 20\log_{10}(k) \tag{2.33}$$

Another important parameter which defines the sharpness of the magnitude response near the maxima, $|H(j\omega_o)|$ is known as pole quality factor Q_o, which is given as follows:

$$Q_o = \frac{\omega_o}{2R_e(p_1)} = \left[\frac{R_e(p_1)^2 + I_m(p_1)^2}{2R_e(p_1)}\right] \tag{2.34}$$

Whereas the depth of $|H(j\omega_z)|$ is defined by the *zero quality factor* Q_z, given as follows:

$$Q_z = \frac{\omega_z}{2R_e(z_1)} = \left[\frac{R_e(z_1)^2 + I_{im}(z_1)^2}{2R_e(z_1)}\right] \tag{2.35}$$

In most cases, $R_e(z_1) = 0$, which means $Q_z \to \infty$ and $\omega_z = I_{im}(z_1)$. This results in infinite (ideally) attenuation at ω_z. From equation (2.34), $R_e(p_1)$ can be expressed as follows, which is shown in Figure 2.13.

$$R_e(p_1) = \omega_o/2Q_o \tag{2.36}$$

Combining equations (2.34) and (2.31a), we get the following relation.

$$I_m(p_1) = \omega_o \left(1 - 1/4Q_o^2\right)^{1/2} \tag{2.37}$$

This is also shown in Figure 2.13, and from equations (2.36) and (2.37), we get:

$$\left\{R_e^2(p_1) + I_m^2(p_1)\right\}^{1/2} = \omega_o \tag{2.38}$$

which means that for all values of Q_o, the location of pole p_1 will lie on a circle with radius ω_o. For $Q_o = 0.5$, the poles became real; whereas for high Q_o, the poles are close to the imaginary axis.

2.6 Responses of Second-order Filter Sections

It is important to note that the zero ω_z can be anywhere on the s plane while deciding the nature of the filter, namely, low pass (LP), high pass (HP), and band pass (BP), and pole frequency ω_o and pole quality factor Q_o are the main design parameters. It is the value of ω_o which differentiates between the pass band and stop band of the LP and HP filters or decides the center frequency of the BP or BR (band reject) filters. The value of Q_o does have an effect on the gain response of the LP and HP sections at ω_o, but it is most significant in deciding the quality of BP or BR filters. Significance of ω_o and Q_o will be illustrated in detail in the next sections.

2.7 Second-order Low Pass Response

When $a_1 = a_2 = 0$, in equation (2.28), the expressions of the transfer function $H(s)$ in equations (2.28)–(2.30) will change to that of a second-order LP transfer function. Since the constant k is only a magnitude multiplier and does not affect the frequency response, it can be scaled, and the LP transfer function can be written as follows:

$$H_{LP}(s) = \frac{k\omega_o^2}{s^2 + (\omega_o/Q_o)s + \omega_o^2} \tag{2.39}$$

The network analysis is done assuming the input to be sinusoidal (or to be a combination of sinusoidal signals), $s \rightarrow j\omega$. Hence, the magnitude and phase of the LP transfer function shall be as follows:

$$\left|H_{LP}(j\omega)\right| = \frac{k\omega_o^2}{\left[(\omega_o^2 - \omega^2) + (\omega_o/Q_o)^2 \omega^2\right]^{1/2}} \tag{2.40}$$

$$= \frac{k}{\left[\left(1-\omega_n^2\right)+(\omega_n/Q_o)^2\right]^{\frac{1}{2}}} \tag{2.41}$$

$$\varphi_{\mathrm{LP}} = -\tan^{-1}\frac{(\omega/\omega_o Q_o)}{(1-\omega^2/\omega_o^2)} \tag{2.42}$$

$$= -\tan^{-1}\frac{(\omega_n/Q_o)}{(1-\omega_n^2)} \tag{2.43}$$

In equations (2.41) and (2.43), $\omega_n = (\omega/\omega_o)$ is termed as a *normalized frequency*.

Magnitude and phase function of the LP are shown in Figure 2.14(a) and (b), respectively, where the magnitude function is as follows:

$$|H_{\mathrm{LP}}(j0)| = k, \ |H_{\mathrm{LP}}(j\omega_o)| = kQ_o \text{ and } |H_{\mathrm{LP}}(j\infty)| \to 0 \tag{2.44}$$

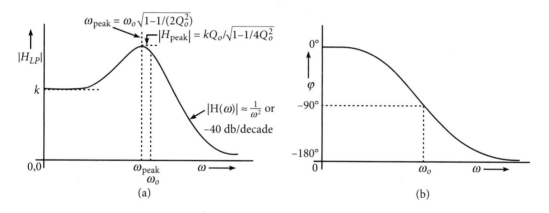

Figure 2.14 (a) Gain variation of a second-order low pass section, and (b) phase variation of the second-order low pass section for all values of Q_o.

These are shown in Figure 2.14(a), where the peak value is obtained by differentiating the magnitude function. Peak value of the transfer function H_{peak}, and the frequency at which it occurs ω_{peak}, are respectively given as follows:

$$H_{\mathrm{peak}} = kQ_o/\left\{1-1/4Q_o^2\right\}^{\frac{1}{2}} \cong kQ_o \tag{2.45}$$

$$\omega_{\mathrm{peak}} = \omega_o\left\{1-1/2Q_o^2\right\}^{\frac{1}{2}} \cong \omega_o \tag{2.46}$$

Approximation in equations (2.45) and (2.46) are satisfactory only with large values of Q_o. For $\omega \gg \omega_o$, rate of drop of the magnitude function is proportional to $(1/\omega^2)$ or -20 dB/dec. The rate of drop for a second-order section is also known as *two-pole roll-off*, as compared to *single-pole roll-off* for a first-order transfer function having only one pole.

For the LP filter, sometimes it is desirable not to have a significant peak in the pass band. However, it is shown in Figure 2.14(a) that relative to $|H_{LP}(j0)|$, H_{peak} is larger by Q_o times, which implies that for avoiding significant peak, Q_o should have a low value (say < 0.9) for the LP filters. For $Q_o = 0.707$, equation (2.46) gives that the peak of the magnitude function shall occur at $\omega = 0$.

It is important to note that while designing LP filters, the usual specifications are given in terms of the *half-power frequency* ω_c, where $|H_{LP}(j\omega_c)|$ is 0.707 times its value at dc $|H_{LP}(j0)|$ (for $Q_o < 0.9$). Since the gain (v_o/v_{in}) falls by a factor of 0.707 at the *half-power frequency* ω_c, it is also called –3 dB frequency. Another required specification for the LP filter design is $|H_{LP}(j0)|$, which decides the gain required by the filter at dc.

For the phase function shown in Figure 2.14(b), value of phase change with ω is as follows:

$$\varphi(0) = 0, \ \varphi(\omega_o) = -90°, \text{ and } \varphi(\omega \to \infty) \to -180° \tag{2.47}$$

Example 2.5: Show that the circuit in Figure 2.15 behaves as a second-order LP function. Design it for $\omega_o = 10$ krad/s and $Q_o = 1/\sqrt{2}$ and $\sqrt{2}$.

Solution: Taking OA as ideal, nodal equations at nodes 2 and 3, respectively, are as follows:

$$V_1(G_1 + G_2 + G_3 + sC_2) - V_{out}G_2 - V_{in}G_1 = 0 \tag{2.48}$$

$$V_1G_3 + V_{out} sC_1 = 0 \tag{2.49}$$

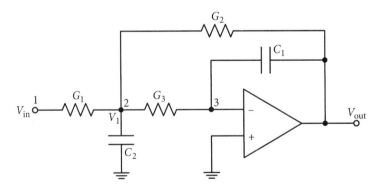

Figure 2.15 *Second-order low pass filter section for Example 2.5.*

Combining equations (2.48) and (2.49), the transfer function is obtained as follows:

$$\frac{V_{out}}{V_{in}} = -\frac{(G_1G_3/C_1C_2)}{s^2 + s\{(G_1 + G_2 + G_3)/C_2\} + (G_2G_3/C_1C_2)} \tag{2.50a}$$

It gives the expressions for ω_o and Q_o as:

$$\omega_o^2 = \frac{G_2 G_3}{C_1 C_2}, \quad Q_o = \frac{C_2}{G_1 + G_2 + G_3} \left(\frac{G_2 G_3}{C_1 C_2} \right)^{1/2} \text{ and } k = \left(\frac{G_1 G_3}{G_2 G_3} \right) \tag{2.50b}$$

Selecting $R_2 = R_3 = 5$ kΩ, with $\omega_o = 10$ krad/s, we get the following from equation (2.50b):

$$C_1 C_2 = 0.04 \times 10^{-14} \tag{2.51a}$$

Corresponding to $Q_o = \sqrt{2}$, selecting $R_1 = 1$ kΩ gives a dc gain of $k = 5$. Required values of the capacitors are obtained from equations (2.50b) and (2.51a) as follows:

$$C_1 = 2.0206 \text{ nF and } C_2 = 0.1974 \text{ } \mu\text{F} \tag{2.51b}$$

Figure 2.16 shows the magnitude responses of the PSpice simulation of the LP filter having Q_o value as $1/\sqrt{2}$ and $\sqrt{2}$. Magnitude response for $Q_o = 1/\sqrt{2}$ does not show any peak and its 3 dB frequency is 1.582 kHz (9.944 krad/s) with a dc gain of 5. However, the response for the corresponding LP filter with $Q_o = \sqrt{2}$, for which, with the same resistance values, the capacitances required are $C_1 = 4.04$ nF and $C_2 = 0.09899$ μF, shows a peak gain of 7.526 at a frequency of 1.378 kHz (8.661 krad/s) in conformity with equations (2.46) and (2.45). Figure 2.16 also shows the corresponding phase responses for the two cases. Though the rate of variation in phase differs, in both the cases, a phase shift of 90° occurs at 1.592 kHz (10.0068 krad/s).

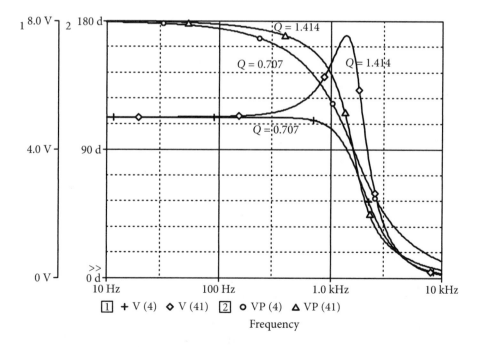

Figure 2.16 Magnitude and phase response of the low pass filter of Figure 2.15 with $Q = \sqrt{2}$ and $1/\sqrt{2}$.

2.7.1   Earthworm seismic data acquisition: application example

The Earthworm System is a seismic network data acquisition and processing system developed by the US Geological Survey in the 1990s [2.3]. The system contained a number of real time electronic seismic wave forms (may be more than 16) that were fed to a multichannel digitizer (consisting of one, two or four 64 channel multiplexer boards).

Like any other data acquisition system, this system also faced the problem of picking up noise. In the beginning, passive filters were used to eliminate/reduce noises. However, passive filters introduced a 24 kΩ impedance between the source and the input. To overcome this limitation, a two-pole LP active filter using a single non-inverting OA as shown in Figure 2.17(a) was employed [2.4]. Quad OA Tl064 provided low impedance while consuming less power. Consumption of less power was an important parameter as a large number of such filters were used in the system.

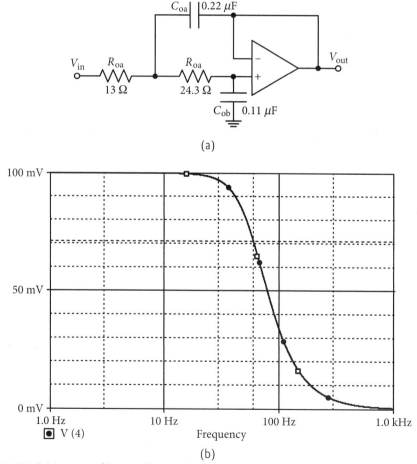

(a)

(b)

Figure 2.17   (a) Low pass filter used in the Earthworm System [2.3]. (b) Simulated response of the low pass filter of Figure 2.17(a).

Figure 2.17(b) shows the simulated response with the 3 dB frequency being 60.17 Hz. It may be noted that even a simple filter can be utilized for major projects.

2.8   Second-order High Pass Response

A biquadratic function can be converted to an HP response when coefficients $a_0 = a_1 = 0$. Equation (2.30) is modified as follows:

$$H_{HP}(s) = \frac{ks^2}{s^2 + (\omega_o / Q_o)s + \omega_o^2} \tag{2.52}$$

Here k is the high frequency gain. The gain function of the HP section is shown in Figure 2.18, which is very similar in nature with the LP response,

$$\omega_{peak} = \omega_o / \left\{1 - 1/2Q_o^2\right\}^{1/2} \cong \omega_o \tag{2.53a}$$

$$H_{peak} = kQ_o / \left\{1 - 1/4Q_0^2\right\}^{1/2} \cong kQ_o \tag{2.53b}$$

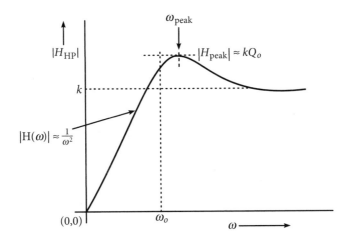

Figure 2.18   Gain variation of a second-order high pass section.

Approximation in equations (2.53) and (2.46) are satisfactory with large values of Q_o. Once again, its gain drops at a rate of –40 dB/dec in the pass band for $\omega \ll \omega_c$, and the gain is 3 dB below the high frequency gain, of $|H_{HP}(\omega \to \infty)| = k$ at the half-power frequency ω_c (for $Q_o < 0.9$). The rate of gain drop is also known as *two-pole roll-off*, as compared to the *single-pole roll-off* for a transfer function having one pole only.

For the HP filter, it is desirable not to have a significant peak in the pass band. However, it is shown in Figure 2.18 that relative to $|H_{HP}(j\infty)|$, H_{peak} is larger by nearly Q_o times, which implies that to avoid significant peak, Q_o should have a low value (say < 0.9). For $Q_o = 0.707$, equation (2.53b) gives that the peak of the magnitude function shall occur at $\omega = \infty$.

It is important to note that while designing an HP filter, the usual specifications are given in terms of the *half-power frequency* ω_c, where the $|H_{HP}(j\omega)|$ is 0.707 times its value at infinity, $|H_{HP}(j\infty)|$ (for $Q_o < 0.9$). Since the gain (v_o/v_{in}) falls by a factor of 0.707 at the *half-power frequency* ω_c, it is called –3 dB frequency. Another required specification for the HP filter design is $|H_{HP}(j\infty)|$, which decides the gain required by the filter at very high frequencies.

Example 2.6: Show that the circuit in Figure 2.19(a) behaves as a second-order HP function. Design it for $\omega_o = 10$krad/s and $Q_o = 1/\sqrt{2}$ and $\sqrt{2}$.

Solution: Taking OA as ideal, the nodal equations at nodes 2 and 3, respectively, are as follows:

$$V_1(sC_1 + G_2 + sC_3 + sC_4) - V_{out}sC_4 - V_{in}sC_1 = 0 \tag{2.54a}$$

$$V_1 sC_3 + V_{out}G_5 = 0 \tag{2.54b}$$

Combining equations (2.54a) and (2.54b), the transfer function is obtained as follows:

$$\frac{V_{out}}{V_{in}} = -\frac{s^2(C_1/C_4)}{s^2 + s\{G_5(C_1 + C_3 + C_4)/C_3C_4\} + (G_2G_5/C_3C_4)} \tag{2.55a}$$

This gives the expressions for ω_o and Q_o as

$$\omega_o^2 = G_2G_5/C_3C_4, \quad Q_o = \frac{C_3C_4}{G_5(C_1 + C_3 + C_4)}\left(\frac{G_2G_5}{C_3C_4}\right)^{\frac{1}{2}} \text{ and } k = (C_1/C_4) \tag{2.55b}$$

Selecting $C_3 = C_4 = 100$ nF, we get the value of $C_1 = 500$ nF for $k = 5$, and the following relation from equation (2.55b) with $\omega_o = 10$ krad/s:

$$G_2G_5 = 10^{-6} \tag{2.56}$$

Corresponding to $Q_o = \sqrt{2}$, using equation (2.55b), and selecting $R_5 = 10$ kΩ, the required values of the resistance $R_2 = 100$ Ω is obtained from equation (2.56).

Figure 2.19(b) shows the magnitude responses of the PSpice simulation of the HP filter having $Q_o = 1/\sqrt{2}$ and $\sqrt{2}$. Magnitude response for $Q_o = 1/\sqrt{2}$ does not show any peak and its 3 dB frequency is 1.602 kHz (10.069 krad/s) with a dc gain of 5.09. However, the response for the corresponding HP filter with $Q_o = \sqrt{2}$, for which, with the same capacitance values, the required resistances are $R_2 = 202$ Ω and $R_5 = 4.949$ kΩ, shows a peak gain of 7.598 at a frequency of 1.8078 kHz (11.363 krad/s) in conformity with equations (2.53a) and (2.53b).

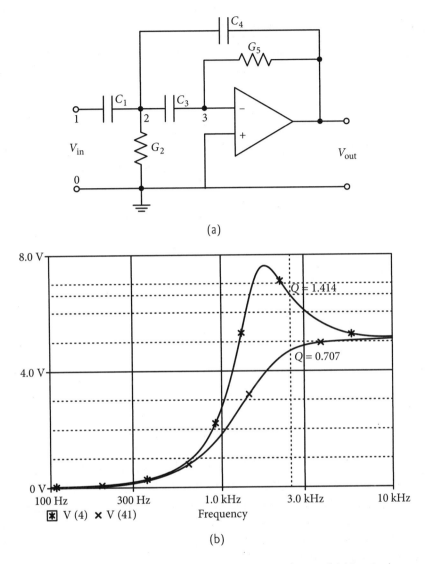

Figure 2.19 (a) Second-order high pass filter circuit for Example 2.6. (b) Magnitude response of the high pass filter of Figure 2.19(a) with $Q = \sqrt{2}$ and $1/\sqrt{2}$.

2.9 Second-order Band Pass Response

When $a_0 = a_2 = 0$, a biquadratic section become a BP section, whose transfer function is given as follows:

$$H_{\text{BP}}(s) = \frac{k(\omega_0 / Q_o)s}{s^2(\omega_o / Q_o)s + \omega_o^2} \tag{2.57}$$

Hence, $\left|H_{\text{BP}}\left(j\omega\right)\right| = \dfrac{ks\omega(\omega_o / Q_o)}{\left[(\omega_o^2 - \omega^2)^2 + (\omega_o / Q_o)^2 \omega^2\right]^{\frac{1}{2}}}$ (2.58)

and $\varphi\left(j\omega\right) = 90^{\circ} - \tan^{-1}\dfrac{(\omega\omega_o / Q_o)}{(\omega_o^2 - \omega^2)}$ (2.59)

Equations (2.58) and (2.59) are the magnitude and phase function of the BP section, which are sketched in Figure 2.20(a) and (b), respectively. Since $H_{\text{BP}}(s)$ has a zero at 0 and ∞, the magnitude function reduces to zero at dc and at infinite frequency; the peak occurs at $\omega = \omega_o$.

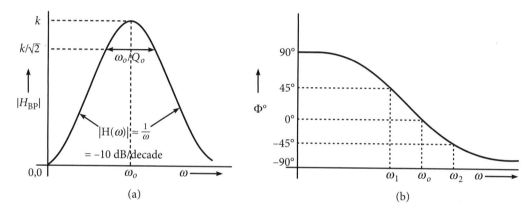

Figure 2.20 (a) Magnitude and (b) phase variation of a second-order band pass section.

The magnitude function drops from the peaks on both sides at a rate of 10 dB/dec with its value becoming 3 dB less than the peak value of k at the half-power frequencies ω_1 and ω_2.

For designing a BP section, important specifications include the bandwidth (BW), the distance between ω_1 and ω_2, or the range of frequencies for which the power output remains more than half of the peak power. The BW, ω_1 and ω_2 are found by putting the square of the magnitude function $\left|H_{\text{BP}}\left(j\omega\right)\right|^2 = (1/2)$. It gives

$$\omega_1, \omega_2 = \omega_o\left[\left\{1 + \left(1/2Q_o\right)^2\right\}^{\frac{1}{2}} \pm \left(1/2Q_o\right)\right]$$ (2.60)

The product and difference of the two frequencies are as follows:

$$\omega_1 \times \omega_2 = \omega_o^2 \text{ and } \left(\omega_2 - \omega_1\right) = \left(\omega_o / Q_o\right) = \text{BW}$$ (2.61)

which means that ω_o is the geometric mean of ω_1 and ω_2 and the BW is inversely proportional to the pole $Q(Q_o)$. Figure 2.21 shows the effect of the value of Q_o on the BP response, which becomes thinner/ sharper as Q increases.

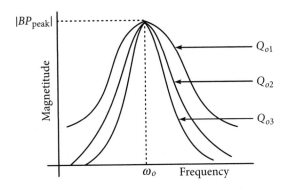

Figure 2.21   Typical response of a band pass filter with varying Q_o $(Q_{o1} < Q_{o2} < Q_{o3})$.

Regarding the phase-function of equation (2.59), it is observed that it is similar to that for the LP case except with the addition of 90° at dc, which means that it asymptotes at −90° for $\omega \to \infty$. Moreover, the values for both the frequencies ω_1 and ω_2 are 45° and −45° from equation (2.59).

Example 2.7: Figure 2.22 shows a single OA based BP filter. Derive its transfer function and compare the response for pole Q value of 2, 5 and 10 at a center frequency of 10 krad/s.

Solution: Considering OA as ideal, with its inverting terminal at virtual ground, the nodal equations at terminal 2 and 3, respectively, are obtained as follows:

$$V_1 (G_1 + G_2 + sC_1 + sC_2) - V_{out} sC_2 - V_{in} G_1 = 0 \tag{2.62}$$

$$V_1 sC_1 + V_{out} G_3 = 0 \tag{2.63}$$

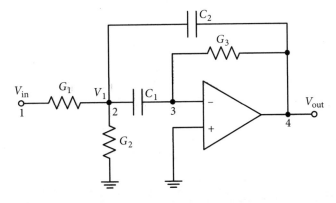

Figure 2.22   A second-order band pass filter section for Example 2.7.

Combining equations (2.62) and (2.63), the transfer function is obtained as:

$$\frac{V_{out}}{V_{in}} = -\frac{s(G_1/C_2)}{s^2 + sG_3\{(C_1+C_2)/C_1C_2\} + \{G_3(G_1+G_2)/C_1C_2\}} \tag{2.64}$$

It gives the expressions for ω_o and Q_o as

$$\omega_o^2 = \frac{G_3(G_1+G_2)}{C_1C_2} \text{ and } Q_o = \frac{\left\{\left(\dfrac{C_1C_2}{G_3}\right)(G_1+G_2)\right\}^{1/2}}{(C_1+C_2)} \tag{2.65}$$

Selecting equal values for capacitors $C_1 = C_2 = 0.005\ \mu F$, for $\omega_o = 10\text{krad/s}$, equation (2.65) provides the following element values.

For $Q_o = 2$, $R_1 = R_2 = 10\ k\Omega$ and $R_3 = 80\ k\Omega$.

For $Q_o = 5$, $R_1 = R_2 = 4\ k\Omega$ and $R_3 = 200\ k\Omega$.

For $Q_o = 10$, $R_1 = R_2 = 2\ k\Omega$ and $R_3 = 400\ k\Omega$.

Figures 2.23 and 2.24 show the magnitude and phase response for the aforementioned three cases; the respective center frequencies, bandwidth, and quality factor obtained through PSpice simulation is as follows:

$f_o = 1.587$ kHz, bandwidth BW = 790 Hz, resulting in $Q_o = 2.008$.

$f_o = 1.578$ kHz, bandwidth BW = 313.7 Hz, resulting in $Q_o = 5.033$.

$f_o = 1.567$ kHz, bandwidth BW = 162.9 Hz, resulting in $Q_o = 10.1$.

As the responses and the resulting parameters show, the circuit works very well at this frequency range.

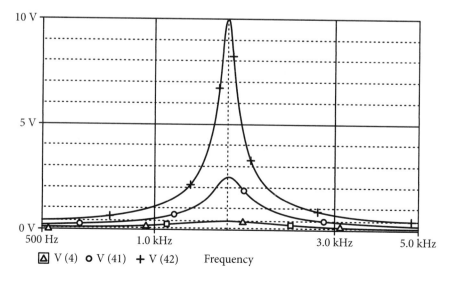

Figure 2.23 Magnitude response of the band pass filter of Figure 2.22 with $Q = 2, 5$ and 10.

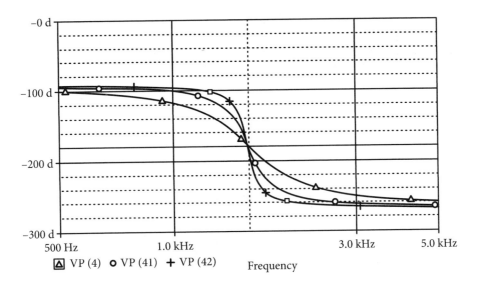

Figure 2.24 Phase response of the band pass filter of Figure 2.22 with $Q = 2$, 5 and 10.

2.10 Band Reject (BR) Response

A BR response, which passes all signals except those falling in certain band of frequencies, is obtained by putting $a_1 = 0$ in the biquadratic function of equation (2.28). It results in the following:

$$H_{BR}(s) = \frac{a_2 s^2 + a_0}{s^2 + (\omega_o / Q_o)s + \omega_o^2} \tag{2.66}$$

$$= \frac{K(s^2 + \omega_z^2)}{s^2 + (\omega_o / Q_o)s + \omega_o^2} \tag{2.67}$$

Here, $K = |H_{BR}(j\omega)|$ is the gain as $\omega \to \infty$ and the rejection band of frequencies is centered at $\omega = \omega_z$ as the numerator has a zero at ω_z. It is the value of Q_o which determines the rate of change of the BR response beyond ω_z, as well as the amount of bump in the response.

The BR filter is also known as *notch filter* because of the shape of the magnitude characteristics. However, depending on the relative value of ω_z in comparison to ω_o, notch filter is called a *symmetrical notch, high pass notch* (HPN) or a *low pass notch* (LPN) for $\omega_o = \omega_z$, $\omega_o > \omega_z$ and $\omega_o < \omega_z$, respectively. The three types of notch responses are shown in Figure 2.25(a), (b), and (c). Here the bump in HPN or LPN occurs at $\omega = \omega_{peak}$. Expressions for the frequency ω_{peak} and the maxima of the transfer function for the LPN and HPN, which occurs at ω_{peak} are given as:

$$\omega_{\text{peak}} = \omega_o \sqrt{\left[1 + \frac{1}{\left|\left\{1-(\omega_o\omega_z)^2\right\}\right|2Q_o^2}\right]}$$

(2.68)

$$|H_{\text{BR}}(j\omega)|_{\max} = KQ_o|\{1-(\omega_z/\omega_o)^2\}|$$

(2.69)

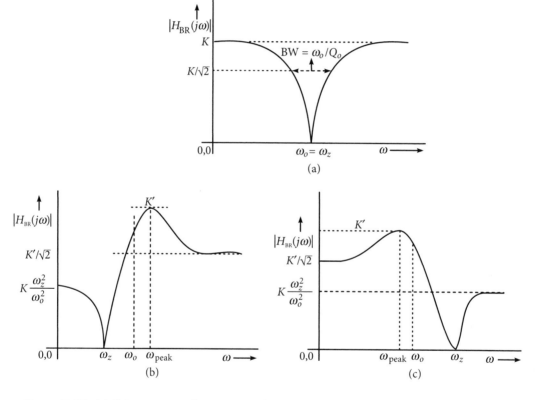

Figure 2.25 (a) Gain response of a symmetrical notch, (b) high pass notch, and (c) low pass notch with $K' = |H_{\text{BR}}(j\omega)|_{\max}$.

Bandwidth of the BR filter is same as that for the BP filter:

$$(\text{BW})_{\text{BR}} = \omega_o/Q_o$$

(2.70)

2.11 Second-order All Pass Response

An all pass (AP) filter has constant magnitude response for all frequencies. For this type of filter to be realized, coefficients of the biquadratic section are selected in such a way that the transfer function becomes:

$$H_{AP}(s) = K \frac{s^2 - (\omega_o / Q_o)s + \omega_o^2}{s^2 + (\omega_o / Q_o)s + \omega_o^2} \tag{2.71}$$

Hence, for sinusoidal input

$$H_{AP}(j\omega) = K \frac{(\omega_o^2 - \omega^2) - j\omega(\omega_o / Q_o)}{(\omega_o^2 - \omega^2) + j\omega(\omega_o / Q_o)} \tag{2.72}$$

Here, $H_{AP}(j\omega)$ has to remain constant for all frequencies and the phase and delay of the AP filter are obtained as follows

$$\varphi_{AP}(\omega_o) = -2\tan^{-1} \frac{\omega(\omega_o / Q_o)}{(\omega_o^2 - \omega^2)} \tag{2.73}$$

$$D_{AP}(\omega_o) = 2\left(\frac{\omega_o}{Q_o}\right) \frac{(\omega_o^2 + \omega^2)}{\left\{(\omega_o^2 - \omega^2)^2 + \omega^2 \left(\frac{\omega_o}{Q_o}\right)^2\right\}} \tag{2.74}$$

Figure 2.26 shows the variation of phase of the AP filter for a certain value of Q_o along with its magnitude response and Figure 2.27 shows the variation of one-half delay for a few values of Q_o. It is observed that for $Q_o = 1/\sqrt{3}$, the delay become *maximally flat*, whereas for $Q_o > 1/\sqrt{3}$, the delay variations have a peak.

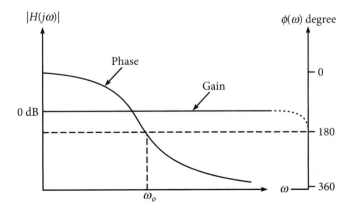

Figure 2.26 Variation of phase and gain response of a second-order all pass filter for a certain value of Q_o.

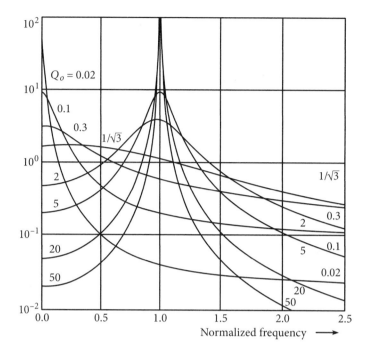

Figure 2.27 One-half delay of second-order all pass filter as a function of Q_o.

In the second-order BR as well as AP filters, finite zeroes are to be realized for which realization methods are a bit different than LP, HP, and BP types. It is for this reason that simulation examples for notch and AP filters shall be taken up at a later stage.

2.12 Effect of Operational Amplifier's Pole on Biquads

Finite frequency dependent gain of the OA, represented by the single-pole roll-off model of equation (1.17) introduces one extra pole for the first-order filter section. In fact, it introduces as many extra poles as the order of the filter section. It affects the filter characteristics by changing all its important parameters like gain, cut-off/ pole frequency and rate of fall of the signal in the stop band; each with varying degree. Amount of variation in the parameter depends on the filter specifications (values of the required gain, pole frequency and pole Q), finite value of the gain–bandwidth product of the OA, and on the method (structure) used for the realization of the filter, like generating biquads using the two-integrator loop method, or any single amplifier generating biquad method, and then cascading them using these biquads in a multiple feedback structure or using direct forms of realizations for higher-order filters. Hence, the effect of OA's poles shall be taken up in later chapters along with different methods of filter realization and the corrective steps applied to overcome the deviations occurring in the filter parameters.

References

[2.1]   Van Valkenburg, M. E. 1976. *Introduction to Modern Network Synthesis*. India: Wiley Eastern Limited.

[2.2]   Trainham, R., M. O. Neill, and I. J. McKenna. 2015. 'An Analog Filter Approach to Frequency Domain Fluorescence Spectroscopy,' *Journal of Fluorescence* 25: 1801.
https://doi.org/10.1007/s10895-015-1669-z

[2.3]   Johnson, Carl E., Alex Bittenbinder, Barbara Bogaert, Lynn Dietz, and Will Kohler. 1995. 'Earthworm: A Flexible Approach to Seismic Network Processing,' *IRIS News Letter* 14 (2): 1–4.

[2.4]   Jenson, E. G. 2000. 'A Filter Circuit Board for the Earthworm Seismic Data Acquisition System.' Open-file Report 00-379. California: US Geological Survey.

Practice Problems

2-1   Find the transfer function for the circuit shown in Figure P2-1. Calculate and verify the frequency at which its gain changes by 3 dBs from dc level using PSpice, with $R_1 = R_2 = 10\ k\Omega$ and $C = 1\ nF$. Consider the OA as near ideal.

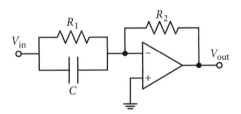

Figure P2.1 Figure for Problem 2-1 and 2-3.

2-2   Find the transfer function for the circuit shown in Figure P2-2 Calculate 3 dB frequency and test the circuit using PSpice with $R_1 = R_2 = 10\ k\Omega$ and $C = 1\ nF$. Consider the OA as near ideal.

2-3   Repeat the problem 2-1 with the bandwidth of the OA as (a) 100 krad/s, (b) 50 krad/s, and (c) 25 krad/s. Find percentage error in the gain at the 3-dB frequency level for the three cases and compare with the case when OA was considered near ideal in the problem 2-1.

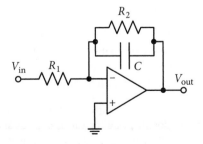

Figure P2.2 Figure for Problem 2-2 and 2-4.

2-4   Repeat the problem 2-2 with the bandwidth of the OA as (a) 100 krad/s, (b) 50 krad/s, and (c) 25 krad/s. Compare the gain at the 3-dB frequency for the three cases and find percentage error in it with the case when OA was considered near ideal in the problem 2-2.

2-5   Find the transfer function for the circuit shown in Figure P2-3 Calculate the peak magnitude and find the frequency at which it occurs using PSpice with $R_1 = R_2 = 10$ kΩ and $C_1 = C_2 = 2$ nF. Consider the OA as near ideal.

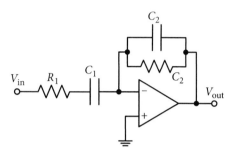

Figure P2.3 Figure for Problem 2-5 and 2-6.

2-6   Repeat the problem 2-5 with the bandwidth of the OA as (a) 100 krad/s, (b) 50 krad/s, and (c) 25 krad/s. Compare the frequency at which peak gain occurs and obtain percentage error in the result for the three cases with the case when OA was considered near ideal in the problem 2-5.

2-7   Find the transfer function for the circuit shown in Figure P2-4. Test the circuit using PSpice with $R_1 = R_2 = 10$ kΩ and $C = 0.5$ nF and find the frequency at which gain drops by 3 dBs. Consider the OA as near ideal.

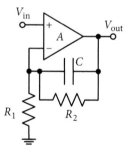

Figure P2.4 Figure for Problem 2-6 and 2-7.

2-8   Repeat the problem 2-7 with the bandwidth of the OA as (a) 100 krad/s, (b) 50 krad/s and (c) 25 krad/s. Compare the result for the three cases with the case when OA was considered ideal in the problem 2-7 while finding error in the frequency at which gain falls by 3 dBs.

2-9   Figure P2.5 shows a second-order passive RLC filter. (a) Derive its transfer function and mention the type of response given by the filter section. (b) Find the values of poles and zeroes when $R = 500$ Ω, $L = 10$ mH and $C = 0.04$ μF. (c) Calculate the parameters ω_o, Q_o and dc gain. (d) If the magnitude response has a peak, then what is the value of the voltage gain and at which frequency does it occur?

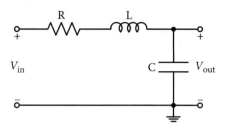

Figure P2.5 Figure for Problem 2-9.

2-10   For the circuit of Figure P2.5, calculate the value of Q_o in each case if R changes from 500 Ω to 250 Ω, 100 Ω and 50 Ω. Find the location of poles on the complex frequency variable plane and show that the poles lie on a semi-circle. What is the radius of the semi-circle?

2-11   Check whether peak in the magnitude response of the circuit in Problem 2-9 occurs at ω_o or not. Justify the location of the peak.

2-12   Repeat Problem 2-9 if location of inductor and the capacitor are interchanged.

2-13   Use the circuit of Figure 2.15 to design a second-order LP filter with the following specifications: cut-off frequency $f_o= 15.9$ kHz, $Q = 2.5$ and dc gain of zero dB. (b) Test the magnitude and phase with PSpice/EWB while using 741 type OA.

2-14   Calculate magnitude and phase of the LP filter with following specifications: $f_o = 1.59$ kHz and $Q = 2.5$. for frequencies $0.25 \times f_o$, $0.5 \times f_o$, f_o, $1.5 \times f_o$, $2 \times f_o$ and compare it with the simulated response.

2-15   Verify equations (2.45) and (2.46) for the LP filter of Problem 2-13 by comparing the parameters by obtaining theoretically and from the simulated response.

2-16   Derive the voltage ratio transfer function for the circuit shown in Figure P2.6. What kind of response is available from it?

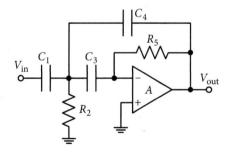

Figure P2.6 Figure for Problem 2-16.

2-17   Design the circuit of Figure P2.6 for critical frequency of 7.95 kHz and $Q_o = 2.5$ and test the magnitude and phase response.

2-18   Verify equations (2.55a) and (2.55b) for the filter section in Problem 2-17 theoretically and from the practical/simulated test results.

2-19   Repeat Problem 2-14 for the filter of Figure P2.6.

2-20   (a)   Determine suitable element values for the realization of the BP filter shown in Figure 2.22 for realizing pole $Q_0 = 5$ and a complex pole pair lying on a circle of radius = 50 krad/s.

   (b)   Determine the peak gain.

   (c)   Determine the spread in element values.

   (d)   Determine error in complex pole radius and Q_0 when OA has $B = 500$ krad/s.

2-21   Redesign the circuit in Figure 2.22 for $\omega_0 = 40$ krad/s and $Q_0 = 10$.

   (a)   Calculate and verify the simulated value of the filter bandwidth while using ideal OA.

   (b)   Repeat (a) for OA with $B = 400$ krad/s.

   (c)   Calculate and verify the phase shift of the filter at 3 dB frequencies.

2-22   Derive equations (2.73) and (2.74).

Chapter 3

Magnitude Approximations

3.1 Introduction

Filters are generally classified in the frequency domain in terms of the amplitude and phase response of their transfer function; though sometimes they are expressed in the time domain as well. The typical characteristics of an ideal LPF (low pass filter) in terms of its variation of attenuation with frequency shown in Figure 1.6(a) is redrawn in Figure 3.1. The transition of the filter from being a pass band to being a stop band occurs abruptly at $\Omega = 1$.

It is well known that the transfer function of an ideal filter, in which transition between pass band and stop band is instant, is physically realizable only by using an infinite number of elements [3.1]. For a practically realizable filter, the transfer function is always expressed by a *real rational function H(s)*, which is a ratio of polynomials in complex variable, $s = (\sigma + j\omega)$ as already given in Section 1.2.1 and repeated here as equation (3.1).

$$H(s) = \frac{N(s)}{D(S)} = \frac{a_m s^m + a_{m-1} s^{m-1} + \ldots + a_2 s^2 + a_1 s + a_0}{b_n s^n + b_{n-1} s^{n-1} + \ldots + b_2 s^2 + b_1 s + b_0} \tag{3.1}$$

In a real rational transfer function, coefficients a_i and b_j are real numbers and the degree of the numerator and the denominator is m and n, respectively. Moreover, the degree of the denominator, n, should be more than or equal to the numerator degree, m, for the physical realization of the transfer function using finite number of elements to be feasible. The condition $n \geq m$ is necessary because ideal filters are non-causal and, therefore, cannot be implemented practically.

To realize a practical form of an LPF, shown as approximated LPF characteristics using a dotted line in Figure 3.1, values of the coefficients a_i and b_j in equation (3.1) are to be determined. The next step will be to find the topology of the filter and values of element to be used, applying the coefficients of equation (3.1). Not only are different methods available for the realization of an arbitrary transfer function, but different forms of approximating the magnitude or phase of the transfer function are also available. Some classical methods of approximating the magnitude function are discussed in this chapter.

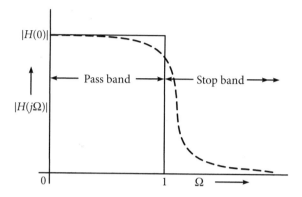

Figure 3.1   Magnitude characteristics of an ideal normalized low pass filter shown by solid line, and that of a practical or real filter shown by dotted line.

The procedure of magnitude approximation which begins by comparing an ideal LPF mathematically, with that of an approximated response is discussed in Section 3.2. One of the most commonly used approximations, namely the maximally flat Butterworth approximation and the design of a Butterworth approximation based LPF is studied in Section 3.3. Also included here is the utilization of a circuit structure in the form of a ladder, called a *lossless ladder*, containing only inductors and capacitors. Equal-ripple approximation is another very important class of magnitude approximation, whose sub-classifications – Chebyshev approximation, inverse Chebyshev approximation and Cauer approximation – have been found to realize filter sections rather economically. In the rest of the chapter, we will describe the development of prototype LPFs using these approximations. Examples have been included of filters of average level order ($n \sim 5,6$) filters. An example of a maximally flat pass band with finite zeros, the significance of which shall be seen later, is also included.

3.2   Magnitude Approximations

Response of the LPF shown by the dotted line in Figure 3.1 represents an approximation to the ideal LPF in terms of the magnitude of the transfer function. In the pass band region, gain of the transfer function is close to the ideal value at low frequencies; the gain reduces to a low value in the stop band region with a finite slope. In practice, the dotted line of the approximated response can take shapes other than the monotonic drop. Other important types of gain variation are discussed in brief in the following sections. Approximation can also be performed for the phase response of the ideal filter which shall be discussed in Chapter 4. In all the magnitude approximations of LPF, the transition band is finite instead of the abrupt transition from the pass band to stop band of the ideal filter. This means that there will be some deviation from the ideal, and hence, some error in the response. However, the amount of intentionally made error, shown in Figure 3.2, can be bounded, as the response has to remain restricted within the shaded region. The maximum allowable attenuation in the pass band is α_{max} and the minimum allowable attenuation in the stop band is α_{min}. The transition band separating the pass and stop band extends from ω_1 to ω_2. Depending on the specifications of the LPF in terms of $\alpha_{max}, \alpha_{min}, \omega_1$ and ω_2, the next step is to find the topology of a network and the element values which satisfy these specifications. It is important to note the term *normalized angular frequency* $\Omega = (\omega/\omega_c)$; by convention, this means normalized cut-off frequency $\Omega_c = 1$.

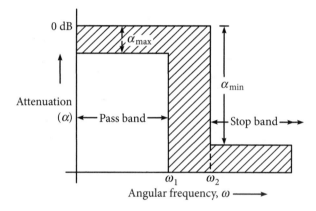

Figure 3.2   Approximated low pass characteristics lie within the shaded region.

Initially, an LP prototype transfer function is considered with all transmission zeros (zeros of the numerator) at infinity, that is, $N(s) = 1$; this is also commonly known as *all pole function*. A number of solutions can be obtained from the general amplitude function of the transfer function. Let us consider the magnitude squared transfer function: $|H(j\omega)|^2$:

$$|H(j\omega)|^2 = \frac{N(j\omega)N(-j\omega)}{D(j\omega)D(-j\omega)} = \left|\frac{N(j\omega)}{D(j\omega)}\right|^2 = \frac{A(\omega^2)}{B(\omega^2)} \tag{3.2}$$

$$= \frac{1}{|D(j\omega)|^2} \rightarrow \frac{1}{B(\omega^2)} \text{ for } N(s) = 1 \tag{3.3}$$

Here, $|H(j\omega)|^2$ is an even rational function for which $|H(j\omega)|$ must be close to $|H(j0)|$ within the frequency range $0 \leq \omega \leq \omega_1$ in the pass band and close to zero for $\omega \geq \omega_2$ in the stop band. Using suitable frequency normalization with respect to pass band edge frequency ω_1, the normalized pass band edge frequency shall be $\Omega = (\omega/\omega_1)$. Hence, the pass band range is up to $\Omega = 1$, and since $N(s)$ has been selected as unity, $|H(j0)| = 1$ for all values of n. The function $|H(j\omega)|$ now modifies into a normalized function $|H(j\Omega)|$. For a mathematical understanding, it is preferable to express $|H(j\Omega)|^2$ in terms of another rational function $|K(j\Omega)|$, such that:

$$|H(j\Omega)|^2 = \frac{1}{1+|K(j\Omega)|^2} \tag{3.4}$$

From equation (3.4), the following relation is obtained:

$$|K(j\Omega)|^2 = \{1/|H(j\Omega)|^2\} - 1 = |D(j\Omega)|^2 - 1 \tag{3.5}$$

Expression for the nth order magnitude squared function modifies from equation (3.3) as given in equation (3.6):

$$|H_n(j\Omega)|^2 = \frac{1}{B_{2n}\Omega^{2n} + B_{2n-2}\Omega^{2n-2} + \ldots + B_4\Omega^4 + B_2\Omega^2 + 1} \tag{3.6}$$

Therefore, the nth order function, $|K_n(j\Omega)|^2$ of equation (3.5) will transform to the following:

$$|K_n(j\Omega)|^2 = B_{2n}\Omega^{2n} + B_{2n-2}\Omega^{2n-2} + \ldots + B_4\Omega^4 + B_2\Omega^2 \tag{3.7}$$

This means that the squared magnitude of the characteristic function is a polynomial in Ω. It is the nature of $|K_n(j\Omega)|^2$ which give different forms of approximations for the ideal LPF (and as a consequence for other types of filter sections also) like maximally flat, Chebyshev, inverse Chebyshev or Cauer type.

3.3   Maximally Flat – Butterworth Approximation

A maximally flat response means that at $\Omega = 0$, not only is its slope (or its first derivative) zero, but the maximum number of derivatives are also equal to zero [3.2]. This stated condition requires that in equation (3.7), the maximum derivatives of K_n are zero, as shown in equation (3.8):

$$\frac{d^k |K_n(j\Omega)|^2}{d^k(\Omega^2)}\Big|_{\Omega=0} = 0 \text{ for } k = 1, 2, \ldots, n-1 \tag{3.8}$$

It means that we are required to make $B_{2n-2} = \ldots = B_4 = B_2 = 0$, resulting in the following expression, where ε is a characteristic term (a significant term effecting approximation):

$$|K_n(j\Omega)|^2 = B_{2n}\Omega^{2n} = \varepsilon^2\Omega^{2n} \tag{3.9}$$

Hence, for a maximally flat response, the magnitude of the squared nth order transfer function is expressed as:

$$|H_n(j\Omega)|^2 = \frac{1}{1+\varepsilon^2\Omega^{2n}} \rightarrow |H_n(j\Omega)| = (1+\varepsilon^2\Omega^{2n})^{-1/2} \tag{3.10}$$

Response given by equation (3.10) is shown in Figure 3.3. Its magnitude decreases monotonically, and at $\Omega = 1$, the loss becomes $10 \log_{10}(1 + \varepsilon^2)$ dB, as its magnitude drops from $|H_n(0)|$ to $|H_n(0)|/(1 + \varepsilon^2)^{0.5}$. Therefore, the expression for the maximum specified loss of α_{max} in the pass band shall be as follows:

$$\alpha_{max} = 10 \log_{10}(1 + \varepsilon^2) \tag{3.11}$$

This gives an important relation for the characteristic term ε as:

$$\varepsilon = \left(10^{0.1\alpha_{max}} - 1\right)^{\frac{1}{2}} \tag{3.12}$$

In the normalized magnitude form of the LP function with maximum flatness at dc ($\Omega = 0$), when $\varepsilon = 1$, it is also called the *Butterworth approximation*; the characteristics being very similar, the terms *maximally flat* and *Butterworth approximation* are sometimes used synonymously.

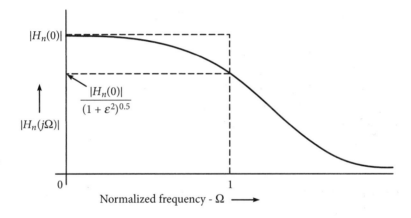

Figure 3.3   Maximally flat normalized low pass response.

For the Butterworth approximation if $\varepsilon = 1$, attenuation at the edge of the pass band, obtained from equation (3.11), is simply:

$$\alpha_{max} = 3 \text{ dB} \tag{3.13}$$

Substituting $\varepsilon = 1$ in equation (3.10), the nth order frequency de-normalized Butterworth response is obtained using the following relation, as mentioned here:

$$\left|H_n(j\omega)\right|^2 = \frac{1}{(1 + \omega^{2n})} \tag{3.14}$$

As all pole responses were selected in the beginning with $N(s) = 1$, the Butterworth response has zeros only at $\omega = \infty$. The response, shown in Figure 3.4, has the following important properties as well.

i.   Magnitude of the transfer function at $\omega = 0$ is unity for all values of n.

ii.   For all values of n, magnitude $|H_n| = 1/\sqrt{2}$ at $\Omega = 1 (\omega = \omega_c)$, corresponding to the attenuation of 3 dBs.

iii.   In the stop band, for $\omega > \omega_c$ $(\Omega > 1)$, $|H_n|$ decreases at the rate of $20n$ dBs per decade.

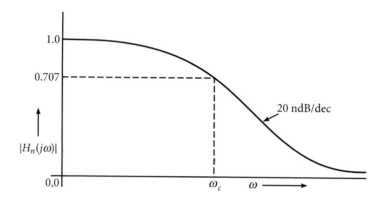

Figure 3.4   Butterworth response having loss of 3 dB at cutoff frequency ω_c.

The transfer function $H(s)$ of equation (3.1) will have n poles. In order to find poles with the Butterworth approximation ω is replaced by (s/j) in equation (3.14). Hence, the poles can be obtained by the roots in the left half-plane of the following relation:

$$D(s)D(-s) = 1 + (-s^2)^n \tag{3.15}$$

These poles have been found to be located on a semicircle in the s-plane whose value (location) can be evaluated from the following:

$$S_k = -\sin\left(\frac{2k-1}{n}\right)\pi \pm j\cos\left(\frac{2k-1}{n}\right)\pi \tag{3.16}$$

The pole locations for $k = 1, 2, \ldots, n$ (up to $n = 8$) are shown in Table 3.1. Coefficients of the Butterworth polynomial $D(s)$ can be obtained from the following recursive relation, with $b_0 = 1$.

$$b_k = b_{k-1}\left\{\cos\left(\frac{k-1}{n}\pi\right)\right\}\Big/\left\{\sin\left(\frac{k}{n}\pi\right)\right\} \tag{3.17}$$

Table 3.2 shows the calculated values of the coefficient b_k up to $n = 8$. Coefficient values in Table 3.2 and for any larger values of n can be calculated from equation (3.17).

Table 3.1   Pole locations for the Butterworth responses

$n = 2$	$n = 3$	$n = 4$	$n = 5$	$n = 6$	$n = 7$	$n = 8$
−0.7071068	−0.5000000	−0.3826834	−0.8090170	−0.2588190	−0.900968	−0.1950903
±j.7071068	±j0.8660254	±j0.9238795	±j0.587785	±j0.9659258	±j.4338837	±j0.9807853
	−1.0000000	−0.9238795	−0.3090170	−0.7071068	−0.2225209	−0.5555702
		±j0.3826834	±j0.9510565	±j0.7071068	±j.9649279	±j0.8314696
			−1.0000000	−0.9659258	−0.6234898	−0.8314696
				±j0.2588190	±j.7818315	±j0.5555702
					−1.0000000	−0.9807853
						±j0.1950903

Table 3.2   Coefficients of the Butterworth polynomial $B_n(s) = s^n + \sum\limits_{k=0}^{n-1} b_k s^k$

n	b_0	b_1	b_2	b_3	b_4	b_5	b_6	b_7
2	1.000	1.4142136						
3	1.000	2.0000000	2.0000000					
4	1.000	2.6131259	3.4142136	2.6131259				
5	1.000	3.2360680	5.2360680	5.2360680	3.2360680			
6	1.000	3.8637033	7.4641016	9.1416202	7.4641016	3.8637033		
7	1.000	4.4939592	10.0978347	14.5917939	14.5917939	10.0978347	4.4939592	
8	1.000	5.1258309	13.137071	21.846151	25.688355	21.846151	13.1370712	5.1258309

3.3.1   Design of low pass Butterworth filter

For designing an LPF (low pass filter), specifications are given in different ways. For example, along with the value of cutoff frequency ω_c (for which $\varepsilon = 1$), $\alpha_{\min}$ is given beyond the stop band corner frequency ω_2. Alternatively, specification can also be given in terms of $\alpha_{\max}$ up to the corner frequency of the pass band ω_1 and $\alpha_{\min}$ beyond the stop band corner frequency ω_2, as shown in Figure 3.2. In order to get a suitable topology and the values of elements used in it, pole locations for the Butterworth response or coefficients of the Butterworth polynomial are to be obtained using Table 3.1 and 3.2, respectively. However, to get either of the values, the order n is to be determined first; the other variable ε has already been given a value of unity for the Butterworth response – if $\varepsilon \neq 1$ for the general maximally flat response, it has to be calculated from the specifications.

At the pass band corner and stop band corner, respectively, we can write:

$$\alpha_{\max} = 10\log_{10}\left(1 + \varepsilon^2 \omega_1^{2n}\right) \tag{3.18}$$

$$\alpha_{\min} = 10\log_{10}\left(1 + \varepsilon^2 \omega_2^{2n}\right) \tag{3.19}$$

From equations (3.18) and (3.19), we get the following expressions:

$$\varepsilon^2 \omega_1^{2n} = \left(10^{0.1\alpha_{\max}} - 1\right) \tag{3.20}$$

$$\varepsilon^2 \omega_2^{2n} = \left(10^{0.1\alpha_{\min}} - 1\right) \tag{3.21}$$

Dividing equation (3.21) by equation (3.20), we get:

$$(\omega_2 / \omega_1)^{2n} = \frac{10^{0.1\alpha_{\min}} - 1}{10^{0.1\alpha_{\max}} - 1} \tag{3.22}$$

Taking log on both sides of equation (3.22), the expression for the degree n is obtained as follows:

$$n = \frac{\log\left[(10^{0.1\alpha_{\min}} - 1) / (10^{0.1\alpha_{\max}} - 1)\right]}{2\log(\omega_2 / \omega_1)} \tag{3.23}$$

Solution of equation (3.23) yields the value of n which should be able to satisfy the given filter specifications. In almost all cases, it is not possible to obtain integers for the calculated value of $n - n$ then has to be rounded off to the next higher integer value for obvious reasons.

To utilize the large amount of data available for the Butterworth response, in terms of pole locations and transfer function for any order n of the filter [1.2], it is useful to find the normalized cutoff frequency ω_{CB} at which attenuation is 3 dB (Table 3.1 and Table 3.2 are small subsets of such information). For $\alpha = 3$ dB, replacing ω_1 by ω_{CB} in equation (3.18) means:

$$3 = 10\log\left(1 + \varepsilon^2 \omega_{CB}^{2n}\right) \tag{3.24}$$

or $\omega_{CB} = [(10^{0.3} - 1)/\varepsilon^2]^{1/2n}$ (3.25)

Equation (3.25) is an important relation between Butterworth and maximally flat responses.

3.3.2   Use of lossless ladder

In a large number of cases while realizing active filters, the starting point is a passive structure. Though different passive structures are available, one of the most used structures is a doubly terminated lossless ladder. Hence, the topic of lossless ladders and their utilization is important and a matter of serious study. In this section, we will discuss the basics of lossless ladders in order to understand their use in developing an all-pole LPF structure. In its most simple form, a terminated lossless ladder is as shown in Figure 3.5(a). The ladder consists of only inductors and capacitors connected in a ladder form with input and output terminating resistances. This ladder structure has been studied extensively for its utilization in realizing passive filters with different magnitude approximations like Butterworth, Chebyshev, or Cauer. Element values for the ladder structure for all the common approximation methods have been calculated and made available for filter orders starting from $n = 2$ to higher n values. Figure 3.5(b) and (c) show the structure of a lossless ladder. The element values for Butterworth approximated filters of order $n = 2$ to 8 for the ladders shown in Figure 3.5 are presented in Table 3.3. The last element in Figure 3.5(b) and (c) differs depending on if n is odd or even for all pole LPFs. The ladder will be either minimum capacitor or minimum inductor when n is odd as the number of inductors will be one more than a capacitor or vice versa. When n is even, the number of inductors and capacitors will be equal and their total will be equal to n as a doubly terminated ladder is a *canonic* structure using the minimum number of dynamic elements.

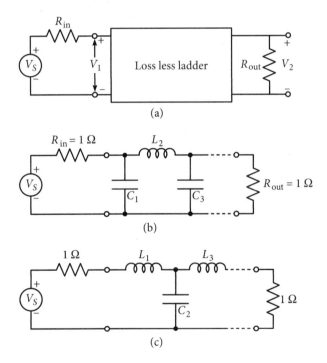

Figure 3.5   (a) Basic structure of a doubly terminated lossless ladder; (b) and (c) Two normalized forms of lossless ladders.

In the normalized low pass doubly terminated ladder, the terminating resistors $R_{in} = R_{out} = 1\ \Omega$ and the frequency normalization is assumed to be done with respect to its 3 dB frequency ω_c. If ω_c is not at 3 dB frequency, a different de-normalizing frequency is to be used, which is given by equation (3.25) as discussed earlier.

Table 3.3   Element values for a doubly terminated lossless ladder for an all-pole LPF using Butterworth approximation

n	C_1	L_2	C_3	L_4	C_5	L_6	C_7	L_8
2	1.414	1.414						
3	1.000	2.000	1.000					
4	0.7654	1.848	1.848	0.7654				
5	0.6180	1.618	2.000	1.618	0.618			
6	0.5176	1.414	1.932	1.932	1.414	0.5176		
7	0.4450	1.247	1.802	2.000	1.802	1.247	0.445	
8	0.3902	1.111	1.663	1.962	1.962	1.663	1.111	0.3902
n	L_1	C_2	L_3	C_4	L_5	C_6	L_7	C_8

Example 3.1: Find the order of the maximally flat LPF which will satisfy the following specifications. Also find the corresponding transfer function.

α_{max} = 1dB, α_{min} = 40dB, ω_1 = 2000 rad/s, and ω_2 = 6000 rad/s.

Solution: To find the order of the filter and its transfer function, equations (3.12) and (3.23) are used for calculating ε and n, respectively.

$$\varepsilon^2 = (10^{0.1} - 1) = 0.25892 \rightarrow \varepsilon = 0.50884$$

$$n = \frac{\log\left[(10^{0.1 \times 40} - 1)/(10^{0.1 \times 1} - 1)\right]}{2\log(6000/2000)} = 4.807$$

which is rounded to the next integer, n = 5.

For the fifth-order Butterworth filter, the values of the pole locations from Table 3.1 gives the following frequency-normalized transfer function.

$$H(S) = \frac{N(S)}{D(S)} = \frac{1}{(S+1)(S^2 + 0.618036S + 1)(S^2 + 1.6186S + 1)} \tag{3.26}$$

$$= \frac{1}{S^5 + 3.236S^4 + 5.236S^3 + 5.236S^2 + 3.236S + 1} \tag{3.27}$$

For using equations (3.24) and (3.25), which are valid for the Butterworth response (and not for a maximally flat response), S shall be replaced by $(j\omega_{CB})$. Hence, using equation (3.25), the normalized cutoff frequency is as follows.

$$\omega_{CB} = \{(10^{0.3} - 1)/0.25892\}^{0.5 \times 5} = 1.144$$

The de-normalized cutoff frequency is given as:

$$\omega_c = \omega_{CB} \times \omega_1 \cong 1.144 \times 2000 = 2288 \text{ rad/s}$$

Hence, equations (3.26) and (3.27) can be modified to the following for the de-normalized frequency

$$H(s) = \frac{1}{(s + 2288)(s^2 + 1414s + 2888^2)(s^2 + 3701.98s + 2288^2)} \tag{3.28}$$

$$= \frac{1}{s^5 + 7.408 \times 10^3 s^4 + 2.741 \times 10^7 s^3 + 6.2714 \times 10^{10} s^2 + 9.11478 \times 10^{13} s + 6.27018 \times 10^{16}} \tag{3.29}$$

Obviously, the next step is to find an active network topology containing the suitable active devices and the values of the passive elements used. One of the most commonly used architecture employs operational amplifiers (OAs) as the active device along with resistances and capacitances (forming the active RC structure). A large variety of procedures are available which lead to the active RC topology and the passive element values for the transfer function given in the form of equations (3.26) to (3.29). These procedures will be discussed later after studying other forms of approximations.

In this section, we make use of Table 3.3 and the lossless ladder of Figure 3.5. For $n = 5$, if the minimum inductor configuration of Figure 3.5(b) is used, the structure's normalized element values from Table 3.3 will be as follows:

$$C_1 = C_5 = 0.618\text{F}, \ C_3 = 2.0 \text{ F and } L_2 = L_4 = 1.618\text{H}$$

De-normalization of the elements is done by using a frequency scaling factor of 2288 rad/s and an impedance scaling factor of 1 kΩ. The de-normalized element values are as follows:

$$C_1 = C_5 = 0.2701 \ \mu\text{F}, C_3 = 0.8741 \ \mu\text{F}, \ L_2 = L_4 = 0.707 \text{ H and } R_{in} = R_{out} = 1 \text{ k}\Omega.$$

The passive ladder shown in Figure 3.6 is simulated and the magnitude response is shown in Figure 3.7. At 318.018 Hz (1998.97 rad/s), attenuation was found to be 0.997 dB and at 955.85 Hz (6008 rad/s), attenuation was 41.9 dBs – an excellent response. The cutoff frequency was found to be at 364.15 Hz (2288.9 rad/s) against the theoretical value of 2288 rad/s. The phase response of the passive filter is also shown in Figure 3.7; it has a phase shift of 180° at the cutoff frequency.

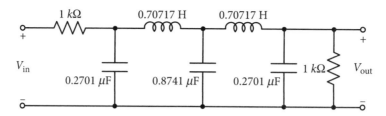

Figure 3.6   Fifth-order Butterworth doubly terminated de-normalized lossless ladder for Example 3.1.

Active realization of this passive fifth-order filter shall be taken up in Chapter 10 using the cascade technique.

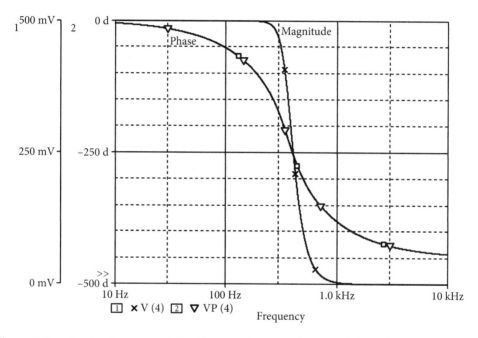

Figure 3.7   Simulated response of the fifth-order low pass Butterwoth filter shown in Figure 3.6 for Example 3.1.

3.4   Equal-ripple Approximations

It is often desirable to obtain a faster attenuation rate beyond the pass band corner frequency – as fast as is practically and economically feasible with lesser number of elements. In the maximally flat Butterworth response, the order of the filter is n and hence, the number of elements used becomes large in order to achieve larger attenuation. Hence, to improve on the value of n, the condition of being maximally flat in the pass band can be dropped. The magnitude characteristic is allowed to ripple between a series of maxima and minima. Ripples can be obtained only in the pass band or stop band, or in both, resulting in following further classifications.

3.5   Chebyshev Approximation

The Chebyshev approximation of a magnitude function is obtained when ripples of equal height appear in the pass band of the transfer function along with a sharp decrease in the gain beyond it. To get such an approximation, the characterizing function of equation (3.5) is selected in the normalized frequency range of $0 \le \Omega \le 1$ as:

$$\left|K\left(j\Omega\right)\right|^2 = \varepsilon^2 C_n^2\left(\Omega\right) = \varepsilon^2 \cos \mathrm{h}^2\left\{n \cos \mathrm{h}^{-1}\left(\Omega\right)\right\} \tag{3.30}$$

Once again, ε is a real constant which is less than 1 and the Chebyshev polynomials are evaluated from the following recursive relation:

$$C_n\left(\Omega\right) = 2\Omega C_{n-1}\left(\Omega\right) - C_{n-2}\left(\Omega\right) \tag{3.31}$$

where $C_0(\Omega) = 1$ and $C_1(\Omega) = \Omega$ and for $\Omega \ge 1$, Chebyshev polynomial is given as follows:

$$C_n(\Omega) = \cos \mathrm{h}\{n \cos \mathrm{h}^{-1}(\Omega)\} \tag{3.32}$$

The amplitude response of the Chebyshev approximation can be obtained from equations (3.2), (3.4), and (3.30). For example, Figure 3.8 shows such an approximation for $n = 4$ (not to the scale), where ripples are shown for an even value of n. For odd values of n, the ripple height remains the same depending on the value of ε; however, at $\Omega = 0$, the function magnitude $|H(0)| = 1$ and for even value of n, $|H(0)| = (1 + \varepsilon^2)^{-\frac{1}{2}}$.

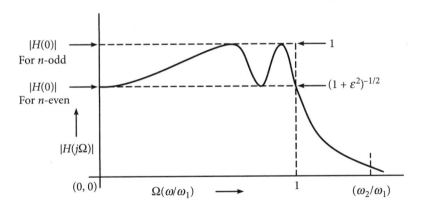

Figure 3.8   Magnitude function of a normalized Chebyshev approximation for order $n = 4$.

3.5.1   Low pass Chebyshev filter design

In order to design a low pass Chebyshev filter, we proceed like the Butterworth case, with ω_1 and ω_2 being the pass band corner frequency and stop band corner frequency, respectively.

At the pass band corner frequency ω_1, use of equations (3.4), (3.10), and (3.30) gives the following expression for maximum attenuation:

$$\alpha_{max} = 10\log\left\{1 + \varepsilon^2 C_n^2\left(\omega_1\right)\right\} \tag{3.33}$$

Since at $\omega = \omega_1$ ($\Omega = 1$), $C_n^2\left(\omega_1\right) = 1$, from equation (3.33), we get:

$$\varepsilon^2 = \left(10^{0.1\alpha_{max}} - 1\right) \tag{3.34}$$

and at normalized ω_2, that is, at $\left(\dfrac{\omega_2}{\omega_1}\right)$ or $\left(\dfrac{\omega_s}{\omega_p}\right)$, α_{min} being the minimum attenuation reached in stop band, its expression is obtained in the same way as equation (3.33) was obtained: use of equation (3.19) gives the expression for minimum attenuation in the stop band α_{min} as:

$$\alpha_{min} = 10\log\left\{1 + \varepsilon^2 C_n^2\left(\omega_2 / \omega_1\right)\right\} \tag{3.35}$$

Use of equation (3.32) modifies the expression for α_{min} as follows:

$$\alpha_{min} = \{10\log\{[1 + \varepsilon^2 \cosh^2 \{n \cosh^{-1}(\omega_2/\omega_1)\}] \tag{3.36}$$

Substituting ε^2 from equation (3.34) in equation (3.36), we get:

$$\cosh\left(n \cosh^{-1}\left(\frac{\omega_2}{\omega_1}\right)\right) = \left(\frac{10^{0.1\alpha_{min}} - 1}{10^{0.1\alpha_{max}} - 1}\right)^{0.5} \tag{3.37}$$

which gives the expression for the order n for the Chebyshev case as follows:

$$n = \frac{\cosh^{-1}\sqrt{\left[(10^{0.1\alpha_{min}} - 1) / (10^{0.1\alpha_{max}} - 1)\right]}}{\cosh^{-1}(\omega_2 / \omega_1)} \tag{3.38}$$

However, a more convenient form of expression is given in equation (3.39) if cosh function is replaced with the natural log function:

$$n \cong \frac{l_n\left[4(10^{0.1\alpha_{min}} - 1) / (10^{0.1\alpha_{max}} - 1)\right]^{1/2}}{l_n\left[(\omega_2 / \omega_1) + ((\omega_2 / \omega_1)^2 - 1)^{1/2}\right]} \tag{3.39}$$

The value of n obtained from equation (3.39) has to be rounded up to the next integer. For order n, analysis has given the location of the left half pole the required transfer function as:

$$s_k = \sigma_K + j\Omega_k \tag{3.40}$$

where,

$$\sigma_k = -\sin h(a)\sin\frac{(2k-1)}{n}\frac{\pi}{2} \qquad (3.41a)$$

$$\Omega_k = j\cos h(a)\cos\frac{(2k-1)}{n}\frac{\pi}{2} \quad k = 0, 1, 2\ (2n-1) \qquad (3.41b)$$

$$a = (1/n)\ \sinh^{-1}\ (1/\varepsilon) \qquad (3.42)$$

It is observed that the poles lie on an ellipse in the complex frequency s plane and substituting ε and n in equation (3.42) and equation (3.41) gives the location (values) of poles. There are extensive tables available that provide the location of poles for various combinations of ε and n. Table 3.4 is a subset of such a table for ε = 0.5dB, 1.0 dB, and 2.0 dB only up to n = 6.

In Table 3.4, the second-order factor for the Chebyshev function in terms of α and β, that is, $(s^2 + 2\alpha s + \alpha^2 + \beta^2)$ is given. It results in the pole frequency $\omega_o = (\alpha^2 + \beta^2)^{1/2}$ and the pole quality factor $Q = (\alpha^2 + \beta^2)^{1/2}/(2\alpha)$. However, in general, the pole frequency and the pole quality factor in terms of the real and imaginary parts of the pole are given as follows:

$$\omega_{ok} = \left(\sigma_k^2 + \omega_k^2\right)^{1/2}, Q_k = \left(\omega_{ok}/2\sigma_k\right) \qquad (3.43)$$

Table 3.4 Pole locations for the Chebyshev approximation, $s = (-\alpha + j\beta)$

N	α_{max} = 0.5 dB		α_{max} = 1 dB		α_{max} = 2 dB	
	α	β	α	β	α	β
1	2.8628	0	1.9625	0	1.3076	0
2	0.7128	1.0040	0.5489	0.8951	0.4019	0.8133
3	0.3132	1.0219	0.2471	0.9660	0.1845	0.9231
	0.6265	0	0.4942	0	0.3689	0
4	0.1754	1.0163	0.1395	0.9834	0.1049	0.9580
	0.4233	0.4209	0.3369	0.4073	0.2532	0.3968
5	0.1120	1.0116	0.0895	0.9901	0.0675	0.9735
	0.2931	0.6252	0.2342	0.6119	0.1766	0.6016
	0.3623	0	0.2895	0	0.2183	0
6	0.0777	1.0085	0.0622	0.9934	0.0470	0.9817
	0.2121	0.7382	0.1699	0.7272	0.1283	0.7187
	0.2898	0.2702	0.2321	0.2662	0.1753	0.2630

Example 3.2: Determine the pole location for the Chebyshev response for n = 3 and α_{max} = 0.5 dB.

Solution: From equation (3.34): $\varepsilon^2 = (10^{0.05} - 1) = 0.122$, $\varepsilon = 0.3493$

The value of the parameter a from equation (3.42) is obtained as follows:

$$a = (1/3)\sinh^{-1}(1/0.3493) = 0.5913$$

which gives sin (ha) = 0.6264 and cos (ha) = 1.18. The location of poles is obtained from equation (3.40) and (3.41) as follows:

$$s_1 = -0.6264, s_2, s_3 = -0.3132 \pm j1.0219 \tag{3.44a}$$

Therefore, the denominator of the transfer function shall be as follows:

$$D(s) = (s + 0.6264)(s^2 + 0.6264s + 1.1424) \tag{3.44b}$$

Example 3.3: Find the order of the Chebyshev LPF for the following specifications. Also find the corresponding transfer function.

$$\alpha_{max} = 0.5\,dB, \ \alpha_{min} = 40\,dB, \omega_1 = 2000\frac{rad}{s} \text{ and } \omega_2 = 6000\,rad/s \tag{3.45}$$

Solution: Using equation (3.39), order n is evaluated as follows:

$$n = \frac{l_n\left[4(10^4 - 1)/(10^{0.05} - 1)\right]^{1/2}}{l_n\left[(6000/2000) + (6000/2000)^2 - 1)^{1/2}\right]} = 3.6 \tag{3.46}$$

This needs to be rounded up to the next integer as 4.

Pole locations can be found as in Example 3.3 or directly using Table 3.4, which are as follows:

$$s_1, s_2 = -0.1754 \pm j1.0163 \text{ and } s_3, s_4 = -0.4233 \pm j0.4209 \tag{3.47}$$

Hence, the normalized transfer function shall be given as follows:

$$H(s) = \frac{0.3377}{\left(s^2 + 0.3508s + 1.0636\right)\left(s^2 + 0.8466s + 0.3563\right)} \tag{3.48}$$

For an even-order transfer function $H(0) = H(1) = \alpha_{max} = 0.5$ dB or 0.944 (normalized), the numerator in $H(s) = (0.944 \times 1.0636 \times 0.3563) = 0.3577$. The obtained transfer function can be realized by direct form synthesis or as a cascade of two second-order non-interactive filter sections. However, its frequency level needs to be de-normalized with respect to 2000 rad/s. The de-normalized transfer function will be as follows:

$$H(s) = \frac{0.3577 \times 2000^2}{\left(s^2 + 0.3508 \times 2000s + 1.0636 \times 2000^2\right)\left(s^2 + 0.8466 \times 2000s + 0.3563 \times 2000^2\right)} \tag{3.49}$$

As in the case of a Butterworth approximated filter, the doubly terminated lossless ladder is also a starting point for active filters when the Chebyshev approximation is used. However, in this case as the corner frequency depends on the ripple width, separate tables are needed for element values for different values of ripple widths. With reference to the lossless ladders of Figure 3.5(b) and (c), Table 3.5 is a small subset containing some commonly used data. For filter requirements not appearing in Table 3.5, we can either consult literature [1.2] or element values can be derived. It is important to note that normalized $R_{in} = 1\Omega$, but it is not equal to R_{out} for even n; its expression is given as $R_{out} = \left\{1 + 2\varepsilon^2 \pm 2\varepsilon\sqrt{(1 + \varepsilon^2)}\right\}R_{in}$.

Table 3.5 LPF element values for Chebyshev approximated response

n	C_1	L_2	C_3	L_4	C_5	L_6	C_7	L_8	R_{out}
				(a) Ripple width = 0.1 dB					
2	0.84304	0.62201							0.73781
3	1.03156	1.14740	1.03156						1.00000
4	1.10879	1.30618	1.77035	0.81807					0.73781
5	1.14681	1.37121	1.97500	1.37121	1.14681				1.00000
6	1.16811	1.40397	2.05621	1.51709	1.90280	0.86184			0.73781
7	1.18118	1.42281	2.09667	1.57340	2.09667	1.42281	1.18118		1.00000
8	1.18975	1.43465	2.11990	1.60101	2.16995	1.58408	1.94447	0.87781	0.73781
				(b) Ripple width = 0.5 dB					
3	1.5963	1.0967	1.5963						1.0000
5	1.7058	1.2296	2.5408	1.2296	1.7058				1.0000
7	1.7373	1.2582	2.6383	1.3443	2.6383	1.2582	1.7373		1.0000
				(c) Ripple width = 1 dB					
3	2.0236	0.9941	2.0263						1.0000
5	2.1349	1.0911	3.0009	1.0911	2.1349				1.0000
7	2.1666	1.1115	3.0936	1.1735	3.0936	1.1115	2.1666		1.0000
	L_1	C_2	L_3	C_4	L_5	C_6	L_7		

3.6   Inverse Chebyshev Approximations

Instead of a maximally flat response in the pass band, having equal ripples in it enables us to realize an active filter having the same specification with a lesser order network; hence, an equal ripple response is more economical than a flat response. It is expected that equal ripples in the stop band structure may further improve response realization. Such an approximation is known as an *inverse Chebyshev approximation*. Further, if there are equal ripples in both the pass band and the stop band responses, it is known as an *elliptic or Cauer approximation*. First, considering the inverse Chebyshev approximation, we can see that allowable attenuation at the edge of the stop band is α_{min}. It is obvious that this may serve no useful purpose for further reduction in α_{min} with frequency, as shown in Figure 3.9. For this kind of approximation, its magnitude function $|H_n(j\Omega)|$ is given by the following relation using equations (3.4) and (3.30):

$$|H_n(j\Omega)|^2 = \frac{1}{1+1/\left[\varepsilon^2 c_n^2(1/\Omega)\right]} \qquad (3.50)$$

where its stop band edge frequency Ω_s is normalized to 1 as shown in Figure 3.9. The significant difference between the inverse Chebyshev and the Chebyshev function is that the frequency normalization in the inverse Chebyshev case is done with respect to the stop band edge frequency (ω_s or ω_2), whereas normalization was done with respect to the pass band edge frequency (ω_p or ω_1) in the Chebyshev approximation. Since $C_n(1) = 1$ for all values of n, the magnitude squared function at $\Omega = 1$ is given as follows:

$$|H_n(j1)|^2 = \frac{1}{\left(1+1/\varepsilon^2\right)} \to |H_n(j1)| = \varepsilon / \left(1+\varepsilon^2\right)^{\frac{1}{2}} \qquad (3.51)$$

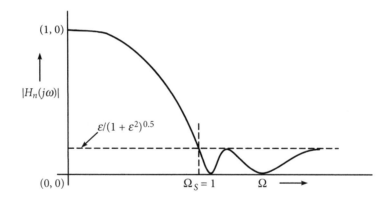

Figure 3.9   Magnitude function variation in a normalized inverse Chebyshev approximation.

The magnitude given by equation (3.51) is the upper limit of the inverse Chebyshev function in the stop band extending from $\Omega = 1$ to ∞, as shown in Figure 3.9. The nature of the magnitude of the ripples in the stop band is the same as it was in the pass band of the Chebyshev function. The number of maxima and minima are also equal to the order of the inverse Chebyshev function as in the pass band. To find the nature of variation of magnitude in the pass band, investigation has to be done at $\Omega = 0$ or $\Omega \ll 1$.

As $C_n(1/\Omega) \approx 2^{n-1}(1/\Omega)^n$ for $\Omega \ll 1$ $\qquad (3.52)$

$$|H_n(j\Omega)|^2 \cong \frac{1}{1+1/\varepsilon^2\left\{(2^{n-1})(1/\Omega^n)\right\}^2} = \frac{1}{1+\Omega^{2n}/(\varepsilon^2 2^{2n-2})} = \frac{1}{1+(\Omega/\Omega_k)^{2n}} \qquad (3.53)$$

where $\Omega_k = (\varepsilon 2^{n-1})^{1/n}$ (3.54)

The nature of equation (3.53) is the same as that of the maximally flat function of equation (3.10), which means that pass band of the inverse Chebyshev response is a maximally flat type of response.

3.6.1 Design of an inverse Chebyshev filter

In the previous section, it was shown that for the inverse Chebyshev response, ripples in the stop band extend from $\Omega = 1$ to ∞, where $\Omega = 1$ corresponds to the edge of the stop band and the function remains maximally flat in the pass band. This means that for an LPF, attenuation specifications will be as shown in Figure 3.10, where α_{max} is the allowable attenuation in the pass band; the attenuation in the stop band has to be at least α_{min}. Hence, the attenuation (in dB) given by the following equation (3.55) can be used.

$$A = -20 \log|H(j\Omega)|$$ (3.55)

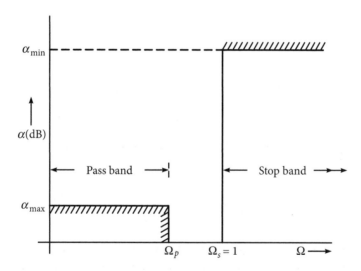

Figure 3.10 Attenuation characteristics of a low pass inverse Chebyshev function.

The equation gives the minimum value of attenuation, α_{min} as follows:

$$\alpha_{min} = -10 \log|H(j\Omega)|^2 = 10 \log(1 + 1/\varepsilon^2)$$ (3.56)

It gives the expression for the constant ε as:

$$\varepsilon = (10^{0.1\alpha_{min}} - 1)^{-1/2}$$ (3.57)

As in Figure 3.10, limit of the attenuation is α_{max} at the edge of the pass band Ω_p. Hence, using equation (3.50), we get the expression for α_{max} in terms of Ω_p and the de-normalized pass band edge frequency as given here.

$$\alpha_{max} = 10\log\left[1+1/\left\{\varepsilon^2 C_n^2\left(1/\Omega_p\right)\right\}\right] \rightarrow \varepsilon^2 C_n^2\left(1/\Omega_p\right) = \left(10^{0.1\alpha_{max}}-1\right) \tag{3.58}$$

Substituting ε from equation (3.57) in equation (3.58),

$$C_n^2(1/\Omega_p) = \frac{10^{0.1\alpha_{max}}-1}{10^{0.1\alpha_{min}}-1} \tag{3.59}$$

As $\Omega_p < 1$ and equation (3.59) is applicable for the pass band, use of equation (3.30) for the value of $C_n^2(\Omega)$ gives:

$$\cos h\left[n\cos h^{-1}(1/\Omega_p)\right] = \frac{(10^{0.1\alpha_{min}}-1)^{\frac{1}{2}}}{(10^{0.1\alpha_{max}}-1)} \tag{3.60}$$

Solving for the order of the function n,

$$n = \frac{\cos h^{-1}\left[(10^{0.1\alpha_{min}}-1)/(10^{0.1\alpha_{max}}-1)\right]^{\frac{1}{2}}}{\cos h^{-1}(1/\Omega_p)} \tag{3.61}$$

If the de-normalized pass band edge frequency is Ω_1 rad/s and the stop band edge frequency is Ω_2 rad/s, $\Omega_p = \left(\dfrac{\Omega_1}{\Omega_2}\right)$, then and equation (3.61) can be modified as:

$$n = \frac{\cos h^{-1}\left[(10^{0.1\alpha_{min}}-1)/(10^{0.1\alpha_{max}}-1)\right]^{\frac{1}{2}}}{\cos h^{-1}(\Omega_2/\Omega_1)} \tag{3.62}$$

Equation (3.62) is the same as that for the Chebyshev function given in equation (3.38); this means that for the same specifications, the order of the inverse Chebyshev function will be the same as that for the Chebyshev response. Value of the parameter ε will also be the same; however, a major difference between the two responses is that the frequency normalization is done with respect to the stop band edge frequency Ω_S in the inverse Chebyshev function.

To find the transfer function of the inverse Chebyshev function, either an appropriate Table like 3.4 can be used, or the location of poles and zeros needs to be determined as follows.

The magnitude squared function of equation (3.40) can be expressed as follows:

$$\left|H_n\left(j\Omega\right)\right|^2 = \frac{\varepsilon^2 C_n^2(1/\Omega)}{1+\varepsilon^2 C_n^2(1/\Omega)} = \frac{z(s)z(-s)}{p(s)p(-s)}\bigg|_{s=j\Omega} \tag{3.63}$$

Hence, zeros are found in the stop band for $\Omega > 1$, when $C_n^2(1/\Omega)$ is expressed through trigonometric functions, and equalized to zero, that is, $C_n^2(1/\Omega_k) = 0$, or:

$$\cos n \cos^{-1}(1/\Omega_k) = 0 \tag{3.64}$$

Equality in equation (3.64) is valid when k is odd (it equals 1 when k is even). Then, with $\varphi_k = n \cos^{-1}(1/\Omega_k)$, we get:

$$\cos n\varphi_k = 0 \text{ when } n\varphi_k = k(\pi/2). \tag{3.65}$$

Equation (3.65) gives

$$\cos^{-1}(1/\Omega_k) = \varphi_k = k\pi/2n \tag{3.66}$$

Therefore, zero frequencies are obtained as follows:

$$\Omega_k = \sec(k\pi/2n) \text{ for } k = 1, 3, 5, \ldots, n \tag{3.67}$$

We will now find the pole location of the inverse Chebyshev function. It can be observed that the denominator of equation (3.63) is the same as that for the Chebyshev function, with a difference that Ω is now replaced by $(1/\Omega)$. This means that to find the pole location for the inverse Chebyshev function, we can first determine the Chebyshev poles using equations (3.40)–(3.42) and then take its reciprocal. However, the value of ε to be used shall be the one obtained for the inverse Chebyshev function using equation (3.47). It is observed that the pole quality factor, Q for the inverse Chebyshev remains the same as that for the Chebyshev function.

Example 3.4: Find the order of the inverse Chebyshev filter for the following specifications. Also find the corresponding transfer function.

$$\alpha_{max} = 1 \, \text{dB}, \ \alpha_{min} = 40 \, \text{dB}, \omega_1 = 2000 \, \text{rad/s and } \omega_2 = 6000 \, \text{rad/s}.$$

Solution: By calculation, the order n of the filter is 3.36; this can be approximated to 4.

Zeros of the transfer function shall be found using equation (3.67) with $\Omega_{zk} = \sec(k\pi/2 \times 4)$. Hence, for $k = 1$,

$$\Omega_{z1} = \sec(\pi/8) = 1.08239 \text{ and for } k = 3, \ \Omega_{z2} = \sec(3\pi/8) = 2.6131 \tag{3.68a}$$

The de-normalized value of the zeros is as follows:

$$z_1 = 6000 \times \Omega_{z1} = 6494.34 \, \text{rad/s} \text{ and } z_2 = 6000 \times \Omega_{z2} = 15678.7 \, \text{rad/s} \tag{3.68b}$$

The first step to finding the pole values is to find the pole of the Chebyshev function with ε obtained using equation (3.47) as follows:

$$\varepsilon = (10^{0.1 \times 40} - 1)^{-1/2} = 0.01 \tag{3.69}$$

Using equation (3.42),

$$a = \left(\frac{1}{4}\right) \sin \mathrm{h}^{-1}(1/0.01) = 1.32458 \tag{3.70}$$

This yields sin h a = sin h 1.32458 = 1.74733 and cos h a = cos h 1.32458 = 2.0132

Hence, the real parts of the pole can be obtained as follows:

$$\sigma_1 = -\sin \mathrm{h}a \times \{\sin(1/4)(\pi/2)\} = -1.74733 \sin(\pi/8) = -0.66867 \tag{3.71a}$$

$$\sigma_2 = -\sin \mathrm{h}a \times \{\sin(3/4)(\pi/2)\} = -1.6143 \tag{3.71b}$$

$$\sigma_3 = -\sin \mathrm{h}a \times \{\sin(5/4)(\pi/2)\} = -1.6143 \tag{3.71c}$$

$$\sigma_4 = -\sin \mathrm{h}a \times \{\sin(7/4)(\pi/2)\} = -0.66867 \tag{3.71d}$$

The imaginary components can be obtained from equation (3.41b) as:

$$\Omega_k = j\cos \mathrm{h}a \times \cos(2k-1)(\pi/2n)$$

$$\Omega_1 = j2.0132 \cos(\pi/8) = j1.8599, \; \Omega_2 = j2.0132 \cos(3\pi/8) = j0.7704$$

$$\Omega_3 = j2.0132 \cos(5\pi/8) = -j0.7704, \; \Omega_4 = j2.0132\cos(7\pi/8) = -j1.859 \tag{3.72}$$

For the Chebyshev function, if we know the value of the real and imaginary parts of the pole, its magnitude and the quality factor shall be given as follows:

$$\Omega_{0k} = \left(\sigma_k^2 + \Omega_k^2\right)^{1/2} \text{ and } Q_{kC} = \left(\Omega_{0k}/2\sigma_k\right) \tag{3.73}$$

Then, the pole location for the inverse Chebyshev response case, $p_k = x_k + jy_k$ is given by:

$$p_k = \frac{(\sigma_k - j\Omega_k)}{(\sigma_k^2 + \Omega_k^2)} \tag{3.74}$$

This gives the magnitude and quality factor for the inverse Chebyshev response as follows:

$$|p_k| = \left(x_k^2 + y_k^2\right)^{1/2} = \left(1/\Omega_{0kIC}\right) \text{ and } Q_{kIC} = Q_{kC} \tag{3.75}$$

Using equations (3.73)–(3.75), we get the following parameters:

$$\Omega_{01} = \left(\sigma_1^2 + \Omega_1^2\right)^{\frac{1}{2}} = \left\{(-0.66867)^2 + (1.8599)^2\right\}^{\frac{1}{2}} = (3.90646)^{\frac{1}{2}} = 1.9764 \qquad (3.76a)$$

$$\Omega_{02} = \left\{(-1.6143)^2 + (0.7704)^2\right\}^{\frac{1}{2}} = (3.19948)^{\frac{1}{2}} = 1.7887 \qquad (3.76b)$$

$$\Omega_{03} = \left\{(-1.6143)^2 + (-0.7704)^2\right\}^{\frac{1}{2}} = \Omega_{02} \text{ and } \Omega_{04} = \Omega_{01} \qquad (3.76c, d)$$

$$Q_{1C} = \frac{\Omega_{01}}{2\sigma_1} = \frac{1.9764}{2 \times 0.66867} = 1.47786, Q_{2C} = \frac{\Omega_{02}}{2\sigma_2} = \frac{1.7887}{2 \times 1.6143} = 0.554, \qquad (3.77a, b)$$

$$Q_{3C} = \frac{\Omega_{02}}{2\sigma_1} = Q_{2C} \text{ and } Q_{4C} = \frac{\Omega_{01}}{2\sigma_2} = Q_{1C} \qquad (3.77c, d)$$

$$p_1 = \frac{\sigma_1 - j\Omega_1}{\sigma_1^2 + \Omega_1^2} = \frac{-0.66867 - j1.8599}{\Omega_{01}^2} = -0.17117 - j0.47611 \qquad (3.78a)$$

$$p_2 = \frac{\sigma_2 - j\Omega_2}{\sigma_2^2 + \Omega_2^2} = \frac{-1.6143 - j0.7704}{\Omega_{02}^2} = -0.50455 - j0.2408 \qquad (3.78b)$$

$$p_3 = \frac{\sigma_2 - j\Omega_2}{\Omega_{02}^2} = \frac{-1.6143 + j0.7704}{\Omega_{02}^2} = -0.50455 + j0.2408 \qquad (3.78c)$$

$$p_4 = \frac{-0.66867 + j1.8599}{\Omega_{01}^2} = -0.17117 + j0.47611 \qquad (3.78d)$$

$$\Omega_{01IC} = (x_1^2 + y_1^2)^{\frac{1}{2}} = \left\{(.17117)^2 + (.47611)^2\right\}^{\frac{1}{2}} = 0.5059 \qquad (3.79a)$$

$$\Omega_{02IC} = \left\{(.50455)^2 + (.2408)^2\right\}^{\frac{1}{2}} = 0.55906 \qquad (3.79b)$$

$$\Omega_{03IC} = \Omega_{02IC}, \Omega_{04IC} = \Omega_{01IC} \qquad (3.79c, d)$$

Instead of following the steps from equations (3.74) to (3.79), the pole location for the inverse Chebyshev function can also be found by taking the inverse of the pole locations of the Chebyshev response obtained from equations (3.71)–(3.73), while using the value of ε obtained using equation (3.47); the quality factors remain the same.

To obtain the transfer function of the inverse Chebyshev response, the pole-pair is associated with the zero nearer to it. Hence, the transfer function is obtained as follows:

$$H(s) = \frac{\left\{s^2 + \Omega_{z1}^2\right\}\left\{s^2 + \Omega_{z3}^2\right\}}{\left[\left\{s^2 + \dfrac{\Omega_{01IC}}{Q_{1IC}}s + \Omega_{01IC}^2\right\}\left\{s^2 + \dfrac{\Omega_{03IC}}{Q_{3IC}}s + \Omega_{03IC}^2\right\}\right]}$$

$$= \frac{\left(s^2 + 1.1757\right)\left(s^2 + 6.8283\right)}{(s^2 + 0.3423s + 0.2559)(s^2 + 1.0091s + 0.31255)} \qquad (3.80)$$

Since the calculation of poles and zero was done in the normalized form with normalization done with respect to the stop band edge frequency, the transfer function is also to be de-normalized with respect to it. Finally, the filter can be realized by using any of the cascade or direct form synthesis procedures. If we wanted to realize the filter as two second-order sections, it can be done by using two notch filters. Since notch filter realization shall be studied later, this transfer function shall also be taken up later.

3.7   Cauer or Elliptic Approximation

The use of Chebyshev or inverse Chebyshev approximation result in an economical or optimal filter section rather than a filter with maximally flat response. It was expected that equal ripples in both the pass band and the stop band will further decrease the required order n of the filter section for the same specification; this assumption was indeed shown to be correct by William Cauer [3.1]. Such an approximation, shown in Figure 3.11, is called a *Cauer approximation* or *elliptic approximation*. As the solutions for this approximation lead to elliptic functions that are not easy to solve, exhaustive tables and design graphs are used instead of solving the elliptic functions.

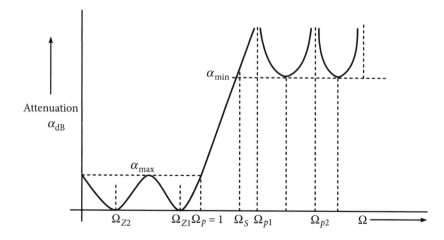

Figure 3.11 Variation of attenuation for a typical Cauer or elliptic response.

In the previous magnitude approximations, different expressions were assigned to the term $|K(j\Omega)|$ of equation (3.4) so that the pass band would be maximally flat, or have equal ripples in the stop band or pass band. To have equal ripples in both the frequency bands, the characteristics function $K(S)$ is selected to be a ratio of polynomials whose poles and zero lie on the imaginary axis of the s plane. For $K(S)$ to be such a new function of $E_n(S)$ of order n, let $E_n(S) = N(S)'/D(S)'$. Then equation (3.4) can be written as:

$$|H_n(j\Omega)|^2 = 1/\left[1 + \varepsilon^2 E_n^2(\Omega)\right]$$

$$= \frac{D'(j\Omega)D'(-j\Omega)}{D'(j\Omega)D'(-j\Omega) + \varepsilon^2 N'(j\Omega)N'(-j\Omega)} \tag{3.81}$$

In equation (3.81), ε is multiplied with $E_n(S)$ as its value is not unity in maximally flat and equal ripple approximations, instead of being unity in case of Butterworth approximation.

This means that the poles of $E_n(\Omega)$ will be the zeros of $|H_n(j\Omega)|$. Analysis of the function assumes that frequency is normalized at the edge of the pass band, that is, $\Omega_p = 1$ and $E_n(\Omega = 1) = 1$. Then, maximum attenuation at the pass band edge from equation (3.81) shall be:

$$\alpha_{max} = 10\log_{10}\left(1 + \varepsilon^2\right) \tag{3.82}$$

This equation gives the same expression for ε which was obtained for the maximally flat or the Chebyshev approximation of equations (3.11) and (3.33), respectively. Further, in order to have equal ripples in the stop band (including at the stop band edge frequency Ω_s with attenuation of α_{min}, it is required that $|E_n| = \pm F$. Hence, in the stop band, the expression of the minimum attenuation shall be:

$$\alpha_{min} = 10\log_{10}(1 + \varepsilon^2 F^2) \tag{3.83}$$

Substitution of ε from equation (3.82) in equation (3.83) results in the expression for F which is a familiar expression for the maximally flat as well as Chebyshev response vide equations (3.23) and (3.39), respectively, in connection with the evaluation of the filter order n.

3.7.1 Design of Cauer filters

The first step in the design of a Cauer filter is to find its order from the same four specifications: α_{max}, α_{min}, Ω_p, and Ω_s. While finding n has been rather straightforward and simple in the approximation methods discussed so far, for the Cauer approximation, is the calculations become quite involved and requires solution of elliptic functions. One way out from this complexity is by using the fact that, invariably, the value of n obtained through calculations is not an integer and, therefore, the next higher integer value is selected. This approximation amounts to a bit of over-designing; however, it is customary that the given values of the specifications can be marginally changed. It allows using a lesser complex graphic process in which n is obtained from a set of curves drawn for the variation of E_n with respect to Ω_s. However, the approximation also requires finding an expression for E_n that needs to be a rational function meeting the requirements of the given specifications. Alternatively, we can use the following simpler method [3.3].

From the given specifications, a *modulator constant q* is calculated from the following relation:

$$q = u + 2u^5 + 15u^9 + 150u^{13} \tag{3.84a}$$

Here, $u = \dfrac{1-(1-k^2)^{\frac{1}{4}}}{\left\{2(1+(1-k^2)^{\frac{1}{4}}\right\}}$ and the selectivity factor,

$$k = \Omega_p/\Omega_s \tag{3.84b}$$

The next step is to find the *discrimination factor D* from the following relation:

$$D = \left(10^{0.1\alpha_{min}} - 1\right) / \left(10^{0.1\alpha_{max}} - 1\right) \tag{3.85}$$

Then, the order of the elliptic filter n is obtained from the following relation:

$$n = \{\log 16D/\log(1/q)\} \tag{3.86}$$

In equation (3.86), the obtained value may not be an integer; the value then has to be rounded off to the next higher integer. Due to the change in the value of n to the next higher integer value, the actual α_{min} in the stop band is changed to the following:

$$\alpha_{min} = 10\log\left(1 + \frac{10^{0.1\alpha_{max}} - 1}{16q^n}\right) \tag{3.87}$$

Obviously, we should ascertain that the value of α_{min} obtained from equation (3.87) satisfies the specifications of the design.

Example 3.5: For the following specifications, find the order of an elliptic filter:

$$\alpha_{max} = 1\ \text{dB}, \alpha_{min} = 40\ \text{dB}, \omega_1 = 2000\ \text{rad/s and } \omega_2 = 6000\ \text{rad/s}$$

Solution: Selectivity factor, $k = 2000 / 6000 = 1/3$. Using equations (3.84)–(3.86) for q, u, D and n, we get the following:

$$u = 0.5\frac{1-(1-1/9)^{\frac{1}{4}}}{(1+(1-1/9)^{\frac{1}{4}}} = 0.00736$$

$$q = 0.00736 + 2(0.00736)^5 + 15(0.00736)^9 + 150(0.00736)^{13} \cong 0.00736$$

Value of the discrimination factor D is obtained as:

$$D = (10^{0.1 \times 40} - 1)/(10^{0.1 \times 2} - 1) = 9999/0.2589 = 38621$$

Then, the order of the filter is obtained from the equation (3.86):

$$n = \log(16 \times 38621)/\log(1/0.00736) = 2.714$$

Hence, the selected value of $n = 3$.

It may be noted that for the same specifications, the required filter order was 5 for the Butterworth approximation, 4 for equal-ripple filters. In practice, the difference becomes more prominent when there is a narrower transition band or selectivity factor with a large value.

The actual minimum stop band attenuation with $n = 3$ from equation (3.87) will be as follows:

$$\alpha_{min} = 10\log\left\{1 + \frac{10^{0.1} - 1}{16(.00736)^3}\right\} = 46.08\,\mathrm{dB}$$

In this expression, the obtained theoretical value of α_{min} is well under control. After finding the order of the filter, the normalized transfer function is to be obtained. Once again, the solution requires elliptic functions, which are quite complex. Algorithms have been developed for the purpose; however, the simpler option is to use the available design tables. In the vast literature pertaining to filters, these tables and the data arranged in the tables have been presented in different ways. Only the specific table corresponding to the stated specifications is to be used to get the location of poles and zeros and the transfer function.

An alternate method for finding the order of the elliptic LPF is to use nomographs. Figure 3.12 is such a nomograph, wherein the attenuation at the pass band edge frequency is α_{max} (normalized to 1 rad/s) and α_{min} is the attenuation at some (normalized) stop band edge frequency Ω_s. To use the nomograph in Figure 3.12, a straight line is drawn through the specified α_{max} and α_{min}, shown as points 1 and 2 in Figure 3.13. Intersection of this line with the ordinate of the nomograph determines point 3. A horizontal line is drawn from point 3 until it meets a vertical line drawn from point 4 which corresponds to the specified frequency Ω_s. The resulting intersection at point 5 decides the order of the required elliptic filter. Almost every time, point 5 lies between two curves corresponding to the loci of the filter orders; the higher value of the filter order is selected for obvious reasons.

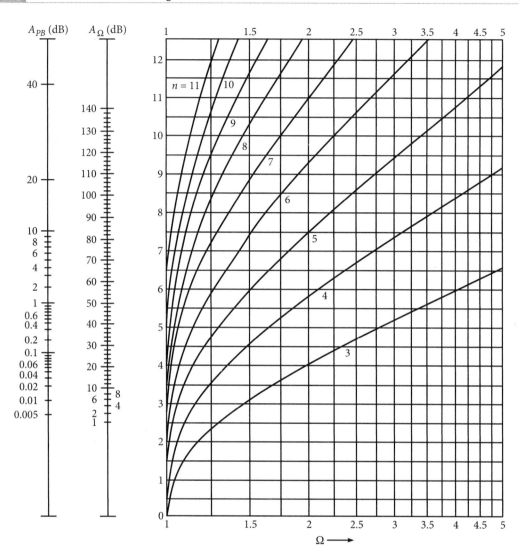

Figure 3.12 A nomograph for determining the order of an elliptic magnitude function.

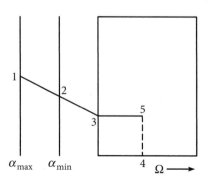

Figure 3.13 Method of using the nomograph in Figure 3.12.

Normalized element values of a doubly terminated lossless ladder can be obtained from other tables. Figure 3.14(a) and (b) show these ladders and Tables 3.6 and 3.7 show the values of the elements for these ladders. Obviously, these ladders and tables are useful when some direct form of filter synthesis is used instead of the cascade form. However, if it is preferred to use the cascade form of synthesis, poles and zeros, and then the transfer function can be found through analyzing the ladder itself.

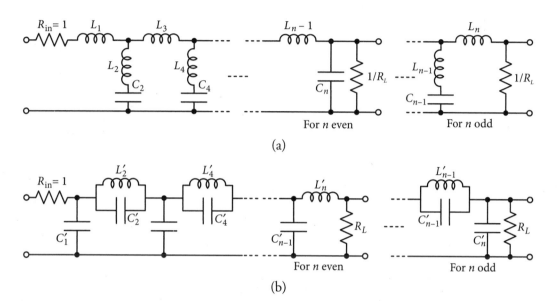

Figure 3.14 (a) Network configuration for Table 3.6 and Table 3.7. (b) Alternate configuration.

Example 3.6: Find a doubly terminated lossless ladder which gives an elliptic response while satisfying the following specifications. Verify the response using PSpice.

$$\alpha_{max} = 0.1 \text{ dB } \alpha_{min} = 54 \text{ dBs, } \omega_p = 100 \text{ krad/s and } \omega_s = 150 \text{ krad/s}$$

Solution: Normalizing the frequency by 100 krad/s, we get $\Omega_s = 1.5$ rad/s. Using the nomograph of Figure 3.12, intersection of the line from the given attenuations of 0.1 dB and 54 dB and the vertical line for $\Omega_s = 1.5$ rad/s falls between $n = 5$ and 6; hence, the required filter order will be 6. For $n = 6$ and $\Omega_s = 1.5$ rad/s, Table 3.7 gives the normalized values of the elements as:

$$R_{in} = R_L = 1 \text{ } \Omega, L_1 = 0.86595 \text{ H}, L_2 = 0.18554 \text{ H}, L_3 = 1.43106 \text{ H}, L_4 = 0.33007 \text{H},$$
$$L_5 = 1.28253 \text{ H}, C_2 = 1.27403 \text{ F}, C_4 = 1.27255 \text{ F}, C_6 = 1.03317 \text{ F}$$

Using frequency scaling, elements are de-normalized by a factor of 100 krad/s and an impedance scaling factor of 10^4. The de-normalized element values are as follows:

$$R_{in} = R_L = 10 \text{ k}\Omega, L_1 = 0.086595 \text{ H}, L_2 = 0.018554 \text{ H}, L_3 = 0.143106 \text{ H}, L_4 = 0.33007 \text{ H},$$
$$L_5 = 0.128253 \text{ H}, C_2 = 1.27403 \text{ nF}, C_4 = 1.27255 \text{nF}, C_6 = 1.03317 \text{ nF}$$

Table 3.6 Element values of a doubly terminated ladder for elliptic filters with pass band ripples of 1 dB

n	ω_s	K_p	L_1	C_2	L_2	L_3	C_4	L_4	L_5	C_6	L_6	L_7
3	1.05	8.134	1.05507	.25223	3.28904	1.05507						
	1.10	11.480	1.22525	.37471	1.94752	1.22525						
	1.20	16.209	1.42450	.52544	1.11977	1.42450						
	1.50	25.176	1.69200	.73340	.48592	1.69200						
	2.00	34.454	1.85199	.85903	.22590	1.85199						
4	1.05	11.322	.63708	.35277	2.41039	1.11522	1.39953					
	1.10	15.942	.80935	.54042	1.40015	1.18107	1.45001					
	1.20	22.293	1.00329	.77733	.79634	1.26621	1.49217					
	1.50	34.179	1.25675	1.11431	.34362	1.38981	1.53225					
	2.00	46.481	1.40677	1.32367	.15960	1.46762	1.55071					
5	1.05	24.135	1.56191	.67560	.83449	1.55460	.26584	3.31881	.88528			
	1.10	30.471	1.69691	.77511	.58827	1.79892	.39922	1.98907	1.12109			
	1.20	38.757	1.82812	.87005	.38720	2.09095	.56347	1.16672	1.38094			
	1.50	53.875	1.97687	.97694	.18824	2.49161	.79362	.51950	1.71889			
	2.00	69.360	2.05594	1.03392	.09152	2.73567	.93561	.24486	1.91939			
6	1.05	29.133	1.07458	.80116	.81300	.92735	.51753	1.71498	.92186	1.60511		
	1.10	36.680	1.22059	.94235	.57746	1.10900	.75718	1.05819	1.01676	1.64682		
	1.20	46.571	1.37146	1.08633	.38284	1.32610	1.05110	.63354	1.12484	1.68498		
	1.50	64.661	1.55425	1.25876	.18779	1.62529	1.46557	.28655	1.26961	1.72482		
	2.00	83.221	1.65661	1.35450	.09179	1.80860	1.72376	.13586	1.35729	1.74424		
7	1.05	40.926	1.82156	.86343	.42668	1.67632	.34381	2.60271	1.23696	.46779	1.63392	1.22362
	1.10	49.816	1.91040	.92662	.30705	1.93579	.48016	1.68753	1.55276	.59277	1.10699	1.41994
	1.20	61.422	1.99168	.98474	.20446	2.22804	.64444	1.04856	1.92724	.73012	.70551	1.62539
	1.50	82.588	2.07882	1.04761	.10016	2.61372	.87393	.48973	2.44021	.90483	.33349	1.87717
	2.00	104.268	2.12329	1.07993	.04884	2.84446	1.01638	.23538	2.75306	1.00567	.16034	2.01924
n	ω_s	K_p	C_1'	L_2'	C_2'	C_3'	L_4'	C_4'	C_5'	L_6'	C_6'	C_7'

1.0-dB pass band ripple

Table 3.7 Element values of a doubly terminated ladder for elliptic filters with pass band ripples of 0.1 dB

n	ω_s	K_p	L_1	C_2	L_2	L_3	C_4	L_4	L_5	C_6	L_6	L_7
3	1.05	1.748	.35550	.15374	5.39596	.35550						
	1.10	3.374	.44626	.26993	2.70353	.44626						
	1.20	6.691	.57336	.44980	1.30805	.57336						
	1.50	14.848	.77031	.74561	.47797	.77031						
	2.00	24.010	.89544	.93759	.20697	.89544						
4	1.05	3.284	.00442	.17221	4.93764	1.01224	.84445				0.1 dB pass band ripple	
	1.10	6.478	.17279	.32758	2.30986	1.04894	.89415					
	1.20	12.085	.37139	.56638	1.09294	1.11938	.92440					
	1.50	23.736	.62815	.94009	.40730	1.24711	.93518					
	2.00	36.023	.77554	1.17646	.17957	1.33473	.93382					
5	1.05	13.841	.70813	.76630	.73572	1.12761	.20138	4.38116	.04985			
	1.10	20.050	.81296	.92418	.49338	1.22445	.37193	2.13500	.29125			
	1.20	28.303	.91441	1.06516	.31628	1.38201	.60131	1.09329	.52974			
	1.50	43.415	1.02789	1.21517	.15134	1.63179	.93525	.44083	.81549			
	2.00	58.901	1.08758	1.29322	.07317	1.79387	1.14330	.20038	.97720			
6	1.05	18.727	.44177	.71651	.90905	.83142	.36274	2.44680	.80463	.99857		
	1.10	26.230	.57630	.88798	.61282	.97304	.59060	1.35666	.94305	1.01381		
	1.20	36.113	.70984	1.06266	.39136	1.15974	.87407	.76185	1.09176	1.02462		
	1.50	54.202	.86595	1.27403	.18554	1.43106	1.27235	.33007	1.28253	1.03317		
	2.00	72.761	.95131	1.39297	.08926	1.60132	1.51866	.15421	1.39521	1.03621		
7	1.05	30.470	.91937	1.07659	.34220	1.09623	.40518	2.20850	.84335	.50342	1.51827	.41098
	1.10	39.357	.98821	1.16726	.24374	1.27743	.59720	1.35681	1.04029	.67881	.96669	.58282
	1.20	50.963	1.05029	1.24872	.16124	1.48377	.82869	.81542	1.28723	.87428	.58918	.75395
	1.50	72.129	1.11593	1.33554	.07857	1.75687	1.15174	.37160	1.63827	1.12502	.26822	.95588
	2.00	93.809	1.14910	1.37979	.03822	1.92026	1.35221	.17692	1.85664	1.27023	.12694	1.06720
n	ω_s	K_p	C_1	L'_2	C_2	C_3	L'_4	C_4	C_5	L'_6	C_6	C_7

Figure 3.15(a) shows the magnitude response of the passive ladder with ordinates on the linear scale and Figure 3.15(b) shows the magnitude on the log scale (magnitude on the log scale is shown to get a better view of both the pass band and the stop band). The simulated pass band edge frequency is 15.92 kHz (100.06 krad/s) and the stop band edge frequency is 23.86 kHz (149.977 krad/s) giving the normalized Ω_s = 1.5 rad/s. Maximum attenuation in the pass band 0.100 dB and the minimum attenuation in the stop band is 54.2 dB.

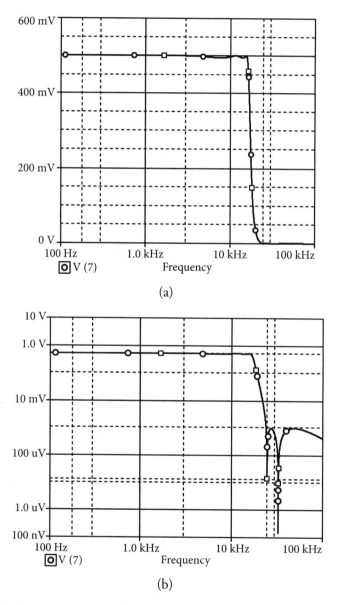

(a)

(b)

Figure 3.15 (a) Magnitude response of the elliptic filter of Example 3.6 with ordinates on the linear scale. (b) with ordinates on the log scale.

3.8 Maximally Flat Pass Band with Finite Zeros

It has been observed that in the maximally flat Butterworth response case, it takes a higher filter order to satisfy the same specifications compared to other approximations. This limitation caused by the slower transition from pass to stop band in the maximally flat Butterworth response is minimized by the introduction of finite zeroes in the otherwise all-pole LPF, where all the zeroes were at infinity.

If finite zeroes are introduced in the maximally flat response, instead of $N(s) = 1$, $N(s)$ will be a polynomial in terms of the complex frequency s in equation (3.3), and the degree of the polynomial will depend on the number of the zeroes to be added. Equation (3.5) will become modified while using equations (3.2) as:

$$|K(j\Omega)|^2 = \frac{B(\Omega^2)}{A(\Omega^2)} - 1 = \frac{B(\Omega^2) - A(\Omega^2)}{A(\Omega^2)} \tag{3.88}$$

However, for the response to remain maximally flat, equation (3.88) has to remain satisfied even when transmission zeroes are introduced. This condition implies that the following relation is satisfied for as many derivatives as possible:

$$\frac{d^k \left(|K(j\Omega)|^2 \right)}{d^k(\Omega^2)} = \frac{d^k}{d^k(\Omega^2)} \left\{ \frac{B(\Omega^2) - A(\Omega^2)}{A(\Omega^2)} \right\} = 0 \tag{3.89}$$

Application of chain rule, differentiation of equation (3.89) gives following relation:

$$B_{2i} = A_{2i} \text{ for } i = 0, 1, ..., (n-1) \tag{3.90}$$

Hence, in equation (3.2) or equation (3.88), $A(\Omega^2)$ is selected in such a way that the desired zeroes are realized and the denominator is the sum of $A(\Omega^2)$ and $B_{2n}\Omega^{2n}$. With a modified transfer function, at $\Omega = 1$, and $|H_n(j1)|^2$ with equation (3.10), we get:

$$|H_n(j1)|^2 = \frac{A(1)}{A(1) + B_{2n}} = \frac{1}{1 + \varepsilon^2} \rightarrow B_{2n} = \varepsilon^2 A(1) \tag{3.91}$$

The transfer function can now be found by multiplying $H_n(j\Omega)$ with $H_n(-j\Omega)$, and substituting $S = j\Omega$:

$$H_n(j\Omega)H_n(-j\Omega)|_{j\Omega=S} = \frac{A(-S^2)}{A(-S^2) + (-1)^n * B_{2n}S^{2n}} \tag{3.92}$$

It is important to note that in an all pole transfer function, the gain drops at a rate of $20n$ dB per decade, but after the addition of finite zeroes, the rate of fall of gain in the transition band increases but the rate of fall of gain at higher frequencies will be at the rate of $20(n - m)$ dB per decade.

The following example is provided to help understand the design process while introducing finite zeroes in a flat pass band transfer function.

Example 3.7: In a maximally flat LPF, it is desired that the dc gain of the filter remains as unity and its gain drops by 1 dB at 20 krad/s. Introduce transmission zeroes at 40 krad/s and 50 krad/s to increase the rate of fall of attenuation in the transition band and find its transfer function.

Solution: As gain is dropping by 1 dB at 20 krad/s, this is taken as the normalizing frequency. Then the transmission zeroes will become $\Omega = 2$ and 2.5, respectively, and with the dc gain as unity, the normalized transfer function is obtained using equation (3.92):

$$H_n(j\Omega)*H_n(-j\Omega) = \frac{\left\{\left(\frac{\Omega_n}{2}\right)^2 - 1\right\}^2 \left\{\left(\frac{\Omega_n}{2.5}\right)^2 - 1\right\}^2}{\left[\left(\frac{\Omega_n}{2}\right)^2 - 1\right]^2 \left\{\left(\frac{\Omega_n}{2.5}\right)^2 - 1\right\}^2 + B_{2n}\Omega^{2n}} \tag{3.93}$$

The numerator is of degree 4 and for a minimum rate of fall of attenuation of 40 dB at *higher frequencies*, the denominator should have degree $n = 6$; hence, B_{12} is to be determined for equation (3.93).

The value of B_{12} can be found from equation (3.91), as at $\Omega = 1$, the gain drops by 1 dB, and we can write:

$$|H_6(j1)|^2 = 1 / \left\{ 1 + \frac{B_{12}}{\left(\frac{3}{4}\right)^2 \left(\frac{5.25}{6.25}\right)^2} \right\} \tag{3.94}$$

which is equal to 1 dB of attenuation or we can calculate that the output will drop by a factor of $10^{-(1/10)} = 0.7943$; hence, comparing this equation with equation (3.94), we get $B_{12} = 0.102767$. The value of B_{12} is substituted in equation (3.94), and while S is replaced by $j\Omega$, the transfer function is obtained from the following:

$$H_6(S)H_6(-S) = \frac{\left(\left(\frac{S}{2}\right)^2 + 1\right)^2 \left(\left(\frac{S}{2.5}\right)^2 + 1\right)^2}{\left(\left(\frac{S}{2}\right)^2 + 1\right)^2 \left(\left(\frac{S}{2.5}\right)^2 + 1\right)^2 + 0.102767 S^{12}} \tag{3.95}$$

The numerator has four roots, which are easily identifiable; however, the root finder is used to find 12 roots of the denominator. Along with a multiplying factor of 64.229, the following roots are obtained:

$$\pm 1.256 \pm j0.41, \pm 0.762 \pm j0.932, \text{ and } \pm 0.238 \pm j1.085 \tag{3.96}$$

In equation (3.96), the roots are in all the quadrants. Selecting the roots on the left half of the s plane, for a real rational function, the following factors (normalized) become available.

$$S^2 + 2.511S + 1.744, S^2 + 1.524S + 1.449 \text{ and } S^2 + 0.475S + 1.234 \tag{3.97}$$

As result of the root multiplying factor of 64.229, the numerator coefficient will become 4 × 6.25/(64.229)$^{\frac{1}{2}}$ = 3.119. The resulting normalized transfer function will be as follows:

$$H_6(S) = \frac{3.119\left\{\left(\frac{S}{2}\right)^2 + 1\right\}\left\{\left(\frac{S}{2.5}\right)^2 + 1\right\}}{\left(S^2 + 2.511S + 1.744\right)\left(S^2 + 1.524S + 1.449\right)\left(S^2 + 0.475S + 1.234\right)} \tag{3.98}$$

Realization of the transfer function of equation (3.98) using cascade technique will be discussed in Chapter 10.

References

[3.1]   Kuo, F. F. 1966. *Network Analysis and Synthesis*. New York: Wiley.

[3.2]   Butterworth, S. 1930. 'On the Theory of Filter Amplifiers,' *Experimental Wireless/ Wireless Engineer* 7: 536–41.

[3.3]   Cauer, W. 1958. *Synthesis of Linear Communication Networks*. New York: McGraw Hill.

[3.4]   Zverev, A. I. 1967. *Handbook of Filter Synthesis*. New York: Wiley.

Practice Problems

3-1   (a) Find the pole location and the coefficients of the Butterworth polynomial for order $n = 5, 8$ and 10. Compare the answers for $n = 5$ and 8 from Table 3.1 and 3.2.

      (b) Plot the pole values calculated in part (a) on the s plane.

      (c) Factorize the Butterworth polynomials found in part (a) using a root finder or any other method.

3-2   (a) Determine the transfer function for an LP filter having a maximally flat magnitude characteristic, which is 2 dB down at 2 rad/s and 32 dBs down at 7.5 rad/s.

      (b) Find a doubly terminated lossless ladder realization for the LP filter.

      (c) Find the element values for the LP ladder filter with its 3 dB frequency at 3.4 kHz.

      (d) Simulate the ladder structure used and verify the results.

3-3   Design an LP filter for the following specifications

a_{max}, dB	a_{min}, dBs	ω_1, rad/s	ω_2, rad/s
1.0	40	2000	4500

    (a)   Determine the degree n of the required maximally flat magnitude response.

    (b)   Determine the location of poles on the s plane.

    (c)   Find the quality factor of each pole.

    Determine the actual loss $\alpha_{max}(\omega_1)$ and $\alpha_{min}(\omega_2)$ at the pass band and the stop band edge frequencies.

3-4   Repeat problem 3-3 for the following specifications:

    $\alpha_{max} = 2.0$ dBs, $\alpha_{min} = 50$ dBs, $\omega_1 = 2000$ rad/s and $\omega_2 = 6000$ rad/s.

3-5   Repeat problem 3-3 for the following specifications:

    $\alpha_{max} = 1.0$ dB, $\alpha_{min} = 30$ dBs, $\omega_1 = 2000$ rad/s and $\omega_2 = 3600$ rad/s.

3-6   Repeat problem 3-3 for the following specifications:

    $\alpha_{max} = 2.0$ dBs, $\alpha_{min} = 40$ dBs, $\omega_1 = 2000$ rad/s and $\omega_2 = 5000$ rad/s.

3-7   Consider the following set of specifications:

    (i) $\alpha_{max} = 0.5$ dB, $\alpha_{min} = 32$ dBs, $\omega_1 = 1500$ rad/s, $\omega_2 = 3600$ rad/s

    (ii) $\alpha_{max} = 1.0$ dB, $\alpha_{min} = 25$ dBs, $\omega_1 = 2000$ rad/s, $\omega_2 = 7000$ rad/s

    (a)   Find the required value of order n of the LP filter with maximally flat response.

    (b)   Determine the actual attenuation at the edge of the pass band and stop band.

    (c)   Determine the attenuation at $2.5\omega_1$ and $5\omega_1$.

3-8   Determine the Chebyshev polynomial $C_4(\Omega)$, $C_5(\Omega)$ and $C_6(\Omega)$, using equation (3.31)

3-9   Determine the pole location for the Chebyshev response for:

    (a) $n = 5$ and $\alpha_{max} = 0.5$ dB,

    (b) $n = 5$ and $\alpha_{max} = 1.0$ dB,

3-10   (a)   Determine the order n, the pole location and the transfer function of an LP filter having 1.0 dB ripple width from 0 to 2.5 rad/s and a maximum of 30 dB attenuation beyond 5.0 rad/s.

    (b)   Find a resistance terminated lossless ladder for the filter realization in part (a).

3-11   Find the transfer function, ω_0 and Q values for the following specifications with the help of Table 3.4.

    (i) $\alpha_{max} = 1$ dB, $n = 6$,

    (ii) $\alpha_{max} = 2$ dB, $n = 5$,

    (iii) $\alpha_{max} = 0.5$ dB, $n = 4$

3-12   A sixth-order LP Chebyshev filter was realized with three options $-\alpha_{max} = 0.5$ dB, 1 dB and 2 dBs. Determine a relationship between ripple width and respective quality factor. Which option shall be preferred and why? (Use Table 3.4)

3-13   In a Chebyshev filter of order 5, (de-normalized) $\omega_{CB} = 1$ Krad/s and $\alpha_{max} = 1.0$ dB. Determine: (a) the value of ε, (b) the value of the pass band edge frequency ω_1, (c) the value of $|H(j\omega)|^2_{min}$, (d) the frequencies of the peaks in pass band, and (e) the frequencies of the valleys in the pass band. (f) Accurately sketch the magnitude response, using only a calculator for the necessary calculations. Use a vertical scale in dBs and a linear radian frequency scale.

3-14   An anti-aliasing filter is needed for an A to D converter working at a sampling rate of 6000 samples/s. Hence, the anti-aliasing filter is to have a minimum attenuation of 60 dBs at 3 kHz using a Chebyshev filter.

(a) If $\omega_1 = 5$ krad/s and $\alpha_{max} = 1$dB, what is the required minimum order?

(b) If $\alpha_{max} = 1$dB and $n = 7$, what is the maximum value of ω_2?

(c) If $\omega_1 = 5\pi$ krad/s and $n = 7$, what is the minimum value of α_{max}.

3-15   Determine the transfer function and give numerical values of poles for part (c) of Problem 3-14. What shall be the value of center frequency and pole-Q of the second order sections?

3-16   (a) Find the required order for a maximally flat magnitude function which is down 1 dB at 1 rad/s and down 34 dBs at 1.5 rad/s.

(b) Repeat part (a) for an equal-ripple pass band filter.

(c) Repeat part (a) for an equal-ripple stop band filter.

3-17   For the attenuation characteristics shown in Figure P3.1, find the attenuation α at the frequency which is 2.5 times the pass band edge frequency.

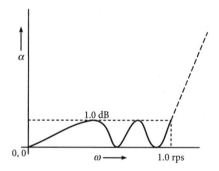

Figure P3.1

3-18   Specifications of an inverse Chebyshev function as shown in Figure P3.2 are as follows:

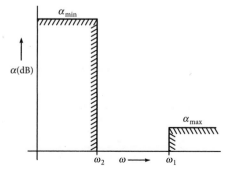

Figure P3.2

$\alpha_{max} = 0.5$ dB, $\alpha_{min} = 20$ dBs, $\omega_1 = 36$ krad/s and $\omega_2 = 80$ krad/s

(a) Determine the order of the filter

(b) Determine the location of poles and zeros.

(c) Determine the frequency of the peaks and the valleys in the stop band.

(d) Find the transfer function satisfying the specifications in terms of the product of second-order (a first-order also if needed) sections.

3-19   Repeat sections (a) to (d) of Problem 3-18 for the following specifications:

$\alpha_{max} = 0.5$ dB, $\alpha_{min} = 30$ dBs, $\omega_1 = 2$ krad/s and $\omega_2 = 3.45$ krad/s

3-20   Find order of an elliptic HP filter using two alternate methods for the following specifications:

(i) $\alpha_{max} = 1.0$ dB, $\alpha_{min} = 30$ dBs, $\omega_1 = 80$ krad/s and $\omega_2 = 50$ krad/s

(ii) $\alpha_{max} = 0.1$ dB, $\alpha_{min} = 20$ dBs, $\omega_1 = 30$ krad/s and $\omega_2 = 15$ krad/s

3-21   Find the passive ladder structures for the elliptic filters of Problem 3-20, with (a) inductors and capacitors in series occurring in the shunt branches, and (b) inductor and capacitors in parallel occurring in the series branches of the networks.

3-22   Find practically suitable values of the elements while integrating for the filters obtained in Problem 3-21 and test the circuits using PSpice.

3-23   Find order of an elliptic filter using two alternate methods for the following specifications:

$\alpha_{max} = 1.0$ dB, $\alpha_{min} = 50$ dBs, $\omega_1 = 20$ krad/s and $\omega_2 = 24$ krad/s.

Find passive ladder structures for the obtained filter with (a) inductors and capacitors in series occurring in the shunt branches, and (b) inductor and capacitors in parallel occurring in the series branches of the networks. Find the actual minimum attenuation in the stop band in both cases.

3-24   It is desired that the dc gain of the maximally flat LP filter with the specifications given in Problem 3-7(ii) remains unity when a zero is introduced at 2750 rad/s. Find the modified transfer functions.

Chapter 4

Delay: Approximation and Optimization

4.1 Introduction

While studying magnitude approximations in Chapter 3, real rational functions were selected in order to achieve desirable attenuation in the stop band and pass band. Not much attention was paid on the unintentional occurrence of phase shift or delay. It has been observed that this delay is neither the same for different approximations nor is it constant. In fact, filter delay is highly dependent on the type of approximation, order of the filter and its parameters. As an example, normalized delay for a fourth-order maximally flat, Chebyshev, inverse Chebyshev and Cauer or elliptic filter, with equal order of pass band attenuation in each case, is shown in Figure 4.1. It can be observed that the delay performance in the maximally flat case is the closest to the ideal requirement of constant delay with frequency. Performance is the worst for the elliptic filter and intermediate for the Chebyshev and the inverse Chebyshev cases. It is important to note that though Figure 4.1 is only a representative sketch, the nature of the functions remain the same: maxima of the delay occur near the pass band edge and relative maxima are dependent on filter specifications in the pass and stop bands.

It is important to have the group delay as constant or the phase response of the filter as linear with frequency. This requirement for filters is necessary in order to process the magnitude and because in many applications, the amount of delay needs to be controlled at the same time. For example, in a telephonic conversation through satellite, generally, the communication channel is active one way and the listener speaks only when the speaker from the other side stops. However, if the delay is too large, two-way conversation will become difficult and sometimes overlapping.

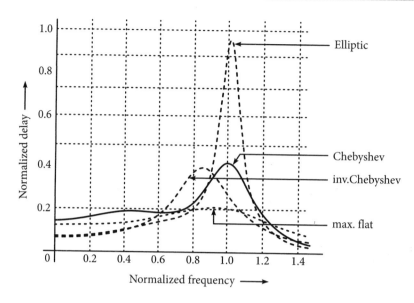

Figure 4.1 Comparison between delay performances of fourth-order elliptic, Chebyshev, inverse Chebyshev, and maximally flat responses.

From this brief discussion, it is clear that in order to avoid distortion due to different phase/group delay in a filter, an ideal transfer function should have its delay constant with frequency.

In Section 4.2, delay is defined. Its relation with the phase change in a filter section and the normalized transfer function of a filter realizing delay are also discussed in detail. As Bessel polynomials are used in realizing such transfer function, these polynomials are reviewed and the Bessel filter design shown in the next section. One of the main functions of Bessel filters is *delay equalization*, that is, trying to make as much uniform delay as possible in the pass band. Such an optimization is done through first-order, second-order, or still higher-order filters; the utilization of these filters is shown in Sections 4.5 to 4.7.

4.2 Delay

If a sine wave with $V_{in} = V_m \sin(\omega t + \varphi)$ is applied to a network (say a filter section) which contains frequency-dependent reactance elements like capacitors or inductors, its output magnitude and phase changes depending on the characteristics of the filter. However, if the magnitude of the output voltage V_{out} is the same as the input, it can be written as:

$$V_{out} = V_{in} \sin(\omega t + \varphi + \theta)$$ (4.1a)

$$= V_{in} \sin(\omega(t - D) + \theta)$$ (4.1b)

In equation (4.1), D is obviously the delay, which is associated with phase shift in the following relation:

$$D = -(\varphi/\omega) \tag{4.2}$$

In equation 4.2, D (in seconds) is a frequency dependent term. Hence, if a signal comprising components having different frequencies is applied to a filter section, these signal components will be delayed by different amounts. Figure 2.14(b) of Chapter 2, redrawn here as Figure 4.2, shows how the phase changes from the input to the output for a second-order LPF section. Delay D can be evaluated at different frequencies employing the expression in equation (2.42), which represents Figure 4.2. Delay can be evaluated not only for LPFs, but for any type of filter from its phase variation plot. It is important to note that a signal comprising a large number of components will produce output signals with a different amount of phase shift/delay resulting in distorted signals. The problem can be solved by observing that the delay D is inversely proportional to the frequency ω. Hence, if the delay, expressed as a derivative of phase shift (equation (4.3)), is a constant, all components of a signal shall be delayed by the same amount. In that case the final output signal will not be distorted, but only delayed by certain amount.

$$D(\omega) = d\varphi/d\omega \tag{4.3}$$

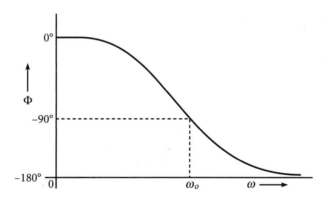

Figure 4.2   Change in phase from input to output for a second-order low pass section.

It is to be noted that the delay $D = -(\varphi/\omega)$ in equation (4.2) is called *phase delay* and the delay given by equation (4.3) is called *group delay*. In practice, the term $D(\omega)$ of equation (4.3) is called *delay* and φ is termed *phase shift*. Hence, in this book, delay $D(\omega)$ will be considered the derivative of phase with frequency.

The output voltage in equation (4.1) and its input in phasor form are represented in the following equation.

$$V_{out} = V \angle (\varphi - \omega D) \text{ and } V_{in} = V \angle \varphi \tag{4.4a}$$

Hence, the transfer function of the filter will be as follows:

$$H(j\omega) = V_{out}(j\omega)/V_{in}(j\omega) = 1.0e^{-j\omega D} \tag{4.4b}$$

Normalizing equation (4.4b) with a frequency $\omega_d = (1/D)$, $H(j\omega)$ will become $e^{-j\omega/\omega_d} = e^{-j\Omega}$. Generalizing the transfer function in the complex frequency s plane by replacing $j\omega$ with s, the desired transfer function shall be as follows:

$$H(s) = V_{out}(s)/V_{in}(s) = e^{-s} \qquad (4.5)$$

A number of methods are available for obtaining a real rational function for the physical realization of the transfer function of equation (4.5) with normalized delay $D = 1$ s. However, the most commonly used method is the *Bessel–Thomson (BT) method*, employing the Bessel function [4.1, 4.2]. Therefore, before discussing the BT method of filter design, the Bessel function and some of its basic properties are discussed in brief.

4.3 Bessel Polynomial

Using the known identity $e^{-s} = \sin h\, s + \cos h\, s$, the transfer function in equation (4.4) can be expressed in terms of hyperbolic functions. As the denominator of $H(s)$ is the sum of $\sin h\, s + \cos h\, s$, it can be expressed in terms of the series expansion of hyperbolic functions: an infinite polynomial in the s plane. It was shown by Storch (1954) that the infinite series for the hyperbolic can be truncated after the nth term and the truncated terms give the maximally flat delay at $\omega = 0$ [4.3]. In addition, such a denominator polynomial was obtained with a class of Bessel polynomials given by the following recursive formula.

$$B_n = (2n-1)B_{n-1} + s^2 B_{n-2} \qquad (4.6)$$

In general, the BT function is formed in such a way that the selected numerator results in $H_n(j0) = 1$. Hence, the form of the transfer function is as follows:

$$H_n(s) = B_n(0)/B_n(s) \qquad (4.7)$$

From equation (4.6), the Bessel polynomial for order n can easily be found. For example, the following relations are given up to $n = 4$.

$$B_0 = 1,\ B_1 = s + 1,\ B_2 = s^2 + 3s + 3,\ B_3 = s^3 + 6s^2 + 15s + 15,$$
$$B_4 = s^4 + 10s^3 + 45s^2 + 105s + 105. \qquad (4.8)$$

We know that there are different methods available for filter realization. However, in a commonly used approach, the filter design uses pole frequency ω_o and pole Q, which are related with the pole location of the transfer function. In the present case, it is required to factorize the Bessel polynomial in order to find the location of poles to find its roots. Unfortunately, there is no simple method to find the polynomial's roots and computer methods have to be employed. Data tables giving roots of the polynomial, location of poles and resultant pole frequency ω_o

and pole Q have been made available for different values of n. Table 4.1 gives the values of ω_o and pole Q for $n = 1$ to 6, which has been obtained from the calculated values of the roots of the Bessel polynomial. Roots of the Bessel polynomial can easily be obtained from root finding programs. Along with Table 4.1, Tables 4.2 and 4.3 give the factored form of $B_n(s)$ for $n = 1$ to 6. Pole-Q and pole location are also given.

Table 4.1 Roots of the Bessel polynomial $B_n(s) = 0$ for $n = 1$ to 6

N	Roots					
1	−1.000000					
2	−1.500000	$\pm j0.866025$				
3	−2.322185	−1.8389073	$\pm j1.754381$			
4	−2.896210	$\pm j0.867234$	−2.1037894	$\pm j2.657418$		
5	−3.646738	−3.3519561	$\pm j1.742661$	−2.3246743	$\pm j3.571022$	
6	−4.248359	$\pm j0.867509$	−3.7357084	$\pm j2.626272$	−2.5159322	$\pm j4.492673$

Table 4.2 Factored form of $B_n(s)$ for $n = 1$ to 6

n	$B_n(s)$	
1	$s + 1$	1
2	$s^2 + 3s + 3$	3
3	$(s^2 + 3.67782s + 6.45944)(s + 2.232219)$	15
4	$(s^2 + 5.79242s + 9.14013)(s^2 + 4.20758 + 11.4878)$	105
5	$(s^2 + 6.70391s + 14.2725)(s^2 + 4.64943s + 18.15631)(s + 3.64674)$	945
6	$(s^2 + 8.49672s + 18.8013)(s^2 + 7.74142s + 20.85282)(s^2 + 5.03186s + 26.51402)$	10,395

Table 4.3 ω_o, pole-Q and real pole σ for the transfer function $H_n(s)$ of equation (4.7)

n	ω_o, and pole-Q are in pairs, real pole $(-\sigma)$ is a single entry			$B_n(s)$
2	1.732;0.577			3
3	2.542;0.691	2.322		15
4	3.023;0.522	3.389;0.806		105
5	3.778;0.564	4.261;0.916	3.647	945
6	4.336;0.510	4.566;0.611	5.149;1.023	10,395

Before moving on to the design method of BT filters, a brief comparison with Butterworth filters is useful. Figure 4.3 shows the variation of the normalized delay D_n with ω up to $n = 10$ in BT filters. It is observed that D_n is maximally flat at $\omega = 0$ with magnitude unity. The delay remains flat for even higher values of ω for larger values of n.

Figure 4.4 shows the attenuation variation of Butterworth filters ($n = 4$ and 8) and the BT approximation filter which also leads to low pass characteristics. Rate of attenuation for both the responses at high frequencies is 20 n dB/dec, but BT response has a much greater transition band and attenuation is at a much slower rate at lower frequencies.

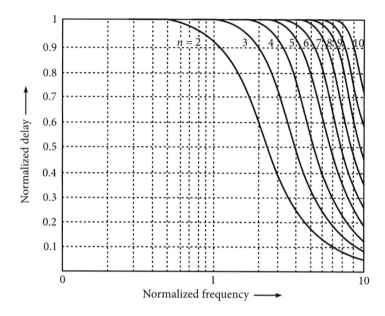

Figure 4.3 Normalized delay in Bessel–Thomson filters of order 2 to 10.

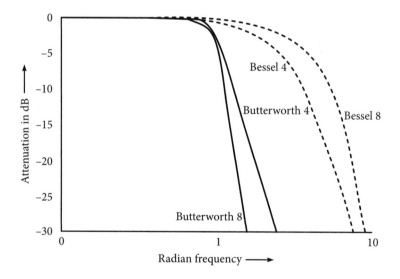

Figure 4.4 Comparison between Bessel–Thomson and Butterworth characteristics for filter of order 4 and 8.

4.4   Design of Bessel Filters

While designing magnitude-approximated filters, it is necessary to determine the values of the attenuation in the pass band, the stop band, and the band edge frequencies. However, for the delay-approximated filter, the amount of permissible deviation in delay compared to the ideal value $D = 1$, from $\omega = 0$ to a certain frequency (say ω_1) is the standard form of specification. However, for the Bessel filter, no closed form solution is available to find the order of the filter for a given delay deviation. Using computer programs, design curves (Figure 4.5(a)) have been obtained showing permissible error (in percent) with normalized frequency as function of the order of the BT filter. Such design curves are used to find the required order of the BTF for the permissible maximum delay error. If, along with the delay requirements, there is constraint on the magnitude as well, curves in Figure 4.5(b) are also used: these curves show the relation between permissible attenuation and order n. After finding the filter order (selecting the greater n obtained from the two specifications), computer program generated pole values or pole frequency ω_0 and pole-Q of second-order sections are obtained. Filters are then realized using a suitable network employing any selected approach.

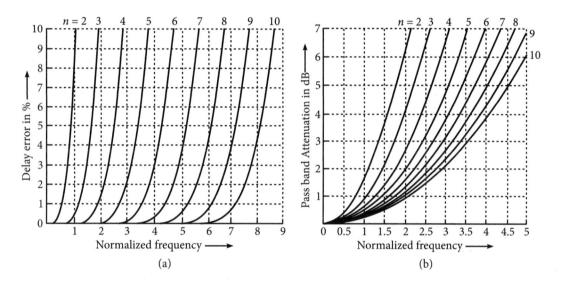

Figure 4.5   Design curves for obtaining the degree n of the Bessel–Thomson filter to meet (a) the maximum delay error and (b) permissible attenuation in pass band.

It can be observed from Figure 4.4 that the use of BT approximation for satisfying attenuation specification requires higher-order filters compared to the case when it is used for approximating delay specification. The problem becomes worse, that is, the required order of the filter becomes larger, if the transition band is narrow or the stop band has to have larger attenuation. It is because of this reason, and the involved process of finding filter networks, that the use of BT approximation only is limited. Other alternatives, for example, the use of a cascading magnitude-approximated filter section and a delay equalizer, are preferred. Such alternatives are discussed in the next section.

Example 4.1: Find the order and the transfer function of a delay filter which can provide delay $D = 200$ μs with a permissible error of 10% in delay at the frequency $\omega_1 = 18$ krad/s. The permissible loss in magnitude at a frequency of $\omega_2 = 10$ krad/s is 2.0 dBs.

Solution: The order of the BTF shall be found using Figures 4.5(a) and (b). For the utilization of these figures, normalized frequency Ω_c is to be found. Its value is obtained as: $\Omega_c = 1/D = 1/200 \ \mu s = 5 \ krad/s$.

Hence, normalized pass band edge frequency (for delay) $\omega_1 = 3.6$ and normalized pass band edge frequency (for magnitude) $\omega_2 = 2$. For normalized $\omega_1 = 3.6$ with 10% allowable deviation in delay required, order is $n = 5$ from Figure 4.5(a). For normalized $\omega_2 = 2$, the required filter order from Figure 4.5(b) is $n = 5$. Therefore, the selected order of the BTF is 5 for which the transfer function is as follows from Table 4.2.

$$H(s) = 945/\{(s^2 + 6.7039s + 14.2725)(s^2 + 4.64943s + 18.1563)(s + 3.64674)\}$$

$$= 945/(s^5 + 15s^4 + 105s^3 + 420s^2 + 945s + 945) \tag{4.9}$$

Example 4.2: Design a delay filter with specifications as given in Example 4.1.

Solution: A fifth-order filter will need one first-order and two second-order sections. From equation (4.9), pole of the first-order section is at $s = -3.6467$. A simple OA–RC (operational amplifier with resistors and capacitors) circuit shown in Figure 4.6(a) will have the following transfer function:

$$H_0(\Omega) = (1/RC)/(s + 1/RC) \tag{4.10}$$

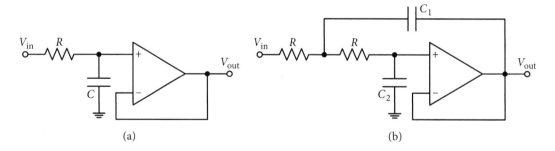

(a)   (b)

Figure 4.6 (a) Active first-order low pass as a first-order passive filter followed by a buffer and (b) a second-order low pass section.

For normalized 3 dB frequency of 1 rad/s and a dc gain of unity, the elements will be as follows:

$$C = 1 \ F \ and \ R = 0.2742 \ \Omega \tag{4.11}$$

For the remaining two second-order sections, the LP circuit shown in Figure 4.6(b) is selected (any other second-order circuit can also be selected), for which the transfer function is given as follows:

$$H(s) = \frac{(G_1 G_2 / C_1 C_2)}{s^2 + \left(\dfrac{G_1 + G_2}{C_1}\right)s + \dfrac{G_1 G_2}{C_1 C_2}} \tag{4.12}$$

From equation (4.12), the following expressions of the parameters are easily obtainable:

Voltage gain at dc equals unity   (4.13a)

$$\omega_o = (\frac{G_1 G_2}{C_1 C_2})^{1/2} \text{ and } \frac{\omega_o}{Q} = \frac{G_1 + G_2}{C_1}$$ (4.13b)

From Table 4.1 or Table 4.3 value of the required two sets of parameters for $H_1(s)$ and $H_2(s)$ are as follows:

$$\omega_{o1} = 3.778, Q_1 = 0.564 \text{ and } \omega_{o2} = 4.261, Q_2 = 0.916$$ (4.14)

Selecting $C_1 = 1F$ and $C_2 = 0.1$ F, equation (4.13b) gives normalized conductance values for the two respective circuits as follows:

$$G_{11} = 6.4776, G_{12} = 0.22034 \text{ and } G_{21} = 4.2109, G_{22} = 0.44075$$ (4.15)

To bring the resistances to a practical level, impedance scaling factor of 10^3 is used. For frequency normalization, the scaling factor of Example 4.1 (5×10^3) is used. The respective element values for the two second-order sections are as follows:

$$R_{11} = 0.1543 \text{ k}\Omega, R_{12} = 4.538 \text{ k}\Omega, C_{11} = 0.2 \text{ }\mu\text{F and } C_{12} = 0.02 \text{ }\mu\text{F}$$ (4.16a)

$$R_{21} = 0.23747 \text{ k}\Omega, R_{22} = 2.2688 \text{ k}\Omega, C_{21} = 0.2 \text{ }\mu\text{F and } C_{22} = 0.02 \text{ }\mu\text{F}$$ (4.16 b)

Similarly, for $H_0(s)$ values of elements after de-normalization are as follows:

$$R = 0.2742 \text{ k}\Omega \text{ and } C = 0.02 \text{ }\mu\text{F}$$ (4.17)

The complete fifth-order delay filter is shown in Figure 4.7: here the three sections are connected in cascade. Figure 4.8 shows the simulated group delay response of the filter, where its value at low frequencies is 200 µs. It gives the delay D at 2.873 kHz (18.085 krad/s) as 185.3 µs:

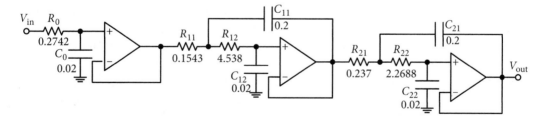

Figure 4.7   Fifth-order delay filter for Example 4.2. All resistors are in kΩ and capacitors are in µF.

Value of D is well within 10% of the specification $D = 200$ µs up to 18 krad/s. Figure 4.8 also shows the magnitude response, where the voltage gain is 0.866 at 1.272 kHz (8 krad/s). Hence, attenuation is 1.24 dB, which is also satisfied being less than 2 dB of specification.

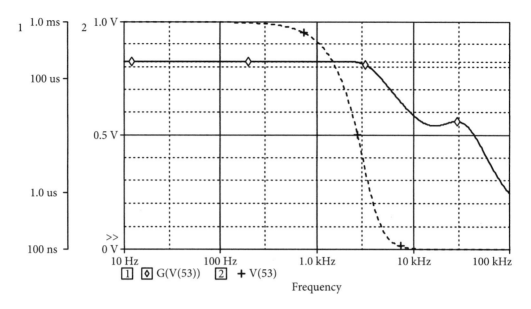

Figure 4.8   Magnitude and delay variation of the fifth-order BT filter for Example 4.2.

4.5   Delay Equalization

By now, the advantages and limitations of magnitude-approximated and delay-approximated filters are obvious. With magnitude-approximation filters, it is not possible to obtain constant delay at all frequencies; whereas for delay/phase-approximated filters, not only is the magnitude roll-off very slow, its design is also rather involved. Simultaneous satisfaction of magnitude and delay mostly requires extremely high-order filters; hence, they are expensive if only BTF is used. In many applications, especially for signals having a sharp transition in the time domain, like a pulse or a pulse train, not only should the magnitude be retained, equal delay is necessary for a very wide range of frequencies. If a sufficiently high-order filter is not used, the pulse output will be distorted. One preferred solution to the problem, as suggested in the last section, is to cascade a magnitude-approximated filter, which satisfies the magnitude response, with an AP (all pass) filter that least affects the magnitude with frequency but controls delay as per the requirement. The simple arrangement is shown in Figure 4.9. Figure 4.10 shows a typical combination of delay provided by a magnitude-approximated filter and an AP filter that is almost constant in the pass band range. As phase or delay equalizers generally use first- or second-order AP sections, their arrangement will be discussed in the following sections.

4.6   Delay Equalization Using First-order Sections

For less challenging cases of delay equalization, a first-order section with the following transfer function can be employed.

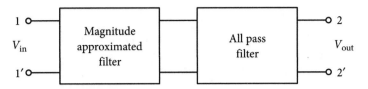

Figure 4.9   Cascading a magnitude-approximated filter and an all pass filter.

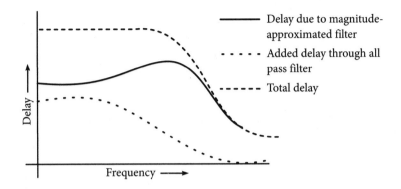

Figure 4.10   Obtaining a constant delay over most of the pass band with the help of an all pass filter.

$$H_{AP1} = k\frac{s - a_0}{s + a_0} \tag{4.18}$$

For sinusoidal inputs, magnitude and phase for the first-order transfer function for equation (4.18) are as follows:

$$|H_{AP1}(j\omega)| = k \text{ and } \varphi_1 = -2\tan(\omega/a_0) \tag{4.19}$$

Corresponding delay is obtained as:

$$D_1 = -(d\varphi_1/d\omega) = (2/a_0)/\{1 + (\omega/a_0)^2\} \tag{4.20}$$

Magnitude of the delay D_1 at $\omega = 0$ is $(2/a_0)$ and goes on decreasing monotonically with ω as given by equation (4.20) and illustrated in the next example.

Example 4.3: Figure 4.11(a) shows a first-order AP section realizing one real pole and one real finite zero. Its transfer function is as follows:

$$H_{AP1}(s) = \frac{V_{out}}{V_{in}} = \frac{s - (R_2/R_1R_3C)}{s + (1/R_3C)} \tag{4.21}$$

In order to equalize the magnitude of real pole and zero at $s = |a_0|$, a convenient choice is $R_1 = R_2 = R_3 = 1\,\Omega$ and $C = 1/a_0$. Frequency scaling by 10^4 and impedance scaling by 10^3 is done to bring component values within the practical range: $C = 0.1\ \mu\text{F}$ and $R_1 = R_2 = R_3 = 1\ \text{k}\Omega$; the values are also shown in Figure 4.11(a). Simulated response for the AP filter is shown in Figure 4.11(b), and some of the delay values with respect to frequency are shown in Table 4.4.

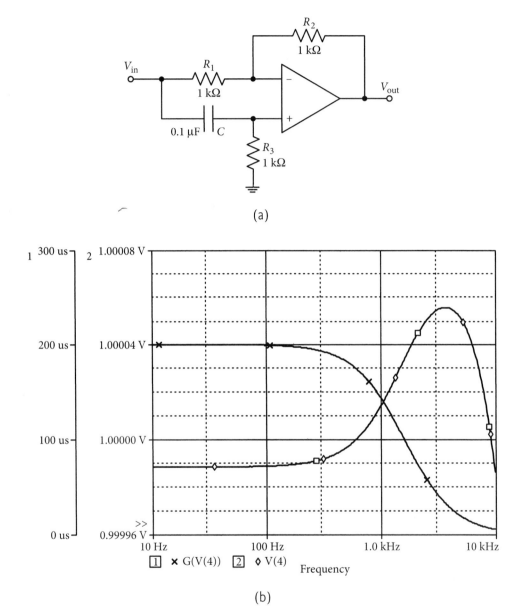

(a)

(b)

Figure 4.11 (a) A first-order all pass section for Example 4.3. (b) Variation of delay and magnitude for the first-order all pass filter of Figure 4.11(a).

Table 4.4 Variation of delay for the first-order AP filter of Figure 4.11(a)

Freq(Hz)	10	100	200	400	600	800	1 k	1.3 k	1.5 k	1.8 k	2 k	3 k	5 k
Delay(μs)	200.3	199.5	197	188.3	175.1	160.1	143.7	120.2	105	87.5	78.5	44.5	18.7

The shape of the delay curve of the first-order AP filter verifies the statement made earlier that it decreases monotonically with frequency. It is thus obvious that the delay equalization through the use of a first-order AP section is limited to the cases where delay to be added is highest at $\omega = 0$ and decreases with ω. Moreover, there is only one parameter a_0, which decides the amount of introduced delay; hence, delay optimization is less flexible and if delay optimization is done without using a computer program, it will require some trial and error. In this example, as the selected RC product $= 10^{-4}$, theoretical value of delay $D = 0.20$ ms; the simulated result is in conformity with this value.

Example 4.4: Design a third-order LP filter using Chebyshev approximation with a 0.5 dB ripple in the pass band and plot its delay response. Use a frequency scaling factor of 10^4 and an impedance scaling factor of 10^3. Connect a suitable first-order AP filter in cascade with the LP filter and observe its effect on the delay differential of the overall composite filter.

Solution: Pole locations of a third-order Chebyshev filter with 0.5 dB pass band ripples are obtained from Table 3.3, which are as shown here:

$$s_1 = -0.62565, \; s_{2,3} = -0.3132 \pm j1.0219 \tag{4.22}$$

The normalized transfer function expression is obtained from equation (4.22) as:

$$H_{ch0} = \frac{0.6265}{(s+0.6265)} \frac{1.1424}{(s^2 + 0.6264s + 1.1424)} \tag{4.23}$$

The transfer function H_{ch0} is realized as a cascade of the following first-order and second-order functions.

$$H_{ch1} = \frac{0.6265}{(s+0.6265)} \tag{4.24}$$

$$H_{ch2} = \frac{1.1424}{(s^2 + 0.6264s + 1.1424)} \tag{4.25}$$

For the first-order section, the circuit shown in Figure 4.6(a) is used and comparison of equation (4.10) with equation (4.24) gives normalized element values as

$$R = 1.0 \; \Omega \text{ and } C = 1/0.6265 \text{ F} \tag{4.26}$$

For the second-order section, the circuit shown in Figure 4.6(b) is used and comparison of the equation (4.25) with equations (4.12) and (4.13) gives the following normalized element values:

$$C_1 = 1 \text{ F}, \; C_2 = 0.05 \text{ F}, \; R_1 = 9.027 \; \Omega \text{ and } R_2 = 1.939 \; \Omega \tag{4.27}$$

A frequency scaling by 10^4 and an impedance scaling by 10^3 are applied on both the first- and second-order functions, and the cascaded third-order filter circuit is shown in Figure 4.12. Its magnitude and delay response are shown in Figure 4.13 and 4.14, respectively. The peak delay at pass band edge frequency is 0.376 ms and the lowest delay at 665 Hz is 0.194 ms. Hence, the delay differential in the designed third-order filter is 0.182 ms or 49.4% which is expected to be reduced by cascading a first-order AP filter of the type shown in Figure 4.11(a). Since we are not using a computer program to calculate the parameters of the AP filter, we employ trial and error. Three different values of the normalized parameter $a_0 = 0.8$, 1.0 and 1.25 are tried for the transfer function of equation (4.18). For the AP of Figure 4.11(a), normalized resistances R_1, R_2, and R_3 of 1 Ω each with $C = 1.25$ F, 1 F, and 0.8 F, respectively, for the three cases of a_0 are used. Using the same frequency and impedance scaling factors of 10^4 and 10^3 on the RC elements, the respective de-normalized a_0 and introduced delays at dc is easily obtained. The overall delay responses in the three cases are simulated using PSpice which are also shown in Figure 4.14. There is an improvement in the delay differential as shown in Table 4.4. Obviously, $a_0 = 1.25$ gives a better result; however, the total delay is still not constant because of the limitation of the first-order section in having only one parameter that controls the shape of the added delay.

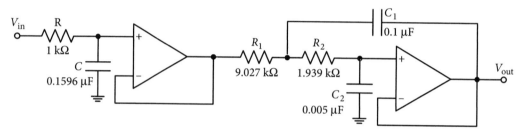

Figure 4.12 Third-order low pass Chebyshev filter for Example 4.4.

Table 4.5 Effect of using a first-order AP section in Example 4.4

Normalized capacitance value (F)	De-normalized a_0	Introduced delay (ms)	Delay differential (%)
1.25	0.8×10^4	0.25	22
1.0	1.0×10^4	0.20	25
0.8	1.25×10^4	0.16	29.5

4.7   Delay Equalization Using Second-order Sections

The transfer function of a second-order AP section in terms of pole frequency ω_o and pole-Q are given as:

$$H_2(s) = k\frac{s^2 - (\omega_o / Q)s + \omega_o^2}{s^2 + (\omega_o / Q)s + \omega_o^2}$$

(4.28)

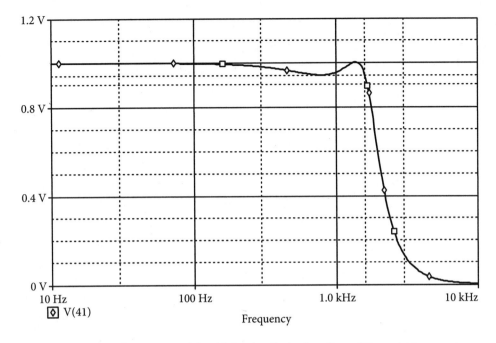

Figure 4.13 Magnitude response of the third-order Chebyshev filter of Figure 4.12.

Obviously, the gain constant k of equation (4.28) needs to be unity if the gain of the overall filter section is to remain unchanged. The important advantage of the second-order section is in its flexibility as there are two parameters, ω_o and Q, available. We can vary the shape of the delay versus frequency curve of the AP section more freely. Using normalized frequency $s_n = (s/\omega_o)$ and $k = 1$, the normalized AP function will be modified as follows:

$$H_2(s_n) = \frac{s_n^2 - s_n(\omega_n/Q) + 1}{s_n^2 + s_n(\omega_n/Q) + 1} \tag{4.29}$$

Replacing s_n by $j\omega_n$, the phase shift of the second-order function is calculated as:

$$\varphi_2(\omega_n) = -2\tan(\frac{\omega_n/Q}{1-\omega_n^2}) \tag{4.30}$$

It corresponds to the delay in the following way:

$$D_2(\omega_n) = -\frac{d\varphi_2(\omega_n)}{d\omega} = \{\frac{d\varphi_2(\omega_n)}{d\omega_n}\}(\frac{d\omega_n}{d\omega})$$

$$= \frac{1}{\omega_o}\left\{\frac{(2/Q)(1+\omega_n^2)}{(1-\omega_n^2)^2 + (\omega_n/Q)^2}\right\} \tag{4.31}$$

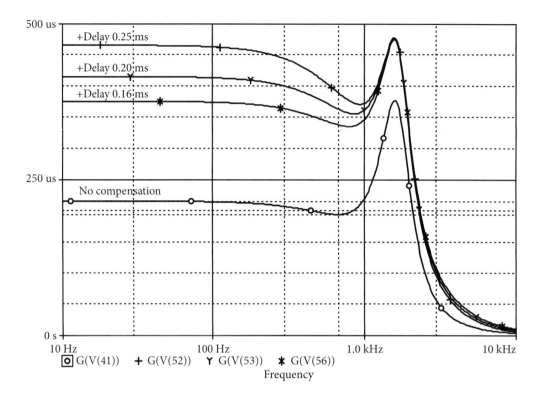

Figure 4.14 Delay in a third-order low pass Chebyshev filter of Example 4.4; the delay with three different all pass filters cascaded to it is also shown.

As mentioned earlier and shown by equation (4.31), variation of D_2 depends on ω_o and Q. Value of the delay at dc is obtained from equation (4.31) as:

$$D_2(0) = (2/\omega_o Q) \tag{4.32}$$

Maximum of the delay curve obtained by taking the derivative of equation (4.31) and equating it to zero results in the following:

$$D_{2,\mathrm{max}} \cong (4Q/\omega_o) \text{ for } \omega_n \cong 1 \tag{4.33}$$

The maxima occur at the frequency

$$\omega_{n,\mathrm{max}} = \{-1 + (4 - Q^{-2})^{0.5}\}^{0.5} \cong 1 \text{ for } Q > 1 \tag{4.34}$$

Equations (4.32) and (4.33) indicate that a larger value of pole Q means a lower delay value at dc and a higher maximum delay.

Based on the selected values of ω_n and Q, the resulting second-order section is then cascaded with the magnitude-approximated filter section. Once again, we use trial and error for simpler cases or a computer program for more demanding cases. More than one second-order AP section will be used if a single section is not able to satisfy the requirement of the delay. Any appropriate circuit employing operational amplifier(s), operational transconductance amplifiers (OTAs) or some other active devices can be used to realize second-order AP section(s).

If a second-order LP section is used instead of an APF, its expression of phase shift, delay, delay at dc, and maximum delay, corresponding to the equations (4.30) to (4.33), are as given below:

$$\varphi_{2LP}(\omega_n) = -\tan\left(\frac{\omega_n / Q}{1 - \omega_n^2}\right) \tag{4.35a}$$

It corresponds to the delay as:

$$D_{2LP}(\omega_n) = -\frac{d\varphi_{2LP}(\omega_n)}{d\omega} = \left\{\frac{d\varphi_{2LP}(\omega_n)}{d\omega_n}\right\}\left(\frac{d\omega_n}{d\omega}\right) = \frac{1}{\omega_o}\left\{\frac{(1/Q)(1+\omega_n^2)}{(1-\omega_n^2)^2 + (\omega_n / Q)^2}\right\} \tag{4.35b}$$

Value of the delay at dc: $D_{2LP}(0) = (1/\omega_o Q)$ \hfill (4.35c)

Maximum of the delay curve $D_{2LP,max} \cong (2Q/\omega_o)$ for $\omega_n \cong 1$ \hfill (4.35d)

Example 4.5: Obtain the delay response of the fifth-order passive Chebyshev filter having a corner frequency of 10 krad/s (1.59 kHz) and a ripple width of 0.5 dB. Apply the delay optimization using first- and second-order AP sections.

Solution: A fifth-order Chebyshev filter is shown in Figure 4.15(a) (its design and simulation will be done in Chapter 5).

Figure 4.15(b) shows the delay caused by the fifth-order Chebyshev filter having 0.5 dB ripple width in the pass band. It shows a peak delay of 1.064 ms at the pass band corner frequency of 10 krad/s (1.59 kHz): this is normal in Chebyshev filters. At low frequency near dc, delay is considerably small having a value of 0.42 ms; it also shows a minimum of ~0.375 ms at around 445 Hz. The ratio of maxima to minima in the pass band is 2.837, which needs to be reduced.

A first-order AP section as shown in Figure 4.11(a) (which essentially has the highest delay at dc) is to be cascaded to a filter as shown in Figure 9.5(a). To introduce a delay of 0.64 ms at dc (the approximate delay of the passive LP filter cascaded with the AP section would be expected to become 1.62 ms; very close to the peak value of its delay at pass band edge). Equation (4.20) gives the characteristic constant of the first-order AP filter:

$$a_0 = (2/D) = 3125 \tag{4.36}$$

Since for the first-order filter shown in Figure 4.11(a), $a_0 = (1/RC)$, with an impedance scaling factor of 10^3, element values are $C = 0.32\ \mu F$ and $R = 1.0\ k\Omega$. The simulated delay response of the designed first-order AP filter (AP1-1) is shown in Figure 4.16. The filter has a maximum delay of 0.64 ms at 10 Hz and as expected, the delay drops monotonically with frequency. The designed first-order AP filter was cascaded with the Chebyshev filter and the resulting circuit is simulated for delay which is also shown in Figure 4.15. This filter has a total delay of 1.06 ms at 10 Hz, and as a result of the sum of the delay of the Chebyshev filter and the delay of the first-order AP section, the maximum delay at pass band edge is 1.12 ms. A minimum delay of 0.591 ms occurs at 1.22 kHz, resulting in a delay differential of 41.3%. This shows an improvement in the differential delay due to the filter of 64.6%. Further improvement in the delay needs to be achieved by the use of an additional second-order AP section.

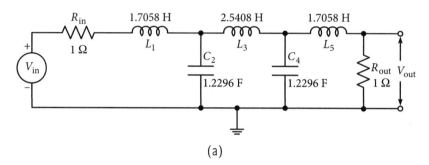

(a)

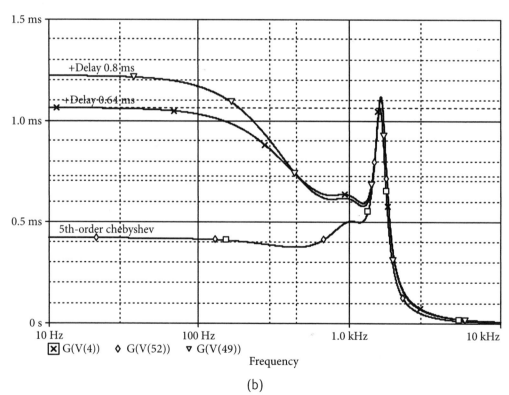

(b)

Figure 4.15 (a) Fifth-order passive Chebyshev approximated filter for Example 4.5. (b) Delay response of the given passive fifth-order filter and total delay when AP1 and AP2 are cascaded with the passive filter.

However, before using second-order AP, another first-order APF (AP1-2) with delay of 0.8 ms at dc is cascaded, (for AP1-2, the only change from AP1-1 is that $C = 0.4$ μF; resistances remain the same) which makes the total delay at dc as 1.22 ms. In this case, a dip occurs with a delay value of 0.576 ms as shown in Figure 4.15, resulting in a differential delay of 52.7%; which means that this alternate of AP1-2 is worse.

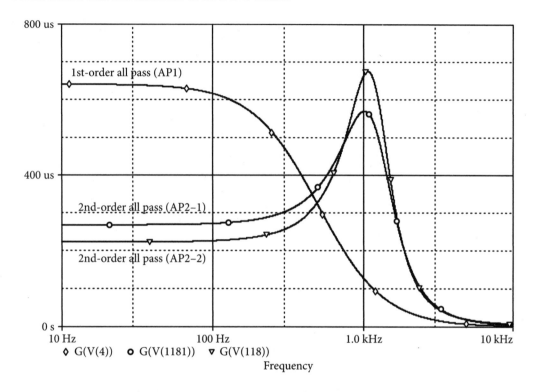

Figure 4.16 Variation of delay for the first-order AP filter (AP1) and two second-order AP filters used in Example 4.5.

For the second-order AP section, we select $\omega_o = 7.5$ krad/s (corresponding to 1200 Hz). For the desired additional delay of 0.53 ms by the second-order AP section at ω_o (an approximation looking at the delay curve after cascading the first-order AP1), equation (4.31) can be used. However, an approximate value of pole-Q can be obtained from equation (4.33) as follows:

$$Q = (0.53 \times 10^{-3} \times 7.5 \times 10^3/4) = 0.99375$$

To realize the second-order AP filter with $\omega_o = 7.5$ krad/s and $Q = 0.99375$, the circuit shown in Figure 8.13(a) was used (any other second-order AP circuit can also be used). Design of the circuit is discussed in Chapter 8, and use of equation (8.31) gives $R = 5$ kΩ, $QR = 4.968$ kΩ and $C = 26.667$ nF; for the equation (8.33b), $\beta = 2.014$, $\gamma = 0$ and $k = 1$ gives $R_f = R_\alpha = 5$ kΩ, $R_\beta = 2.484$ kΩ. Delay introduced due to the second-order AP2-1 circuit is shown in Figure 4.16. The designed AP2-1 section was also cascaded and the overall circuit is shown in Figure 4.17.

The values of the elements of the second-order AP are also shown in Figure 4.17 and the filter's delay response is obtained through PSpice simulation.

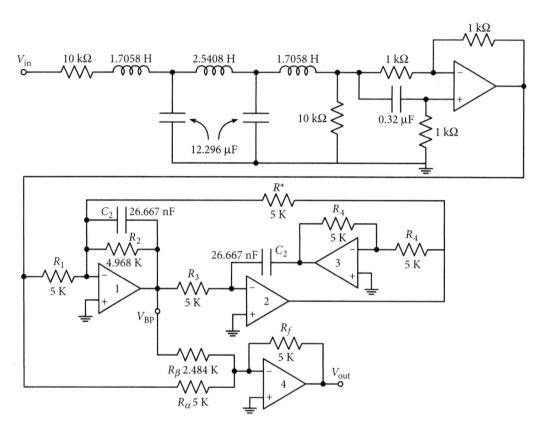

Figure 4.17 Passive fifth-order filter being delay compensated with a first-order and a second-order all pass filter.

The magnitude response of the composite filter is shown in Figure 4.18 along with the magnitude response after cascading with AP1-1 and AP2-1. It is important to note that the response magnitude remains unaffected.

The delay response of the compensated filter after cascading with AP1-1, having $D(0) = 0.64$ ms, as well as second-order sections AP2-1 (case I) is shown in Figure 4.19. It improves the delay differential at dc (1.33 ms) and the peak delay at pass band edge frequency (1.42 ms). The peak delay and the delay at dc are close; a minimum delay of 1.06 ms occurs at nearly 520 Hz, and the ratio of minimum to maximum delay in the pass band is 1.33. The delay curve still has depressions at around 520 Hz and around 1320 Hz; but the delay differential has improved to 25.3%.

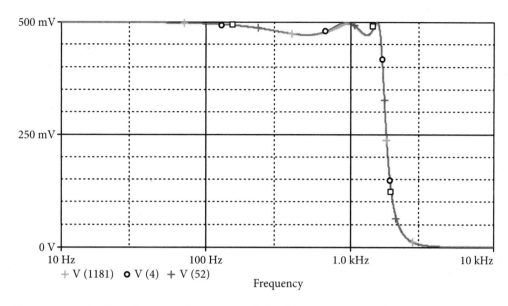

Figure 4.18 Unaffected magnitude response of the delay compensated filter of Example 4.5 along with the magnitude response before cascading with AP1 and AP2-1.

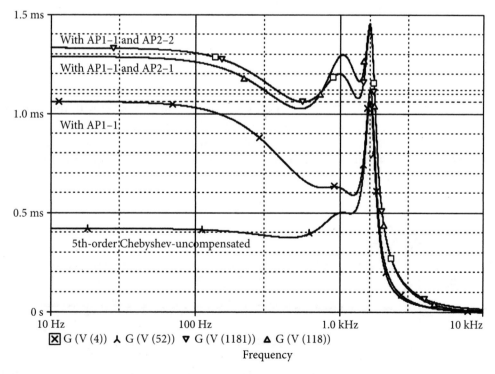

Figure 4.19 Variation of the delay for the uncompensated fifth-order Chebshev filter and delay after compensation with first- and second-order AP filters.

A second attempt was made with another second-order section AP2-2 with $Q = 1.2$ (case II), for which enhanced delay by 0.64 ms is spread over a frequency range of around 1200 Hz, (another possible choice for a second-order section can be that which has a different value of ω_o). The modified filter was realized with new element values ($QR = 6$ kΩ, $R_\beta = 3$ kΩ) and the delay response was simulated (Figure 4.19). Delay of the Chebyshev filter shown in Figure 4.15 is also included here for comparison sake. It shows the ratio of the maximum to minimum delay as 1.418 in case II since the maximum value of the delay is 1.45 ms and minimum is 1.0223 ms. Case I with $Q = 0.993$ appears to be slightly better. Obviously some more trials may give even better results without resorting to computer usage.

Example 4.6: Obtain a maximally flat filter for the following specifications and use AP filter(s) to improve its delay response.

$$\alpha_{max} = 1 \text{ dB}, \; \alpha_{min} = 40 \text{ dBs}, \; \omega_1 = 2000 \text{ rad/s and } \omega_2 = 6000 \text{ rad/s} \qquad (4.36)$$

Solution: In Chapter 3, Example 3.1, order of the filter was 5 for the aforementioned specifications and a passive filter structure with element values was given in Figure 3.6. The filter in active form is shown in Figure 4.20 with all element values (its design in active form shall be taken up later in Chapter 10). Figure 4.21 shows the delay due to the Butterworth filter which remains nearly constant for 1.416 ms up to 100 Hz and the peak delay of 2.25 ms near the pass band edge. Since the ratio of maxima to minima (1.56) is not very high, utilization of the first-order AP filter was considered adequate. The AP filter section shown in Figure 4.11(a) with $D(0) = 1.2$ ms and 1.0 ms, having all resistances as 1 kΩ and capacitance as 0.6 μF and 0.5 μF, respectively, were used. Delay of the compensated circuits is also shown in Figure 4.21. Case II with $D(0) = 1$ ms gives sufficiently good results as the maxima to minima ratio reduces to nearly 1.169 from 1.56. However, case I is slightly better with the maxima to minima ratio a bit less at 1.133. The example shows that the delay introduced in maximally flat filters is much less compared to equal-ripple filters; hence, it is easier to compensate their delay. When it is important to have uniform delay, then an overall compensated filter using a maximally flat approximation becomes economical rather than using an equal-ripple base filter, though its order may be less.

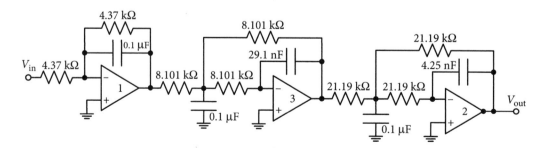

Figure 4.20 Fifth-order Butterworth filter circuit for Example 4.6.

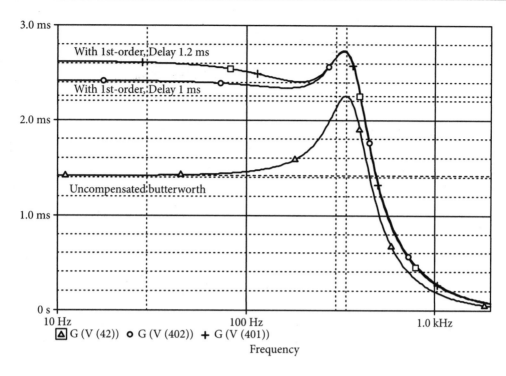

Figure 4.21 Variation of delay in a fifth-order Butterworth filter, and with delay compensation for Example 4.6.

Practice Problems

4-1 While looking at the phase response of a filter, the following phase versus frequency observations were taken. Calculate the phase delay at each frequency and the group delay at each frequency differential.

Frequency (Hz)	197.956—202.076	396.314—403.081	598.358—601.744
Phase (degrees)	71.885—74.558	−68.942—(−73.489)	−150.554—(−151.338)

4-2 Using equation (4.6), find the Bessel polynomial for filter order $n = 5$ and 6, and verify its roots given in Table 4.1.

4-3 Calculate the delay in a normalized fifth-order maximally flat and Chebyshev filter at the pass band edge frequency. In both cases, the voltage gain drops by 1 dB at the pass band edge frequency.

4-4 Find the order and the transfer function of a delay filter which can provide a delay of 0.5 ms, with a permissible delay error of 8% at 8.5 krad/s. Will the required order of the filter increase if permissible loss in magnitude is 3 dBs at 5 krad/s?

4-5 Design a delay filter which satisfies the specifications in Problem 4-4 and check the design through simulation.

4-6   Find the order and the transfer function of a delay filter for which permissible loss in magnitude is 2 dB at 6 krad/s; the filter should provide a delay of 0.25 ms with permissible delay error of 5% at 9 krad/s.

4-7   Repeat Problem 4-5 for the specifications in Problem 4-6.

4-8   Design a BT filter which provides 120 μs delay. Attenuation error is not expected to be more than 2 dBs up to $\omega = 12$ krad/s and the delay should stay below 6% in the frequency range below 12 krad/s. Realize the filter as a cascade of sections having maximally flat approximations and test using PSpice.

4-9   Obtain the delay response of a third-order passive Chebyshev filter having corner frequency of 10 krad/s and ripple width of 1.0 dB, and apply delay optimization using first- and second-order AP sections.

4-10   Design a BT delay filter with practical value of elements, which has a 4% deviation at the normalized frequency $\omega = 4$ krad/s and almost 1.5 dB attenuation at $\omega = 2.1$ krad/s. The filter has to provide 500 μs delay.

4-11   Obtain a maximally flat filter for the following specifications and use AP filter(s) for improving its delay response.

$\alpha_{max} = 0.5$ dB, $\alpha_{min} = 32$ dBs, $\omega_1 = 1500$ rad/s, $\omega_2 = 3600$ rad/s

4-12   Using the circuit of Figure 4.11(a), design a first-order AP filter having maximum delay of 10 ms. Employ practical values of components, test the circuit and find the frequency where introduced delay reduces to 9 ms.

4-13   Design a third-order maximally flat filter with 3 dB gain loss at a frequency of 1 krad/s. Find its delay at 10 Hz, 100 Hz, 500 Hz and 1 kHz. What is the ratio of maximum to minimum delay in the pass band? Cascade a first-order AP filter having maximum delay equal to the delay differential of the maximally flat filter between delays at 10 Hz and 1 kHz. What is the value of the ratio of the maximum to minimum delay in the pass band for the composite filter?

4-14   Repeat Problem 4-13 for (i) for a third-order maximally flat filter and (ii) a third-order equal-ripple filter having 1 dB ripple width.

4-15   Derive equation (4.35c) and find the delay at 1 kHz in a second-order filter if the value of the pole-Q and normalized ω_n is (i) 2 and 0.5, (ii) 2 and 0.75, (iii) 5 and 0.5.

4-16   Derive equation (4.35d) and find the value of maximum delay for each case in Problem 4-15; also find the frequency at which maximum delay occurs.

Chapter 5

Frequency and Impedance Transformations

5.1   Introduction

In Chapter 3, magnitude approximations were studied for the low pass (LP) response. Specifications for the LP response contained maximum attenuation in the pass band α_{max}, minimum attenuation in the stop band α_{min}, and corner frequencies of the pass band Ω_p and stop band Ω_s (or the selectivity factor $= (\Omega_p/\Omega_s)$); for a normalized LP filter, the value of the selectivity factor becomes Ω_s. Such approximation methods are not commonly available for other types of filter responses like high pass (HP), band pass (BP), or band reject (BR). However, this is not too much of an issue as *frequency transformations* are available, which can convert all the important characteristics of an LP filter to that of any other type of filter response and vice-versa. This process of using frequency transformations is a longer procedure compared to direct approximation of other types of responses, but it has some basic advantages. Instead of using different approximation procedures for the different types of filters, extensively available charts and tables for the LP response are used for the maximally flat-Butterworth, Chebyshev, inverse Chebyshev, and the elliptic forms of approximations. The values of poles and zeros, the expression of the transfer function and structure in ladder form with element values are available for small and large order n of the LP filter. The procedure of frequency transformation is lengthy because it involves conversion of the specifications of the *filter to be designed* (FTD) in terms of a corresponding *low pass prototype* (LPP). After designing the LPP, its transfer function is then converted back to that of the FTD.

Transformation of an LPP to the prototype HP, BP, and BR is described in Sections 5.3 to 5.5; the respective transformation factors are also described. Besides these transformations, the level of impedance has to be changed to control the values of the passive components that are allowed to be used in a practical circuit. Impedance scaling and conversion of an LPP to an LP filter of some other frequency is studied in Section 5.2.

For convenience and to avoid confusion between the frequency axes of the FTD and LPP, different symbols are used. For the LPP, the complex frequency variable is expressed in capital letters $S = \Sigma + j\Omega$; small letters $s = \sigma + j\omega$ are used for the FTD. To transform the transfer function of the LPP, $H_{LP}(S)$ in terms of the transfer function of either HP: $T_{HP}(s)$, BP: $H_{BP}(s)$, or band elimination (stop): $H_{BE}(S)$ functions, we need to find an appropriate functional relation as follows:

$$\Omega = f(\omega) \tag{5.1}$$

The function $f(\omega)$ has to be selected in such a way that the approximated magnitude function of LPP, $|H_{LP}(j\Omega)| = |H_{LPP}\{f(j\omega)\}| = |H_{HP}(j\omega)|$ (say) for HPF.

It is important to note that transformation through equation (5.1) affects only the frequency axis. The magnitude on the y-axis is not affected; therefore, the amount of variation of gain in the pass and stop band will remain the same.

There is an alternate method of converting the LPP magnitude response to other responses known as the *network transformation* method. It will be shown that this method is more convenient as it can use the available extensive networks and element values of doubly terminated LPP ladders for any arbitrary specifications.

5.2 Frequency and Impedance Scaling

We have discussed *normalized* and *de-normalized* frequency earlier. Study of approximation can be in both forms, but doing it in normalized form is comparatively easy. Changing from one frequency level to another is called *frequency scaling*; changing the frequency level from unit frequency to another frequency (usually higher) is called *frequency de-normalization*. In this section, we will express frequency and impedance scaling in a formal way and also observe their effect on the location of poles and zeros of the transfer function of the prototype filter.

5.2.1 Frequency scaling

The simplest form of frequency transformation is a frequency scaling operation which is given in terms of the frequency scaling parameter ω_o as follows:

$$S = (s/\omega_o) \rightarrow s = \omega_o S \tag{5.2}$$

The transformation in equation (5.2) converts an LPP response to another LP response at a different frequency level: from normalized to de-normalized, with S being considered as a normalized frequency.

Use of the transformation equation (5.2) changes a numerator factor $(S - z_i)$ to $(s - \omega_o z_i)$ and a denominator factor $(S - p_j)$ to $(s - \omega_o p_j)$ in the factorized form of an LPP transfer function. This means that poles and zeros for the new LPF are simply multiplied by the frequency transformation factor ω_o. Hence, expressions for the new zeros and poles will be as follows:

$$(z_i)_s = (\omega_o z_i)_S, \ (p_j)_s = (\omega_o p_j)_S \tag{5.3}$$

While transforming an LPP to another LP, the normalization frequency and its mirror image in the negative x-axis, $\pm\Omega$ converts to the frequency $\pm\omega_o\Omega$. However, the transfer function retains the same magnitude. Hence, as shown in Figure 5.1(a) and (b), magnitude of the LPP and LP are equal, that is, $|T_{LPP}(j\Omega)| = |H_{LP}(j\omega)|$ and the pass band edge frequency $\Omega_p = 1$ rad/s and stop band edge frequency Ω_S rad/s gets converted to $\omega_p = \omega_o$ rad/s and $\omega_s = \omega_o\Omega_S$ rad/s, respectively.

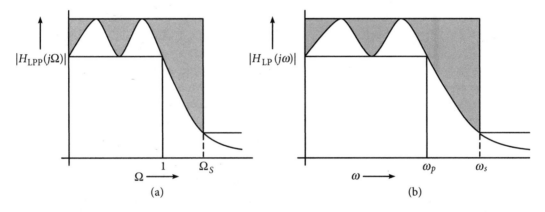

Figure 5.1   Transformation of frequency level from (a) low pass prototype to (b) another low pass.

A filter section is realized either using only passive elements or including active elements. It is only the inductors and capacitors which are frequency dependent; hence, only the values of these elements are affected during frequency transformations. Use of the frequency transformation equation (5.2) converts the inductor impedance of LPP, $Z_L = j\Omega L$ ohms to $Z'_L = (j\Omega L / \omega_o)$ ohms and the capacitor impedance of LPP, $Z_c = 1/j\Omega C$ ohms to $Z'_C = (\omega_o / j\Omega C)$ ohms. Thus, the value of the inductance and the capacitance are divided by the factor ω_o, as illustrated in Figure 5.2(a) and (b).

5.2.2   Impedance scaling

After designing a filter, the circuit configuration is to be selected. Element values of the selected circuit depend on the selected architecture, specifications of the filter and the frequency range of operation. The resulting circuit has to be realized in either discrete form or in an integrated circuit (IC) form. For either form of practical realization, element values should be in a practical range. For example, in IC form, capacitance values should be as small as possible, preferably below the nF range and the resistances should be of the order of or less than a few kilo ohms range. To convert the element values of the designed circuit to within the practically desirable range, *impedance scaling* is almost essential, where all impedances of the network are scaled by a common factor say k. It is important to note that while performing impedance scaling the voltage ratio transfer function of the circuit is not affected because it is dimensionless.

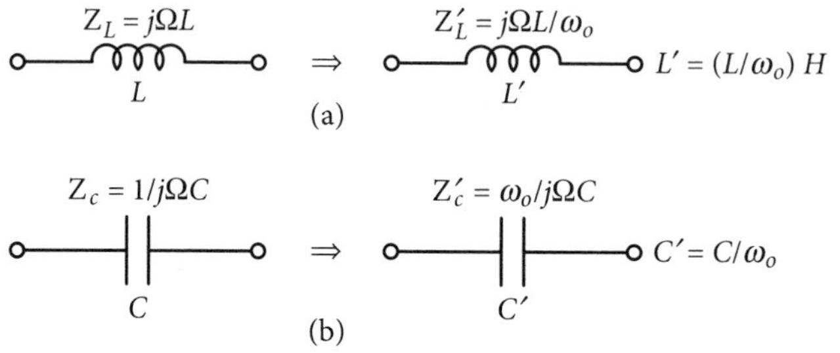

$$Z_L = j\Omega L \qquad\qquad Z_L' = j\Omega L / \omega_o$$

$$L \qquad\qquad\qquad L' \qquad\qquad L' = (L/\omega_o)\ H$$

$$(a)$$

$$Z_c = 1/j\Omega C \qquad\qquad Z_c' = \omega_o/j\Omega C$$

$$C \qquad\qquad\qquad C' \qquad\qquad C' = C/\omega_o$$

$$(b)$$

Figure 5.2   Change in the value of (a) inductor and (b) capacitor, due to the frequency transformation.

Similar to the frequency scaling, during impedance scaling also, the initial circuit whose impedance level is to be changed is called the *normalized impedance circuit* (NIC) and the circuit after the impedance scaling is called the *de-normalized impedance circuit* (DIC). Hence, impedance scaling operation is expressed as:

$$z(\text{DIC}) = k \times Z(\text{NIC}) \tag{5.4}$$

Application of equation (5.4) changes the impedance level of resistor (R), inductance (L), capacitance (C), transconductance gain coefficient (G_m) and transresistance gain coefficient (R_n). The respective changed expressions for the DIC are as follows.

$$kR,\ Z(\omega kL),\ Z\left(1/\omega \frac{C}{k}\right),\ Z\left(\frac{G_m}{k}\right),\ \text{and}\ ZkR_n \tag{5.5}$$

These changed expressions result in a change in the respective circuit element values:

$$r = kR\ \Omega,\ l = kL\ H,\ c = C/k\ F,\ g_m = G_m/k\ \text{mho, and}\ r_n = kR_n\ \Omega \tag{5.6}$$

Example 5.1: Apply frequency scaling by a facto $\omega_o = 10$ krad/s and impedance scaling factor $k = 10^5$ to the LPP passive ladder structure of a seventh-order Chebyshev filter with a 1 dB pass band ripple width (shown in Figure 5.3(a)), and find the element values after the scaling.

Solution: In the original ladder structure, values of the elements are:

$$R_1^* = R_2^* = 1\Omega, C_1^* = C_7^* = 2.1666F, C_3^* = C_5^* = 3.0936F,$$
$$L_2^* = L_6^* = 1.1115H,\ L_4^* = 1.735H \tag{5.7a}$$

Application of frequency scaling will not change the resistor values, but inductances and capacitances will be divided by 10^4; hence, the frequency scaled elements are as follows:

$R_1 = R_2 = 1 \ \Omega, \ C_1 = C_7 = 0.21666 \ \text{mF}, \ C_3 = C_5 = 0.30936 \ \text{mF},$
$L_2 = L_6 = 0.11115 \ \text{mH}, \ L_4 = 0.1735 \ \text{mH}$ (5.7b)

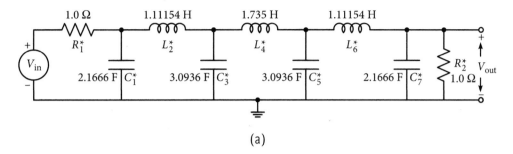

(a)

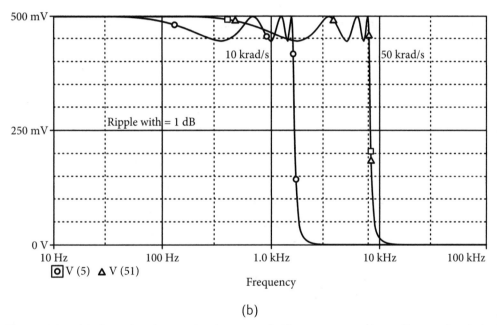

(b)

Figure 5.3   (a) Seventh-order passive low pass ladder structure with I dB pass band ripple.
(b) Transformation from low pass prototype to low pass responses at higher frequencies.

Impedance scaling by a factor of 10^5 will increase the values of the resistors and inductors but decrease the values of the capacitances. Hence, the final element values will be:

$R_1 = R_2 = 100 \ \text{k}\Omega, \ C_1 = C_7 = 2.1666 \ \text{nF}, \ C_3 = C_5 = 3.0936 \ \text{nF},$
$L_2 = L_6 = 11.115 \ \text{H}, \ L_4 = 17.35 \ \text{H}$ (5.7c)

The original passive ladder had a pass band edge frequency ω_p of 1.0 rad/s. After the frequency transformation, the design value of ω_p becomes 10 krad/s. The PSpice simulated value of ω_p from the response shown in Figure 5.3(b) is 9.995 krad/s (1.59 kHz). The ladder was frequency transformed again by a factor of 50 krad/s; the response in this case is also shown in Figure 5.3(b). ω_p is 49.78 krad/s (7.92 kHz).

5.3 Low Pass to High Pass Transformations

Most of the time, the magnitude function of a filter $|H(j\omega)|$ is sketched in the first quadrant, that is, where the frequency remains positive, though the function $|H(j\omega)|^2$ spreads on to both quadrants. As the magnitude function $H(j\Omega)$ is an even function $|H(j\Omega)| = |H(-j\Omega)|$, it gets reflected on the negative x-axis. For example, Figure 5.4 shows a sketch for a maximally flat response for the complete range of frequency, that is, from $-\infty$ to $+\infty$. For the normalized frequency response of an LP filter, pass band extends in the range $|\Omega| \le 1$ and the stop band ranges from Ω_s to ∞ and from $-\Omega_s$ to $-\infty$. To convert the normalized LPP of Figure 5.4 to an HP response, the pass band should range from $\omega = 1$ to ∞ and $\omega = -1$ to $-\infty$, and the stop band from $-\omega_s$ to $+ \omega_s$, respectively, as shown in Figure 5.5. Study of the two figures suggests the form of frequency transformation from an LPP to an HP *FTD*. Zeros of the LPP at $\Omega = \infty$ and $-\infty$ are to be converted so that they are at $\omega = 0$ for the HP filter. Comparison between Figures 5.4 and 5.5 suggests that the pass band of the LP ($-1 \le \Omega \le +1$) needs to be converted to a pass band of HP as ($+1 \le \Omega \le -1$). Such a transformation is obtained by selecting:

$$s = 1/S \text{ or } S = 1/s \tag{5.8}$$

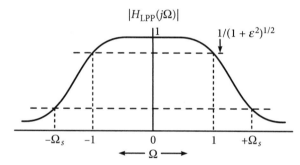

Figure 5.4 Even function response of a maximally flat low pass function.

While working on a transfer function in terms of the complex frequency variable s, it is needed to replace $j\Omega$ by S and $j\omega$ by s. Hence, from equation (5.8),

$$j\omega = 1/j\Omega \rightarrow \omega = -(1/\Omega) \text{ or } j\Omega = 1/j\omega = -(j/\omega) \rightarrow \Omega = -(1/\omega) \tag{5.9}$$

Let us consider a normalized second-order LPP with quality factor Q and dc gain K, and corresponding frequency de-normalized transfer function, as given by equation (5.10) below:

$$H_{LPP}(S) = K\frac{1}{S^2 + \left(\frac{1}{Q}\right)S + 1} = K\frac{\omega_o^2}{S^2 + \left(\frac{\omega_o}{Q}\right)S + \omega_o^2} \tag{5.10a}$$

Application of the transformation equation (5.8) shall lead to the following corresponding HP transfer functions:

$$H_{HP}(s) = K\frac{s^2}{s^2 + \left(\frac{1}{Q}\right)s + 1} = K\frac{s^2}{s^2 + \left(\frac{1}{\omega_o Q}\right)s + \frac{1}{\omega_o^2}} \tag{5.10b}$$

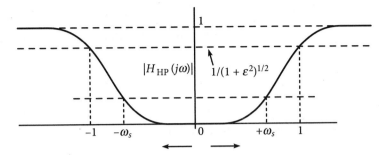

Figure 5.5   Even function response of a maximally flat normalized high pass function.

Obviously, it will retain the same quality factor Q but pass band edge frequency ω_o shall become $(1/\omega_o)$ for the HP filter.

Based on the aforementioned discussion, to obtain a network for a normalized HP filter (HPF) section, the following steps are to be taken.

1. Normalize the specifications of the given HPF by dividing the frequency axis by the pass band edge frequency ω_p so that its pass band is in the normalized frequency range $\omega \geq 1$.
2. Find the normalized stop band edge frequency of the HPF (ω_s/ω_p).
3. Using the transformation relation of equation (5.9) (neglecting the negative sign), obtain the selectivity factor for the LPP (Ω_p/Ω_s).
4. Use any type of magnitude approximation for obtaining the $H_{LPP}(s)$ for the calculated selectivity factor and the given attenuation (or ripples) in pass and stop band.
5. Apply the LP to HP frequency transformation of equation (5.8).

Example 5.2: Design an HPF using LP to HP transformation, with a maximally flat response having the following specifications:

$$\alpha_{min} = 40 \text{ dB},\ \alpha_{max} = 1 \text{dB},\ \omega_s = 1000 \text{ rad/s and } \omega_p = 4000 \text{ rad/s}$$

Also determine the attenuation at 1500 rad/s and 750 rad/s.

Solution: In the first step, specifications of the HPF are normalized by dividing the frequency range by ω_p, so the stop band edge normalized frequency becomes 0.25 rad/s.

Next, the selectivity factor of the normalized LPP $= 1/\Omega_s = 1/0.25 = 4$.

Design of the LPP requires calculation of the factor ε and order of the filter n. Application of equations (3.12) and (3.23), respectively, gives:

$$\varepsilon^2 = (10^{0.1\alpha_{max}} - 1) = (10^{0.1} - 1) = 0.258 \tag{5.11}$$

$$n = \frac{\log\{(10^4 - 1)/(10^{0.1} - 1)\}}{2\log 4} = 3.79 \tag{5.12}$$

Therefore, order of the LPP will be 4. Use of Table 3.1 gives the location of pole for $n = 4$ as:

$$p_{1,2} = 0.3826836 \pm j0.9238795 \text{ and } p_{3,4} = -0.9238795 \pm j0.3826836$$

The normalized transfer function of the LPP is obtained as shown here:

$$H_{LPP}(S) = \frac{1}{(S^2 + 0.7653668\,S + 1)(S^2 + 1.847749\,S + 1)} \tag{5.13}$$

Applying LP to HP transformation of equation (5.8) on the transfer function $H_{LPP}(S)$, the transfer function of the fourth-order normalized HP becomes:

$$H_{HP}(s) = \frac{s^4}{(s^2 + 0.7653668s + 1)(s^2 + 1.847749s + 1)} \tag{5.14}$$

There are several options to synthesize equation (5.14). In one option, the section is broken into two second-order sections with transfer functions H_1 and H_2 (given in the following equations), which will be cascaded and then frequency and impedance scaling shall be applied.

$$H_1(s) = \frac{s^2}{(s^2 + 0.7653668s + 1)} \tag{5.15}$$

$$H_2(s) = \frac{s^2}{(s^2 + 1.847749s + 1)} \tag{5.16}$$

A single amplifier second-order filter section shown in Figure 5.6 has the following expression in equation (5.17). It is used to realize the transfer functions H_1 and H_2.

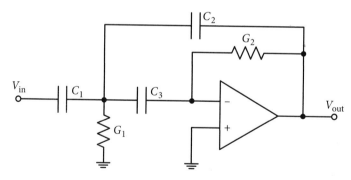

Figure 5.6   A single amplifier, second-order high pass filter.

$$\frac{V_{out}}{V_{in}} = -\frac{(C_1/C_3)s^2}{s^2 + G_2 \dfrac{(C_1 + C_2 + C_3)}{C_2 C_3}s + \dfrac{G_1 G_2}{C_2 C_3}} \tag{5.17}$$

To maintain high frequency gain as unity, comparing equations (5.15) and (5.17), and assuming

$C_1 = C_3 = 1$, we get:

$$G_2 = (1/G_1), \text{ and with } G_2 \frac{3C}{C^2} = 0.7653 \rightarrow R_2 = 3.919\Omega, \text{ and } R_1 = 0.2551\Omega \qquad (5.18a)$$

Frequency normalization is to be done with respect to the 3 dB frequency. To convert the pass band edge frequency of 4000 rad/s to 3 dB frequency, from equation (3.25):

$$\omega_{CB} = \{(10^{0.1 \times 3} - 1)/0.2589\}^{1/2 \times 4} = 1.183 \qquad (5.18b)$$

Hence, the frequency scaling factor will be $4000/1.183 = 3381.2$ rad/s. If all the three capacitors are selected as 0.1 μF, which is a convenient practical value, the impedance scaling factor shall be $10^7/3381 = 2957$. Using this impedance scaling factor, we get $R_{11} = 754\ \Omega$ and $R_{12} = 11.588$ kΩ.

In the same way, comparing equation (5.17) with the transfer function H_2 of equation (5.16), element values are $C_{12} = C_{22} = C_{32} = 0.1$ μF, $R_{12} = 1.797$ kΩ and $R_{22} = 4.803$ kΩ. The cascaded fourth-order HP circuits with the element values used are shown in Figure 5.7(a); their PSpice simulation is shown in Figure 5.7(b).

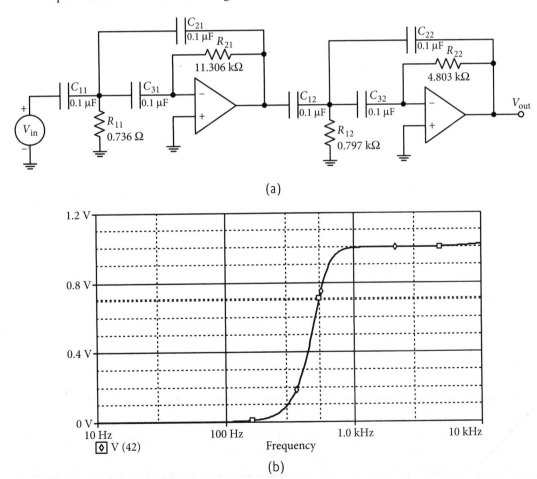

(a)

(b)

Figure 5.7   (a) Fourth-order high pass maximally flat filter for Example 5.2. (b) Magnitude response of the fourth-order high pass filter of Figure 5.7(a) for Example 5.2.

Simulated 3 dB frequency is 539.6 Hzor 3391 rad/s, high frequency gain is unity, pass band edge frequency (4000 rad/s) attenuation is 42.6 dB, and stop band edge frequency (1000 rad/s) attenuation is 0.972 dB. Attenuation at 1500 rad/s and 750 rad/s is 28.2 dB and 52.4 dB, respectively. The observed parameters are very close to the design values as specifications are very well satisfied.

Low Pass to High Pass Network transformation: LP to HP frequency transformation can also be applied directly on a network of a LPP. For a network which contains, resistors, inductors, capacitors and active device, frequency trans-formation shall affect only the inductors and capacitors. Use of equation (5.8) shall transform these components as shown in figure 5.8. Impedance of inductance (SL) gets changed to capacitive impedance having capacitor value of (1/L) and correspondingly capacitive impedance (1/SC) changes to inductive impedance with inductance value (1/C). This conversion process is often used instead of dealing in pole/zero reciprocation. While performing network transformation it is important to note that such a conversion shall be done on the elements, when the LP transfer function is normalized to get a normalized HPF with the pass band edge frequency $\Omega_p = 1$. Later, frequency scaling shall be performed on the normalized HPF for a desired pass band edge frequency; pole-Q shall remain unchanged from the LPF.

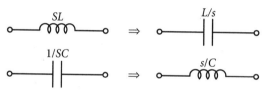

Figure 5.8   Application of an LP to HP frequency transformation on inductive and capacitive impedances.

5.4   Low Pass to Band Pass Transformation

Figure 5.9(a) shows the frequency response of a normalized LP function, approximated in maximally flat form with its 3 dB frequency at $\Omega = 1$ and normalized stop band edge frequency Ω_s. Application of a suitable frequency transformation should give a band pass (BP) response as shown in Figure 5.9(b) converting the 3 dB frequency of the LPF to the lower and upper cut-off frequencies and the stop band frequency gets converted to the two stop band frequencies ω_{s1} and ω_{s2} of the BPF. The pole frequency at $\Omega = 0$ is converted to the normalized center frequency $\omega_o = 1$. The LP response in Figure 5.9(a) was shown only in the first quadrant; whereas for the rational transfer function spread over the whole frequency range of $-\infty$ to $+\infty$, Figure 5.10(a) shows the LP response in inverse Chebyshev form. Its pass band ranges from $\Omega = -1$ to $+1$, which is to be transformed to the pass band of the BPF, extending from frequency ω_1 to ω_2 for positive frequencies. Obviously, the center frequency of the BPF where its magnitude is maximum will lie within the frequency range $\omega_1 \leq \omega \leq \omega_2$; this equals 1.0 for the normalized frequency BPF with $\omega_1 < 1$ and $\omega_2 > 1$. For the LPP, there will be a pole at $\Omega = 0$ and zero at $\Omega = \pm\infty$, whereas for the transformed BP, there will be zeros at $\omega = \pm 1$ and a pole at $\omega = 0$ and $\pm\infty$. For such a conversion, the following function will be sufficient.

$$\Omega = \frac{1}{(BW)}\frac{(\omega-1)(\omega+1)}{\omega} = \frac{1}{BW}\frac{(\omega^2-1)}{\omega} \tag{5.19}$$

In equation (5.19), the term BW has been included to normalize and adjust the slope of the function as explained in the following text. For the LPP, its pass band edge frequency $\Omega = 1$. Hence, from equation (5.19), we get:

$$\Omega = 1 = \frac{1}{BW}\frac{(\omega^2-1)}{\omega} \rightarrow \omega^2 - \omega * BW - 1 = 0 \tag{5.20}$$

It has been shown that the pass band edge frequency of the LPP ($\Omega = 1$) has been transformed as the pass band edge frequencies of the transformed BPF: $-\omega_1$ and ω_2. As these frequencies $-\omega_1$ and ω_2 should be the solution of equation (5.20), we can express the following.

$$(\omega + \omega_1)(\omega - \omega_2) = 0 \rightarrow \omega^2 - \omega(\omega_2 - \omega_1) - \omega_1\omega_2 = 0 \tag{5.21}$$

It is important to note that the scheme is more useful in passive structures employing both inductors and capacitors. In active-RC circuits it is not preferable as such, because it will convert capacitors as inductors, which shall have to be simulated using active-RC circuits.

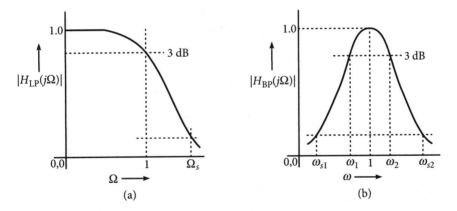

Figure 5.9   Application of a normalized low pass magnitude response transformation to convert it to a normalized band pass response.

Comparing equations (5.20) and (5.21), we get the following expression:

BW = $\omega_2 - \omega_1$, and the product of the normalized pass band edge frequencies,

$$\omega_1 \times \omega_2 = 1 \tag{5.22}$$

Hence, BW = $(\omega_2 - \omega_1)$, introduced in equation (5.19), is the bandwidth of the BPF and its normalized center frequency $\omega = 1$ is the geometric mean of the pass band edge frequencies, ω_1 and ω_2.

Now, multiplying equation (5.19) by j

$$jΩ = \frac{j}{BW}\left(\frac{ω^2 - 1}{ω}\right) = -\frac{1}{BW}\frac{ω^2 - 1}{jω} \tag{5.23}$$

With $S = jΩ$ and $s = jω$, we can write equation (5.23) as follows:

$$S = \frac{1}{BW}\frac{s^2 + 1}{s} = Q(s + 1/s) \tag{5.24}$$

Hence, Q, referred to as the *quality factor* is defined as the center frequency ($ω = 1$) of the BPF divided by the bandwidth (BW).

For a normalized BPF with center frequency as $ω_o$ instead of 1.0, equations (5.19), (5.22) and (5.23) will be modified as follows:

$$Ω = \frac{1}{BW}\frac{ω^2 - ω_o^2}{ω} \tag{5.25}$$

$$ω_1 \times ω_2 = ω_o^2 \tag{5.26}$$

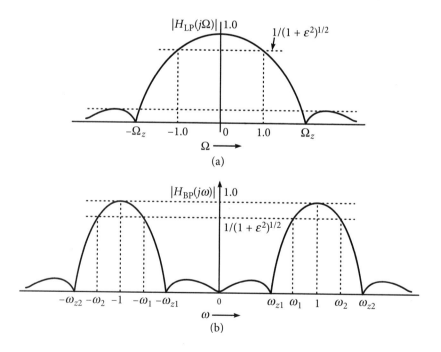

Figure 5.10 (a) A normalized low pass response in inverse Chebyshev approximated form transformed to (b) a normalized band pass response.

$$S = Q\left(\frac{s}{ω_o} + \frac{ω_o}{s}\right) \tag{5.27}$$

With quality factor Q as:

$$Q = \omega_o/(\omega_2 - \omega_1) \tag{5.28}$$

Either equation (5.19) representing the normalized transformation factor or the de-normalized transformation factor of equation (5.25) can be used to convert LPP to BP functions. However, with transformation, the order of the BP function becomes double of the order of the LPP. For example, a third-order LPP became a sixth-order BP, and so on. Hence, the application of the frequency transformation will change the transfer function. Denominator of the transfer function can be expressed as before, either in a polynomial form or in terms of second-order factors. When the denominator is expressed in polynomial form, any direct form of synthesis procedure can be adopted. Alternatively, when the denominator is in the factorized form, a variety of methods employing second-order sections, including a cascade of second-order networks can be used. However, for the factorization, location of poles of the transformed BP function has to be found out as will be discussed here.

A first-order LP function with a real pole at $S = -\Sigma_r$ results in two complex poles in the normalized BP functions as follows:

$$p_{1,2} = -(\Sigma_r/2Q) \pm j\{1 - (\Sigma_r/2Q)^2\}^{1/2} \tag{5.29}$$

While arriving at the result in equation (5.29), it is assumed that $2Q > \Sigma_r$, so that the poles p_1 and p_2 are complex.

With the pole of the first-order LPP being at $-\Sigma_r$, its transfer function $H_{\mathrm{LPP}}(S) = 1/(S + \Sigma_r)$ will be converted to a second-order transfer function for which the transformed BP section will be as follows:

$$H_{\mathrm{BP}}(s) = 1/(s - p_1)\,(s - p_2) = 1/\{s^2 + (\Sigma_r/Q)s + 1\} \tag{5.30}$$

Somewhat more complex is the case when the LPP has complex conjugate poles $(-\Sigma \pm j\Omega)$, which are converted to four poles for the BP function. These four roots appear in the following conjugate complex pair form.

$$\left\{s^2 + (\omega_{o1}/q_1) + \omega_{o1}^2\right\}\left\{s^2 + (\omega_{o2}/q_2) + \omega_{o2}^2\right\} \tag{5.31}$$

Here, ω_{o1}, ω_{o2} are the pole frequencies and q_1, q_2 are the pole quality factors of the two second-order BP sections. Because of the nature of the complex conjugate poles of the LPP function and the transformation factor, equation (5.31) possess the following properties:

$$\omega_{o1} \times \omega_{o2} = 1 \text{ and } q_1 = q_2 = q \tag{5.32}$$

This means that the two normalized pole frequencies are reciprocal of each other and symmetrical about $\omega_o = 1$; the pole quality factor are equal in value.

If the LPP has the following second-order transfer function

$$H_{\mathrm{LPP}}(S) = 1/\left\{S^2 + (\Omega_o/Q) + \Omega_o^2\right\} \tag{5.33}$$

Utilization of the properties in equation (5.32) helps in finding the expression for the pole frequencies ω_{o2} and ω_{o1} which have been shown to be:

$$\omega_{o2} = \frac{1}{\omega_{o1}} = \frac{q\Omega_o}{2Q^2} + \frac{1}{2}\left(\frac{\Omega_o^2}{Q^2} - \frac{1}{q^2}\right)^{1/2} \tag{5.34}$$

From equation (5.34), ω_{o1} and ω_{o2} can be found once value of q is known, which can be obtained from the following:

$$q^2 = \frac{Q^2}{\Omega_o}\left[\left(\frac{2Q^2}{\Omega_o} + \frac{\Omega_o}{2}\right) \pm \left\{\left(\frac{2Q^2}{\Omega_o} + \frac{\Omega_o}{2}\right)^2 - 1\right\}^{1/2}\right] \tag{5.35}$$

Restriction in equation (5.35) is that only the plus sign of the square root is taken to obtain the positive value of ω_{o2}.

Example 5.4: An LPP has pole pairs at $S = \Sigma \pm j\Omega = -1.0 \pm j1.0$. Find the location of poles and values of the pole-Q for a transformed BPF.

Solution: For the given pole location, transfer function of the LPP shall be:

$$H_{\mathrm{LPP}}(S) = 1/(S^2 + 2S + 2) \tag{5.36}$$

Hence, the normalized pole frequency and the pole-Q of the LPP are $\sqrt{2}$ and $1/\sqrt{2}$, respectively. Using the LPP to BP transformation factor of equation (5.24), the transfer function of the normalized BPF shall be obtained as follows (for ($Q = 1/\sqrt{2}$)):

$$T_{\mathrm{BP}}(S) = \frac{1}{Q^2\left(\frac{s^2+1}{s}\right)^2 + 2Q\frac{s^2+1}{s} + 2} = \frac{2s^2}{s^4 + 2\sqrt{2}s^3 + 6s^2 + 2\sqrt{2}s + 1} \tag{5.37}$$

To find the location of the four poles of the BPF, equation (5.35) is used to find q (for ($Q = 1/\sqrt{2}$)) as:

$$q^2 = \frac{1}{(2)\sqrt{2}}\left[\left(\frac{2}{2\sqrt{2}} + \frac{\sqrt{2}}{2}\right) \pm \left\{\left(\frac{2}{(2)\sqrt{2}} + \frac{\sqrt{2}}{2}\right)^2 - 1\right\}^{\frac{1}{2}}\right] = 0.5(1 + 1/\sqrt{2}) = 0.85355$$

The equation gives $q = 0.9293$; hence, from equation (5.34):

$$\omega_{o2} = \frac{0.9293\sqrt{2}}{2} \times 2 + \frac{1}{2}\left(2 \times 2 - \frac{1}{0.85355}\right)^{\frac{1}{2}} = 2.155, \text{ and } \omega_{o1} = \frac{1}{2.155} = 0.464 \tag{5.38}$$

Location of the poles for the BP section are given as:

$$s_{1,2} = -\omega_{o1}\left\{\frac{1}{2q} \pm j\left(1 - \frac{1}{4q^2}\right)^{1/2}\right\}$$

$$= -0.464\left\{\frac{1}{2*0.9293} \pm j\left(1 - \frac{1}{4q^2}\right)^{1/2}\right\} = -0.2496 \pm j0.3901 \qquad (5.39a)$$

$$s_{3,4} = -1.159 \pm j1.8121 \qquad (5.39b)$$

Obviously, roots of the denominator in equation (5.37) shall yield the pole values as obtained in equation (5.39).

5.4.1 Design steps for transformation to BPF

To design a BPF with the requisite specifications, the following steps are to be taken if an LPP to BP frequency transformation is used.

i.   Calculate the pole frequency ω_o of the BPF. If it is not given in direct form, it may be obtained from the pass band edge frequencies as $\omega_o = (\omega_{p1} \times \omega_{p2})^{1/2}$.

ii. Next, the stop band frequencies of the BPF are made geometrically symmetric with respect to the pole frequency obtained in step (i) through altering ω_{s1} or ω_{s2}.

   However, this choice of alteration in either ω_{s1} or ω_{s2} has to be such that one of these becomes more constrained; the stop band becomes narrow, making the design specification a little more severe. For $\omega_{s1} < (\omega_o^2 / \omega_{s2})$, ω_{s1} is to be assigned a new value as $\omega_{s1} \geq (\omega_o^2 / \omega_{s2})$. Otherwise, a new value assigned to ω_{s2} shall will be calculated from $\omega_{s2} = (\omega_o^2 / \omega_{s1})$.

iii. Using the modified stop band of the BPF, selectivity factor of the LPP is calculated as $\Omega_S = (\omega_{s2} - \omega_{s1})/(\omega_{p2} - \omega_{p1})$.

iv. Parameter ω_o becomes modified due to the changed value in ω_{s1} or ω_{s2} as $\omega_o = (\omega_{s1} \times \omega_{s2})^{1/2}$.

v.   Change in the value of ω_o creates asymmetry in ω_{p1} and ω_{p2} with respect to it. Hence, either ω_{p1} or ω_{p2} is to be constrained similar to the case for stop band. If $\omega_{p1} < (\omega_o^2 / \omega_{p2})$, then ω_{p1} is assigned a new value from $\omega_{p1} \geq (\omega_o^2 / \omega_{p2})$. Otherwise, ω_{p2} is assigned a new value from $\omega_{p2} = (\omega_o^2 / \omega_{p1})$.

vi. The modified selectivity factor of the LPP is now calculated due to the change in the pass band frequency range.

vii. Out of the two selectivity factors obtained in step (iii) and step (vi), the larger one is selected; transformation parameters ω_o and BW are evaluated corresponding to the steps (i)–(iii) or (iv)–(vi), whichever leads to the larger value of the LPP selectivity, as it leads to the lowest order n for the LPP(S).

viii. Any method of approximation can be used and the LPP transfer function $H_{LPP}(S)$ is then obtained using the calculated value of the order n and ripple factor ε.

ix. A transformation factor is used to obtain a BP transfer function by replacing S with $\{Q (s^2 + 1)/s\}$ in $H_{LPP}(S)$.

x. The BP is now realized selecting any suitable synthesis process.

Example 5.5: Using the given unsymmetrical frequency specification of a BPF, calculate the selectivity factor for an LPP from which BPF is to be obtained; pass band frequencies, $\omega_{p1} = 5(2\pi)$ krad/s and $\omega_{p2} = 7.2(2\pi)$ krad/s, stop band frequencies, $\omega_{s1} = 4(2\pi)$ krad/s and $\omega_{s2} = 10(2\pi)$ krad/s.

Solution: Center frequency of the BPF, $\omega_o = (\omega_{p1} \times \omega_{p2})^{1/2} = (10\pi \times 14.4\pi)^{1/2} = 6 \times (2\pi)$ krad/s. With $(\omega_o^2 / \omega_{s2}) = 36 \times (2\pi)^2 / 20\pi = 7.2$ krad/s being less than ω_{s1}, the new value of ω_{s2} shall be $\leq 36(2\pi)^2/8\pi = 18\pi$ krad/s. As $(\omega_{s1}\omega_{s2})^{1/2} = (9 \times 4)^{1/2} 2\pi = 12\pi$, equals the center frequency ω_o, there shall be no change in the pass band edge frequency. Since BW $= (7.2–5)2\pi = 4.4\pi$ krad/s, selectivity factor will be $= (9–4)/(7.2–5) = 2.27$.

Example 5.6: Find the transfer function of a BPF with the following specifications using LP to BP transformation: maximum attenuation of 1 dB between 4 and 9 kHz and minimum attenuation of 40 dBs below 1.5 kHz and beyond 22.5 kHz.

Solution: With the pass band edge frequencies being 4 and 9 kHz, center frequency $f_o = (4 \times 9)^{1/2} = 6$ kHz, and bandwidth $= 5$ kHz; hence, pole-Q $= 1.2$.

First, an LPP is to be obtained. Therefore, the specifications of the BPF are to be transformed for the LPP. All the frequencies are normalized with respect to f_o. It gives lower cut-off frequency $\omega_1 = 0.6667$, upper cut-off frequency $\omega_2 = 1.5$; and their product is unity. Normalized lower stop band edge frequency $\omega_{s1} = (1.5/6) = 0.25$ and upper stop band edge frequency $\omega_{s2} = (22.5/6) = 3.75$. Since product of ω_{s1} and ω_{s2} is not unity but less than ω_o^2, a new value has to be given to ω_{s1}, which is equal to $\omega_o^2/\omega_{s2} = 0.26667$ as mentioned in step (ii) of the design process. With the modified stop band, selectivity of the LPP will become:

$$\Omega_S = (\omega_{s2} – \omega_{s1})/(\omega_{p2} – \omega_{p1}) = (3.75 – 0.26667)/(1.5 – 0.6667) = 4.18 \tag{5.40}$$

Required order of the LPP with Chebyshev approximation (from Chapter 4) will be:

$$n = \frac{\ln\left\{4\left(10^{4.0.} – 1\right)/\left(10^{0.1} – 1\right)\right\}^{1/2}}{\ln\left\{4.18 + \left(4.18^2 – 1\right)^{1/2}\right\}} = \frac{5.9738}{2.1088} = 2.832 \tag{5.41}$$

Since it is to be rounded to the next integer, $n = 3$. Pole location of the third-order Chebyshev filter obtained from Table 3.4 is as follows:

$$S_1 = -0.4942, S_{2-3} = -0.2471 \pm j0.966 \tag{5.42}$$

Normalized transfer function of the third-order LPP will become:

$$H_{\mathrm{LPP}}(S) = \frac{0.4942}{(S + 0.4942)(S^2 + 0.4942S + 1)} \qquad (5.43)$$

$|H_{\mathrm{LPP}}(j\Omega_S = 4.18)|$ from equation (5.43) shows that attenuation is little over 40 dBs, satisfying the requirement.

Numerator of the equation (5.43) is 0.4942 as that will result in $H_{\mathrm{LPP}}(0)$ being unity for the third-order Chebyshev filter. Next, equation (5.27) will be applied on equation (5.43) to get the normalized transfer function of the BPF.

$$H_{\mathrm{BP}}(s) = \frac{0.4942s^3}{1.728s^6 + 1.4236s^5 + 6.6772s^4 + 3.34109s^3 + 6.6772s^2 + 1.4236s + 1.728} \qquad (5.44)$$

The root finder is used to find roots in equation (5.44) which are as follows:

$$s_{1-2} = -0.206 \pm j\,0.979,\ s_{3-4} = -0.142 \pm j\,1.478 \text{ and } s_{5-6} = 0.064 \pm j\,0.67 \qquad (5.45)$$

Hence, equation (5.44) is broken into three second-order sections for which factorization of the denominator gives:

$$1.728\,(s^2 + 0.412s + 1)\,(s^2 + 0.284s + 2.204)\,(s^2 + 0.129s + 0.454) \qquad (5.46)$$

Equation (5.46) along with the numerator in equation (5.44) can be broken into three second-order sections and realized using the cascade method or equation (5.44) can be used for any direct form of synthesis.

5.4.2 Low pass to band pass network transformation

Similar to the LP to HP transformation case, the LP to BP transformation can also be applied directly to the LPP network. An inductor in the LPP having impedance $Z_p(S) = SL_p$ gets converted to a series combination of an inductor and a capacitor as shown in Figures 5.11(a).

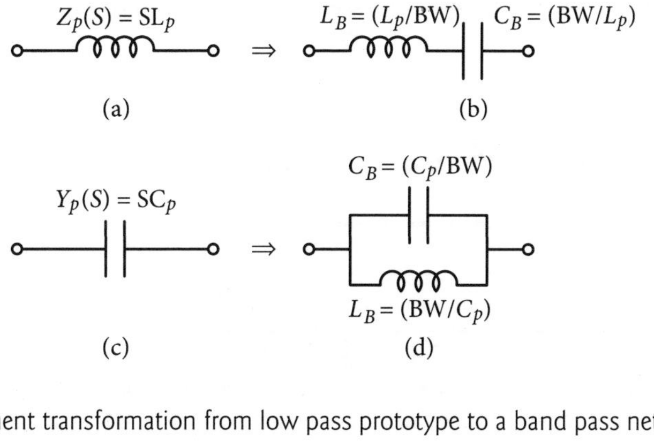

Figure 5.11 Element transformation from low pass prototype to a band pass network.

$$Z_{BP}(s) = \frac{1}{BW}\left(\frac{s^2+1}{s}\right)L_p = \frac{L_p}{BW}s + \frac{1}{\left(\dfrac{BW}{L_p}\right)s} \tag{5.47}$$

Expressions of the resulting series combination of inductor and capacitor, respectively, are: inductor L_p/BW and capacitor $C_B = BW/L_p$ as shown in Figure 5.11(b). Likewise, a capacitor in the LPP having an admittance $Y_p(s) = SC_p$, shown in Figure 5.11(c), gets transformed as follows:

$$Y_B = \frac{1}{BW}\left(\frac{s^2+1}{s}\right)C_p = \frac{sC_p}{BW} + \frac{1}{\dfrac{BW}{C_p}s} \tag{5.48}$$

Equation (5.48) represents a parallel combination of a capacitor (Cp/BW) and an inductor (BW/C_p) as shown in Figure 5.11(d). Resistance being frequency independent, it is not affected by the frequency transformation. Hence, conversion of an LPP network to a BP network can easily be done by using the aforementioned transformations of the inductors and capacitors. It is important to note that the transformed BP network will be a frequency normalized network with center frequency $\omega_o = 1$, which will be applied to passive structures.

Example: 5.7: A third-order Chebyshev approximated normalized LPP ladder structure as shown in Figure 5.12(a) is to be transformed to a BPF through network transformation. Obtain the resulting network and element values of a frequency normalized BP network having normalized BW = 0.1. Also obtain the element values for center frequency $\omega_o = 10^5$ rad/s with an impedance scaling factor of 10^4.

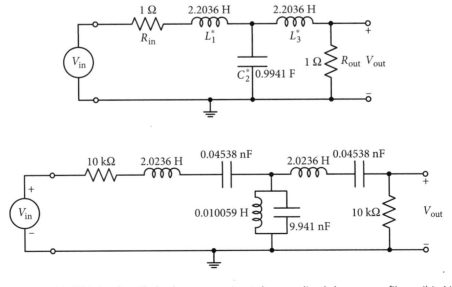

Figure 5.12 (a) Third-order Chebyshev approximated normalized low pass filter. (b) Network transformed de-normalized band pass filter from part (a).

Solution: For the LPP, which is to have a pass band ripple width of 1 dB, element values are as follows:

$$R_{in} = R_{out} = 1\Omega, L_1^* = L_3^* = 2.0236 \text{H and } C_2^* = 0.9941 \text{ F}$$

With BW = 0.1, using equations (5.47) and (5.48), $L_1^* = L_3^*$ changes to a series combination of $l_{1,3} = 20.236$ H and $c_{1,3} = 0.04941$ F, and C_2^* transforms to a parallel combination of $c_2 = 9.941$ F and $l_2 = 0.10059$ H.

Application of frequency translation from $\omega_o = 1$ to 10^5 rad/s and impedance scaling by 10^4 converts elements to the following:

$$L_1 = L_3 = 2.0236 \text{ H}, C_1 = C_3 = 0.04438 \text{ nF}, L_2 = 10.059 \text{ mH}, C_2 = 9.941 \text{ nF},$$
$$\text{and } R_{in} = R_{out} = 10 \text{ k}$$

Figure 5.12(b) shows the transformed BP network with element values. Figure 5.13 shows its PSpice simulated response. The response keeps the nature of variation of output voltage very well with the simulated center frequency being 15.917 kHz; the ripple width in the pass band is 1.15 dB. The lower and upper 3 dB cut-off frequencies are 15.104 kHz and 16.771 kHz, giving BW = 1.667 kHz, and resulting in a pole Q of 9.548.

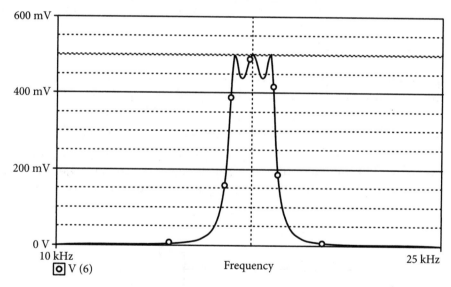

Figure 5.13 Simulated response of the band pass filter of Figure 5.12(b) obtained through network transformation for Example 5.7.

5.5 Low Pass to Band Reject Transformation

Band reject (BR) response being similar in nature to a BP response, it can be transformed from an LPP using a transformation factor similar to the one used for BP transformation:

$$S = \text{BW}\frac{s}{(s^2 + 1)} \tag{5.49}$$

Transformation through equation (5.49) can also be explained in terms of two transformations, one from LPP to HPP and one from HPP to a BP transformation as mentioned in the following two steps, resulting in a BR function.

S is replaced by $(1/S')$, then $\tag{5.50}$

S' is replaced by $\dfrac{1}{\text{BW}}\left(\dfrac{s^2+1}{s}\right)$ $\tag{5.51}$

The BR magnitude response has the following constraints and conditions.

$$\omega_{s1} \times \omega_{s2} = \omega_{p1} \times \omega_{p2} = 1 \tag{5.52}$$

$$\omega_{s2} - \omega_{s1} = \text{BW}\Omega'_s = \text{BW} / \Omega_s \tag{5.53}$$

To realize a BRF, specifications are given in terms of pass band and stop edge frequencies, ω_{p1}, ω_{p2}, ω_{s1} and ω_{s2} and the pass band and stop band attenuations $A_{\max}$ and $A_{\min}$, respectively. Using similar procedure as that for the BP case, selectivity factor of the LPP is found from the expression $\Omega_S = (\omega_{p2} - \omega_{p1})/(\omega_{s2} - \omega_{s1})$. In the same way, frequency specification must be symmetrized with respect to $\omega_o = 1$.

Band reject network transformation: Application of the transformation factor of equation (5.49) with an inductor of the LPP network having admittance $Y_R(S) = 1/(Sl_p)$ becomes an admittance $Y_{BR}(s) = sC_R + 1/sL_R$; here $C_R = 1/\text{BW}l_p$ and $L_R = \text{BW}l_p$. A capacitor c_p in the LPP with impedance $z_c(S) = 1/(Sc_p)$ gets transformed to an impedance function $Z_{BR}(s) = sL'_R + 1/(sC'_R)$ with $L'_R = 1/(c_p\text{BW})$ and $C'_R = (c_p\text{BW})$. This means that like the BP case, inductances and capacitances of the LPP are transformed to a parallel and series combination, respectively, of an inductor and a capacitor in the BR network. Figure 5.14 shows such a transformation of an inductor and a capacitor.

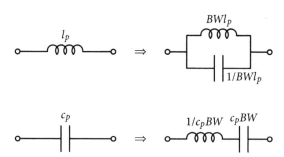

Figure 5.14 Transformation of low pass prototype network elements to normalized network elements of band reject filter.

Practice Problems

5-1 The circuit shown in Figure P5.1 is a prototype filter at 1 rad/s level. Scale the circuit so that it will have a load resistance value of 1 kΩ and the parallel LC branch will resonate at 10 kHz.

5-2 What will be the value of resistance scaling factor k_R and frequency scaling factor k_ω, for the circuit shown in Figure P5.1, so that the load capacitance will become 10 pF and the inductor will have a value of 5 mH.

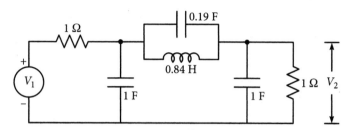

Figure P5.1

5-3 The network shown in Figure P5.2 is to be scaled by increasing the level of impedance by 100 and the level of frequency from 1 rad/s to 10^5 rad/s. Find the element values in the scaled network.

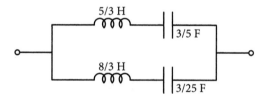

Figure P5.2

5-4 Design an HPF having maximally flat response and the following specifications, using LP to HP transformation:

$\alpha_{min} = 30$ dBs, $\alpha_{max} = 1$ dB, $\omega_s = 1$ krad/s and $\omega_p = 2.6$ krad/s

5-5 Redesign the HPF having the specifications of Problem 5-4 using Chebyshev approximation. Also find the filter attenuation at 1.2 krad/s and test the design.

5-6 (a) Design an HPF with a maximally flat response for which specifications are shown in Figure P5.3 employing LP to HP transformation.

(b) Determine the actual attenuation of the filter at 1800 rad/s and 2200 rad/s.

5-7 An HPF with equal ripples in the pass band is to be designed, employing LP to HP transformation, for which specifications are shown in Figure P5.3.

Design the filter using either a single OA filter circuit of Figure 5.6 or the Sallen–Key section and test the circuit.

Modify the circuit which provides a 10 dB increase in the gain at high frequencies, without employing addition OA.

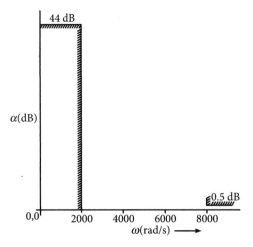

Figure P5.3

5-8   Apply the LP to HP transformation to the following network function $H(s)$, and compare the critical frequencies for both the network functions. What is the inference while comparing the critical frequencies of the LP and the HP functions?

$$H(s) = \frac{2s+1}{s^2 + 4s + 6}$$

5-9   Find the transfer function of the Sallen–Key circuit shown in Figure P5.4. Apply the LP to HP transformation $s \rightarrow 1/s$ and obtain the transfer function and structure of the transformed circuit. Apply impedance scaling factor of 10^3 and frequency scaling factor of 10^4 and simulate the circuit.

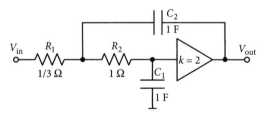

Figure P5.4

5-10   Apply the LP to BP transformation on the LP circuit of Figure P5.5. Find the transfer function of the LP prototype and the transformed network. Determine the value of the pole-Q for the BPF. Use suitable impedance scaling on the BP network such that its center frequency is 10 kHz and test the circuit using PSpice.

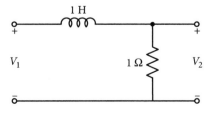

Figure P5.5

5-11   Design a BPF which satisfies the specification shown in Figure P5.6, with attenuation being 0 dB at $\omega = 2700$ rad/s. Construct the circuit with suitable second- (and first-) order sections; maximally flat approximation is to be usedfor the LP prototype.

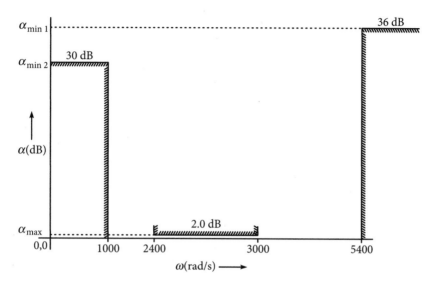

Figure P5.6

5-12   Repeat Problem 5-11 with the equal-ripple approximation used for the LP prototype.

5-13   With a maximally flat response for a BPF, it is desired that the maximum allowable attenuation is 1 dB in the frequency band of 1000 rad/s to 2000 rad/s. Design the BPF with the constraint that only two OAs can be used in the final realization. What shall be the largest obtainable attenuation at a frequency of 6000 rad/s?

5-14   Design a filter with a maximally flat response for which the specifications are: attenuation $= 30$ dB for $0 \leq \omega \leq 500$ rad/s and 4000 rad/s $\leq \omega \leq \infty$, attenuation $= 2$ dB for 1000 rps $\leq \omega \leq 2000$ rps. The mid-band gain is to be 0 dB, and only 0.1 μF capacitors can be used in the final realization.

5-15   Redesign the filter in Problem 5-14 with a pass band having equal ripples.

5-16   The LP prototype shown in Figure P5.7 has a 3dB frequency of 1 rad/s.
   (a) Apply an LP to BP transform so that the BP filter has $Q = 10$ and center frequency $f_o = 1$ kHz. Verify the response using a computer method.
   (b) Convert the LPF to a BRF with band stop width of 0.25 kHz and maximum attenuation at 1 kHz.

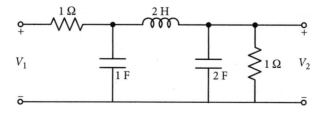

Figure P5.7

5-17   Third-order Chebyshev approximated normalized LP prototype structure of Figure 5.12(a) is to be transformed to a BS filter through network transformation. Obtain resulting network and element values for normalized bandwidth of 0.1 for the BS filter. Also obtain element values for center frequency of 10^5 rad/s and after using impedance scaling factor of 10^4.

Verify the response for the passive BS filter.

5-18   Repeat Problem 5-13 but employ Chebyshev approximation.

Chapter 6

Sensitivity of Active Networks

6.1 Introduction

In the previous chapters, comparatively simple methods have been discussed for the realization of active OA (operational amplifier) RC filters. Many more methods of filter synthesis will be discussed later, providing many more filter circuit configurations. Among this large number of available circuits, the choice of a 'best' filter circuit may depend on the specific user's requirements. However, every application ideally requires a practical filter for which performance parameters, like ω_o (center frequency or cut-off frequency) and quality (pole-Q), are needed exactly as designed, and expected to remain invariant with use in varying environment. However, in practice, the user is satisfied if the parameters remain within such limits that do not make the filter impractical. Though there are different reasons during fabrication which cause deviations in the performance parameters, there is one factor which is common to all circuits at design stage. This factor can be termed as the first consideration in connection with these deviations. It is studied under *sensitivity* and is due to the following reasons.

i.   Design of a filter circuit assumes active and passive elements to be ideal, whereas in every practical fabrication process, the nominal value of the passive element has statistical variations around its mean value. In general, sophisticated, higher level fabrication processes reduce the parameter variations; elements are said to have smaller *tolerance*. However, for such advanced processes, filter fabrication cost will go up.

ii. The values of both active and passive components change with change in operating conditions like change in temperature, humidity and supply voltage. Some chemical changes due to aging also affect the element values.

Whatever be the reason for the difference between the practical element value(s) and the original design value(s), performance of the filter gets affected, and it is said that the filter performance parameters are *sensitive* to the elements used. Sensitivity studies of filter circuit parameters provide the information whether a particular circuit will meet the given specifications under likely tolerances of the elements or not. The studies also help in establishing filter stability and hence the filter's utility in the long term.

Study of *sensitivity* begins with single-element (incremental) sensitivity, that is, the effect of change in a single-element on a certain filter parameter. Evaluation of incremental sensitivity is very important as it gives significant information about the filter. Additionally, it is also widely used for finding other advanced form of sensitivities. Hence, study of single-element sensitivity is taken up first in Section 6.2 in detail. For most applications, this study suffices the requirements for filter design. However, other significant sensitivity factors like *transfer function sensitivity* and *sensitivity of second-order*, which are important because of the greater utility of second-order filter sections, is taken up in Sections 6.3 and 6.4. Further, sensitivity of higher-order filters and advanced topics such as multi-parameter sensitivity are discussed in brief towards the end of the chapter.

6.2 Single-element (Incremental) Sensitivity

Every single parameter of a filter, say pole frequency, quality factor, poles and zeros of the transfer function depend on the parameters of the active devices and the values of the passive components and their tolerances. Let P be a performance parameter of a filter, and x be one of the elements (a passive component or parameter of an active device) which may cause change in the parameter P; we can express this relation as $P = f(x, s)$. s has been added as P is also a function of the complex frequency. However, we shall restrict our study to $P = f(x)$ only (for the sake of keeping the expressions simple); this means that we are operating at a fixed frequency or over a small band of frequency, where a small frequency change has little or no effect on the value of the element x.

Generally, the process of finding a change in P due to a change $\Delta x = x - x_o$ in the element x is done through Taylor's series expansion of P around the nominal value x_o of the element x, as shown in Figure 6.1:

$$P(x) = P(x_o) + \left.\frac{\partial P(x)}{\partial x}\right|_{x_o} dx + \frac{1}{2}\left.\frac{\partial^2 P(x)}{\partial x^2}\right|_{x_o} (dx)^2 + \ \ldots\ldots \tag{6.1}$$

If it is assumed that the change Δx in Figure 6.1 is small and at the nominal point x_o, the curvature showing the variation of P due to the variation in x is also not very large, then the second- and higher-order derivative terms in equation (6.1) can be neglected. This omission leads to the following expression for absolute change in the parameter P:

$$\Delta P(x_o) = P(x_o + dx) - P(x_o) \cong \left.\frac{\partial P(x)}{\partial x}\right|_{x_o} dx \tag{6.2}$$

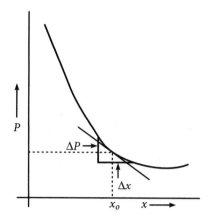

Figure 6.1   Small change in the parameter P, shown as ΔP, due to a small change Δx in x at the nominal value of the element as x_o.

In many cases, it is not very useful to find the absolute change in the parameter ΔP given by equation (6.2); instead, the useful term is the relative change in P, which is given as:

$$\frac{\Delta P(x_o)}{P(x_o)} \cong \frac{x_o}{P(x_o)} \left.\frac{\partial P(x)}{\partial x}\right|_{x_o} \left(\frac{dx}{x_o}\right) \tag{6.3}$$

Part of the right-hand side of equation (6.3), which is given in equation (6.4), and expressed as S_x^P, is known as *sensitivity* of the parameter P with respect to the element x at its nominal value x_o. Sensitivity can also be expressed in the natural log form of the same equation.

$$S_x^P = \left.\frac{x_o}{P(x_o)}\frac{\partial P(x)}{\partial x}\right|_{x_o} = \left.\frac{\partial P/P}{\partial x/x}\right|_{x_o} = \left.\frac{d(\ln P)}{d(\ln x)}\right|_{x_o} \tag{6.4}$$

This expression of single-element sensitivity in equation (6.4), in which the amount of deviation in x has been assumed to be small, also known as *incremental sensitivity*, is extremely useful while analyzing the sensitivity of electronic circuits including filter circuits. Once the value of the sensitivity S_x^P is known, the *relative change* or *variability* of the parameter P can be determined from the relative change in an element x, as follows.

$$\frac{\Delta P}{P} \cong S_x^P \frac{dx}{x} \tag{6.5}$$

As we know, the parameter P can be the pole frequency ω_o, pole-Q, transfer function $H(s)$ or its poles and zeros. Hence, if one wants to express the parameters' sensitivities with respect to (say) a resistor R, then the expression for the sensitivities will simply be:

$$S_R^{\omega_o} = \frac{R}{\omega_o}\frac{\partial \omega_o}{\partial R} = \frac{d(\ln \omega_o)}{d(\ln R)} \tag{6.6a}$$

$$S_R^Q = \frac{R}{Q}\frac{\partial Q}{\partial R} = \frac{d(\ln Q)}{d(\ln R)} \tag{6.6b}$$

$$S_R^{H(s)} = \frac{R}{H(s)}\frac{\partial H(s)}{\partial R} = \frac{d(\ln H(s))}{d(\ln R)} \tag{6.6c}$$

It is obvious from equation (6.5) that it is always desirable to have the sensitivity as small as possible in order to have a better option of a smaller relative change in the parameter P. At the same time, it is important to note that a larger sensitivity is acceptable with respect to those elements which are very stable. This is because the product of a larger sensitivity and components of smaller variability will result in an allowable smaller variability in the parameter P.

In general, evaluation of sensitivity is not difficult, especially for lower-order filter circuits. Though the obtained sensitivity figures with respect to a single-element do provide a fair assessment of the stability of the filter, evaluation of sensitivity figures with respect to all the active and passive individual elements (with the remaining elements considered as constants) do not give a complete picture; the reason being that in each case, sensitivity evaluation is done at the nominal value of that element, whereas the parameter P depends on other elements also. Hence, if the nominal values of other elements change, then the simple incremental sensitivity evaluation will not remain correct. Of course, the amount of incorrectness will depend upon the changes in the nominal values of the other elements. An accurate evaluation is done under the topic of multi-parameter (or multi-element) sensitivity evaluation, where account is taken of the fact that a network parameter depends on many elements which can simultaneously change by varying amounts. At this stage, consideration of simultaneous change in the parameters can also be done in a simplistic way. For example, if parameter $P = f(x_1, x_2, \ldots., x_n)$, then the likely total change in P is found as:

$$\Delta P = \Delta P(x_1, x_2, \ldots., x_n) = \frac{\partial P}{\partial x_1}dx_1 + \frac{\partial P}{\partial x_2}dx_2 + \ldots\ldots + \frac{\partial P}{\partial x_n}dx_n \tag{6.7}$$

The relative change in the parameter P can be written as:

$$\frac{\Delta P}{P} = \frac{x_1}{P}\frac{\partial P}{\partial x_1}\frac{dx_1}{x_1} + \frac{x_2}{P}\frac{\partial P}{\partial x_2}\frac{dx_2}{x_2} + \ldots\ldots. + \frac{x_n}{P}\frac{\partial P}{\partial x_n}\frac{dx_n}{x_n}$$

$$= S_{x_1}^P\frac{dx_1}{x_1} + S_{x_2}^P\frac{dx_2}{x_2} + \ldots\ldots + S_{x_n}^P\frac{dx_n}{x_n} \tag{6.8}$$

which means that the total relative change is the sum of individual sensitivities multiplied with the relative change in the elements.

It was mentioned earlier that performance parameters are functions of the complex frequency s; hence, the sensitivity expression is also a function of s. While finding the sensitivity for a filter section, the proper frequency range should be kept in mind, as the sensitivity value may get changed at different frequency levels.

Example 6.1: Find the incremental sensitivity of the pole frequency and the quality factor for the RLC filter section shown in Figure 6.2. Also find the total relative change in ω_o and Q if the inductor and capacitor change by -5% and the resistors change by 8%.

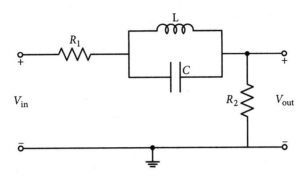

Figure 6.2   A simple RLC filter section for Example 6.1.

Solution: The transfer function of the filter section of Figure 6.2 is obtained as follows

$$H(s) = \frac{V_{out}}{V_{in}} = \frac{R_2}{R_1 + R_2} \frac{(s^2 + 1/LC)}{s^2 + \{1/C(R_1 + R_2)\} + 1/LC} \tag{6.9}$$

Expression of the pole frequency ω_o and the quality factor Q are as follows

$$\omega_o = 1/\sqrt{(LC)}, \text{ and } Q = (R_1 + R_2)\sqrt{(C/L)} \tag{6.10}$$

Use of equations (6.6) and (6.7) on equation (6.10) gives the incremental or single-element sensitivity figures as:

$$S_{R_1}^{\omega_o} = S_{R_2}^{\omega_o} = 0, \quad S_L^{\omega_o} = -\frac{1}{2}, \quad S_C^{\omega_o} = -\frac{1}{2}$$

$$S_{R_1}^{Q} = \frac{R_1}{R_1 + R_2}, \quad S_{R_2}^{Q} = \frac{R_2}{R_1 + R_2}, \quad S_C^{Q} = \frac{1}{2}, \quad S_L^{Q} = -\frac{1}{2}$$

It means that any change in the value of R_1 and R_2 does not make any difference in the value of ω_o but it affects Q, which depends on the relative values of R_1 and R_2. For example, if $R_1 = 2R_2 = 2R$, $S_{R_1}^{Q} = \frac{2}{3}$ and $S_{R_2}^{Q} = \frac{1}{3}$; $\pm 1\%$ change in R_1 will cause a change of $\pm\frac{2}{3}\%$ in Q, whereas a change of $\pm 1\%$ in R_2 will cause a change of $\pm\frac{1}{3}\%$ in Q.

Similarly, a change of 1% in L or C will cause a change of -0.5% in ω_o, whereas a 1% change in C will change Q by 0.5%, but a 1% change in L will change Q by -0.5%

The total relative change in ω_o, and Q are computed using equation (6.8) as follows:

$$\frac{\Delta\omega_o}{\omega_o} = S_L^{\omega_o}\frac{\Delta L}{L} + S_C^{\omega_o}\frac{\Delta C}{C} + S_{R_1}^{\omega_o}\frac{\Delta R_1}{R_1} + S_{R_2}^{\omega_o}\frac{\Delta R_2}{R_2} = -\frac{1}{2}\left(\frac{\Delta L}{L} + \frac{\Delta C}{C}\right)$$

$$\frac{\Delta Q}{Q} = S_L^{Q}\frac{\Delta L}{L} + S_C^{Q}\frac{\Delta C}{C} + S_{R_1}^{Q}\frac{\Delta R_1}{R_1} + S_{R_2}^{Q}\frac{\Delta R_2}{R_2} = -\frac{1}{2}\left(\frac{\Delta L}{L} - \frac{\Delta C}{C}\right) + \frac{1}{R_1 + R_2}(\Delta R_1 + \Delta R_2)$$

It can easily be observed for this example that if relative change in L and C are in the same direction, then their effects add in the case of ω_o but cancels each other in the case of Q; the opposite happens if the relative change in L and C occur in the opposite direction. For the given changes in elements, the following will be the values of the important relative changes:

$$\frac{\Delta\omega_o}{\omega_o} = 5\% \text{ and } \frac{\Delta Q}{Q} = 0 + 8\%$$

Example 6.2: Figure 6.3 shows a single amplifier biquad. Its transfer function is as shown in equation (6.11) while OA is considered as ideal. (a) Find all the incremental sensitivities for the important parameters of the biquad and (b) find the variability in Q for small changes in the active parameter.

$$H(s) = \frac{V_o(s)}{V_{in}(s)} = -\frac{(1/aR_1C_1)s}{s^2 + \left\{\dfrac{1}{R_2C_1} + \dfrac{1}{R_2C_2} - \dfrac{K}{R_1C_1(1-K)}\right\}s + \dfrac{1}{R_1R_2C_1C_2}} \tag{6.11}$$

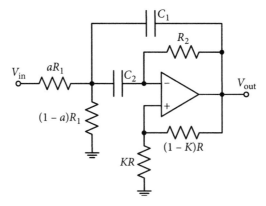

Figure 6.3 Delyiannis–Friend single amplifier biquad with Q-enhancement.

Solution: If there was no positive feedback in the circuit, the non-inverting terminal of the OA would have been connected to the ground and the value of K would have become zero. If in the beginning it is assumed that $C_1 = C_2 = C$, the transfer function in equation (6.11) will be modified as:

$$H(s) = -\frac{H_m(1/aR_1C)s}{s^2 + (2/R_2C)s + 1/C^2R_1R_2} \qquad (6.12)$$

which can be written in the standard format of a second-order filter having parameters ω_o, Q and mid-band gain H_m as:

$$H(s) = -\frac{H_m(\omega_o/Q_o)s}{s^2 + (\omega_o/Q_o)s + \omega_o^2} \qquad (6.13)$$

Comparison of equations (6.12) and (6.13) gives expressions for the important parameters: Center frequency ω_o:

$$\omega_o = 1/C\sqrt{(R_1R_2)} \qquad (6.14a)$$

Quality factor without positive feedback Q_o, and mid-band H_m are as follows:

$$Q_o = \omega_o CR_2/2 = 0.5\sqrt{(R_2/R_1)} \qquad (6.14b)$$

$$H_m = R_2/2aR_1 = 2Q_o^2/a \qquad (6.14c)$$

For the given circuit with positive feedback expression, ω_o remains unchanged, but Q is enhanced to Q_o, and its expression is obtained from the following equations:

$$\left(\frac{\omega_o}{Q}\right) = \frac{1}{R_2C}\left\{2 - \frac{K}{(1-K)}\frac{R_2}{R_1}\right\} \qquad (6.15)$$

With $K/(1-K) = \delta$, we get $Q = \pm Q_o/(1-2\delta Q_o^2)$ (6.16a)

where, $Q_o = \dfrac{1}{2}\sqrt{C_1R_2/C_2R_1}$ (6.16b)

The expression of the mid-band gain is:

$$H_B = \frac{H_m}{1 - 2\delta Q_o^2}\left(\frac{1}{1-K}\right) = \frac{H_m}{1-K}\frac{Q}{Q_o} \qquad (6.17)$$

(a) Using equations (6.6) and (6.14), we get the sensitivities as

$$S_{R_1}^{\omega_o} = \frac{R_1}{\omega_o}\frac{\partial \omega_o}{\partial R_1} = -\frac{1}{2}, \quad S_{R_2}^{\omega_o} = -\frac{1}{2} \text{ and } S_C^{\omega_o} = \frac{C}{\omega_o}\frac{\partial \omega_o}{\partial C} = -1$$

Sensitivity $S_C^{\omega_o} = 1$ is a little misleading because capacitors C_1 and C_2 were assumed equal. It is important to note that while doing the sensitivity calculations, the general expression should be used without any specific ratio between element values as it was done in this case. Hence, the correct value will be obtained from the general expression of $\omega_o^2 = 1/R_1 R_2 C_1 C_2$ as $S_{C_1,C_2}^{\omega_o} = -\frac{1}{2}$.

All the passive sensitivities are $(-1/2)$, which is a theoretical minimum for an active filter based on the product of two RC time constants. Designers try to design filters with ω_o sensitivities as close to these minimum values as possible. As far as active sensitivity of ω_o is concerned, in this case, $S_K^{\omega_o} = 0$, which is highly desirable.

Before finding Q – sensitivities, it is desirable to note that the second-order filter section employs R_1, R_2, C_1, C_2, and a potential divider involving resistor R for deciding the value of K, whereas there are only two filter parameters, ω_o, and Q; H_B can be controlled independently by the factor a. Therefore, some assumptions have to be taken for the element values. One such assumption has already been taken in the form of having the capacitors C_1 and C_2 as equal; this is very attractive from the point of view of integrated circuit fabrication. Let another assumption be that $\sqrt{(R_2/R_1)} = r$; then, the expression for Q will become:

$$Q = \frac{Q_o}{1 - 2\delta Q_o^2} = \frac{\left(\frac{1}{2}\right)\sqrt{(R_2/R_1)}}{1 - 2(K/1-K)(1/4)(R_2/R_1)} = \frac{r}{2 - r^2 K/(1-K)} \tag{6.18a}$$

With $C_1 = C_2 = C$, Q is independent of C_1 and C_2; hence, the sensitivity of Q with respect to these capacitors is zero. Sensitivity of Q with respect to C_1 and C_2 can separately be obtained from the general expression of Q obtained from equation (6.11).

Sensitivity of Q with respect to K is obtained now as:

$$S_K^Q = \frac{K}{Q}\frac{\partial Q}{\partial K} = \frac{K}{Q}\frac{\partial}{\partial K}\frac{r}{\{2 - r^2 K/(1-K)\}} = \frac{K}{Q}\frac{\partial}{\partial K}\frac{r(1-K)}{2(1-K) - r^2 K} \tag{6.18b}$$

Alternatively, expressing Q from equation (6.18a) in terms of the ratio of polynomials as $Q = N(s)/D(s)$ will give:

$$S_K^Q = \frac{K}{N(s)}\frac{\partial(1-K)r}{\partial K} - \frac{K}{D(s)}\frac{\partial\{2(1-K) - Kr^2\}}{\partial K}$$

$$= \frac{Kr^2}{(1-K)\{2(1-K) - Kr^2\}} = \frac{Kr}{(1-K)^2}Q \tag{6.19}$$

Obviously, Q sensitivity with respect to K will depend on the selected value K (or r) and the value of K (or r) will be dependent on the specified value of Q. It is obvious that sensitivity will shoot to very high values with the value of K nearing unity.

(b) Calculation of variability in Q: For a selected value $r = 1$, the obtained value of K for $Q = 10$ is $(19/29)$ from equation (6.19). Hence, from equation (6.19): $S_K^Q = 55.1$.

This is rather a large value for active Q sensitivity. A small 0.5% of change in the value of K means a large -27.55% change in the value of Q. Worse, if K becomes $(2/3)$, for a mere increase of $(1/57)$, $Q \rightarrow$ infinity and the network become unstable. Practically, it is very difficult, with this choice of element (or the value of K) to set K accurately; the circuit becomes almost non-workable.

Instead of using equation (6.19), the percentage change in Q can also be calculated directly from the expression of Q in equation (6.18a); for the same value of $r = 1$, we get:

$$Q = (1 - K)/(2 - 3K) \tag{6.20}$$

Equation (6.20) shows that for a change of $+0.5\%$ in K (from 19/29), the change in the value of Q will be $+38.5\%$, and if K changes by -0.5%, Q will change by -21.4%.

It is significant to note that variations in Q obtained through different methods of calculations have big differences. The reason behind this is the fact that while deriving the sensitivity definition in equation (6.4), it is assumed that at the nominal point, change in the element with respect to which sensitivity is being calculated is small, whereas in this example, it is not so. This tells us that accuracy of calculating the variability of a performance parameter depend on the smallness, or otherwise of the rate of change of the element x at the nominal point x_0.

6.2.1   Semi-relative sensitivity

In Example 6.1, calculation of the single-element incremental sensitivity was not involved because the expressions of ω_o and Q were simple, and the parameters were in simple relations with the elements. In reality, all the cases are not so simple and even evaluation of single-element sensitivity becomes involved and cumbersome, or the value of sensitivity becomes infinite. Instead of a direct application of equation (6.4) using the involved expressions of the parameters, alternatives are available, which are in fact derived from the definition of equation (6.3) and (6.4). Some of such relations are as follows:

$$S_x^{P_1 P_2} = S_x^{P_1} + S_x^{P_2} \tag{6.22a}$$

$$S_x^{P_1/P_2} = S_x^{P_1} - S_x^{P_2} \tag{6.22b}$$

With P being a function of y as $P(y)$ and y being a function of x as $y(x)$, then:

$$S_x^P = S_y^P \times S_x^y \tag{6.22c}$$

Additionally, when k and n are constant, the following relations are useful:

$$S_x^{1/P} = S_{1/x}^P = -S_x^P \text{ and } S_x^{kP} = S_x^P \tag{6.23a, b}$$

$$S_x^{P^n} = nS_x^P \text{ and } S_x^{kx^n} = n \qquad (6.24\text{a, b})$$

$$S_x^{k+P} = \frac{k}{k+P}S_x^P \text{ and } S_x^{\sum P_i} = \frac{1}{\sum P_i}\sum\left(P_iS_x^{P_i}\right) \qquad (6.25\text{a, b})$$

There are some cases where it is the absolute change in P, rather than the relative value which is desired; hence, the value of sensitivity itself is not important. For example, if S_x^P is to be evaluated near (or at) a nominal point $P \to 0$, then its value will tend to infinity. This result is not very useful for practical purposes. Hence, in such cases, instead of finding S_x^P directly, the following *semi-relative sensitivity* measure is evaluated:

$$Q_x^{P(x)} = x\frac{dP}{dx} \qquad (6.26)$$

It will be observed that $Q_x^{P(x)}$ is very useful in many cases as will be shown soon.

6.3   Transfer Function Sensitivity

Expression of a general transfer function introduced in Chapter1 is repeated here:

$$H(s) = \frac{N(s)}{D(s)} = \frac{a_m s^m + a_{m-1}s^{m-1} + \ldots\ldots + a_1 s + a_0}{b_n s^n + b_{n-1}s^{n-1} + \ldots\ldots + b_1 s + b_0} \qquad (6.27)$$

Both the numerator and the denominator in equation (6.27) are functions of the elements used in the construction of the circuit for which the transfer function is obtained. Hence, their coefficients are also functions of these elements. Use of equation (6.22b) on equation (6.27) gives the following relation for the transfer function sensitivity which simplifies its evaluation considerably:

$$S_x^{H(s)} = S_x^{N(s)} - S_x^{D(s)} = x\left(\frac{1}{N(s)}\frac{\partial N(s)}{\partial x} - \frac{1}{D(s)}\frac{\partial D(s)}{\partial x}\right) \qquad (6.28)$$

Most of the time, more than one coefficient in $N(s)$ as well as in $D(s)$ depends on an element x. It results in the modification of equation (6.28) as:

$$S_x^{H(s)} = \frac{x}{N(s)}\sum\frac{\partial a_i}{\partial x}s^i - \frac{x}{D(s)}\sum\frac{\partial b_i}{\partial x}s^i \qquad (6.29)$$

Obviously, the two summations in equation (6.29) will comprise only those terms for which a_i or b_i, respectively, depend on element x; this leads to the understanding of the transfer function sensitivity from a different angle. When the numerator and denominator of the transfer function are factorized, zeros and poles also deviate from their nominal positions when any element gets changed. A relation between the amount of shift in any pole or zero location and the resultant behavior of the transfer function can be obtained and studied. If $H(s)$ given in equation (6.27) is factorized in terms of poles and zeros and its natural log is obtained, we get:

$$\ln H(s) = \ln k + \sum_{i=1}^{m} \ln(s - z_i) - \sum_{i=1}^{n} \ln(s - p_i) \tag{6.30}$$

Here, along with the poles and zeros, the coefficient $k = (a_m/b_n)$ as well may be a function of active parameters and the passive elements. A derivative of equation (6.30) is taken and then both sides of the equation are multiplied with x, which results in the following important relation in terms of semi-relative sensitivity expressions of equation (6.26):

$$S_x^{H(s)} = S_x^k - \sum_{i=1}^{m} \frac{Q_x^{z_i}}{(s - z_i)} + \sum_{i=1}^{n} \frac{Q_x^{p_i}}{(s - p_i)} \tag{6.31}$$

Equation (6.31) clearly shows that any shift in the location of a single pole or zero will affect the transfer function and its sensitivity becomes high at frequencies close to a pole or zero of $H(s)$. But for physical frequencies, with $s = j\omega$, the transfer function sensitivity tends towards infinity when the transmission zero is on the $j\omega$-axis. In addition, for $s = j\omega$, with larger Q values, complex pole pairs are very near to the $j\omega$-axis, resulting in $|(j\omega - p_i)|$ becoming small and the sensitivity of $H(s)$ becoming high. It is illustrated in Figure 6.4 with the help of the location of poles of an eighth-order Chebyshev filter with $\alpha_{max} = 0.5$ dB.

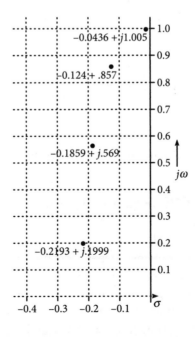

Figure 6.4   Pole location in the second ordinate for an eighth-order Chebyshev filter with $\alpha_{max} =$ 0.5 dB.

Use of equation (6.31) is not very desirable as factorization of a polynomial of even order 4 (and more) is extremely difficult (without a root finder).

Significant relations are obtained in the context of the transfer function sensitivity if $H(s)$ is expressed in terms of its magnitude and phase as:

$$H(j\omega, x) = |H(j\omega, x)| e^{j\varphi(\omega, x)} \tag{6.32}$$

Using the relation (6.22a), and later equation (6.26) for the phase part, we get

$$S_x^{H(j\omega)} = S_x^{|H(j\omega)|} + jQ_x^{\varphi(\omega)} \tag{6.33}$$

A significant inference from equation (6.33) is that the real part of the transfer function is the magnitude sensitivity:

$$\mathrm{Re}\left\{ S_x^{H(j\omega)} \right\} = S_x^{|H(j\omega)|} \tag{6.34}$$

and the imaginary part of the transfer function sensitivity is the semi-relative phase sensitivity:

$$\mathrm{Im}\left\{ S_x^{H(j\omega)} \right\} = Q_x^{\varphi(j\omega)} \tag{6.35}$$

The right-hand side of equation (6.31) is the partial fraction expansion of $S_x^{H(s)}$ in equation (6.28); hence, $Q_x^{z_i}$ and $Q_x^{p_i}$ are the residues. Hence, equalizing the mentioned equations for $s \to p_i$, the last term in equation (6.32) will dominate, resulting in the following relation:

$$\lim_{s \to p_i} x \left(\frac{1}{N(s)} \frac{\partial N(s)}{\partial x} - \frac{1}{D(s)} \frac{\partial D(s)}{\partial x} \right) = \lim_{s \to p_i} \frac{Q_x^{p_i}}{(s - p_i)} \tag{6.36}$$

Since p_i is the pole of $H(s)$, which is obtained through factorizing the denominator, the term $\dfrac{1}{D(s)} \dfrac{\partial D(s)}{\partial x}$ dominates $\dfrac{1}{N(s)} \dfrac{\partial N(s)}{\partial x}$; an important inference.

6.4   Second-order Filter Sensitivities

The sensitivity expressions developed so far are valid for both active and passive filters of any order. Values of the sensitivities obtained depend on the type of structure used, and these values helps in comparing the circuits from the point of view of sensitivity. However, there are certain relations which allow us to get useful information about second-order filters (with $Q > 0.5$). Since second-order active filters form an important entity in the design of any higher-order filter, this area of sensitivity study forms an useful and important topic.

Transfer function of a general biquadratic function is written as:

$$H(s) = \frac{a_2(s-z_1)(s-z_2)}{s^2+(\omega_o/Q_o)s+\omega_o^2} = \frac{a_2 s^2 + a_1 s + a_0}{s^2+(\omega_o/Q_o)s+\omega_o^2} \qquad (6.37)$$

Since for all practical purposes, active filters are used with $Q > 0.5$, poles are complex conjugate and their expression, from equation (6.37), is obtained as:

$$p_1, \; p_1^* = -\frac{\omega_o}{2Q_o} \pm j\omega_o(1-1/4Q_o^2)^{\frac{1}{2}} \qquad (6.38)$$

In order to determine an important relation for the sensitivities of biquadratic sections with respect to an element x, we first proceed with the conjugate poles considering ω_o and Q_o to be functions of x by taking the derivative of p_1 as:

$$\frac{\partial p_1}{\partial x} = -\frac{\partial \omega_o}{\partial x}\left\{\frac{1}{2Q} - j(1-\frac{1}{4Q^2})^{\frac{1}{2}}\right\} - \omega_o \frac{\partial}{\partial x}\left\{\frac{1}{2Q} - j(1-\frac{1}{4Q^2})^{\frac{1}{2}}\right\} \qquad (6.39a)$$

We first multiply both sides of equation (6.39a) with x. Then, multiplying and dividing the first term on the right-hand side by ω_o, and taking the derivative of the second term with respect to x, the following relation is obtained:

$$x\frac{\partial p_1}{\partial x} = -\frac{x}{\omega o}\omega_o\frac{\partial \omega_o}{\partial x}\left\{\frac{1}{2Q} - j(1-\frac{1}{4Q^2})^{\frac{1}{2}}\right\} - \omega_o\left[-\frac{1}{2Q^2} - j\frac{1/2Q^3}{2\left(1-\frac{1}{4Q^2}\right)^{\frac{1}{2}}}\right]x\frac{\partial Q}{\partial x} \qquad (6.39b)$$

Dividing equation (6.39b) by p_1 of equation (6.38), and after a bit of manipulation, we get:

$$\frac{x}{p_1}\frac{\partial p_1}{\partial x} = S_x^{p_1} = S_x^{\omega_o} - jS_x^Q/(4Q^2-1)^{\frac{1}{2}} \qquad (6.40)$$

Following the same procedure, it is shown that, with p_1^*, we get:

$$\frac{x}{p_1^*}\frac{\partial p_1^*}{\partial x} = S_x^{p_1^*} = S_x^{\omega_o} + jS_x^Q/(4Q^2-1)^{\frac{1}{2}} \qquad (6.41)$$

Equations (6.40) and (6.41) show that pole-sensitivity depends on both the sensitivities of ω_o and Q but an important observation is that pole-sensitivity is more dependent on ω_o sensitivity than that of Q. In fact, the position of pole p_1 is $(4Q^2 - 1)^{1/2} \cong 2Q$ times more sensitive to the deviations in ω_o than the deviations in the Q value. It is for this reason that the filter designer needs to care more for the ω_o sensitivity than the Q sensitivity. For a second-order active RC filter, for which ω_o depends on the product of two RC time constants (R_1C_1 and R_2C_2), designers try to get equal to or as close to the ideal ω_o sensitivities of $(-1/2)$ with respect to all the mentioned elements.

From equation (6.36), an important observation was made that while finding the transfer function sensitivity, the dominant term is the one which depends on the denominator $D(s)$, or poles of $H(s)$, and the remaining terms can be neglected. The reason for finding this expression is that the poles lie mostly in the pass band and the zeros of the transfer function lie in the stop band and it does not affect the filter output. Therefore, for most of our discussion on the sensitivity of a biquad, the following modified form of the equation (6.28) is used for finding the sensitivity of the transfer function. Later, for specific cases, the effect of the contribution due to the zeros can be added.

$$S_x^{H(s)} = -\frac{x}{D(s)}\frac{\partial D(s)}{\partial x} = -x\frac{\left(2\omega_o + \dfrac{s}{Q}\right)\dfrac{\partial \omega_o}{\partial x} - \dfrac{s\omega_o}{Q^2}\dfrac{\partial Q}{\partial x}}{s^2 + \left(\dfrac{\omega_o}{Q}\right)s + \omega_o^2} \tag{6.42}$$

Multiplying and dividing the first term in the numerator of the right-hand side term by ω_o in equation (6.42), simplifies it to the following expression:

$$S_x^{H(s)} = \left\{-\left(2\omega_o^2 + s\frac{\omega_o}{Q}\right)S_x^{\omega_o} - s\frac{\omega_o}{Q}S_x^Q\right\}/D(s) \tag{6.43}$$

In order to find the magnitude sensitivity and semi-relative phase sensitivity of the biquad as given in equation (6.43), s is replaced by $j\omega$ in equation (6.43), and the expression is multiplied and divided with the conjugate of the denominator $D(j\omega)$.

$$S_x^{H(j\omega)} = -\frac{\left\{\left(2\omega_o^2 + j\omega\dfrac{\omega_o}{Q}\right)S_x^{\omega_o} - j\omega\dfrac{\omega_o}{Q}S_x^Q\right\}\left\{\left(\omega_o^2 - \omega^2\right) - j\dfrac{\omega\omega_o}{Q}\right\}}{\left\{\left(\omega_o^2 - \omega^2\right) + j\dfrac{\omega\omega_o}{Q}\right\}\left\{\left(\omega_o^2 - \omega^2\right) - j\dfrac{\omega\omega_o}{Q}\right\}}$$

$$= -\frac{\left\{2\omega_o^2\left(\omega_o^2 - \omega^2\right) + \dfrac{\omega^2\omega_o^2}{Q^2}\right\}S_x^{\omega_o} - \dfrac{\omega^2\omega_o^2}{Q^2}S_x^Q}{\left(\omega_o^2 - \omega^2\right)^2 + \omega^2\omega_o^2/Q^2} + j\frac{\dfrac{\omega\omega_o}{Q}\left(\omega_o^2 + \omega^2\right)S_x^{\omega_o} + \dfrac{\omega\omega_o}{Q}\left(\omega_o^2 - \omega^2\right)S_x^Q}{\left(\omega_o^2 - \omega^2\right)^2 + \omega^2\omega_o^2/Q^2} \tag{6.44}$$

The right-hand side is divided in the numerator and denominator by ω_o^4, equation (6.44) is now modified in terms of $\omega_n = (\omega/\omega_o)$.

$$S_x^{H(j\omega_n)} = -\frac{\left\{2\left(1-\omega_n^2\right)+\dfrac{\omega_n^2}{Q^2}\right\}S_x^{\omega_o} - \dfrac{\omega_n^2}{Q^2}S_x^Q}{\left\{\left(1-\omega_n^2\right)^2+\dfrac{\omega_n^2}{Q^2}\right\}} + j\frac{\omega_n}{Q}\frac{\left(1+\omega_n^2\right)S_x^{\omega_o}+\left(1-\omega_n^2\right)S_x^Q}{\left\{\left(1-\omega_n^2\right)^2+\dfrac{\omega_n^2}{Q^2}\right\}} \tag{6.45}$$

The real part of equation (6.45) may be written as:

$$S_x^{|H(j\omega_n)|} = S_{\omega_o}^{|H(j\omega_n)|}S_x^{\omega_o} + S_Q^{|H(j\omega_n)|}S_x^Q \tag{6.46}$$

where, $S_{\omega_o}^{|H(j\omega_n)|} = \dfrac{\left\{2\left(1-\omega_n^2\right)+\dfrac{\omega_n^2}{Q^2}\right\}}{\left\{\left(1-\omega_n^2\right)^2+\dfrac{\omega_n^2}{Q^2}\right\}}$ and $S_Q^{|H(j\omega_n)|} = \dfrac{\dfrac{\omega_n^2}{Q^2}}{\left\{\left(1-\omega_n^2\right)^2+\dfrac{\omega_n^2}{Q^2}\right\}}$ \hfill (6.47)

The imaginary part of equation (6.45) may be written as:

$$Q_x^{\varphi(\omega_n)} = Q_{\omega_o}^{\varphi(\omega_n)}S_x^{\omega_o} + Q_Q^{\varphi(\omega_n)}S_x^Q \tag{6.48}$$

where, $Q_{\omega_o}^{\varphi(\omega_n)} = \dfrac{\omega_n}{Q}\dfrac{\left(1+\omega_n^2\right)}{\left[\left(1-\omega_n^2\right)^2+\dfrac{\omega_n^2}{Q^2}\right]}$ and $Q_Q^{\varphi(\omega_n)} = \dfrac{\omega_n}{Q}\dfrac{\left(1-\omega_n^2\right)}{\left\{\left(1-\omega_n^2\right)^2+\dfrac{\omega_n^2}{Q^2}\right\}}$ \hfill (6.49)

For a given value of Q, the determination of the terms given in equation (6.47) provides information about the overall magnitude sensitivity of a biquad using equation (6.46); sensitivity of ω_o and Q should have been calculated for the biquad with respect to the element x. Such calculations are very important since for most of the filters, specifications are given in terms of the required magnitude in the pass and stop bands. Moreover, the functions given in equation (6.47) also provide useful information if these terms are plotted as shown in Figures 6.5(a) and (b), respectively, for a few different values of Q.

For example, it can be observed from Figure 6.5(a) that the maximum of the Q magnitude sensitivities of $|H(j\omega)|$ is unity for any value of $Q.S_Q^{|H(j\omega_n)|} = 1$ and it occurs at $\omega_n = 1$; its variation with frequency is very similar to the response of a band pass (BP) filter having midband gain equal to unity.

However, the shape of the term $S_{\omega_o}^{|H(j\omega_n)|}$ is a bit complicated, as shown in Figure 6.5(b). For large values of Q, it is given as:

$$\max\left\{S_{\omega_o}^{|H(j\omega_n)|}\right\} \cong (Q-1)\,\text{at}\,\omega_{n2} \cong \left(1+\frac{1}{2Q}\right) \tag{6.50a}$$

$$\min\{S_{\omega_o}^{|H(j\omega_n)|}\} \cong (-Q-1) \text{ at } \omega_{n1} \cong \left(1 - \frac{1}{2Q}\right)$$ (6.50b)

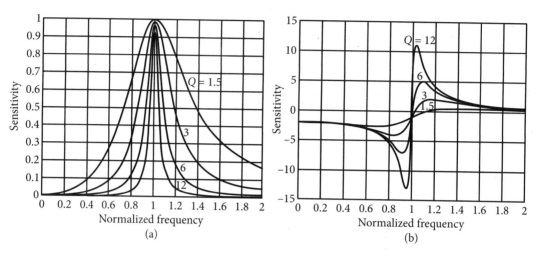

Figure 6.5 Sensitivity of the magnitude and phase function of the second-order transfer function for a few values of Q: (a) $S_Q^{|H(j\omega_n)|}$; (b) $S_{\omega_o}^{|H(j\omega_n)|}$.

From Figure 6.5(b) and equation (6.50), it is clear that the magnitude maxima (and the minima) directly depend on the value of Q and these are located close to $\omega_n = 1$ (in the pass band and close to its edge). Comparison of the maxima in the two figures, Figures 6.5(a) and (b) confirms that the effect of the first term in equation (6.46) is more pronounced, nearly $(4Q^2 - 1)^{1/2}$ or around $(2Q)$ times. It again shows that ω_o sensitivities are to be kept at minimum and nearly Q times less than Q sensitivities if their effect is to be made comparable. The peaks occur as shown by equation (6.50a, b) at a distance of $(\omega_n/2Q)$ from ω_n, which means approximately at 3 dB frequencies.

The phase sensitivity of the transfer function is also important. Not only does it evaluate deviation in phase output in a filter whose specification is given in terms of magnitude attenuation, but it is the phase in terms of which specifications are mentioned while designing phase equalizers or all pass filters. Hence, for observation of phase deviation, equation (6.48) needs to be studied.

Other than keeping $S_x^{\omega_o}$ small, it is observed that error in both magnitude and phase increases with increase in the design value of Q. This implies that filter design is easy with smaller values of Q.

6.5 Sensitivity Considerations for High-order Active Filters

In the last section, sensitivity considerations for a general second-order section were discussed. The study has its own importance, as the information obtained is not only for a standalone second-order section, but is also useful while designing higher-order filters employing a few section-order sections. Before taking up the sensitivity consideration of higher-order filters employing section-order sections, an important issue needs to be considered.

In Section 6.3, it was observable from equation (6.31) that the transfer function sensitivity is very high in the vicinity of either a pole or a zero of a transfer function. This means that for any higher-order transfer function, with high selectivity (having complex–conjugate poles close to the $j\omega$-axis), its sensitivity will be high throughout the pass and stop band. Hence, designing a high-order filter will result, in general, in a high transfer function sensitivity. This means that in fabricated elements with practically obtained variability, deviations in the transfer function are likely to go beyond acceptable limits, making the design impractical.

The problem has been overcome in different ways as will be seen in the following sections.

6.5.1   Simulation of LC ladder method

Equation (6.28) shows that the overall sensitivity of a transfer function depends on the difference term $\left\{ \dfrac{1}{N(s)} \dfrac{\partial N(s)}{\partial x} - \dfrac{1}{D(s)} \dfrac{\partial D(s)}{\partial x} \right\}$. If a circuit topology is such that the two terms here are equal or near equal, it will result in a zero or very small transfer function sensitivity. Such a methodology has been employed using a doubly terminated lossless ladder structure as shown in Figure 6.6 in a block form. It has been shown mathematically that for such a lossless ladder structure, the magnitude sensitivity $S_x^{|H(j\omega)|}$ of equation (6.34) becomes zero, if it is designed for maximum power transfer condition. However, in some other studies, the transfer function sensitivities are shown to be not exactly zero but small in the pass band region.

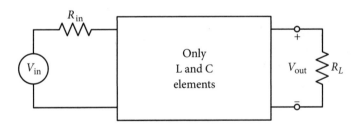

Figure 6.6   A doubly terminated lossless ladder structure in block form.

It needs to be kept in mind that though the sensitivities remain small in the pass band while using doubly terminated lossless ladders, the sensitivities with respect to inductors and capacitors can be large in the transition and stop bands.

After it was proved mathematically that a doubly terminated lossless ladder designed for maximum power transfer possesses low sensitivities in the pass band, they have been extensively used in the passive RLC filter design. Their modified versions in active forms have also been used in large numbers. Such methods include either active simulation of active components or operational simulation of passive components (Chapter 9).

6.5.2 Sensitivity in cascade design

The direct realization of higher-order filters means that all poles and zeros are functions of almost all the elements used, which means that each of the term $S_x^{z_i}$ and $S_x^{p_i}$ is finite. Instead of using the doubly terminated ladder structure, an alternate method of reduction in the transfer function sensitivities comes in the form of cascade realization (Chapter 10) of higher-order filters. In this method, any element x affects only one pole pair and corresponding zero(s). Sensitivity of other pole pairs and zero(s) is zero with respect to this particular element x. Under such a condition, equation (6.31) reduces to the following for a second-order section:

$$S_x^{H(s)} = S_x^K + \frac{(Q_x^{p_1})^*}{(s - p_1^*)} + \frac{Q_x^{p_1}}{(s - p_1)} - \frac{Q_x^{z_1}}{(s - z_1)} - \frac{(Q_x^{z_1})^*}{(s - z_1^*)} \tag{6.51}$$

If the second-order function represented by poles and zeros of equation (6.51) is an all pole function (or the effect of sensitivity of zeros and the gain constant K is not affective in the pass band), equation (6.51) reduces to the following.

$$S_x^{H(s)} = \frac{(Q_x^{p_1})^*}{(s - p_1^*)} + \frac{Q_x^{p_1}}{(s - p_1)} \tag{6.52}$$

Equation (6.52) informs that the transfer function sensitivity with respect to the element x near pole pair p_1, p_1^* depends only on this particular second-order function in the cascade formation. For the rest of the second-order functions in the cascade, their sensitivity with respect to x is zero. This result has led to the useful cascade design methodology of higher filter design discussed in Chapter 10. Of course, each second-order section to be connected in cascade needs to be non-interactive with its transfer function.

Since the overall sensitivity of the transfer function $H(s)$ depends on the sensitivity of each one of the constituent second-order sections, it is important that each section is designed with optimum sensitivities with respect to the elements used in that section so that variability of $H(s)$ is minimum.

6.6 Multi-parameter Sensitivity

Incremental sensitivity (or single-parameter sensitivity) has been found to be very useful as a large amount of information becomes available through it. Most of the sensitivity studies discussed so far is sufficient for comparing filter circuits. However, incremental sensitivity-based comparison has to be applied with caution, remembering the assumptions made, which include that effective change in elements is small, the nominal point of the element x is fixed and the rest of the components remain unchanged. In practice, these assumptions might not be true. Element tolerances are different and almost all elements may vary simultaneously, either by the same amount or differently. Under such conditions, single-element sensitivity measures will not give precise information; they may even mislead by providing incorrect information depending on the amount of the magnitude of approximations. Therefore, it becomes important to study multi-parameter sensitivity measures. Unfortunately, because of the large number of varying elements, many of them inter-dependent as well, finding multi-parameter sensitivity becomes very involved and a high amount of computations are required, which necessitates the use of computers and related software.

Practice Problems

6-1   Derive the voltage ratio transfer function for the circuit shown in Figure P6.1, and show that it s realizes a BP response. Determine the incremental sensitivity of the parameters ω_0, Q and the mid-band gain H with respect to the elements used.

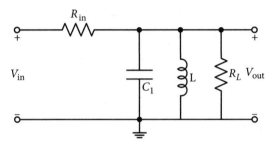

Figure P6.1

(a) Also find the changes in the parameters ω_0, Q and H, when the inductor changes by +5%, the capacitance changes by – 6% and resistors change by +10%.

(b) If all the elements are likely to change in the positive or negative direction by the same amount as in part (a), then calculate the worst case percent deviation in the filter parameters.

6-2   Repeat Problem 6-1 for the circuit displayed in Figure P6.2 after showing that it realizes a band stop response. The concerned filter parameters are ω_0, Q and attenuation α in the stop band.

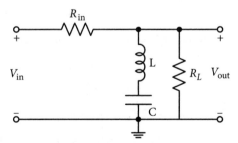

Figure P6.2

6-3   (a) Design a BP filter using the circuit diagram shown in Figure 6.3 with no positive feedback, while assuming that $\sqrt{(R_2 / R_1)} = 10$, for $\omega_0 = 10^4$ rad/s, and $Q = 5$. What shall be the value of mid-band gain H_m for $C_1 = C_2$. Find the sensitivities of the parameters ω_0, Q and H_m with respect to the elements R_1, R_2, C_1 and C_2.

(b) What are the changes in the parameters when positive feedback is introduced with (i) $K = 1/10$, (ii) $K = 1/20$.

(c) Find the sensitivities for the parameters ω_o, Q and H_m with respect to the passive elements and K.

(d) With the given values of K in part (b), find the variability in Q using two approaches. Which one out of the two processes is more accurate?

6-4   Repeat Problem 6-3 for $\omega_o = 10^4$ rad/s, $Q = 2$ and $H_m = 12.5$, with the assumption that $\sqrt{(R_2/R_1)} = 5$.

6-5   Prove the relations in equations (6.22)–(6.25) using equation (6.4), with P_i being a function of x, and n and k being constant.

6-6   Find the transfer function $H(s)$ for the circuit shown in Figure P6.3 and determine the expressions for the parameter ω_o and Q

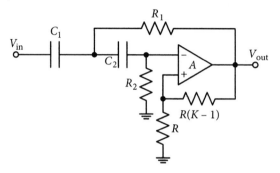

Figure P6.3

Find the sensitivities of ω_o, Q and $H(j\omega)$ with respect to all the passive and active elements. It would be better if the sensitivity expressions are given in terms of ω_o and Q

6-7   For the circuit shown in Figure P6.3, $\omega_o = 5 \times 10^3$ rad/s and $Q = 20$.

(a) Show by calculation that the maximum of the sensitivity of $|H(j\omega)|$ with respect to Q is unity.

(b) Find the maximum and minimum sensitivity of $|H(j\omega)|$ with respect to ω_o.

(c) Find the magnitude and phase sensitivities of $|H(j\omega_n)|$ at $\omega_n = 0.25, 0.5, 1.0, 1.25$ and 1.5.

6-8   Obtain the voltage ratio transfer function for the circuit shown in Figure P6.4, and find the sensitivity of Q with respect to all the passive and active elements for:

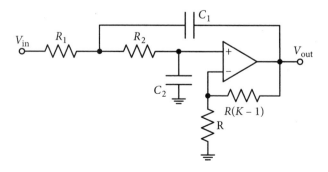

Figure P6.4

(i) $R_1 = 1\Omega$, $R_2 = 1\Omega$, $C_1 = 4Q$, $C_2 = 1/4Q$, $K = 1$.

(ii) $R_1 = 1\Omega$, $R_2 = 2Q\,\Omega$, $C_1 = 1$, $C_2 = 1/2Q$, $K = 2.5$.

(iii) $R_1 = 1\Omega$, $R_2 = 1\Omega$, $C_1 = 1$, $C_2 = 1$, $K = 3 - 1/Q$.

Evaluate the sensitivity values found for $Q = 20$ and discuss the results.

6-9   Derive the transfer function for the circuit shown in Figure P6.5, while considering the OA as ideal and find the sensitivity expression for ω_o and Q with respect the resistors R_1 and R_2.

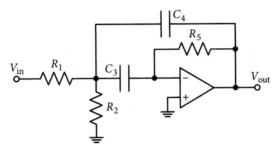

Figure P6.5

6-10   Design the filter in Figure P6.5 for $\omega_o = 10$ krad/s and $Q = 10$, and determine the displacement in the pole location if (a) all resistors and capacitors increase by 5%, and (b) all resistor values increase and all capacitor values decrease by 5%.

6-11   For the filter circuit of Figure 6.3, (a) discus the effect of selecting the parameter $r = 4$ and (b) design the circuit for $\omega_o = (3.4 \times 2\pi)$ krad/s.

6-12   Find the transfer function sensitivity for the Tow–Thomas biquadratic filter circuit shown in Figure P6.6 with respect to the passive components R_1 and C_1.

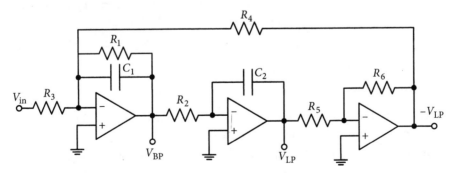

Figure P6.6 Tow–Thomas biquad.

Chapter *7*

Single Amplifier Second-order Filters

7.1 Introduction

Depending on the specifications, the required order of the filter may be small or big. For less challenging requirements, a second-order or even a first-order section may be adequate. However, higher-order sections may become necessary in other situations. For higher-order filter requirements, either the direct form approach is used, or a combination of second-order second sections is used which are connected in a cascade form or in a multiple-loop feedback form. Hence, the study of and realization methods of second-order sections assumes importance as they are used either stand-alone or form the building blocks for higher-order filters. Since the overall performance of the higher-order filter depends on the individual second-order building blocks, it is very important that these building blocks have desirable and attractive properties; the following are some such requirements.

(c) Low sensitivities of the two important filter parameters, pole frequency ω_o and pole-Q, with respect to the circuit elements are highly desirable.

(d) Utilization of a lesser number of passive elements (active elements as well, though in present discussion, it is assumed that only one active component is used) makes the realization economical in discrete, as well as in integrated form.

(e) Values of the passive components should be in the practical range of integrated circuit (IC) fabrication, and the component spread is preferred to be as small as possible in order to make them attractive in IC form.

(f) Independent tunability of the parameters ω_o and Q is important as analog filters require some tuning.

(g) To properly connect the second-order blocks in cascade or in multiple feedback system, their input impedance should be as high as possible and the output impedance as low as possible at all working frequencies. Otherwise, it can change the characteristics of the overall higher-order filter.

A large number of second-order active building blocks are available with each one claiming advantages over the others. Study of the development procedure of such sections might help the readers to gain more insight towards selecting the optimal section or help in designing new circuits/improving upon the existing ones.

Hence, we begin the study of development of second-order sections using a single passive feedback in Section 7.2; the effect of using different types of passive structures in the feedback is also included. Versatility of the configuration is shown to be improved with multiple feedbacks in Section 7.3. Certain constraints of the single input are overcome in differential-mode input. Next, general active RC feedback is studied in Section 7.5 to gain more understanding of the factors responsible for controlling the sensitivities. It was found that the well-known circuits of Sallen–Key and Delyiannis–Friend, which will be studied in Sections 7.7 and 7.8, are, in fact, special cases of the general active RC feedback structure discussed in Section 7.5.

7.2   Single-feedback Basic Biquadratic Section

A simple and common configuration for generating a biquadratic (biquad) section is shown in Figure 7.1. It uses a single-ended operational amplifier (OA) with its non-inverting input grounded, a passive RC two-port (three terminals) section (N_p) in the negative feedback path and another RC two-port section in the feed-forward path (N_z). Let the two-port sections be represented by the admittance parameter y_z and y_p as follows:

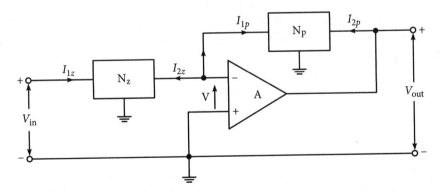

Figure 7.1   Basic biquad configuration using single feedback with a single-ended operational amplifier.

$$\begin{bmatrix} I_{1z} \\ I_{2z} \end{bmatrix} = \begin{bmatrix} y_{11z} & -y_{12z} \\ -y_{21z} & y_{22z} \end{bmatrix} \begin{bmatrix} V_{in} \\ V \end{bmatrix}$$

$$(7.1)$$

$$\begin{bmatrix} I_{1p} \\ I_{2p} \end{bmatrix} = \begin{bmatrix} y_{11p} & -y_{12p} \\ -y_{21p} & y_{22p} \end{bmatrix} \begin{bmatrix} V \\ V_{out} \end{bmatrix} \tag{7.2}$$

With the inverting input terminal of the OA at the virtual ground potential, and its input current neglected being small, summation of currents at the inverting terminal gives:

$$I_{2z} + I_{1p} = 0 \rightarrow -y_{21z}\,V_{in} + y_{22z}\,V + y_{11p}\,V - y_{12p}\,V_{out} = 0 \tag{7.3}$$

If the OA is considered ideal, $A \rightarrow \infty$, $V = 0$ and the transfer function is obtained as:

$$(V_{out}/V_{in}) = (-y_{21z}/y_{12p}) \tag{7.4}$$

The networks N_z and N_p being RC-only networks, their natural frequencies lie on the negative real axis of the complex frequency s plane. If the two networks are selected in such a way that their natural frequencies cancel each other, then they will not affect the overall transfer function. In that case, if the transfer function is represented as a ratio of two polynomials in s as in equation (7.5), then selection of an arbitrary polynomial $Q(s)$ representing the natural frequencies of the two RC networks shall be given as in equation (7.6).

$$(V_{out}/V_{in}) = \{N(s)/D(s)\} \tag{7.5}$$

$$y_{21z} = -\{N(s)/Q(s)\} \text{ and } -y_{12p} = -\{D(s)/Q(s)\} \tag{7.6}$$

It is obvious from equation (7.5) and (7.6) that the zeros of the transfer function are decided only by the zeros of the transfer admittance y_{21z} of the feed-forward network. Since an RC network without inductors can have transmission zeros anywhere in the complex frequency plane except on the positive real axis, all types of stable transfer functions like LP (low pass), HP (high pass), BP (band pass) or BR (band reject) can be realized except an odd order all pass function. Again, from equation (7.5) and (7.6), poles of the transfer function are decided by the zeros of the transfer admittance y_{12p}, with the same location restrictions.

We can use the following general expression for a normalized biquad:

$$H(s) = \frac{N(s)}{s^2 + (\omega_o / Q) + \omega_o^2} \tag{7.7}$$

From the aforementioned discussion, it is obvious that realization of the transfer function reduces to the synthesis of the two RC two ports N_z and N_p. Conventional methods of RC synthesis are to be used for the realization of the RC networks. However, if $N(s)$ and $D(s)$ contain complex roots, then the synthesis is a little involved. For the second-order admittance functions, Table 7.1 can also be used, though the choice is not limited to this table.

Table 7.1   Short circuit transfer admittances of some common RC two ports sections

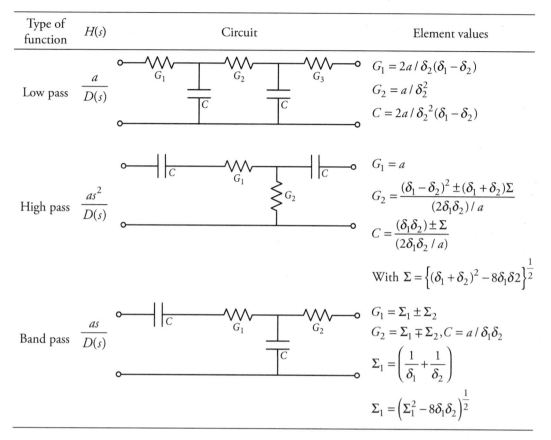

Type of function	$H(s)$	Circuit	Element values
Low pass	$\dfrac{a}{D(s)}$		$G_1 = 2a/\delta_2(\delta_1 - \delta_2)$ $G_2 = a/\delta_2^2$ $C = 2a/\delta_2^2(\delta_1 - \delta_2)$
High pass	$\dfrac{as^2}{D(s)}$		$G_1 = a$ $G_2 = \dfrac{(\delta_1 - \delta_2)^2 \pm (\delta_1 + \delta_2)\Sigma}{(2\delta_1\delta_2)/a}$ $C = \dfrac{(\delta_1\delta_2) \pm \Sigma}{(2\delta_1\delta_2/a)}$ With $\Sigma = \left\{(\delta_1 + \delta_2)^2 - 8\delta_1\delta2\right\}^{\frac{1}{2}}$
Band pass	$\dfrac{as}{D(s)}$		$G_1 = \Sigma_1 \pm \Sigma_2$ $G_2 = \Sigma_1 \mp \Sigma_2, C = a/\delta_1\delta_2$ $\Sigma_1 = \left(\dfrac{1}{\delta_1} + \dfrac{1}{\delta_2}\right)$ $\Sigma_1 = \left(\Sigma_1^2 - 8\delta_1\delta_2\right)^{\frac{1}{2}}$

Example: 7.1 Realize a second-order BPF (band pass filter) section with a pole-Q of 5, a pole frequency of 5 kHz and a center band gain of 20 dB using a single-ended OA based configuration.

Solution: The following will be the frequency-normalized ($\omega_o = 1$) transfer function of a second-order BPF while using the configuration of Figure 7.1.

$$\frac{V_{out}}{V_{in}} = \frac{-hs}{s^2 + bs + 1} \tag{7.8}$$

$(\omega_o/Q) = b$, gives $b = 1/5$, and with $s = j\omega$, the gain magnitude at the center band of 20 dBs gives $20 \log(h/b) = 10$, or $h = 2$. Assuming a tertiary polynomial $Q(s) = (s + \delta)$, for the N_z block, we get the parameter y_{21z} as:

$$y_{21z} = -hs/(s + \delta) \tag{7.9}$$

A suggested circuit for realizing equation (7.9) is shown in Figure 7.2(a) for which:

$$y_{21z} = (-s/R)/(s + 1/RC) \tag{7.10}$$

Comparing equations (7.9) and (7.10), we get:

$$R = 1/h = 0.5 \ \Omega \text{ and } C = h/\delta \tag{7.11}$$

To cancel the arbitrary polynomial $Q(s)$ of equation (7.6) and to find the two-port N_p, it is assumed that:

$$RC = C_2 R_1 R_2/(R_1 + R_2) \tag{7.12}$$

With selected $Q(s)$ and using the denominator of the transfer function in equation (7.8), $-y_{12p}$ can be written in compact and, then in expanded form as:

$$-y_{12p} = \frac{s^2 + bs + 1}{s + \delta} = s + (b - \delta) + \frac{(1 + \delta^2 - \delta b)}{(s + \delta)} \tag{7.13}$$

Equation (7.13) represents a parallel combination of a capacitor, a resistor and a bridged-T network circuit as shown in Figure 7.2(b). Expressions for the elements shown in the figure are compared with the denominator in equation (7.8); it gives, $a_2 = 1$, $a_1 = b$ and $a_0 = 1$. The bridged-T network is simplified by choosing $\delta = b = 0.2$, which results in R_3 being infinite. With R_3 open circuited, the T network has the following expression for y_{12}:

$$y_{12} = -\frac{(1/R_1 R_2 C_2)}{s + \frac{1}{C_2}\left(\frac{1}{R_1} + \frac{1}{R_2}\right)} \tag{7.14}$$

A parallel combination of the admittance of a T network and capacitance C_1 gives the admittance of the feedback network $-y_{12p}$ as:

$$-y_{12p} = \frac{s^2 + s\frac{(R_1 + R_2)}{R_1 R_2}\frac{1}{C_2} + \frac{1}{R_1 R_2 C_1 C_2}}{\frac{s}{C_1} + \frac{R_1 + R_2}{R_1 R_2 C_1 C_2}} \tag{7.15}$$

Further, from equation (7.13) and Figure 7.2(b), with the following element values and relations ($C_1 = a_2 = 1$ F, $a_1 = b = 0.2$, $a_0 = 1$ and $C = h/\delta = 10$ F), the obtained normalized resistance values are:

$$G_1 = G_2 = 2 \ (1 - 0.2 \times 0.2 + 1 \times 0.2 \times 0.2)/0.2 = 10 \text{ Mho} \rightarrow R_1 = R_2 = 0.1 \ \Omega \tag{7.16a}$$

Using equation (7.12) and $R = 0.5 \ \Omega$, we get the value of $C_2 = (4/\delta^2) = 100$ F. Dividing equation (7.10) by equation (7.15), (y_{21z} by $-y_{12p}$), we obtain the transfer function of the filter for which the normalized pole frequency, $\omega_o = 1/\sqrt{R_1 R_2 C_1 C_2}$ = 1 rad/s. De-normalization with frequency

$\omega_o = 2\pi \times 5 \times 10^3$rad/s will give the element values as:

$$C_1 = (1/\pi \times 10^4) = 31.8 \ \mu F, \ C_2 = 3.18 \ mF \ and \ C = (10/\pi \times 10^4) = 318 \ \mu F \qquad (7.16b)$$

Further, for realizing resistance and capacitance values appropriate for IC fabrication, the values obtained can be impedance scaled (say) by a factor of 10^4. The resulting element values will be as follows:

$$C_1 = 31.8nF, \ C_2 = 318 \ nF, \ C = 31.8 \ nF \ and \ R_1 = R_2 = 1 \ k\Omega; \ R = 5 \ k\Omega \qquad (7.17a)$$

The complete second-order BP section using bridged-T network in feedback path is now shown in Figure 7.2(c). Figure 7.2(d) shows the network's PSpice simulated response having a mid-band gain of 9.975 at a center frequency of 4.86 kHz. Bandwidth being 945.3 Hz results in pole-Q = 5.14.

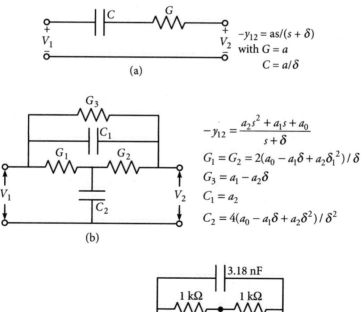

(a)

$-y_{12} = as/(s + \delta)$
with $G = a$
$C = a/\delta$

(b)

$$-y_{12} = \frac{a_2 s^2 + a_1 s + a_0}{s + \delta}$$
$G_1 = G_2 = 2(a_0 - a_1\delta + a_2\delta_1^2)/\delta$
$G_3 = a_1 - a_2\delta$
$C_1 = a_2$
$C_2 = 4(a_0 - a_1\delta + a_2\delta^2)/\delta^2$

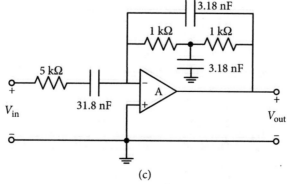

(c)

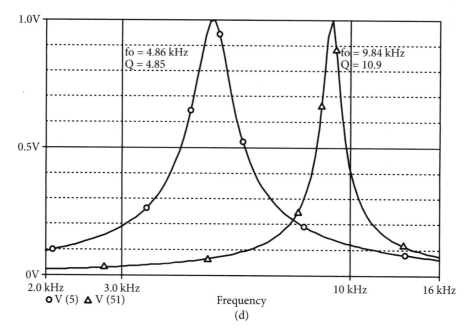

Figure 7.2 (a) An admittance function realization for Example 7.1. (b) An admittance function realization using bridged-T network. (c) Second-order band pass filter section for Example 7.1. (d) Magnitude responses of a band pass filter using a bridged-T network for Example 7.1.

Another BPF for a center frequency of 10 kHz with a mid-band gain of 10 and $Q = 10$ is designed using the same structure. The required element values in this case are as follows:

$$C_1 = 1 \text{ nF}, C_2 = 0.1 \text{ nF}, C = 40 \text{ nF and } R_1 = R_2 = 7.96 \text{ k}\Omega; R = 159.2 \text{ k}\Omega \qquad (7.17b)$$

The simulated response is also shown in Figure 7. 2(d); value of the center frequency is 9.12 kHz; mid-band gain is 9.84 and $Q = 10.9$.

It is to be noted that the element for the LP, HP and BP responses can be obtained directly from the relations given in Table 7.2.

Table 7.2 Element values for the multi-loop feedback structure

Elements shown in Figure 7.5 (explained later)	Element values of a low pass section	Element values of a high pass section	Element values of a band pass section
Y_1	$G_1 = b / \sqrt{b_0}$	$C_1 = b$	$G_1 = b$
Y_2	$C_2 = \dfrac{(2b_0 + b)}{b_1 \sqrt{b_0}}$	$G_2 = \dfrac{b_0(2+b)}{b_1}$	$G_2 = \dfrac{2b_0}{b_1} - b$
Y_3	$G_3 = \sqrt{b_0}$	$C_3 = 1$	$C_3 = 1$

Contd.

Y_4	$G_4 = G_3$	$C_4 = C_3$	$C_4 = C_3$
Y_5	$C_1 = \dfrac{b_1 \sqrt{b_0}}{(2b_0 + b)}$	$G_5 = \dfrac{b_1}{(2 + b)}$	$G_5 = \dfrac{b_1}{2}$

7.2.1 Signal conditioning modules: application example

Signal conditioning modules (SCMs) are an essential part of circuits for measuring process control variables, such as temperature, pressure, strain, and so on. SCMs are subject to a number of noise signals which are both electrically and magnetically induced. Elimination of noise from the mixed signal requires appropriate filtering.

One such module, namely, the *DSCA 43 DIN rail analog input module, 4-way isolation, multiple LP, anti-aliasing filter* is discussed here as an application example [7.1] of a single feedback biquadratic configuration. The module requires LPF with a 4 Hz, 3 dB frequency at a roll-off of 120 dBs per decade. Such a filter attenuates the 50/60 Hz noise by 100 dBs.

To get the required roll-off, order of the LPF can be calculated: it is found to be six. Hence, from Table 3.1, the pole locations of the sixth-order Butterworth characteristics are as follows:

$$-0.2588 \pm j0.9659, -0.7071 \pm j0.7071, -0.9659 \pm j0.2588. \tag{7.18}$$

Since it will be an all pole filter, the following will be the transfer functions of the three second-order sections that will be connected in cascade.

$$H_1(s) = 1/(s^2 + 0.5176s + 1) \tag{7.19a}$$

$$H_2(s) = 1/(s^2 + 1.41s + 1) \tag{7.19b}$$

$$H_1(s) = 1/(s^2 + 1.9318s + 1) \tag{7.19c}$$

A number of circuits are available for realizing second-order LPFs. Figure 7.1 shows the structure of the single feedback with a single input op amp circuit that will be used for this application example. The following is a standard form of the normalized transfer function of a second-order LPF:

$$\frac{V_{out}}{V_{in}} = -\frac{h_{lp}}{\left(s^2 + bs + 1\right)} \tag{7.20}$$

For the first LP section, using equation (7.19), $b = 0.5176$ and with dc gain of unity, $h_{lp} = 1$.

Following the procedure of Example 7.1, a tertiary polynomial $Q(s) = (s + \delta)$. The following will be the parameter y_{21z} for the block N_z:

$$y_{21z} = -a/(s + \delta) \tag{7.21a}$$

A suggested circuit for realizing equation (7.21a) is shown in Figure 7.3(a), for which:

$$G = 2a/\delta \text{ and } C = 4a/\delta^2 \tag{7.21b}$$

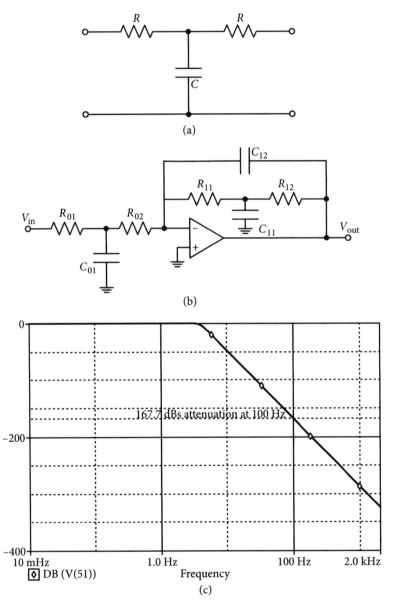

(a)

(b)

167.7 dBs attenuation at 100 Hz

DB (V(51))

(c)

Figure 7.3   (a) A suggested circuit for the realization of equation (7.21a). (b) A second-order low pass section obtained while employing a single feedback biquadratic section. (c) Simulated magnitude response of the sixth-order filter shown in Figure 7.3(b).

To cancel the arbitrary polynomial $Q(s)$ of equation (7.21a) and find the two-port N_p, the bridged-T network of Figure 7.2(b) is selected. With $a_2 = 1$, $a_1 = b = 0.5176$ and $b_o = 1$, we get $C_1 = 1$ F, $C_2 = \dfrac{4}{0.5176^2} = 14.936$ F. The resistance values are: $R_1 = R_2 = \delta/2 = 0.2588$, $R_3 = \infty$, with $\delta = b$.

For the T network of Figure 7.3(a), $R_{01} = \delta/2b = 0.2588$ and $C_{01} = 4b/\delta^2 = 4/0.5166^2 = 14.936$ F.

The component values are de-normalized using an impedance scaling factor of 10 kΩ and frequency scaling factor of 24.143 rad/s (4 Hz). The element value for the first second-order section is given in equation (7.22a) and the corresponding first LP section with element symbols is shown in Figure 7.3(b).

$$R_{01} = 2.588 \text{ k}\Omega = R_{11} = R_{12}, \; C_{01} = 59.378 \; \mu\text{F} = C_{11}, \; C_{12} = 3.977 \; \mu\text{F} \tag{7.22a}$$

Similar steps were taken to design the other two second-order sections. De-normalized values of the corresponding elements are as follows:

$$R_{02} = 7.705 \text{ k}\Omega = R_{21} = R_{22}, \; C_{02} = 7.954 \; \mu\text{F} = C_{21}, \; C_{22} = 3.977 \; \mu\text{F} \tag{7.22b}$$

$$R_{03} = 9.659 \text{ k}\Omega = R_{31} = R_{32}, \; C_{03} = 4.262 \; \mu\text{F} = C_{31}, \; C_{32} = 3.977 \; \mu\text{F} \tag{7.22c}$$

The three sections are cascaded with $H_1(s)$ being the first section with minimum pole-Q followed by $H_2(s)$ and $H_3(s)$. The simulated magnitude response of the cascaded filter is shown in Figure 7.3(c). Its 3-dB frequency occurs at 3.9989 Hz and with an attenuation of 47.726 dBs at 10 Hz and 167.73 dBs at 100 Hz, the obtained roll-off is 120.3 dBs per decade.

7.2.2   Bridged twin-T RC network

In a number of applications, it becomes necessary for a BP section to be more selective. In such cases, the filter requires a high value of pole-Q (or a small value of the coefficient b in equation (7.8)). It was observed in Example 7.1 that even for $Q = 5$, the component spread is high, which is not attractive in IC implementation. Hence, instead of using the bridge-T network shown in Figure 7.2(b), it is advised to use a bridged twin-T network which is shown in Figure 7.4(a). A bridged twin-T network is in effect a parallel combination of an HP-T, an LP-T network and a resistance R^*. For the HP-T network, which comprises C_1, C_2 and R_3, transfer admittance, $-y_{12H}$ is given as:

$$-y_{12H} = \frac{\left\{(C_1 C_2)/(C_1 + C_2)\right\}s^2}{s + \dfrac{1}{R_3(C_1 + C_2)}} \tag{7.23a}$$

However, for the LP-T network comprising R_1, R_2 and C_3, transfer admittance is given as:

$$-y_{12L} = \frac{1/(R_1 R_2 C_3)}{s + \dfrac{1}{C_3}\left(\dfrac{1}{R_1} + \dfrac{1}{R_2}\right)} \tag{7.23b}$$

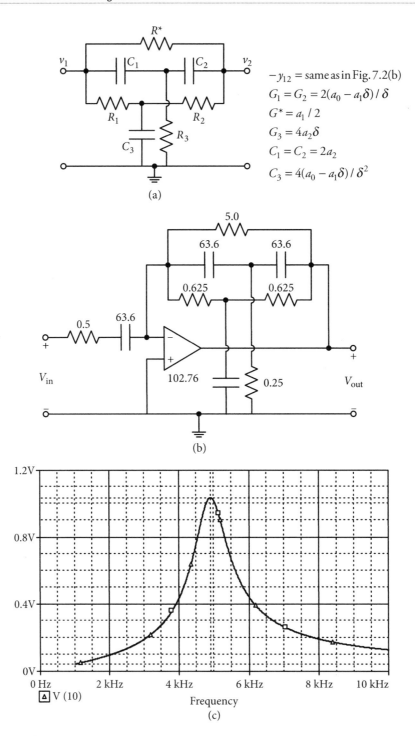

$$-y_{12} = \text{same as in Fig. 7.2(b)}$$
$$G_1 = G_2 = 2(a_0 - a_1\delta)/\delta$$
$$G^* = a_1/2$$
$$G_3 = 4a_2\delta$$
$$C_1 = C_2 = 2a_2$$
$$C_3 = 4(a_0 - a_1\delta)/\delta^2$$

(a)

(b)

(c)

Figure 7.4   (a) A bridged twin-T network. (b) Circuit realization for the band pass filter of Example 7.2. Capacitors are in nF and resistors are in kΩ. (c) Magnitude response of the band pass filter using twin-T feedback for Example 7.2.

If the twin-T network is symmetrical, $R_1 = R_2$ and $C_1 = C_2$, a factor of half appears in the simplified expression of equation (7.23a, b), which forces:

b to be halved → effective R^* is halved. (7.24)

Utilization of the twin-T network in the feedback path of the basic configuration of Figure 7.1 requires the following form of representation of the transfer admittance $-y_{12p}$:

$$-y_{12p} = \frac{s^2}{(s+\delta)} + \frac{(1-\delta b)}{s+\delta} + b$$ (7.25)

Again, expressions of the elements in Figure 7.4(a) and the denominator of equation (7.8) are compared. It is worth noting that the twin-T network is a third-order section, but can be reduced to a second-order section, according to our present requirement by single pole–zero cancellation through the selection of δ as:

$$\delta = \frac{1}{R_3(C_1+C_2)} = \frac{1}{C_3}\left(\frac{1}{R_1}+\frac{1}{R_2}\right) \rightarrow \frac{1}{2R_3C_1} = \frac{2}{C_3R_1}$$ (7.26)

The following example will illustrate the procedure and the advantage in terms of getting a low component spread.

Example 7.2: Redesign the BP section of Example 7.1 while using the twin-T network of Figure 7.4(a).

Solution: With $Q = 5$, $a_1 = b = 0.2$, and as in Example 7.1, $h = 2$, $R = (1/h) = 0.5\ \Omega$ and $C = h/\delta = 2$ F. For δ assumed to be unity and using the expressions for elements in Figure 7.4(a), with $a_0 = 1$, $a_1 = 0.2$ and $a_2 = 1$.

$C = (h/\delta) = 2$ F (7.27a)

$R_1 = R_2 = \{\delta/2(a_0 - a_1\delta)\} = 0.625\ \Omega$ and $R_3 = 1/4a_2\delta = 0.25\ \Omega$ (7.27b)

$R^* = (2/0.2)/2 = 5\Omega$, $C_1 = C_2 = 2a_2 = 2$ F, $C_3 = 4\ (a_0 - a_1\delta)/\delta^2 = 3.2$ F (7.27c)

The component spread comes down significantly for capacitance from 100 to 1.6, though the resistance spread has increased to 20. For $\omega_o = 2\pi \times 5$ krad/s, capacitance values will change as follows:

$C_1 = C_2 = 63.6 \times 10^{-6}$ F, $C_3 = 102.76 \times 10^{-6}$ F and $C = 63.6 \times 10^{-6}$ F (7.27d)

To bring all components in the preferable range for IC integration, impedance scaling by a factor of 10^3 is performed to give the following element values.

$R_1 = R_2 = 625 \; \Omega, \; R_3 = 250 \; \Omega, \; R^* = 5 \; k\Omega, \; C_1 = C_2 = 63.6 \; nF, \; C_3 = 102.76 \; nF$ and $C = 63.6 \; nF$

$$(7.27e)$$

The complete circuit is shown in Figure 7.4(b) and its PSpice simulated response is shown in Figure 7.4(c). Its center frequency is 4.93 kHz, bandwidth is 938 Hz resulting in $Q = 5.25$, and a mid-band gain of 10.33; these results are close to the design specifications with improved capacitor spread.

7.3   Multiple Feedback Single Amplifier Biquad (SAB)

A large number of configurations are available in literature in which a number of two-terminal passive elements are connected in the negative feedback path of an OA. The multiple feedbacks can provide realizations of biquadratic and all pole functions with different levels of advantages and constraints. A widely used multi feedback single amplifier arrangement obtained from a multiple-loop feedback arrangement [7.2], is a double-ladder or Rauch structure shown in Figure 7.5. Application of Kirchhoff's current law (KCL) at nodes 1 and 2 with node 1 at virtual ground gives the following equations, respectively.

$$V_{out}Y_5 + V_2Y_3 = 0 \tag{7.28a}$$

$$(V_{in} - V_2)Y_1 + (V_{out} - V_2)\,Y_4 + (-V_2)Y_3 - V_2Y_2 = 0 \tag{7.28b}$$

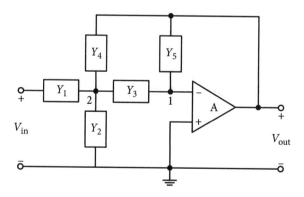

Figure 7.5   Double ladder multiple feedback single amplifier second-order section.

Hence, the voltage ratio transfer function is obtained as follows:

$$\frac{V_{out}}{V_{in}} = \frac{-Y_1Y_3}{Y_3Y_4 + Y_5\,(Y_1 + Y_2 + Y_3 + Y_4)} \tag{7.28c}$$

Selection of elements for the admittances Y_1 to Y_5, either as a resistor or capacitor, can give a large number of circuits. Their selection decides whether the filter would be an LP, HP, BP or any other type. Instead of using a single-element, a combination of RC components can also be used: this provides more versatility to the network. For any specific selected combination of elements, the network can be studied to determine the kind of response that can be obtained and any constraint, such as the achievability of a quality factor, or the level of sensitivity of the parameters with respect to the passive and active elements. Out of the many choices available, if Y_1 and Y_3 are conductances, Y_5 is a capacitor and either Y_2 or Y_4 is a capacitor as shown in Figure 7.6, an LP response with the following transfer function is obtained.

$$\frac{V_{out}}{V_{in}} = -\frac{G_1 G_3}{C_2 C_5 s^2 + C_5 \left(G_1 + G_3 + G_4\right)s + G_3 G_4} \tag{7.29}$$

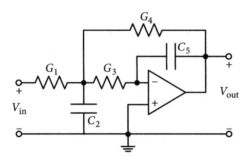

Figure 7.6   Second-order low pass function realization of equation (7.29).

Further showing the versatility of the structure, a BP function can be realized with either Y_1 or Y_3 as a capacitor, and the other one a conductor. If $Y_1 = G_1$ and $Y_3 = sC_3$, Y_4 needs to be a capacitor, Y_5 a conductor and Y_2 can be either a capacitor or conductor (shown in Figure 7.7). Hence, selecting $Y_2 = G_2$, the transfer function shall be that of a BPF:

$$\frac{V_{out}}{V_{in}} = \frac{-G_1 C_3 s}{C_3 C_4 s^2 + G_5 \left(C_3 + C_4\right)s + G_5(G_1 + G_2)} \tag{7.30}$$

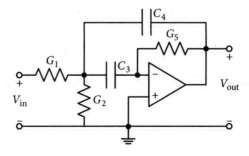

Figure 7.7   Second-order band pass function realization of equation (7.30).

For the realization of an HP function, the numerator will have a term with s^2. Hence, selecting both Y_1 and Y_3 as capacitors, Y_4 also needs to be a capacitor. With Y_2 and Y_5 as conductors, the circuit will be as shown in Figure 7.8(a) with the following HP function.

$$\frac{V_{out}}{V_{in}} = \frac{-C_1 C_3 s^2}{C_3 C_4 s^2 + G_5\left(C_1 + C_3 + C_4\right)s + G_2 G_5}$$
(7.31)

Designing of the filter section can easily be performed using the coefficient matching technique. Consider the HP second-order voltage ratio transfer function:

$$\frac{V_{out}}{V_{in}} = -\frac{h s^2}{b_2 s^2 + b_1 s + b_0}$$
(7.32)

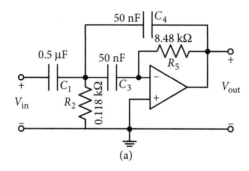

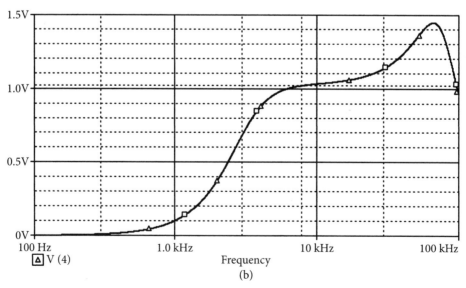

Figure 7.8   (a) Second-order high pass function realization of equation (7.31). (b) Second-order high pass filter response for the circuit shown in Figure 7.8(a) for Example 7.3.

By comparing the coefficients for the HP function of equation (7.31) with equation (7.32), the following design equations can be obtained.

$$h = C_1 C_3, \; b_0 = G_2 G_5 \tag{7.33a-b}$$

$$b_1 = G_5(C_1 + C_3 + C_4), \; b_2 = C_3 C_4 \tag{7.33c-d}$$

Selection of element values usually put a constraint on the performance parameters. For example, if equal value capacitors are used, that is, $C_1 = C_3 = C_4 = C$, with $b_2 = h$, h will also be unity. If we use equal value capacitors, having h as unity is not a severe constraint, rather it is an advantage. However, these capacitors may result in a high spread of resistance; it may also require an additional amplifier to get the required gain (if needed).

Example 7.3: Design a normalized second-order HP Butterworth filter using multiple feedback in a single OA configuration, with a pass band gain of 20 dBs. De-normalize the filter for a cut-off frequency of 3.2 kHz. Use suitable element values and obtain the filter's response.

Solution: The network structure is shown in Figure 7.8(a) and the desired Butterworth function has $h = 10$, $b_1 = \sqrt{2}$ and $b_0 = 1$. As mentioned earlier, many possibilities are available for the selection of element values. Solving equations (7.32) and (7.33) for equal value capacitors and without using an extra amplifier for getting $h = 10$, two possible sets of relations and resulting elements are as follows:

Set 1: With $C_1 = C_3 = C$, $h = 10 = C^2 \rightarrow C = \sqrt{10}$, $b_o = G_2 G_5$, $b_1 = \sqrt{2} = G_5\left(2\sqrt{10} + C_4\right)$ and $b_2 = 1 = \sqrt{10}C_4 \rightarrow C_4 = 1/\sqrt{10}, G_5 = \sqrt{20/21}, G_2 = 21/\sqrt{20}$

Set 2: With $C_3 = C_4 = C$, $b_2 = 1$ $C^2 \rightarrow C = 1$, $h = 10 = C_1$, $b_1 = \sqrt{2} = G_5(10 + 1 + 1) \rightarrow G_5 = \sqrt{2}/12, b_0 = 1 = G_2 \sqrt{2}/12 \rightarrow G_2 = 12/\sqrt{2}$

Element values for the normalized functions are in ohms and farads, respectively. De-normalizing the filter section with $\omega = 2\pi \times 3.2$ krad/s and using an impedance scaling factor of 10^3, value of the elements for the second set are as follows:

$$C_3 = C_4 = 50 \text{ nF}, \; C_1 = 0.5 \text{ μF}, \; R_2 = 0.118 \text{ kΩ and } R_5 = 8.48 \text{ kΩ}$$

Figure 7.8(b) shows the PSpice simulated response of the HP filter. The filter's output flattens at 1.028 volts for an input voltage of 0.1 volt, providing a high frequency gain of 10.028. The cut-off occurs at 3.19 kHz; a very close response to the design. A peaking occurs at higher voltage due to the frequency dependent gain of the OA.

7.3.1   Power line communication: automatic meter reading

Automatic meter reading (AMR) is one of the most well-known applications of power line communication (PLC). PLC is now considered an optimal solution to provide communication between a residential or industrial consumer and power distributers. However, the frequency response of the electric grid for a certain consumer is different from any other consumer due to various reasons; because of this reason, a system level solution is required.

The European regulatory committee responsible for allowing the communication requirement (CENELEC) has provided five different frequency bands and stipulated the maximum transmission and distribution levels when transmitting data over the power line. The frequency range for the signal transmission of signals has been divided into five bands, and the frequency range for the distribution company use and their licensees is from 9 kHz to 95 kHz. A fourth-order LP Butterworth filter having a cut-off frequency of 95 kHz have been employed [7.3]. Though other circuit structures can be used, the multiple feedback topology of Figure 7.6 was selected; two such stages may be connected in cascade.

From Table 3.1, location of poles for a fourth-order Butterworth filter are as follows:

$$s_{1,2} = -0.38268 \pm j0.9238, \; s_{3,4} = -0.9238 \pm j0.38268$$

Hence, the normalized transfer function for the two second-order stages will be as follows:

$$H_1(s) = \frac{2.5}{s^2 + 0.76536s + 1} \text{ and } H_2(s) = \frac{2.0}{s^2 + 1.9576s + 1}$$

To get an overall dc gain of 5 (14 dBs), the numerators of the transfer functions were selected as 2.5 and 2.0, respectively. Both the transfer functions are compared with equation (7.29) and the elements are de-normalized with a frequency scaling factor of 95(2π) krad/s and an impedance scaling factor of 1 kΩ. The resulting element values for the two sections are as follows:

$$R_{11} = R_{31} = 3.135 \text{ k}\Omega, \; R_{41} = 7.839 \text{ k}\Omega, \; C_{21} = 1.674 \text{ nF}, \; C_{51} = 68.1 \text{ pF}$$

$$R_{12} = R_{32} = 1.277 \text{ k}\Omega, \; R_{42} = 2.554 \text{ k}\Omega, \; C_{22} = 1.674 \text{ nF}, \; C_{52} = 0.5132 \text{ nF}$$

The simulated frequency response of the filter is shown in Figure 7.9, having a 3 dB frequency of 93.1 kHz.

Another PLC module for the distribution of signals in the CENELEC band of frequencies is available [7.4]. The module's receiver signal path is shown in Figure 7.10. It consists of a passive HPF with a cut-off frequency of 25 kHz, an active LPF with a cut-off frequency of 125 kHz, with two in-between programmable gain amplifiers (PGAs). The third-order passive HPF with element values is shown in Figure 7.11(a). The third-order LPF introduces 6 dBs of attenuation to keep the dc level at 1.65 V according to the specifications. For the third-order active LP filter, a multiple feedback circuit as shown in Figure 7.6 is used. The value of poles for a third-order Butterworth filter from Table 3.1 are as follows:

$$s_{1,2} = -0.5 \pm j0.866 \text{ and } s_3 = -1.0$$

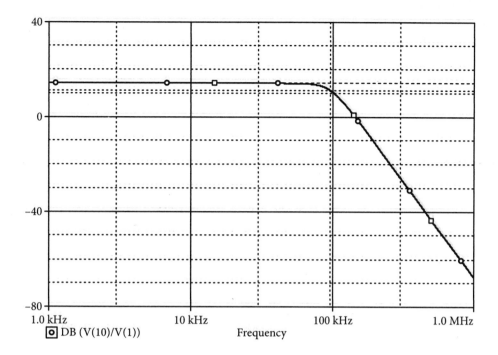

DB (V(10)/V(1))   Frequency

Figure 7.9   Magnitude response of a fourth-order low pass filter for power line communication.

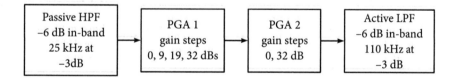

Figure 7.10   Receiver signal path for a PLC module [7.4].

Hence, the transfer function of the LPF with a dc gain of 0.5 (-6dB) will be as follows:

$$H_{LP}(s) = \frac{0.5}{(s^2+s+1)(s+1)} = \frac{0.5}{(s^2+s+1)}\frac{1}{(s+1)}$$

For the first-order LP section, the normalized elements will be $R_0 = 1\Omega$ and $C_0 = 1$ F. Using equation (7.29), the normalized elements for the second-order section are $R_1 = R_3 = 4\Omega$, $R_4 = 2\Omega$, $C_2 = 1$F, $C_5 = 0.125$ F. De-normalization with a frequency scaling factor of $125 \times (2\pi)$ krad/s and an impedance scaling factor of $10^3\ \Omega$ results in the following element values.

$$R_0 = 1\ k\Omega,\ C_0 = 0.12727\ nF,\ R_1 = R_3 = 4\ k\Omega,\ R_4 = 2\ k\Omega,\ C_2 = 0.12727\ nF,\ C_5 = 0.159\ nF$$

The third-order LPF is shown in Figure 7.11(b). The filter needs to remove signals having a frequency above 110 kHz with a 6 dBs in-band attenuation. The simulated frequency response of the HPF and the LPF is shown in Figure 7.11(c) with a 3 dB frequency of 125.3 kHz, in conformity with specifications.

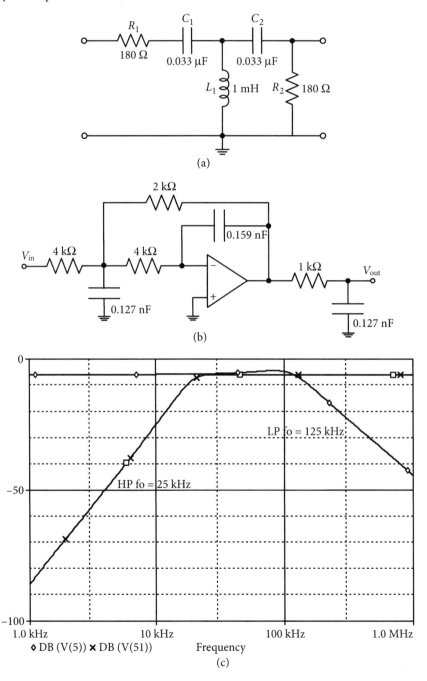

Figure 7.11 (a) Passive high pass filter structure for Figure 7.10 [7.4]. (b) Third-order low pass filter. (c) Responses of the third-order active low pass and passive high pass for CENELEC module.

7.4   Differential Input Single Amplifier Filter Sections

Single amplifier single-ended sections explained in Sections 7.2 and 7.3 are good examples of efficient and economical implementation of second-order sections. However, there are certain general constraints when using these sections, which include their inability to realize complex conjugate zeros or their need for a specific choice of elements. It has been observed that utilization of the differential nature of OAs increase the versatility of the filter section. In the following sub-sections, we will discuss this nature in more detail.

7.4.1   Differential input single feedback biquad

Figure 7.12(a) shows an OA in differential mode with single feedback for realizing the same range of transfer functions as that in the sections explained in Sections 7.2 and 7.3 [7.2]. The open circuit transfer functions of the passive RC networks N_z and N_p are $H_z(s)$ and $H_p(s)$, respectively. With the open-loop gain of the OA being A, we get:

$$V_{out} = A(V_1 - V_2) = A(H_z V_{in} - H_p V_{out})$$

$$(V_{out}/V_{in}) = H_z/(H_p + 1/A) = H_z/H_p \text{ for } A \to \infty \tag{7.34}$$

If the poles of $H_z(s)$ and $H_p(s)$ are chosen in such a way that they cancel each other, the synthesis of the transfer function resolves into the separate synthesis of two-port RC networks. For the normalized second-order case, the following $H_z(s)$ can be realized using a number of simple ladder structures, whereas the $H_p(s)$ of equation (7.36) may be a bridged-T or twin-T structure as explained earlier.

$$H_z(s) = N(s)/(s^2 + bs + 1) \tag{7.35}$$

$$H_p(s) = \frac{s^2 + (1/Q)s + 1}{s^2 + bs + 1} \tag{7.36}$$

For finite A, substituting equations (7.35) and (7.36) in equation (7.34), we get:

$$\frac{V_{out}}{V_{in}} = \frac{1}{1 + 1/A} \frac{N(s)}{s^2 + \left(\frac{1}{Q}\right)\left(1 + \frac{Q_p}{A}\right)s + 1} \tag{7.37}$$

This equation gives the effective selectivity or quality factor as:

$$Q_e = Q/(1 + Q_p/A) \tag{7.38}$$

In equations (7.37) and (7.38), $Q_p = 1/b$, the pole-Q of the denominator from equation (7.36). While using the bridged twin-T network of Figure 7.2(b) for $H_p(s)$, the selected element values are as follows:

$$R_1 = R_2 = R, \ R_3 = (R/2), \ C_1 = C_2 = C, \ C_3 = 2C \text{ and } R^* = (R/a)$$

This gives the transfer function of $H_p(s)$ as:

$$\frac{V_{out}}{V_{in}} = \frac{N(s)}{s^2 + \left(\dfrac{4}{A} + 2a\right)s + (1 + 2a)} \tag{7.39}$$

From equation (7.39), expressions of the center frequency and quality factor, respectively, are as follows:

$$\omega_{oe} = \omega_o \left(1 + 2a\right)^{\frac{1}{2}} \tag{7.40}$$

$$Q_e = \frac{\left(1 + 2a\right)^{\frac{1}{2}}}{\left(2a + 4/A\right)} \cong \frac{1}{2a} \tag{7.41}$$

The small shift in the pole frequency given by equation (7.40) is easily corrected by a pre-distortion of the twin-T design. Structure of the H_z network needs to be in the form shown in Figure 7.12(b); otherwise, the response functions will be restricted. The advantage of using OA in differential mode is that it avoids increase in sensitivity due to interaction between the two-port networks H_z and H_p. Order of sensitivity of the overall network is mainly decided by the choice of the network H_p.

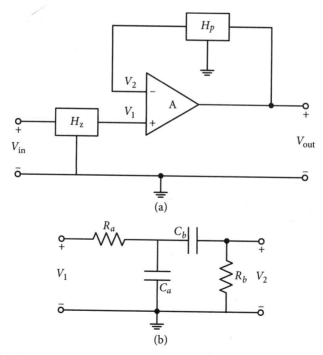

Figure 7.12 (a) Differential input single feedback single operational amplifier biquad. (b) A preferred RC two port network for realizing network H_z shown in Figure 7.12(a).

7.4.2   General differential input single OA biquad

The biquad realizations discussed in the previous sections have the restriction that there are no positive real poles and zero for their transfer functions. To overcome this restriction, additional single-ended infinite gain amplifiers are used. A more economical way to avoid this constraint is to use a general differential input configuration as shown in Figure 7.13. Applying Kirchhoff's current law at the terminals V_p and V_z, we get the following relations for obtaining the transfer function [7.5].

$$(V_p - V_{in})\, y_1 + V_p y_2 + (V_p - V_{out})\, y_3 = 0 \tag{7.42a}$$

$$(V_z - V_{in})\, y_a + V_z y_b + (V_z - V_{out})\, y_c = 0 \tag{7.42b}$$

$$V_{out} = (V_z - V_p)A \tag{7.43}$$

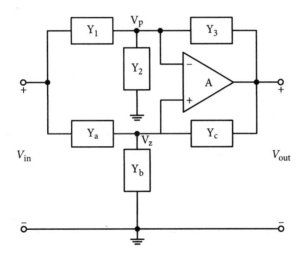

Figure 7.13   General differential input single operational amplifier biquad configuration.

The obtained transfer function for the general differential input single OA is as follows:

$$\frac{V_{out}}{V_{in}} = \frac{y_a\left(y_1 + y_2 + y_3\right) - y_1(y_a + y_b + y_c)}{y_3\left(y_a + y_b + y_c\right) - y_c\left(y_1 + y_2 + y_3\right) + (1/A)(y_1 + y_2 + y_3)(y_a + y_b + y_c)} \tag{7.44}$$

When $A \rightarrow \infty$, the transfer function of equation (7.44) is considerably simplified as shown in equation (7.46), provided the condition presented in equation (7.45) is satisfied.

$$(y_a + y_b + y_c) = (y_1 + y_2 + y_3) \tag{7.45}$$

$$\frac{V_{out}}{V_{in}} = \frac{y_a - y_1}{y_3 - y_c} \tag{7.46}$$

Equation (7.46) ensures that any real transfer function $N(s)/D(s)$ can be realized from the configuration shown in Figure 7.13.

Procedurally, the numerator and denominator are to be divided by a polynomial $Q(s)$ with simple negative real roots, which has order at least one less than that of $N(s)$ or $D(s)$. The problem will then reduce to realizing the driving point admittances (y_1, y_3, y_a and y_c) using RC elements. The rest of the driving point function y_2 and y_b are obtainable using equation (7.45) as follows:

$$y_b - y_2 = (y_3 - y_c) - (y_a - y_1) \tag{7.47}$$

$$= \{D(s) - N(s)\}/Q(s) \tag{7.48}$$

Example 7.4: Realize the following BP transfer function using the configuration of Figure 7.13.

$$H(s) = \frac{s}{s^2 + 0.20s + 1.01} \tag{7.49}$$

Solution: Selecting auxiliary polynomial $Q(s) = (s + 1)$ and an order less than $H(s)$, we get:

$$\frac{N(s)}{Q(s)} = \frac{s}{(s+1)} = \frac{s}{s+1} - 0 = y_a - y_1 \tag{7.50}$$

$$\frac{D(s)}{Q(s)} = \frac{s^2 + 0.2s + 1.01}{(s+1)} = (s+1.01) - \frac{1.81s}{s+1} = y_3 - y_c \tag{7.51a}$$

$$\frac{D(s) - N(s)}{Q(s)} = \frac{s^2 - 0.8s + 1.01}{(s+1)} = (s+1.01) - \frac{2.81s}{(s+1)} = y_b - y_2 \tag{7.51b}$$

Here, y_1 is an open circuit, y_a, y_c and y_2 are a series combination of R and C elements and y_3 and y_b have parallel RC elements. The normalized element values from equations (7.50–51) are as follows:

$R_a = 1\ \Omega$ in series with $C_a = 1$ F, $R_c = 0.5525\ \Omega$ in series with $C_c = 1.81$ F, $R_2 = 0.3558\ \Omega$ in series with $C_2 = 2.81$ F, $R_3 = R_b = 0.99\ \Omega$ in parallel with $C_3 = C_b = 1$ F, respectively.

In order to get the center frequency at 10^4rad/s, the frequency is de-normalized with a factor of $(1.01)^{-\frac{1}{2}} \times 10^4$. Further, an impedance scaling of 10^4 gives the following element values, which are also shown in the filter circuit presented in Figure 7.14(a).

$R_a = 10\text{k}\Omega$, $C_a = 10.0498$ nF, $R_c = 5.525$ kΩ, $C_c = 18.19$ nF, $R_2 = 3.558$ kΩ,
$C_2 = 28.24$ nF, $R_3 = R_b = 9.9$ kΩ, and $C_3 = C_b = 10.0498$ nF. $\tag{7.52}$

Figure 7.14(b) shows the PSpice simulated magnitude and phase response of the BPF. The obtained mid-band gain is 5.0. Center frequency $f_o = 1.595$ kHz and a bandwidth of 318 Hz give the value of pole-Q as 5.017; this confirms the design parameter values. In the phase response, a $0°$ phase shift occurs at 1.6 kHz as per design.

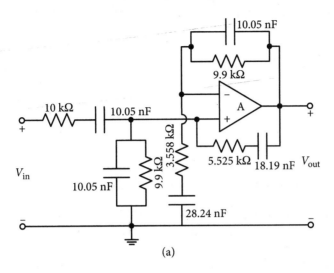

(a)

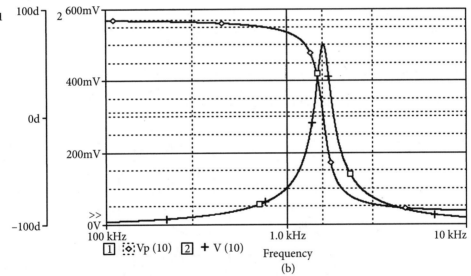

(b)

Figure 7.14 (a) Second-order band pass filter section using a differential mode operational amplifier for Example 7.4. (b) Magnitude and phase response of the differential input single operational amplifier band pass filter depicted in Figure 7.14(a).

Example 7.5: Design a BR filter with the following normalized transfer function using a general differential input single OA biquadratic circuit.

$$H(s) = \frac{N(s)}{D(s)} = \frac{s^2 + 0.25}{s^2 + 0.09s + 0.83}$$

Solution: Selecting the auxiliary polynomial $Q = (s + 1)$, we get:

$$\frac{N(s)}{Q(s)} = \frac{s^2 + 0.25}{s+1} = (s+0.25) - \frac{1.25s}{s+1} \rightarrow y_a = s+0.25 \text{ and } y_1 = 1.25s/(s+1).$$

$$\frac{D(s)}{Q(s)} = \frac{s^2 + 0.09s + 0.83}{s+1} = s+0.83 - \frac{1.74s}{s+1} \rightarrow y_3 = s+0.83 \text{ and } y_c = 1.74s/(s+1)$$

$$\frac{D(s) - N(s)}{Q(s)} = \frac{0.09s + 0.58}{s+1} = 0.58 - \frac{0.49s}{s+1} \rightarrow y_b = 0.58 \text{ and } y_2 = 0.49s/(s+1)$$

Admittances y_1, y_c and y_2 are a series combination of R and C elements, y_a and y_3 are parallel RC elements and y_b is only resistive. The normalized element values are as follows:

R_a = 4.0 Ω, C_a = 1 F, R_b = 1.724 Ω, R_c = 0.5747 Ω, C_c = 1.74 F, R_1 = 0.8 Ω, C_1 =1.25 F, R_2 = 2.0408 Ω, C_2 = 0.49 F, R_3 = 1.2048 Ω and C_3 = 1 F

Frequency de-normalization by a factor of $0.83^{-0.5} \times 10^4$ and an impedance scaling by 10^4 will result in a notch frequency of $(\sqrt{0.25}\sqrt{}/\sqrt{0.83}) \times 10^4$ rad/s (873.1 Hz), a dc gain of 0.3012 and a high frequency gain of unity with the following element values, which are shown in Figure 7.15.

R_a = 40 kΩ, C_a = 9.1104 nF, R_b = 17.24 kΩ, R_c = 5.747 kΩ,
C_c = 15.852 nF, R_1 = 8.0 kΩ, C_1 = 11.388 nF, R_2 = 20.408 kΩ,
C_2 = 4.461 nF, R_3 = 12.048 kΩ and C_3 = 9.1104 nF

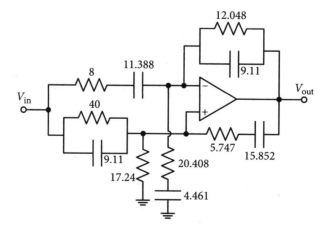

Figure 7.15 Second-order band stop filter using a differential mode operational amplifier for Example 7.5. All capacitors are in nF and resistors in kΩ.

Figure 7.16 shows the PSpice simulated magnitude and phase responses of the BR filter. Its gain at dc is 0.3011(0.3012) and gain at high frequencies is 0.998 (1.0); notch occurs at 875.04

Hz (873.1 Hz) and a peak gain of 7.14 occurs at 1.595 kHz (1.591 kHz). This response is very close to the design.

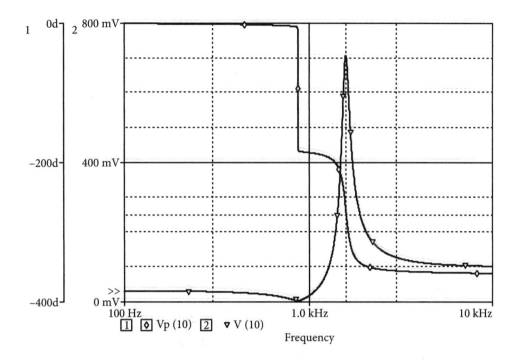

Figure 7.16 Magnitude and phase responses of the band reject filter for Example 7.5.

7.5   General Active RC Feedback Single Amplifier Biquad

In Sections 7.2 to 7.4, single amplifier biquad generation using OAs in single-ended mode and in differential mode were discussed. Obviously, a large number of circuit configurations are available and many more are likely to be found. Such circuits are selected for a particular application depending on their performance characteristics. However, the criterion that is very important for all circuits is the sensitivity factors with respect to the elements used. Study of a general active RC feedback circuit is helpful in gaining an understanding of the factors responsible for increase (decrease) in sensitivities. Such a study also helps in a systematic generation of good quality single amplifier RC biquads. It can be seen that the circuits studied in the previous sections were in effect special cases of such a general case.

Consider a general active RC single amplifier block configuration shown in Figure 7.17. With H_{ij} being the transfer function corresponding to the output and input terminals i and j, we get

$$V_{out} = A(s)[V_{out}\{H_{42}(s) - H_{32}(s)\}] + A(s)[V_{in}\{H_{41}(s) - H_{31}(s)\}] \tag{7.53a}$$

With OA being ideal (infinite gain), the overall transfer function is obtained as:

$$H(s) = \frac{V_{out}}{V_{in}} = \frac{H_{41}(s) - H_{31}(s)}{H_{32}(s) - H_{42}(s)} \tag{7.53b}$$

From equation (7.53b), the following two important inferences are obtained.

(a) For $H(s)$ to be a second-order function, the RC network should also be second-order.

(b) Transmission zeros of the active RC network are determined by the feed-forward path, whereas poles (natural frequencies) are set by the feedback path; this confirms an inference already seen in the previous sections.

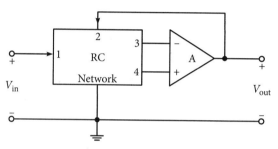

Figure 7.17 General active RC single amplifier configuration for biquad realization.

Theoretically, there is no problem in realizing a transfer function using the configuration shown in Figure 7.17, but it requires a three-port RC network, which is rather involved. In order to make the realization simpler, the network is broken into two-port RC networks as shown in Figure 7.18. With transfer function $H_{ij}(s) = (V_i/V_j)$ notation, output is obtained as follows:

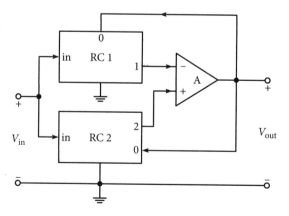

Figure 7.18 General active RC single amplifier configuration from Figure 7.17 with two two-port networks: RC1 and RC2.

$$\frac{V_{out}}{V_{in}} = \frac{H_{2in} - H_{1in}}{(H_{10} - H_{20}) + (1/A(s))} \tag{7.54}$$

$$= \frac{1}{D_{1\text{in}} D_{2\text{in}}} \frac{\left(N_{2\text{in}} D_{1\text{in}} - N_{1\text{in}} D_{2\text{in}}\right) D_{10} D_{20}}{\left(N_{10} D_{20} - N_{20} D_{10}\right)} \quad \text{For } A(s) \text{ nearing infinity} \tag{7.55}$$

Poles of the transfer function, from equation (7.55), are given by the following equation:

$$(N_{10} D_{20} - N_{20} D_{10}) = 0 \tag{7.56}$$

In order to get a second-order polynomial for getting only two roots in equation (7.56), it is essential that the transfer functions H_{10} and H_{20} are selected in such a way that the denominator D_{10} and D_{20} are same and cancel each other. Otherwise, the order of the polynomial will become more than two. In fact, even selecting $D_{10} = D_{20}$ is not sufficient to make the polynomial of equation (7.56) a second-order one since practically there is always some mismatch between the components of D_{10} and D_{20}. Therefore, it is advised to select either RC1 or RC2 as frequency independent. (The same argument is valid for the configuration of Figure 7.12(a)). Selecting RC2 (say) as a purely resistive network as shown in Figure 7.19, $H_{20} = (k-1)/k$. Hence, the poles are now determined from equation (7.57), which is obtained using equation (7.56).

$$H_{10} - \frac{(k-1)}{k} = 0 \rightarrow N_{10} - \frac{k-1}{k} D_{10} = 0 \tag{7.57}$$

In equation (7.57), if $k = 1$, poles of the realized transfer function depend on the numerator polynomial of the negative feedback network in the same way as for the structure shown in Figure 7.5 and Figure 7.12(a). For $k > 1$, some positive feedback is also applied which causes enhancement in the pole -Q, the main RC feedback goes to the inverting terminal of the OA. Hence, this structure is also called *enhanced negative feedback (ENF) circuit.*

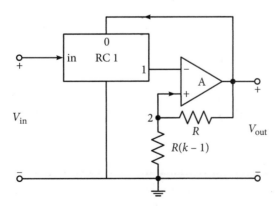

Figure 7.19 Simplified structure from Figure 7.18 while ensuring that it is a second-order active RC network.

Instead of making RC2 resistive, RC1 network can be made resistive with RC2 as a frequency dependent structure. Such a configuration is called an *enhanced positive feedback*

(EPF) circuit. The EPF configuration is shown in Figure 7.20. With $H_{10} = 1/k$, poles are obtained from equation (7.54) using the following relation:

$$\frac{1}{k} - H_{20} = 0 \rightarrow \frac{D_{20}}{k} - N_{20} = 0 \tag{7.58}$$

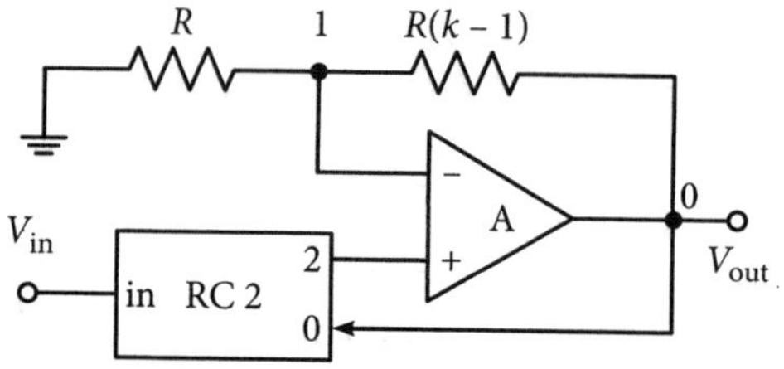

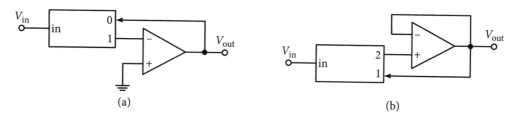

Figure 7.20 Enhanced positive feedback circuit obtained from Figure 7.18.

Special cases of these structures can be obtained with $k = 1$ as well. Figures 7.21(a) and (b) show the modified forms of Figures 7.19 and 7.20, respectively, with $k = 1$; it is essential that these forms are studied to understand the limitations of similar structures.

Figure 7.21 (a) Infinite gain negative feedback (NF), and (b) unity gain positive feedback (PF) configuration.

7.6   Coefficient Matching Synthesis Technique

Before taking up specific cases of general RC feedback discussed in the previous section, a significant method of network synthesis needs to be formalized, though we are already using a method without mentioning it as a procedure.

In the *coefficient matching synthesis technique*, a network is available for which the transfer function is determined by the conventional method. Coefficients in the transfer function are obviously in terms of the elements used in the filter structure. The coefficients of the transfer function are then equated to the function to be realized. This results in a set of simultaneous algebraic equations in terms of normalized (or sometimes de-normalized) values of the components. In general, such equations are lesser in number than the number of elements used. This allows assumption of values for some of the elements, which affects the performance, sensitivity, and component spread. Therefore caution is necessary.

As a technique, the coefficient matching method is particularly useful for second-order (or at most third-order) sections, and has been extensively used. Obviously, the choice of the initial filter structure decides the usefulness, advantages or limitations of the filter; many choices are available for any particular type of filtering action. For example, a large number of circuits can be used in order to obtain a passive RC network in the ENF and EPF configurations. At the same time, as mentioned earlier, the selection of elements or certain element ratios (depending upon the degree of freedom in design) also affects the filter performance. The technique becomes difficult for higher orders, hence, it is seldom used.

7.7   Sallen and Key Biquad

The Sallen and Key biquad circuit, introduced in 1955, was probably the first set of circuits that could realize almost all types of second-order responses [7.6]. Figure 7.22 shows one of these circuits realizing a LP response. Incidentally, it happens to be an ENF circuit where the passive circuit RC1 is realized by R_1, R_2, C_1, and C_2. Obviously, a different combination of these elements or some other passive structure may give another type of response. The transfer function of the circuit in Figure 7.22 can be obtained using equation (7.54) or (7.55). However, a direct application of KCL at nodes 1 and 3 gives the transfer function more easily, as shown here. While considering OA as ideal:

$$V_1 = V_2 \text{ and } V_2 = \{(k-1)/k\} V_{out} \tag{7.59}$$

$$V_1 (G_2 + sC_2) - V_3 G_2 = 0 \tag{7.60}$$

$$V_3 (G_1 + G_2 + sC_1) - V_{in} G_1 - V_1 G_2 - V_{out} sC_1 = 0 \tag{7.61}$$

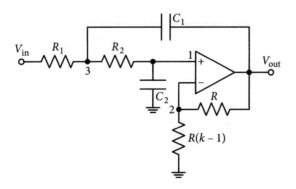

Figure 7.22   Sallen and Key second-order low pass filter section.

Substituting V_3 from equation (7.60) and V_1 and V_2 from equation (7.59) in equation (7.61) and simplifying, we get:

$$\frac{V_{out}}{V_{in}} = \frac{\left(k/k-1\right)\left(G_1G_2/C_1C_2\right)}{s^2 + \left\{\dfrac{\left(G_1+G_2\right)}{C_1} - \dfrac{G_2}{(k-1)C_2}\right\} + \dfrac{G_1G_2}{C_1C_2}}$$ (7.62)

Comparing equation (7.62) with the standard format of a second-order LP section shown again in equation (7.63) and applying the coefficient matching technique:

$$H_{LP}(s) = \frac{h_{lp}\omega_o^2}{s^2 + \left(\omega_o/Q\right)s + \omega_o^2}$$ (7.63)

The following relations are obtained.

$$\omega_o^2 = \frac{G_1G_2}{C_1C_2}$$ (7.64)

$$\frac{\omega_o}{Q} = \frac{G_1+G_2}{C_1} - \frac{G_2}{(k-1)C_2} \rightarrow Q = \frac{(k-1)\left(C_1C_2G_1G_2\right)^{\frac{1}{2}}}{(k-1)\left(G_1+G_2\right)C_2 - G_2C_1}$$ (7.65)

$$h_{lp} = k/(k-1)$$ (7.66)

From equation (7.66), it is obvious that the dc gain h_{lp} shall will always be more than unity. It is not a serious problem but shows one of the limitations connected with dc gain. The issue can be resolved by cascading an amplifier, though this may increase the cost of the filter. Alternatively, if it is necessary to have a dc gain less than that obtained as such, only a resistive potential divider at the input or output can serve the purpose, provided the divider's loading effect is taken care of.

There are three design parameters, h_{lp}, ω_o and Q against four passive elements, G_1, G_2, C_1, and C_2, and coefficient k which affects h_{lp} as well as Q. Hence, there are options available for pre-selecting some element values or their ratios. To make the equations suitable for integration, we can select $C_1 = C_2 = C$, then, equations (7.64)–(7.66) modify as:

$$\omega_o^2 = \left(G_1G_2/C^2\right)$$ (7.67)

$$Q = (k-1)\left(G_1G_2\right)^{\frac{1}{2}}/\left\{(k-1)G_1 + (k-2)G_2\right\}$$ (7.68)

Further, if $G_1 = G_2 = G$, $\omega_o = G/C$ (7.69)

$$Q = (k-1)/(2k-3)$$ (7.70)

Example 7.6: Design an LP Chebyshev second-order filter having a ripple width of 0.5 dB and a corner frequency of 3.18 kHz using the Sallen and Key circuit. It should have maximum gain of unity.

Solution: From Table 3.4, the pole location for the desired specifications is as follows:

$$s = -0.7128 \pm j1.004$$

This gives the normalized transfer function of the second-order filter as:

$$H(s) = \frac{1.5161 \times 0.944}{s^2 + 1.4256s + 1.5161} \qquad (7.71)$$

In the numerator of equation (7.71), 0.944 comes from the fact that for an even order Chebyshev filter, the dc gain drops from normalized unity by the ripple width; which is 0.5 dB in this case.

With $\omega_o = \sqrt{1.5161} = 1.2313$, equation (7.71) gives $Q = 1.2313/1.4256 = 0.8637$

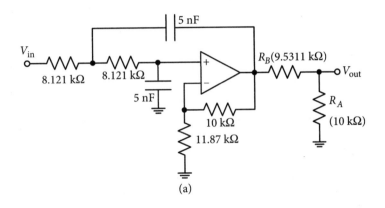

(a)

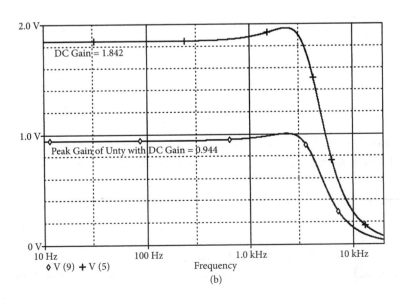

(b)

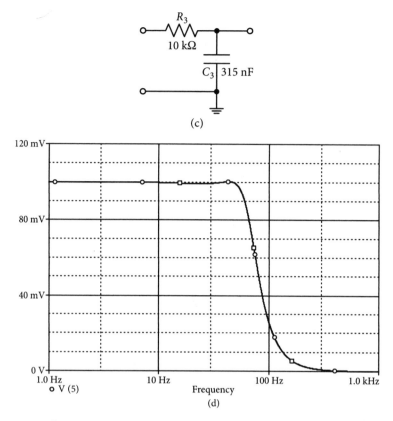

(c)

Figure 7.23   (a) Sallen and Key low pass circuit with element values for Example 7.5. (b) Second-order low pass Chebyshev filter response using the Sallen–Key circuit shown in Figure 7.23(a). (c) First-order filter section component in a moving air vehicle model. [with permission from N. Chattaraj et al. [7.7]]. (d) Simulated response of the filter employed with moving air vehicle.

Selecting normalized $C = 1$, equation (7.67) gives $G_1 \times G_2 = 1.5161$. Further, with $G_1 = G_2 = G = 1.2313$, and using $Q = 0.8637$, the value of k is obtained from equation (7.70) as:

$$0.8637 = \{(k-1)/(2k-3)\} \rightarrow k = 2.18738$$

The value of the dc gain obtained from equation (7.66) is 1.842, and in order to get a dc gain of 0.944 and a maximum gain of unity, a potential divider at the output as shown in Figure 7.23(a) is used with a ratio of R_A and R_B as 1: 0.9513.

De-normalizing the elements with a frequency scaling factor of 3.18 kHz and an impedance scaling factor of 10^4, the element values are:

$C_1 = C_2 = 5$ nF, $R_1 = R_2 = 8.121$ kΩ, $R = 10$ kΩ, $R(k-1) = 11.873$ kΩ, $R_A = 10$ kΩ and $R_B = 9.513$ kΩ

The circuit in Figure 7.23(a) shows these element values, and the simulated response is shown in Figure 7.23(b). The pass band edge frequency is 3.21 kHz and the ripple width is 0.51 dB, both of which are in very close agreement with the design. The effect of the passive potential divider is also clear in the two responses, bringing the final gain to unity.

The component spread is an important consideration in the selection of filters while implementing in integrated form. Another important issue is the sensitivity expressions and values for the passive and active elements used with respect to the design parameters. In the present case, the component spread is ideal as far as R_1, R_2, C_1 and C_2 are concerned. R_A and R_B depend on the amount of reduction in the filter gain (if reduction is required) but the value of k decides the final component spread and it also affects the sensitivity of Q as will be shown here.

Using equations (7.64) and (7.66), the sensitivity of ω_o and h_{lp} is found to be:

$$S^{\omega_o}_{R_1, R_2, C_1, C_2} = -\frac{1}{2}, \, S^{\omega_o}_k = 0$$

$$S^{h_{lp}}_{R_1, R_2, C_1, C_2} = 0, \, S^{h_{lp}}_k = -\frac{1}{k-1} = -0.533$$

From equation (7.65), evaluation of the sensitivity of Q is a bit involved, but for the present case, without losing any generality, from equation (7.70):

$$S^Q_k = \frac{k}{Q} \frac{\partial Q}{\partial k} = \frac{k}{(k-1)(2k-3)} = 1.34$$

which means that for the present design, with $k = 2.187$, Q will change 1.34% for a 1% change in the value of Q. The value of Q-sensitivity is not alarming but the condition will become worse if k reaches values near 1.5 for which the denominator in Q-sensitivity will increase alarmingly.

7.7.1 Micro air vehicle: a case study

Flapping wing micro air vehicles (MAVs) technology has attracted considerable attention in recent times. Advances in micro-electromechanical systems have led to the development of a number of new miniature devices such as infrared sensors and cameras, and MAVs can act as platform for these micro sized systems. Piezoelectrically actuated flapping wing MAVs is an emerging technique. Hence, MAVs based on piezoelectrical actuators are now subjects of intense study. In one such study [7.7], it was suggested that an active filter based unipolar high voltage driver be used to obtain the desired result. Analysis and synthesis of a Chebyshev active LPF was done for this purpose.

The average flapping wing frequency of insects, that were observed for inspiration, lies below the 50 Hz range. Hence, in this case, a flapping frequency range of 30–50 Hz was considered for designing a driver of the piezoelectric actuator.

The evaluated specifications for the LPF in this case were as follows: maximum gain = 1, minimum allowable gain = 0.99, pass band edge frequency $\omega_p = 2\pi(50)$ rad/s. Hence, the required order of the Chebyshev filter was three for which the following was the normalized transfer function.

$$H(s) = \frac{1.77}{(s+1.006)(s^2 + 1.006s + 1.76)} \tag{7.72}$$

The Sallen–Key circuit shown in Figure 7.22 was selected for the second-order section with the following values for the de-normalized elements: $R_1 = R_2 = 10$ kΩ, $C_1 = 633$ nF, $C_2 = 91$ nF. The circuit was followed by an RC circuit shown in Figure 7.23(c), which represents the first-order component of the transfer function; this models the capacitance of the piezoelectric bimorph actuator in series electrical connection.

The simulated response of the filter is shown in Figure 7.23(d).

7.8   Delyiannis–Friend Biquad

The classic and widely reported second-order filter circuit given by Deliyannis [7.9] and Friend [7.10] (D&F circuit) is an excellent example of an EPF configuration. It is convenient to study the circuit first without positive feedback and later introduce the feedback. An example of a second-order BP section was illustrated in Section 2.9 of Chapter 2. It is basically the same as a D&F circuit, except for a small difference in the formation of input resistance, which obviously reflects in some difference in the expressions of center frequency ω_o and pole-Q with the D&F circuit shown in Figure 7.24.

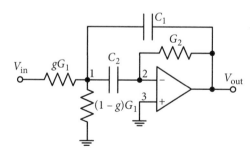

Figure 7.24 Deliyannis and Friend's second-order band pass filter circuit.

Applying KCL at terminals 1 and 2, respectively, we get:

$$V_1(gG_1 + G_1 - gG_1 + sC_1 + sC_2) - V_{in}gG_1 - V_2sC_2 - V_{out}sC_1 = 0 \tag{7.73}$$

$$V_2(sC_2 + G_2) - V_1sC_2 - V_{out}G_2 = 0 \tag{7.74}$$

For the OA open-loop gain being finite:

$$V_{out} = A(V_3 - V_2).$$
(7.75)

If the effect of finite A is to be investigated, equation (7.75) will be used; otherwise, considering OA as ideal, $V_3 = V_2 = 0$. With OA as ideal, equation (7.73) and (7.74) are simplified and combined to obtain the transfer function as:

$$\frac{V_{out}}{V_{in}} = -\frac{g(G_1/C_1)s}{s^2 + s\dfrac{(C_1+C_2)}{C_1C_2}G_2 + \dfrac{G_1G_2}{C_1C_2}}$$
(7.76)

Comparing this equation with the standard format of a second-order BP section of equation (7.77):

$$H_{BP}(s) = \frac{h_{bp}(\omega_o/Q)s}{s^2 + (\omega_o/Q)s + \omega_o^2}$$
(7.77)

The coefficient matching technique gives the relations of important parameters as:

$$\omega_o^2 = (G_1G_2/C_1C_2)$$
(7.78)

$$\frac{\omega_o}{Q} = \frac{(C_1+C_2)}{C_1C_2}G_2 \rightarrow Q = (C_1C_2)^{\frac{1}{2}}(G_1/G_2)^{\frac{1}{2}}/(C_1+C_2)$$
(7.79)

$$h_{bp} = g(1 + C_2/C_1)Q^2$$
(7.80)

Ordinarily, equations (7.78)–(7.80) are used to find element values for the given specifications ω_o, Q and h_{bp}. However, without losing generality, a good choice is $C_1 = C_2 = C$, which modifies these equations as equations (7.78)–(7.80), respectively.

$$\omega_o^2 = G_1G_2/C^2 \rightarrow \omega_o = (G_1G_2)^{\frac{1}{2}}/C$$
(7.81)

$$Q = \tfrac{1}{2}(G_1/G_2)^{\frac{1}{2}}$$
(7.82)

$$h_{bp} = 2gQ^2$$
(7.83)

From equations (7.78)–(7.80), the expressions of the elements are found in terms of the specifications:

$$R_2 = (2Q/\omega_o C), \quad R_1 = 1/2Q\omega_o C, \text{ and } g = h_{bp}/2Q^2$$
(7.84)

Equation (7.84) shows the high resistance spread as $(R_1/R_2) = 1/4Q^2$; the mid-band gain is proportional to Q^2. These characteristics were reflected in Example 2.7; for $Q = 10$ and effective $g = 0.5$, resistance spread was 200 and mid-band gain was 100. If the mid-band gain is to be reduced, a smaller value of coefficient g has to be used.

7.8.1   Enhanced-Q circuit

It is observed that resistance spread can be considerably reduced when the D&F circuit is modified in the form of an EPF configuration as shown in Figure 7.25. Denominator of its transfer function can be formed using equation (7.58). However, there is an alternate method which uses the application of KCL as earlier; with OA taken as ideal:

$$V_2 = V_3 = V_{\text{out}}\left(k-1\right)/k \tag{7.85}$$

$$V_1\left(gG_1 + G_1 - gG_1 + sC_1 + sC_2\right) - V_{\text{in}}gG_1 - V_2 sC_2 - V_{\text{out}} sC_1 = 0 \tag{7.86}$$

$$V_2\left(sC_2 + G_2\right) - V_1 sC_2 - V_{\text{out}} G_2 = 0 \tag{7.87}$$

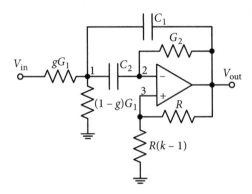

Figure 7.25 Deliyannis and Friend's circuit with Q-enhancement in ENF configuration.

Simplifying the aforementioned three equations, we get the transfer function.

$$H(s) = -\frac{gG_1 ks / C_1}{s^2 + \left\{\dfrac{G_1\left(1-k\right)}{C_1} + \dfrac{C_1 + C_2}{C_1 C_2}G_2\right\}s + \dfrac{G_1 G_2}{C_1 C_2}} \tag{7.88}$$

Once again, we can select $C_1 = C_2 = C$, and the transfer function will simplify as:

$$H(s) = -\frac{gG_1 ks / C}{s^2 + \left\{\dfrac{G_1\left(1-k\right)}{C} + \dfrac{2}{C}G_2\right\}s + \dfrac{G_1 G_2}{C^2}} \tag{7.89}$$

From equation (7.89), the expression for center frequency remains the same; expression of the enhanced $Q(Q_e)$, and gain $h(h_e)$ are obtained as follows:

$$\omega_o = \left(G_1 G_2\right)^{\frac{1}{2}} / C \tag{7.90}$$

$$\frac{\omega_o}{Q_e} = \frac{\left(1-k\right)G_1 + 2G_2}{C} \rightarrow Q_e = \frac{1}{2}\frac{\left(G_1 / G_2\right)^{\frac{1}{2}}}{\frac{1}{2}\left(1-k\right)\left(G_1 / G_2\right)+1} \tag{7.91}$$

$$h_e = \frac{kgG_1}{\left(1-k\right)G_1 + 2G_2} \tag{7.92}$$

Expression of Q_e in equation (7.91) and h_e in equation (7.92) can be written in terms of the quality factor Q from equation (7.82).

$$Q_e = \frac{Q}{2Q^2\left(1-k\right)+1} \tag{7.93}$$

$$h_e = \frac{gk}{\left(1-k\right)+\dfrac{1}{2Q^2}} \rightarrow 2gkQ_e Q \tag{7.94}$$

Equations (7.90) to (7.94) can be used to find the expressions or relations for element values. However, before discussing these relations, we need to determine the constraints, which can also be helpful in the design of the filter. Since coefficient $k > 1$ (when $k = 1$, it does not remain EPF form of the circuit), from equation (7.93), we get that:

$$2Q^2\left(1-k\right)<1 \rightarrow Q < \left(\frac{1+k}{2}\right)^{\frac{1}{2}} \tag{7.95}$$

To keep the resistance ratio small while forming the potential divider, the value of k should not be large, as even with a small value of Q, sufficiently large values of Q_e can be realized, as can be seen from equation (7.93). From equation (7.82), it can be observed that a small value of Q means a small ratio between R_1 and R_2 resulting in smaller resistance spread.

The first step in the filter design is to assume a small arbitrary value for k; then, equation (7.95) is used to find the upper limit for Q. A value for Q is assumed which is less than its upper limit and equation (7.82) is used to get a ratio between G_1 and G_2 and their normalized values. Since the expression for ω_o remains unchanged in the EPF case, equation (7.90) is used to give a nominal value of C for a selected nominal value of G_1 and G_2.

Equation (7.93) now gives the value of k for a specified value of Q_e and the assumed value of Q. The obtained value of k gives resistance values for the potential divider and equation (7.94) gives the value of coefficient g for a specified value of h_e.

Example 7.7: Design a BPF having a central frequency of 20 krad/s, pole-Q = 10 and mid-band gain of 10 employing an enhanced Q circuit. Repeat the design for pole-Q = 20 and a mid-band gain of 20.

Solution: Selecting arbitrarily k = 3, a small value, from equation (7.95):

$$Q < \left(\frac{1+3}{2}\right)^{\frac{1}{2}} = 1.414$$

Hence, we select Q = 1.2, and from equation (7.82):

$$Q = 1.2 = \frac{1}{2}\sqrt{(G_1 / G_2)} \rightarrow \text{for } G_2 = 1, G_1 = 5.76 \tag{7.96}$$

Next, using equation (7.90), for normalized center frequency = 1

$$\sqrt{(G_1 G_2)} = C = 2.4 \tag{7.97}$$

For Q_e = 10, applying equation (7.93) with Q = 1.2:

$$Q_e = 10 = \frac{1.2}{2*1.2*1.2*(1-k)+1} \rightarrow k = 1.30555 \tag{7.98}$$

And then, using equation (7.94), we get:

$$h_e = 10 = 2g \times 1.30555 \times 1.2 \times 10 \rightarrow g = 0.31915, (1-g) = 0.68085 \tag{7.99}$$

Applying a frequency scaling of 20 krad/s and an impedance scaling of 10 kΩ, element values are:

$$R = 10\,\text{k}\Omega, R(k-1) = 3.0555\,\text{k}\Omega, \frac{R_1}{g} = 5.4395\,\text{k}\Omega, \frac{R_1}{1-g} = 2.55\,\text{k}\Omega, R_2 = 10\,\text{k}\Omega, C_1 = C_2 = 12\,\text{nF} \tag{7.100}$$

Using these element values, the circuit shown in Figure 7.25 is simulated and its response is shown in Figure 7.26. Its center frequency is 3.1623 kHz (19.877 krad/s); a bandwidth of 313.22 Hz gives Q_e = 10.096 and the mid-band gain h_e is 10.007.

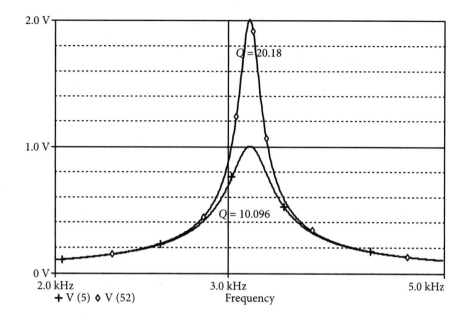

Figure 7.26 Simulated response of the band pass filter, with theoretical $Q = 10$ and $Q = 20$, using the enhanced Q circuit shown in Figure 7.25. Input to the filters was 0.1 V.

Not only is the obtained parameter very close to the design, the resistance spread is less than 4.

Using the same procedure for the second case with $Q_e = 20$ and $h_e = 20$, the obtained value of $k = 1.3264$ and $g = 0.3141$. With capacitor values remaining the same, resistance values are:

$$R = 10\,\mathrm{k\Omega},\ R(k-1) = 3.2638\,\mathrm{k\Omega},\ \frac{R_1}{g} = 5.5266\,\mathrm{k\Omega},\ \frac{R_1}{1-g} = 2.5312\,\mathrm{k\Omega},\ R_2 = 10\,\mathrm{k\Omega} \quad (7.101)$$

The simulated response is also shown in Figure 7.26. The center frequency is the same with realized pole-$Q = 20.18$ and mid-band gain = 20.

References

[7.1]   Dataforth Corporation. *Application Note AN112: Filtering in Signal Conditioning Module, SCMs.*

[7.2]   Bohn, E. V. 1963. *Transform Analysis of Linear Systems.* Massachusetts: Addison-Wesley.

[7.3]   Little, Wayne. 2010. *NCS 5650 PLC Filter Design*, Publication Note Order AND8466/D.

[7.4]   *NXP Semiconductor Design Reference* Manual, Document No. DRM170, May 2016.

[7.5]   Hamilton, T.A., and A. S. Sedra. 1972. 'Some New Configurations for Active Filters,' *IEEE Transactions.* CT-19 (1): 25–33.

[7.6]   Sallen, R. P. and E. L. Key. 1955. 'A Practical Method of Designing RC Active Filters'. *IRE Transactions.* CT-2: 74-85.

[7.7]   Chattaraj, N. and R. Ganguli. 2016. 'Active Filter Drivers for Piezo-actuators for Flapping-wing Micro Air Vehicles'. *International Journal of Micro Air Vehicles* 21–28. {http://us.sagepub-com/en-us/nam/open-access-at-sage}

[7.9]   Deliyannis, T. 1968. 'High-Q Factor Circuit with Reduced Sensitivity'. *Electronic Letters* 4 (26): 577-579.

[7.10]   Friend, J. J. 1970. 'A Single Amplifier Biquadratic Filter Section'. *Digest of Technical Papers.* p. 189.

Practice Problems

7-1   Design and test a second-order BP filter using a single OA and a bridged-T network in its feedback path. Its center frequency is to be 25 krad/s, mid-band gain should be of 20 dBs and pole-Q of 2.5.

7-2   (a) Derive two-port y parameters for the circuit shown in Figure P7.1.

(b) Design and test a second-order LP filter using a bridged-T RC network of Figure 7.3(a) in the feedback path of an OA, and the RC network of part (a). Cut-off frequency of the filter is 100 krad/s and the dc gain 0.5.

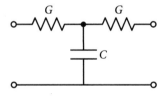

Figure P7.1

7-3   Repeat Problem 7-2 (b) for a maximally flat LP filter for which dc gain is to be unity, and the gain drops by 1 dB at 10 krad/s.

7-4   (a) Derive two-port parameters for the circuit shown in Figure P7.2.

(b) Design and test a second-order HP filter using the bridged-T RC network in the feedback path of an OA, and the RC network of part (a). The cut-off frequency of the filter is 50 krad/s and the high frequency gain is 0.5.

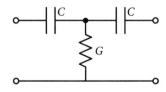

Figure P7.2

7-5   Design and test a second-order BP filter using a single OA and a bridged twin-T network in its feedback path. Its center frequency is to be 10 kHz, mid-band gain 5 and pole-Q 20.

7-6   Using multiple feedback with a single OA, design and test a second-order LP filter with a dc gain of 6 dBs and a cut-off frequency of 100 krad/s.

7-7   Repeat Problem 7-6 for a 1 dB drop occurring at 50 krad/s; the filter should have a dc gain of 12 dBs.

7-8   Using multiple feedback with a single OA, design and test a second-order HP filter with a high frequency gain of unity and an attenuation of 2 dB at a frequency of 60 krad/s.

7-9   Using multiple feedback with a single OA, redesign the BPF with specifications in Problem 7-1.

7-10   Realize and test the following transfer function using the configuration of Figure 7.13.

$$H(s) = \frac{4 \times 10^4 s}{s^2 + 2 \times 10^3 s + 10^8}$$

7-11   Design and test a BR filter for the following transfer function using the general differential input single OA configuration shown in Figure 7.13.

$$H(s) = \frac{s^2 + 1.44}{s^2 + 0.1s + 1.21}$$

De-normalize with a frequency scaling factor of 50 krad/s and a suitable impedance scaling factor to bring all components in a range compatible with integration.

7-12   Redesign and test $H(s)$ of Problem 7-11, if the constant term in the numerator is 0.25.

7-13   Design a second-order AP section using the configuration as in Problem 7-11 such that its phase shift becomes $-180°$ at 25 krad/s.

7-14   Design a Sallen and Key LP second-order filter with the following specifications:

Ripple width in the pass band = 1 dB, pass band edge frequency = 100 krad/s.

Test the circuit with OA having a very large bandwidth, as if OA is ideal.

7-15   Design and test a maximally flat Sallen and Key LP circuit for $f_o = 1.59$ kHz and $Q = 2$. Modify the circuit to get dc gain of unity. Find the incremental sensitivities of ω_0 and Q with respect to the passive elements.

7-16   (a) Replace the passive components in Figure 7.24 by the circuit shown in Figure P7.3, and show that it realized a second-order LP filter. Obtain the expressions of the parameters (a) considering OA as ideal and (b) as non-ideal.

7-17   Design and test a circuit obtained in problem 7-16 (a) with cut-off frequency of 20 krad/s and $Q = 2.5$ using OA 741. Justify the difference in the peak magnitude value with the value of Q.

7-18   (a) Design and test the D&F circuit of Figure 7.24 with the following specifications:

$\omega_0 = 10$ kHz, $Q = 10$ and mid-band gain = 5

(b) Find the incremental sensitivities of the parameters ω_0 and Q with respect to the passive elements.

7-19   Redesign and test the BP filter of Problem 7-18 using the Q-enhancement circuit of Figure 7.25. Compare the component spread with that in Problem 7-18.

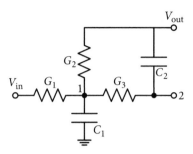

Figure P7.3

7-20   Replace the passive components in Figure 7.24 by the circuit shown in Figure P7.4, and show that it realizes a second-order HP filter. Obtain the expressions of the parameters (a) considering OA as ideal and (b) as non-ideal.

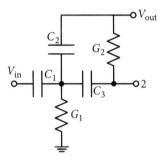

Figure P7.4

7-21   Design and test the circuit obtained in Problem 7-20 (a) with a cut-off frequency of 20 krad/s and $Q = 2.5$ using OA741. Justify the difference in the peak magnitude value with the value of Q.

7-22   Repeat Problem 7-19 using an enhanced-Q D&F circuit with increased value of Q as 20. Also find the sensitivities of the parameters with respect to the elements.

Chapter 8

Multi Amplifier Second-order Filter Sections

8.1 Introduction

In Chapter 7, the basics of realizing second- (or third-) order filter sections using only one active device (OA) were explained. Such circuits are capable of providing any arbitrary second-order function; they are also economical from the point of view of the use of active devices. However, depending on the filter specifications and the configuration chosen, the resulting circuit may not fulfil all the requirements like small number of passive components used and specific spread, sensitivity and variability. It is for this reason that many second-order filter sections use two, three or more OAs: multi amplifier biquads (MABs). Obviously, the intention is to overcome the mentioned limitations of the single amplifier biquad (SAB). In addition, a significant feature of multi amplifier biquadratic sections is their versatility in terms of providing more than one kind of response (like LP and BP) at the output terminals leading to general biquadratic structures.

Almost all MABs use two integrators in a loop, a technique known as the state-variable approach. Based on this technique, an important practical circuit known as the KHN (Kerwin-Huelsman–Newcomb) biquad can be assembled. The scheme explained in detail in Section 8.2 realizes three types of output responses. A direct modification of the scheme, known as Tow–Thomas biquad is studied in Section 8.3. The schemes, being interesting and useful, are further studied while employing active compensation to inverting or non-inverting integrators used in the loop. Active compensation leads to another well-known biquad, the Ackerberg–Mossberg filter, which is studied in Section 8.5. Many schemes have been implemented to utilize these structures and obtain other types of responses as explained in Section 8.6. Another scheme for obtaining a multi-response configuration using a generalized impedance convertor (GIC) is explained at the end of the chapter.

While designing a SAB, it was observed that a frequency-dependent finite gain of the OA results in a deviation in the performance parameters ω_o and pole-Q. To compensate for these deviations, biquads using *composite amplifiers* are also used, in which instead of using only a passive negative feedback, an active feedback network is used. These amplifiers increase the number of OAs used, making it a MAB, though the design itself remains a SAB type.

8.2   State Variable Multi Amplifier Biquad

There are a number of two, three or more amplifier biquad circuits. Almost all of these circuits are based on the *state variable* form of realization technique, first introduced by Kerwin, Huelsman and Newcomb (popularly known as the KHN biquad) [8.1]. The scheme, in its generality uses n integrators for an nth order transfer function, which are then appropriately connected the way integrators are connected in the analog computation method. To realize a second-order section, only two integrators are required along with a summer. Hence, in its basic form, a state variable biquad uses three amplifiers, with three outputs as shown in Figure 8.1. The configuration includes an integrator ($-a_1/s$) with feedback k_1 making it a lossy integrator, a lossless integrator ($-a_2/s$) with feed back factor $-k_2$ and two summers S_1 and S_2. Here, S_2 is used to convert a lossless integrator into a lossy one by combining the lossless integrator with feedback; the summer S_1 is used to complete the feedback loop for the integrators. As there is no element in between the two summers, the summers are generally combined. It is to be noted that both integrators are in inverting mode and use negative loop feedback to ensure stability. With the transfer function of the integrators as ($-a_1/s$) and ($-a_2/s$), the three available transfer functions of the section in Figure 8.1 are obtained from the following equations:

$$V_{o1} = kV_{in} + (-k_2)V_{o3} + k_1 V_{o2} \tag{8.1a}$$

$$V_{o2} = -\frac{a_1}{s}V_{o1} \text{ and } V_{o3} = -\frac{a_2}{s}V_{o2} \tag{8.1b}$$

The obtained transfer functions are:

$$\left(\frac{V_{o1}}{V_{in}}\right) = \frac{ks^2}{D(s)}, \left(\frac{V_{o2}}{V_{in}}\right) = -\frac{ka_1 s}{D(s)}, \left(\frac{V_{o3}}{V_{in}}\right) = \frac{ka_1 a_2}{D(s)} \tag{8.1c}$$

where $D(s) = s^2 + a_1 k_1 s + a_1 a_2 k_2$ \tag{8.1d}

The three outputs given in equations (8.1a)–(8.1c) are HP (high pass), BP (band pass), and LP (low pass), respectively, with their center frequency and pole-Q being decided by equation (8.1d).

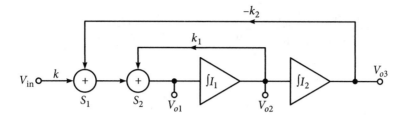

Figure 8.1   Basic two-integrator loop realizing a second-order filter in which summers S_1 and S_2 can be combined.

For the block diagram shown in Figure 8.1, three OAs are to be used, one each for the integrators and one for the combined summer. Another modification can be done by merging the summer in the lossy integrator. However, use of this modification requires the use of differential inputs for the summing of V_{in} and feedbacks for the lossy integrator and the loop. It also makes the adjustment of summing coefficients a bit difficult. To avoid this requirement, all the inputs at the summing integrator $\int I_1$ can become inverting signals. The filter will need a further modification by the addition of an inverter after $\int I_2$ as shown in Figure 8.2. It is to be noted that this configuration gives only BP and LP functions; HP is not available.

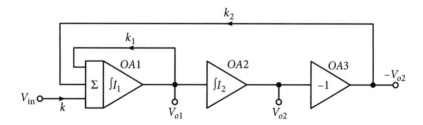

Figure 8.2   Two-integrator loop in modified form from Figure 8.1.

8.3   Tow–Thomas Biquad

A practical implementation of Figure 8.2, given in Figure 8.3, is known as the Tow and Thomas (TT) configuration [8.2,8.3]. Here, OA1 realizes a lossy integrator, OA2 realizes a lossless integrator and OA3 works as an inverter. Initially, considering OAs as ideal with infinite open-loop gain, the realized transfer function can be obtained through application of the Kirchhoff current law (KCL) at the inverting input terminals of OA1 and OA2.

$$(V_{in} - 0)/R_3 + (-V_{o2}/R_4) = (0 - V_{o1})/\{R_1/(1 + sC_1R_1)\} \tag{8.2}$$

$$(V_{o1} - 0)/R_2 = (0 - V_{o2})sC_2 \tag{8.3}$$

From equations (8.2)–(8.3), the following transfer functions are obtained:

$$\frac{V_{o1}}{V_{in}} = -\frac{\left(s/(C_1 R_1)\right)\left(R_1 / R_3\right)}{s^2 + s/(C_1 R_1) + 1/(C_1 C_2 R_2 R_4)} \rightarrow -\frac{h_{bp}\,(\omega_o / Q)s}{s^2 + (\omega_o / Q)s + \omega_o^2} \tag{8.4}$$

$$\frac{V_{o2}}{V_{in}} = \frac{1/C_1 C_2 R_2 R_3}{s^2 + s/(C_1 R_1) + 1/(C_1 C_2 R_2 R_4)} \rightarrow \frac{h_{lp}\omega_o^2}{s^2 + (\omega_o / Q)s + \omega_o^2} \tag{8.5}$$

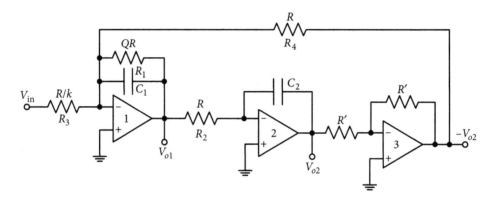

Figure 8.3   Tow–Thomas biquad. A practical implementation of Figure 8.2.

For the BP function of equation (8.4) and the LP function of equation (8.5), the important performance parameters ω_o, pole-Q, mid-band gain h_{bp} at ω_o, and gain at low frequencies or at dc for the LP case h_{lp} are as follows:

$$\omega_o = 1/(C_1 C_2 R_2 R_4)^{1/2} \tag{8.6}$$

$$Q = \sqrt{\left(\frac{C_1}{C_2}\right)}\frac{R_1}{\sqrt{(R_2 R_4)}} \tag{8.7}$$

$$h_{bp} = (R_1/R_3) \text{ and } h_{lp} = (R_4/R_3) \tag{8.8}$$

Six passive components have been used in the biquad and there are only three design parameters. These multiple components give some choice in the selection of component values and flexibility in obtaining the desired parameters. As R_3 appears only in the gain factor terms of equation (8.8), it is used for fixing the dc gain of the LPF (low pass filter) and the mid-band gain of the BPF (band pass filter). Since it is always preferable to have equal valued capacitors, $C_1 = C_2 = C$. Pole-Q can be easily controlled by R_1 even when R_3 and R_4 are equalized; the component spread for resistors normally equals Q. The two resistors R' used with the inverter can be of some arbitrary value close to any one of the other resistors used.

One attractive feature of the TT configuration is its passive sensitivities, which are at their theoretical minimum. If x is a passive component, sensitivity expressions are

$$S_x^{\omega_o} = -(1/2) \text{ and } \left|S_x^Q\right| \le 1. \tag{8.9}$$

Example 8.1: Design a BPF using a Tow–Thomas biquad having a center frequency of 3.4 kHz and pole-Q of 5. The filter's mid-band gain needs to be 20 dBs. Also discuss the obtained LP response.

Solution: With a normalized center frequency of 1 and $Q = 5$, $(\omega_o/Q) = 0.2$. Hence, we can begin with designing a normalized BP response as shown in equation (8.10).

$$H_{bpn}(S) = \frac{0.2h_{bp}S}{S^2 + 0.2S + 1} \tag{8.10}$$

The required mid-band gain being 20 dB or $h_{bp} = 10$, means $(R_1/R_3) = 10$ from equation (8.8). Selecting the normalized capacitors as $C_1 = C_2 = 1$, R_2 and R_4 will also be nominally 1 from equation (8.6). Hence, from equation (8.7), we get $R_1 = 5$ for $Q = 5$. De-normalizing the capacitors with $\omega = 2\pi \times 3.4$ krad/s and using an impedance scaling factor of 10^3 gives the following element values:

$$C_1 = C_2 = 0.0468 \ \mu F, \ R_1 = 5 \ k\Omega, \ R_3 = 0.5 \ k\Omega, \ R_2 = R_4 = R = 1 \ k\Omega \tag{8.11}$$

Figure 8.4 shows the PSpice simulated magnitude response of the BPF with a mid-band gain of 10.828, center frequency of 3.382 kHz. Its bandwidth = 0.62 kHz resulting in a pole-Q of 5.45. The LP response is also available as V_{o2}, as shown in the figure, having a dc gain of 2 and a peak at 3.352 kHz, with peak gain being 10.915.

In another set of responses, the desired center frequency of a BPF was 300 krad/s. $Q = 10$, and mid-band gain was unity. Using the same steps as in the first set, the obtained de-normalized element values are:

$$C_1 = C_2 = 0.01 \ \mu F, \ R_1 = R_3 = 3.333 \ k\Omega, \ R_2 = R_4 = 0.3333 \ k\Omega, \ R' = 5 \ k\Omega$$

The simulated BP response is also shown in Figure 8.4 with $\omega_o = 2\pi \ (44.027) = 276.74$ krad/s, bandwidth of 5.07 kHz which results in $Q = 8.67$ and a mid-band gain of 8.404. The corresponding LP response has a peak gain of 8.76 at 43.78 kHz (275.18 krad/s), and its voltage gain at dc is 1. While obtaining the transfer functions for the TT biquad in equations (8.4) and (8.5), OA were assumed to be ideal with an infinite open-loop gain. It is now well-known that the frequency-dependent gain creates deviations in performance. For example, in the first case, while using 741 type of OAs at $f_o = 3.4$ kHz and $Q = 5$, the simulated values show respective percent errors as 5.87 and 9; error in the mid-band gain was 8.28 percent. In the second case, at $f_o = 47.72$ kHz and $Q = 10$, the respective percent errors were 7.9, 13.3 and 16; a significant amount of error which increases with frequency. If suitable correction is not done, the performance will become impracticable. Hence, passive as well as active compensations are used. In this configuration as well as in many other cases, integrators have often been used, so before moving on to other biquads, it is suggested that we find the deviations caused in ω_o and pole-Q and the methods employed in integrators for the compensation of errors.

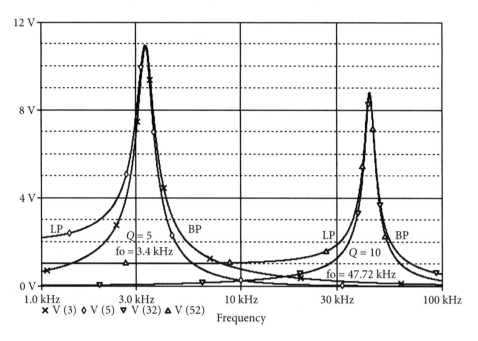

Figure 8.4 Band pass and low pass responses using a Tow–Thomas circuit at low and higher frequency levels for Example 8.1.

8.4 Active Compensation for Inverting Integrators

In Chapter 1, Section 1.8, integrators using OAs were briefly discussed. Figure 1.11(a) showed a lossy inverting integrator using an OA. The same integrator is now drawn in lossless form in Figure 8.5(a), without a feedback resistor. Using the single-pole roll-off model of equation (1.17) for the OA, the ideal integrator gets converted into a lossy integrator as expressed by equation (1.22) and rewritten as equation (8.12).

$$\frac{V_{\text{out}}}{V_{\text{in}}} = -\frac{1}{sCR}\frac{1}{1+\frac{1}{A}\left(1+\frac{1}{sCR}\right)} \tag{8.12}$$

With $A \cong (B/s)$, B being the gain bandwidth product, and with the condition that $(B \times CR) \gg 1$.

$$\frac{V_{\text{out}}}{V_{\text{in}}} = -\frac{1}{sCR(1+s/B)} = -\frac{1}{j\omega CR - \left(\omega^2 CR/B\right)} \text{ for } s = j\omega \tag{8.13}$$

Equation (8.13) gives the quality factor of the inverting integrator as a ratio of imaginary parts to real parts as:

$$Q_I = -\frac{\omega CR}{\left(\omega^2 CR / B\right)} = \left(-B / \omega\right) = -\left|A(j\omega)\right|$$

(8.14)

The integrator quality factor Q_1 is negative and depends on the magnitude of the gain. A larger value of $A(j\omega)$ is better, but as working frequency becomes large, Q_1 becomes smaller. This introduces a frequency-dependent loss in the ideal integrator and therefore, error is introduced in the parameters of such filters which employ the integrator. To overcome the problem, passive or active compensation is used in integrators.

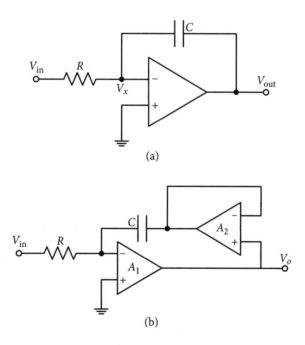

(a)

(b)

Figure 8.5   (a) Inverting integrator. (b) Active compensation for an inverting integrator.

Passive compensation has the advantage of using only one extra passive element; the compensation becomes near ideal. However, the compensation may not be accurate and will be variable with frequency.

An alternate solution is in the form of active compensation, shown in Figure 8.5(b), for which the transfer function is obtained as:

$$\frac{V_o}{V_{in}} = -\frac{1}{\dfrac{sCR}{(1+1/A_2)} + \dfrac{1+sCR}{A_1}}$$

(8.15)

Using simplified single-pole roll-off models for the OAs, $A_1 = (B_1/s)$ and $A_2 = (B_2/s)$. Applying truncation of Taylor's series expansion after the second-order term for $|A| \gg 1$, as we get:

$$\frac{1}{1+1/A} \cong 1 - \frac{1}{A} + \frac{1}{A^2} \tag{8.16}$$

For $s = j\omega$, equation (8.15) yields:

$$\frac{V_o}{V_{in}} = \frac{1}{\text{Re}(\omega) + j\text{Im}(\omega)} \tag{8.17}$$

$$\text{Re}(\omega) = \frac{\omega^2 CR}{B_2}\left(1 - \frac{B_2}{B_1} - \frac{\omega^2}{B_2^2}\right) \tag{8.18}$$

$$\text{Im}(\omega) = \omega CR\left(1 - \frac{1}{B_1 CR} - \frac{\omega^2}{B_1^2}\right) \tag{8.19}$$

For the matched OAs with $B_1 = B_2$, quality factor of the integrator $Q_I = \{\text{Im}(\omega)/\text{Re}(\omega)\}$ is obtained as follows:

$$Q_I = -(B_1/\omega)^3 = -|A(j\omega)|^3 \tag{8.20}$$

8.4.1 Compensation for a non-inverting integrator

In its simplest form, a non-inverting integration is obtained by cascading an inverting integrator and an inverter as shown in Figure 8.6(a). However, there are some other configurations for the non-inverting integrator as well. One such circuit is shown in Figure 8.6(b), for which the transfer function is obtained as:

$$\frac{V_o}{V_{in}} = -\frac{1}{sCR}\frac{1}{\left[\left\{\dfrac{1}{A_1(s)} + \dfrac{1}{sCRA_1(s)} + \dfrac{1}{1+\dfrac{2}{A_2(s)}}\right\}\right]} \tag{8.21}$$

For matched OAs, the quality of the integrator is simplified as:

$$Q_{NI} = +(B/\omega) = +|A(j\omega)| \tag{8.22}$$

It is significant to note that here Q_{NI} is positive, whereas for the inverting integrator, Q_I was negative. This opposite nature of change in quality factor has been found to be useful while designing filters.

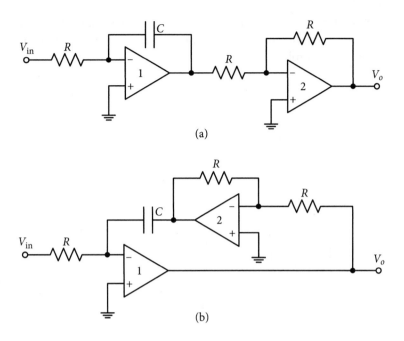

Figure 8.6   (a) A simple method to obtain a non-inverting integrator, and (b) an alternate non-inverting integrator with positive quality factor Q_{NI}.

It is not only the quality of integrators that change; the phase shift also changes. Therefore, compensating circuits of different configurations give different multi-amplifier biquadratic circuits with varied phase responses. Without going into a detailed study of the amount of affected performance due to the frequency-dependent finite open-loop gain of the OAs, let us discuss one prominent circuit which employs active compensation.

8.5   Ackerberg–Mossberg Biquad

Figure 8.7(a) shows the modification of the two-integrator loop of Figure 8.2 using the active compensation circuit of Figure 8.6(b), which was given by Ackerberg and Mossberg (AM) [8.4]. As the basic structure remains the same, the center frequency depends on the same RC product.

The circuit provides an LP response and a BP response. Assuming ideal OAs with $A \to \infty$, the obtained transfer functions are as following:

$$-\frac{V_{LP}}{V_{in}} = \frac{k / R^2 C^2}{s^2 + (s / CRQ) + 1 / R^2 C^2} \tag{8.23}$$

$$\frac{V_{BP}}{V_{in}} = -\frac{ks / CR}{s^2 + (s / CRQ) + 1 / R^2 C^2} \tag{8.24}$$

The quality factor, and the mid-band gain of the BPF = k are controlled by the resistance ratios shown in Figure 8.7(a). For the inverter OA3, equal valued resistances R^* are used; expression of the center frequency is as follows.

$$\omega_o = 1/RC \qquad\qquad (8.25)$$

Using the expressions of equation (8.22) and (8.21), deviations in the quality factor and pole frequency can be obtained in the AM structure.

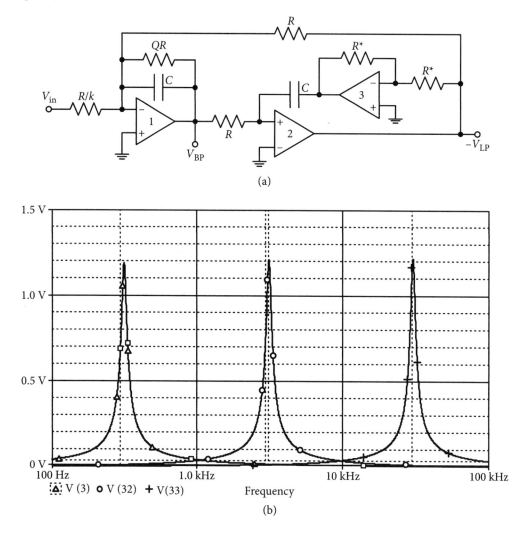

(a)

$\blacktriangle$ V (3) o V (32) + V(33)

Frequency

(b)

Figure 8.7   (a) Ackerberg–Mossberg biquadratic structure. (b) Band pass filter responses with $\omega_o = 2$ krad/s, 20 krad/s and 200 krad/s using the Ackerberg–Mossberg structure.

The coefficient matching technique described in Chapter 7 will be used to find element values, which will be shown in Example 8.2. Once the value of R or C is selected, the rest of the element values are evaluated from equation (8.25).

Example 8.2: Obtain a BP response using the AM structure of Figure 8.7(a) having a center frequency of 2 krad/s (3.1818 kHz), with $Q = 12$ and a mid-band gain of 12. Also obtain responses for center frequencies of 20 krad/s (3.181 kHz) and 200 krad/s (31.818 kHz) and comment on the obtained parameters.

Solution: Case (i) From equation (8.25), center frequency $\omega_o = 1/RC$, so we select $C = 0.1\ \mu$F, $R = 5\ k\Omega$. Therefore, resistance $QR = 60\ k\Omega$, and for mid-band gain of 12, $k = 1$, hence, (R/k) $= 5\ k\Omega$. R^* is arbitrarily selected as $10\ k\Omega$; a value in-between values of other resistors. The circuit is simulated through PSpice and its response is shown in Figure 8.7(b). The measured center frequency is 317.75 Hz (1997.2 rad/s), mid-band gain is 11.926 and with a bandwidth of 26.61 Hz, $Q = 11.94$.

Case (ii) With all resistors remaining the same, the required value of $C = 0.01\ \mu$F for $\omega_o = $ 20 krad/s. The simulated response is also shown in Figure 8.7(b). The measured parameters are $\omega_o = 19.793$ krad/s (3.149 kHz), mid-band gain = 11.738 and bandwidth of 264.3 Hz resulting in $Q = 11.92$.

Case (iii) For $\omega_o = 200$ krad/s, the required capacitor $C = 1$ nF. The simulated response is also shown in Figure 8.7(b). The measured parameters are $\omega_o = 192.15$ krad/s (30.57 kHz), mid-band gain = 12.139 and bandwidth of 2.464 kHz, which results in $Q = 12.36$. Table 8.1 shows the percent error in the simulated parameters for the three cases at different frequencies.

Table 8.1 Percent error in the parameters of filters realized using AM configuration for Example 8.2

	ω_o	Q	mid-band gain
Case (i), $\omega_o = 2$ krad/s	0.135	0.5	0.616
Case (ii), $\omega_o = 20$krad/s	0.375	0.66	0.516
Case (iii), $\omega_o = 200$krad/s	3.92	−3.0	−1.158

A comparison of percent errors in the parameters of the filters designed using Tow–Thomas and AM configurations show a marked improvement in the latter case, especially at higher frequencies. This confirms the utility of the active compensation employed in the AM circuitry.

8.6   Multi-output Biquad Using Summing Amplifier

As mentioned earlier, one of the advantages of using more than one amplifier in a circuit is its versatility in obtaining more than one kind of response simultaneously. For example, in one method, using a summing amplifier, a circuit which has already generated LP and BP responses, can also generate other kind of responses by adding the specific input. The process is illustrated by using structure of Figure 8.2 which shows a two-integrator second-order generating circuit. When the circuit's two outputs are summed with an input using an additional summing amplifier as shown in Figure 8.8, three responses are simultaneously available as V_{o1}, V_{o2} and V_{out}. Employing the relations of equation (8.1), the obtained transfer function is as follows:

$$\frac{V_{out}}{V_{in}} = -\frac{\alpha s^2 + a_1 s\left(\alpha k_1 - \beta k\right) + a_1 a_2 \left(\alpha k_2 - \gamma k\right)}{s^2 + a_1 k_1 s + a_1 a_2 k_2} \tag{8.26}$$

Selection of summing coefficients α, β and γ decides the type of available response at the output V_{out}.

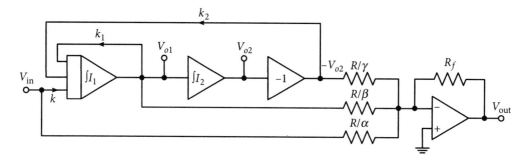

Figure 8.8 Generation of a general biquad using a two-integrator network and a summer.

If the AM circuit shown in Figure 8.7(a) is used to generate a general biquadratic circuit with a summing amplifier and inputs coming from the original input V_{in}, $+V_{LP}$ output at the output terminal of OA3, and V_{BP}, the arrangement will look as shown in Figure 8.9. The output voltage shall be as follows using equations (8.23) and (8.24); obviously any other two-integrator loop circuit can also be used.

$$V_{out} = \alpha V_{in} + \beta V_{BP} + \gamma V_{LP} = \alpha V_{in} - \beta \frac{k\omega_o}{D(s)} s V_{in} - \gamma \frac{k\omega_o^2}{D(s)} V_{in}$$

$$\frac{V_{out}}{V_{in}} = \frac{\alpha s^2 + \left(\alpha - kQ\beta\right)\left(\omega_o / Q\right)s + \left(\alpha - \gamma k\right)\omega_o^2}{s^2 + \left(\omega_o / Q\right)s + \omega_o^2} \tag{8.27}$$

From equations (8.23)–(8.24) and from Figure 8.9, the concerned relations are as follows:

$$\omega_o^2 = \frac{1}{RC}, \quad \alpha = \frac{R_f}{R_\alpha}, \quad \beta = \frac{R_f}{R_\beta} \text{ and } \gamma = \frac{R_f}{R_\gamma} \tag{8.28}$$

Selection of coefficients α, β and γ will decide the numerator terms; this will then decide the type of response at the output. The following examples will illustrate the generation of responses other than BP and LP using the coefficient matching technique.

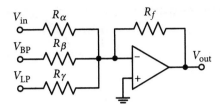

Figure 8.9   Summing amplifier used for obtaining a general biquad.

Example 8.3: Design a notch filter employing an AM circuit and a summing amplifier which should have a notch at 20 krad/s and $Q = 5$.

Solution: In equation (8.27), we select $\alpha = 1$. Hence, for normalized $\omega_o = 1$, $(1 - kQ\beta)$ has to be zero, which gives the coefficients as:

$$(1 - 5k\beta) = 0 \rightarrow \beta = 0.2 \text{ if } k = 1 \tag{8.29}$$

$$1 - \gamma \times 1 = -1 \rightarrow \gamma = 0 \tag{8.30}$$

The selected value of $C = 0.01 \ \mu F$ gives the value of R from equation (8.28) as 5 kΩ for $\omega_o =$ 20 krad/s. Hence, $R_1 = (R/k) = 5$ kΩ, $R_2 = QR = 25$ kΩ, $R_3 = R^* = R_1 = R_\alpha = R_f = R_4 = 5$ kΩ, $R_\beta = 25$ kΩ and R_γ is open as shown in Figure 8.10(a).

Figure 8.10(b) shows the PSpice simulated response of the notch filter having a notch at 3.168 kHz (19.913 krad/s). The cut off frequencies of the filter are 3.5 kHz and 2.866 kHz, resulting a bandwidth of 634 Hz and $Q = 4.996$. The input voltage being 100 mV, the voltage level at the notch drops to 0.98 mV or an attenuation of 40.17 dBs.

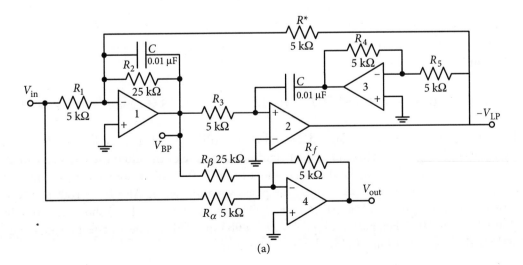

(a)

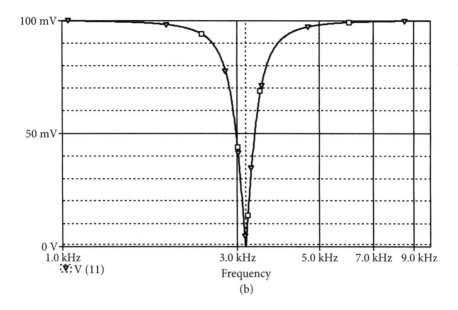

Figure 8.10   (a) Second-order notch filter circuit using an Ackerberg–Mossberg circuit and a summing amplifier. (b) Magnitude response of the notch filter of Figure 8.10(a).

Example 8.4: Design a second-order APF (all pass filter) which has a pole-$Q = 2$ and a $0°$ phase delay at 2 krad/s using an AM circuit and a summing amplifier.

Solution: The first consideration in getting an APF is to fix $\alpha = 1$; then, using equation (8.27):

$$\omega_o^2 = (1 - \gamma k)\omega_o^2 \rightarrow \gamma k = 0 \text{ and } \gamma \text{ is taken as } 0 \tag{8.31}$$

For $\omega_o = 1$, $Q = 2$, and assuming $k = 1$; $(1 - kQ\beta\omega_o) = -1 \rightarrow \beta = 1$ (8.32)

Critical frequency being 2 krad/s, the selected value of capacitor $C = 0.1$ μF, which requires $R = 5$ kΩ. Having obtained the coefficients in equations (8.31)–(8.32), the remaining element values are:

$$C_2 = 0.1 \text{ μF}, C_3 = 0, R_1 = 5 \text{ kΩ}, R_2 = 10 \text{ kΩ}, R_4 = R_5 = R^* = 5 \text{ kΩ and } R_f = R_\alpha = R_\beta = 5 \text{ kΩ}.$$

The circuit diagram of the designed second-order APF with element values is shown in Figure 8.11(a). Figure 8.11(b) shows the magnitude response of the filter: there is a very small variation in magnitude; a dip of 0.13 mV and a rise of 0.042 mV from an average constant value of 100 mV. The figure also shows the phase variation in the APF from $180°$ to $-180°$ with a zero-degree phase shift at 317.68 Hz (1996.8 rad/s). Figure 8.11(c) shows variation in group delay; at 310.6 Hz, peak group delay $D = 4.065$ ms. This is a near perfect response due to the active compensation employed in the AM circuit.

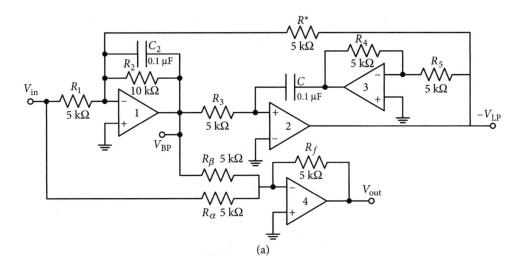

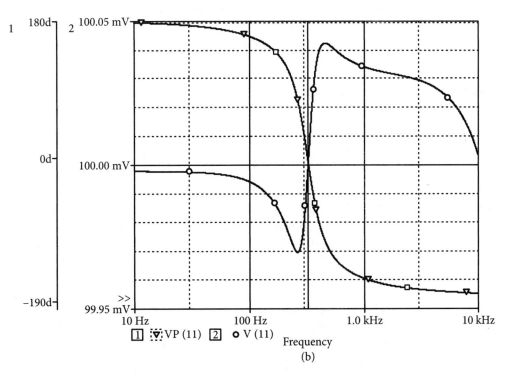

(a)

(b)

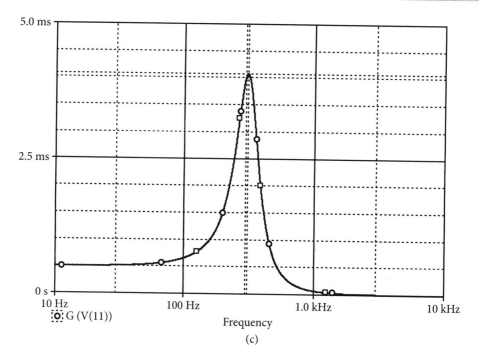

Figure 8.11 (a) Second-order all pass filter circuit using the Ackerberg–Mossberg circuit and a summing amplifier. (b) Magnitude and phase responses of the all pass filter shown in Figure 8.11(a). (c) Group delay response of the all pass filter shown in Figure 8.11(a).

8.6.1   Biquad using modified summation method

In a slightly modified but efficient approach, the additional summing amplifier can be avoided by the application of a weighted input signal at the virtual ground terminals of the two integrators. The advantage of connecting V_{in} in such a way is that the poles given by equation (8.26) are not affected. If this scheme is applied to the AM circuit shown in Figure 8.7(a), the resulting biquad becomes as shown in Figure 8.12. With the OAs considered ideal, the transfer function of the circuit is given as:

$$\frac{V_{out}}{V_{in}} = -\frac{\alpha s^2 + \{s(k-\beta)/CR\} + \gamma/C^2R^2}{s^2 + \{s/(CRQ)\} + 1/C^2R^2} \tag{8.33}$$

All types of responses are easily obtainable by selecting different weighting coefficients:

For LP, $k = \beta = \alpha = 0$, and for HP, $k = \beta = \gamma = 0$ (8.34a-b)

For BP, $\gamma = \beta = \alpha = 0$, and for notch, $k = \beta = 0$, $\alpha < \gamma$ (8.34c-d)

For HP notch, $k = \beta = 0$, $\alpha > \gamma$, and for LP notch, $k = \beta = 0$, $\alpha < \gamma$ (8.34e-f)

Lastly for the AP, $\alpha = \gamma = k = 1$, $\beta = (1 + 1/Q)$ (8.34g)

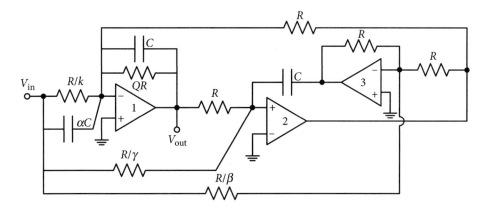

Figure 8.12 Generation of a general biquadratic circuit through the application of a weighted input at the virtual ground terminals of the operational amplifiers using an AM biquad.

The general biquad has many desirable characteristics apart from providing all types of second-order responses. Active and passive sensitivities are found to be low and since pole locations are not changed while connecting the input at the virtual ground terminals, sensitivities remain the same as that for the AM circuit. Center frequency ω_o and pole-Q are independently tuneable with the help of the input resistance connected to the second integrator and the resistor QR, respectively. Component spread is also small as will become obvious from the following examples. Additionally, when parameters are set for deciding the location of zeros or the type of response, it does not affect the pole location.

Example 8.5: Design an HPF having a 3dB cut-off frequency of 20 krad/s and $Q = 2$ using the generalized biquad shown in Figure 8.12.

Solution: From equation (8.33), as $\omega_o = (1/RC) = 20$ krad/s for the general biquadratic filter circuit shown in Figure 8.12, selection of $C = 0.01$ µF gives $R = 5$ kΩ. To find the other component values, we use the condition $k = \beta = \gamma = 0$ from equation (8.34b); it gives, $R_1 = (R/k) = \infty = R_\beta = (R/\beta) = R_\gamma = (R/\gamma)$, $R_2 = QR = 10$ kΩ and $R_3 = R_4 = 5$ kΩ. Selection of R_3 and R_4 is intentionally done to keep as many resistances equal as possible; a good choice in integrated circuits. Capacitor $C_2 = C = 0.01$ µF and $C_3 = \alpha C = 0.01$ µF for $\alpha = 1$. The resulting circuit is shown in Figure 8.13(a) and the simulated response in Figure 8.13(b).

A peak occurs at 3.386 kHz where its voltage gain is 2.073 against the ideal value of 2.0. Voltage gain at high frequencies is unity. The evaluated value of the simulated $\omega_o = 3.386(1 - 1/2 \times 2^2)^{1/2} = 3.167$ kHz = 19.908 krad/s.

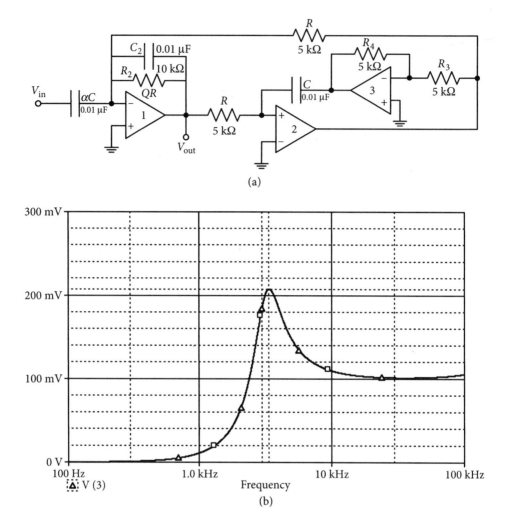

(a)

(b)

Figure 8.13 (a) High pass filter using the Ackerberg–Mossberg biquad and applying the modified summation method for Example 8.5. (b) Response of the HPF shown in Figure 8.13(a).

Example 8.6: Design a HP notch filter having a pole frequency of 2 krad/s, a zero at 1 krad/s and $Q = 2$ using the modified summation method.

Solution: From equation (8.33) for $\omega_o = 2\text{krad/s}$, the selected capacitor $C = 0.01\ \mu\text{F}$ gives $R = 5\ \text{k}\Omega$. For a zero at 1 krad/s, $\gamma^{1/2} = (1/2)$, hence, $(R/\gamma) = (R/0.25) = 20\ \text{k}\Omega$. Since α has to be more than γ for a HP notch, it is taken as 1; therefore, $\alpha C = C = 0.01\ \mu\text{F}$, and $C_2 = C$. With $Q = 2$, other elements will be the same as in Example 8.5, with $k = \beta = 0$. The circuit with element values is shown in Figure 8.14(a) and its PSpice simulated response is shown in Figure 8.14(b).

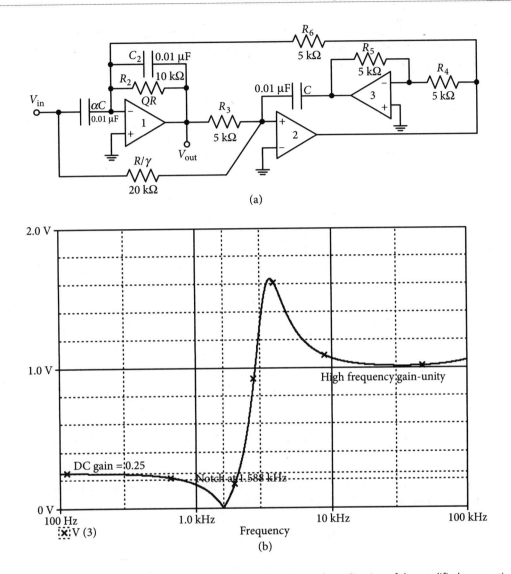

Figure 8.14 (a) Generation of a high pass notch circuit through application of the modified summation method using the Ackerberg–Mossberg biquad. (b) Magnitude response of the high pass notch filter shown in Figure 8.14(a).

The notch occurs at 1.588 kHz with a low frequency gain of 0.25; the peak occurs at 3.565 kHz where due to $Q = 2$, voltage gain is 1.636. At high frequencies, the output voltage levels at unity gain; this verifies all the specifications.

8.6.2   Active noise control: application example

Feedback control systems are used in active noise and vibration control. Such systems can be realized using either digital signal processing or analog signal methodology. Each process has advantages and limitations. There are certain applications where analog feedback control systems are preferred, such as active control of earmuffs [8.5] and similar applications, where the process delay must be small.

In practice, analog control systems include a filter or a bank of filters [8.6]. Parameters of the analog filter are usually adjusted with variable resistance and/or changing capacitor values employing switching arrangements. The switching arrangement is normally done manually. In the application case presented here, microprocessor driven, real-time control of the parameters of the filter bank has already been developed. This circuit discussed here helps in noise control. Without going into the development process, the basic arrangement of the biquadratic filter bank system is shown in Figure 8.15 in block form.

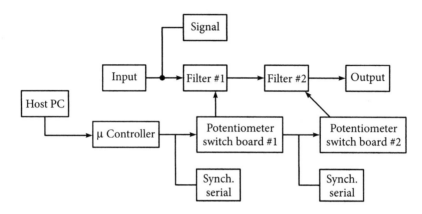

Figure 8.15 The block diagram of a system for active noise control {with permission from M. Antila et al.} [8.6].

As a specific example, equation (8.35) shows a transfer function, which was obtained as a useful transfer function for a feedback control system.

$$H(s) = \frac{1.01s^2 - 70s + 2226065}{s^2 + 272s + 319092} \tag{8.35}$$

To implement the aforementioned transfer function, a general biquadratic circuit, as shown in Figure 8.12, applying the modified summation method was used. Comparing equation (8.35) with equation (8.33) gives:

$$\omega_o^2 = 319092, \alpha = 1.01, \gamma = (2226065/319092) = 6.976, \ Q = (\omega_o/272) = 2.076,$$
and if k is selected as 0.5, $\beta = (70/560.88)+0.5 = 0.6239$ \hfill (8.36)

Applying an impedance scaling factor of 10 kΩ and a frequency scaling factor of $\omega_o = 564.88$ rad/s, element values for the circuit in Figure 8.12 are obtained as:

$R = 10$ kΩ, $R/k = 20$ kΩ, $QR = 20.76$ kΩ, $R/\beta = 16.028$ kΩ, $R/\gamma = 1.433$ kΩ $C = 0.177$ μF and $\alpha C = 0.1788$ μF

The simulated response of the biquadratic filter section is shown in Figure 8.16. Its low frequency gain is 16.87 dB, the high frequency gain is 0.158 dB, the notch frequency is 234.4 Hz and the peak gain of 22.3 dB occurs at 512.4 rad/s (81.5 Hz).

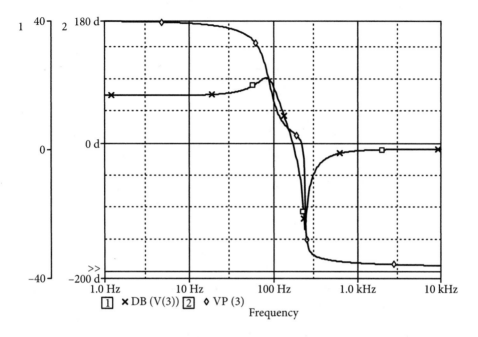

Figure 8.16 Simulated response of a biquadratic filter modeling noise control (equation (8.35)).

8.7 Generalized Impedance Converter Based Biquad

A significant alternative to obtain a multi amplifier biquad is a technique which is based on the use of a generalized impedance converter (GIC) [8.7]. In its basic form, this improvised biquad starts with a passive structure and its grounded inductor is replaced with a GIC. Presently, without going into a detailed description of a GIC, observe a second-order passive BPF structure in Figure 8.17(a) and its conversion to a second-order active filter circuit in Figure 8.17(b). GIC is shown in a dotted rectangle replacing the inductor in Figure 8.17(a). The transfer function of the passive BP filter is:

$$\frac{V_1}{V_{\text{in}}} = \frac{(1/RC)s}{s^2 + (1/RC)s + 1/(LC)} \tag{8.37}$$

Here center frequency ω_o and pole-Q (Q_o) are:

$$\omega_o = (1/LC)^{1/2} \text{ and } Q_o = R(C/L)^{1/2} \tag{8.38}$$

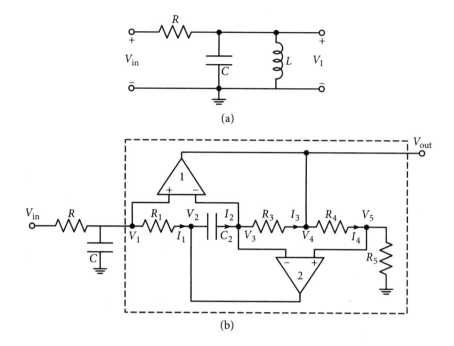

Figure 8.17 (a) A second-order passive band pass passive filter; (b) inductor in part (a) replaced with a generalized impedance convertor shown inside the dotted line.

For the circuit in Figure 8.17(b), with the OAs assumed to be ideal and A→∞, $V_1 = V_3 = V_5$, the following are the current–voltage relations:

$$I_4 = (V_5/R_5), \ V_4 = V_5 + I_4 R_4 = V_5(1 + R_4/R_5) \tag{8.39}$$

$$I_3 = I_2 \rightarrow (V_3 - V_4)/R_3 = (V_5 - V_4)/R_3 = -V_5 R_4/(R_3 R_5) = (V_2 - V_3)sC_2 \tag{8.40}$$

$$I_1 = (V_1 - V_2)/R_1 \ \text{or} \ (V_1/I_1) = s(C_2 R_1 R_3 R_5)/R_4 \tag{8.41}$$

Equation (8.41) confirms that the circuit within the dotted rectangle in Figure 8.17(a) realizes an inductance with its expression given as:

$$L = C_2 R_1 R_3 R_5/R_4 \tag{8.42}$$

Therefore, substituting the expression of L from equation (8.42) into equation (8.37) and V_4 from equation (8.39), the transfer function of the active second-order section of Figure 8.17(b) is obtained as:

$$\frac{V_{out}}{V_{in}} = \frac{(1/RC)(1 + R_4/R_5)s}{s^2 + (1/RC)s + (R_4/CC_2 R_1 R_3 R_5)} \tag{8.43}$$

If output voltage is V_1, transfer function is same as in equation (8.43) but numerator will not have the term $(1 + R_4/R_5)$. So the mid-band gain will be unity in this case. Here, ω_o and Q_o are given by the following relations:

$$\omega_o = (R_4/CC_2R_1R_3R_5)^{1/2} \tag{8.44}$$

$$Q_o = R(C/C_2)^{1/2} (R_4/R_1R_3R_5)^{1/2} \tag{8.45}$$

For ω_o and Q_o, the passive sensitivities with respect to the elements x are calculated as:

$$\left|S_x^{\omega_o}\right| = \frac{1}{2}, \left|S_x^{Q_o}\right| = \frac{1}{2} \text{ and } \left|S_R^{Q_o}\right| = 1 \tag{8.46}$$

Clearly, all passive sensitivities are very low; hence, the proposed circuit employing a GIC enjoys excellent sensitivity figures, provided its active sensitivities are also low. A detailed discussion on GIC sensitivities will be taken up later with the description of the GIC structure. In addition to the very low sensitivities, the GIC based active second-order filter structure has the following advantages, which makes the circuit very attractive.

(a) Component spread is small; most of the passive elements can be made equal.

(b) Parameters ω_o and Q_o and mid-band gain of the BP can be independently tuned.

(c) The circuit is suitable for cascading as it has *infinite* input impedance at the frequency ω_o.

Example 8.7: Design a BPF for a center frequency of 20 krad/s and $Q = 5$ using the GIC based configuration of inductance shown in Figure 8.17(b).

Solution: Selecting equal valued capacitors C and C_2 as 0.01 μF, equation (8.44) gives $R_1 = R_3 = R_4 = R_5 = 5$ kΩ. For $Q = 5$, equation (8.45) gives $R = 25$ kΩ. Using these element values, the filter structure is shown in Figure 8.18(a). The response is simulated using PSpice and shown in Figure 8.18(b).

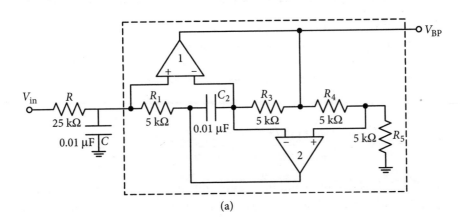

(a)

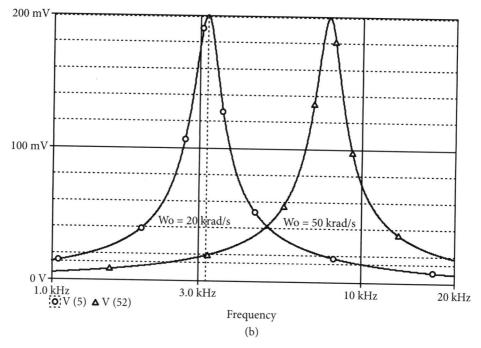

Figure 8.18 (a) Second-order active band pass filter with inductor replaced using a generalized impedance convertor for Example 8.7; (b) responses of the band pass filter in part (a).

The obtained BP response realizes a center frequency as 3.177 kHz (19.969 krad/s) with a voltage gain of 1.991. The bandwidth of the filter is 635 Hz resulting in $Q = 5.003$.

Another BP response was obtained for a higher center frequency of 50 krad/s. The circuit requires capacitors C and C_2 each of 0.004 µF, with all resistors remaining the same, for same values of Q. The simulated response is also shown in Figure 8.18(b) with a center frequency of 7.821 kHz (491.6 krad/s) and a voltage gain of 1.984. Bandwidth of the filter is 1.563 kHz, resulting in the value of Q as 5.004.

8.7.1 General biquad using generalized impedance converter

In addition to the discussed BPFs in the previous section, other types of transfer functions can be realized using the well-known process of *lifting grounded elements completely or partially from the ground* while using GICs. Inclusion of resistance R_7 and splitting of input resistance R and capacitance C are used to provide feedback. Such a configuration is shown in Figure 8.19, with its transfer function obtained as:

$$\frac{V_{out}}{V_{in}} = \frac{\left[\alpha H - \gamma(H-1)\right]s^2 + \left[\beta H - \gamma(H-1)\right]s/QRC + \left(\gamma/R^2C^2\right)}{s^2 + (1/QRC)s + (H-1)/R^2C^2}$$ (8.47)

In equation (8.47), H is the mid-band filter gain, and an appropriate choice of the weighting coefficients, α, β and γ, determines the type of obtained filter response. At this stage, it may be noted that getting a pure LP response is very difficult in this scheme; it is advised to use some other configuration of a GIC.

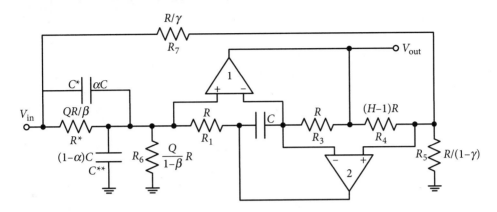

Figure 8.19 Biquadratic circuit obtained using the technique of lifting some elements from the ground partially using GIC circuit of Figure 8.17(b).

Example 8.8: Design a notch filter which should have a notch at 20 krad/s and $Q = 4$ using a GIC based general biquad.

Solution: A convenient choice of a notch filter from equation (8.47) is to assume $H = 2$ and $\gamma = 1$, with which $\omega_o = (1/RC)$. For notch frequency $\omega_o = 20$ krad/s, an easy choice of components is $C_2 = 0.01$ μF and $R = 5$ kΩ. With $H = 2$ and $\gamma = 1$, from equation (4.49), we need to have the following:

$$\alpha H - \gamma(H-1) = 1 \rightarrow \alpha = 1 \text{ and } \beta H - \gamma(H-1) = 0 \rightarrow \beta = 0.5 \qquad (8.48)$$

Application of equation (8.48) gives the following element values:

$$R^* = \frac{QR}{\beta} = 40\,\text{k}\Omega, \; R_6 = \frac{Q}{(1-\beta)}R = 40\,\text{k}\Omega, \; R_7 = \frac{R}{\gamma} = 5\,\text{k}\Omega, \; R_5 = \frac{R}{1-\gamma} = \text{open}$$

$$R_1 = R_3 = R_4 = 5 \text{ k}\Omega, \; C^* = \alpha C = 0.01 \text{ μF and } C^{**} = (1-\alpha)C = 0$$

Figure 8.20(a) shows the circuit structure of the notch filter with element values and Figure 8.20(b) shows the PSpice simulated magnitude response. For the simulated notch which occurs at 3.169 kHz (19.91 krad/s), the output voltage level is 2.396 mV for an input voltage of 100 mV; an attenuation of 32.4 dBs. Its 3 dB bandwidth is 799.8 Hz, resulting in $Q = 3.96$.

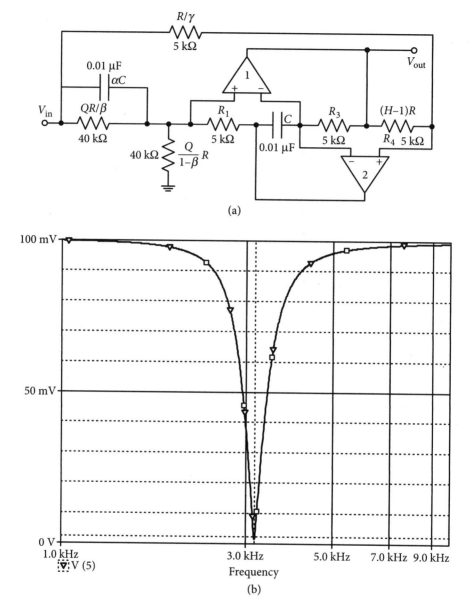

(a)

(b)

Figure 8.20 (a) Notch filter circuit obtained from the GIC circuit of Figure 8.17(b), while lifting some elements from ground partially. (b) Notch filter response for the circuit in Figure 8.20(a).

Example 8.9: Design an APF which has a phase shift of 180° at 40 krad/s and $Q = 2$ using GIC.

Solution: For the multifunctional configuration of Figure 8.19, to give an AP response, we need to have the following conditions from equation (8.47):

$$\alpha H - \gamma(H-1) = 1, H-1 = 1 \rightarrow H = 2 \text{ and } \beta H - \gamma(H-1) = -1 \qquad (8.49)$$

In equation (8.49), if $\gamma = 1$, $\beta = 0$ and $\alpha = 1$. For center frequency of 40 krad/s, if C_2 is selected as 5 nF, R = 5 kΩ, and with $Q = 2$:

$$R^* = \frac{QR}{\beta} = \infty, R_6 = \frac{QR}{1-\beta} = 1 \text{ k}\Omega, R_4 = (H-1)R = 5\text{k}\Omega, R_5 = \frac{R}{1-\gamma} = \infty, R_7 = \frac{R}{\gamma} = 5\text{k}\Omega,$$

$$R_1 = R_3 = 5\text{k}\Omega \text{ and } C^* = C_2 = 5\text{nF}, C^{**} = (1-\alpha)C_2 = 0 \qquad (8.50)$$

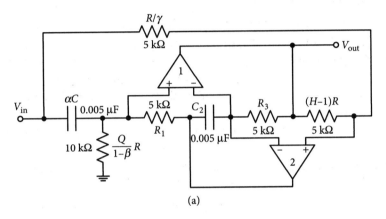

(a)

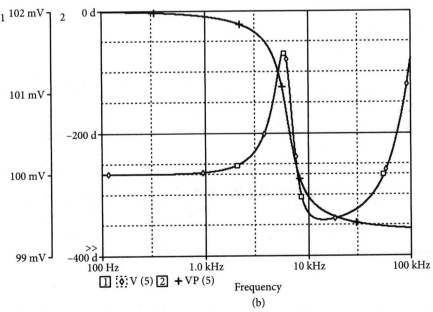

(b)

Figure 8.21 (a) All pass circuit obtained using GIC circuit of Figure 8.17(b). (b) Magnitude and phase responses of the all pass shown in Figure 8.21(a).

Figure 8.21(a) shows the circuit structure of the APF with element values and Figure 8.21(b) shows the PSpice simulated magnitude response. Gain is almost unity; there is a bit of rise having maximum gain of 1.0148 and a small drop for a minimum gain of 0.9943. Figure 8.21(b) also shows the phase response of the APF, having a phase shift of $-180°$ at 6.3 kHz (39.6 krad/s).

References

[8.1]   Kerwin, W. J., L. P. Huelsman, and R. W. Newcomb. 1967. 'State-Variable Synthesis for Insensitive Integrated Transfer Functions,' *IEEE Journal of Solid-State Circuits* SC-2: 87-92.

[8.2]   Tow, J. 1969. 'A Step-by-Step Active Filter Design,' *IEEE Spectrum* 6: 64-8.

[8.3]   Thomas, L. C. 1971. 'The Biquad: Part I-Some Practical Design Considerations; Part II-A Multipurpose Active Filtering System,' *IEEE Transactions on. Circuit Theory* CT-18: 350-7.

[8.4]   Ackerberg, D., and K. Mossberg. 1974. 'A Versatile Active RC Building Block with Inherent Compensation for the Finite Bandwidth of the Amplifier,' *IEEE Transactions on Circuits and Systems* 21: 758.4.

[8.5]   Hall, D. L., and B. Flatua. 1998. 'On Analog Feedback Control for Magneto-strictive Transducer Linearization,' *Journal of Sound and Vibration* 211: 481-94.

[8.6]   Antila, M., K. Hakanen, and J. Kataja. 2002. 'Microcontroller-Driven Analogue Filter for Active Noise Control.' Southampton, UK: *ISVR, ACTIVE.*

[8.7]   Antoniou, A. 1967. 'Gyrators Using Operational Amplifiers,' *Electron Letters* 3: 350–2.

Practice Problems

8-1   Figure P8.1 shows the circuit diagram of a KHN biquad which was shown in Figure 8.1 as a block diagram form. Obtain all the three voltage ratio transfers functions available from the circuit. What kinds of responses are available?

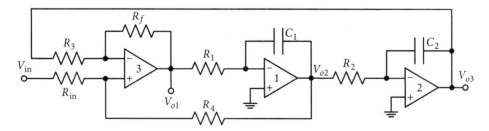

Figure P8.1

8-2   Design and test a second-order BP filter using a KHN circuit having center frequency of 3.4 kHz, $Q_0 = 3$ and mid-band gain of 5.

8-3   Design and test a HP filter using KHN circuit for which attenuation falls by 2 dB at 10 krad/s.

8-4   Design a KHN based second-order LP filter for cut-off frequency of 10 krad/s and $Q = 5$. Use equal value capacitors C_1 and C_2 and equal valued resistors R_1 and R_2. What is the gain of the filter at dc?

8-5   Design and test the second-order Tow–Thomas (TT) BP circuit of Figure 8.3 for the following specifications.

$\omega_o = 20$ krad/s, $Q_o = 10$ and peak gain of 10.

8-6   Derive the transfer function (V_{out}/V_{in}) for the circuit shown in Figure P8.2. Design a filter, using equal value capacitors for the following transfer function:

$$\frac{V_{out}}{V_{in}} = \frac{s^2 + 0.25}{s^2 + 0.2s + 1.21}$$

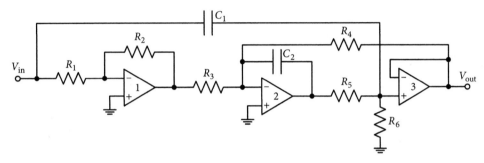

Figure P8.2

8-7   (a)   Derive expressions for the three transfer functions for the Tow–Thomas structure shown in Figure P8.3, with OAs considered ideal.

(b)   Find the incremental sensitivity of the parameters ω_o and Q_o with respect to the passive elements.

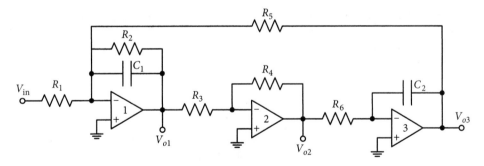

Figure P8.3

8-8   If the OAs are represented by the model $A(s) = B/s$, even lossless integrators become lossy. Figure P8.4 shows the representation of the two-integrator loop biquad in such a case. Show that the normalized transfer function with $\omega_o = 1$ and $k = H(\omega_o/Q)$, where H is the mid-band gain:

$$\frac{V_{out}}{V_{in}} = \frac{\dfrac{H}{Q\tau_1\tau_2}(s\tau_2 + q_2)}{s^2 + \left(\dfrac{q_1}{\tau_1} + \dfrac{q_2}{\tau_2} + \dfrac{1}{Q\tau_1}\right)s + \left(1 + q_1 q_2 + \dfrac{q_2}{Q}\right)\dfrac{1}{\tau_1\tau_2}}$$

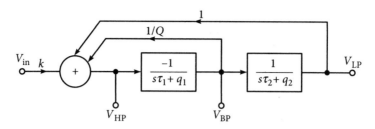

Figure P8.4

8-9   Compare the transfer function in Problem 8-8 with a standard form of BP transfer function and (a) show that the resultant equivalent parameters will be:

$$\omega_{oe} = \left\{ \left(1 + q_1 q_2 + \frac{q_2}{Q} \right) \frac{1}{\tau_1 \tau_2} \right\}^{1/2}$$

$$Q_{oe} = Q \frac{\omega_{oe} \tau_1}{1 + Q(q_1 + q_2 \tau_1 / \tau_2)}, \quad H_e = H \frac{1}{1 + Q(q_1 + q_2 \tau_1 / \tau_2)}$$

(b)   What is the magnitude of a parasitic zero in the transfer function?

8-10   Design a TT BP filter shown in Figure 8.3 for $\omega_o = 60$ krad/s and $Q_o = 15$ with mid-band gain H being unity. Estimate the deviation in the parameters of the filter if OAs have bandwidth $= 0.5 \times 10^6$ rad/s. Verify the results using PSpice simulation.

8-11   Repeat Problem 8-10 for $\omega_o = 120$ krad/s, $Q = 20$ and mid band gain $H = 1$.

8-12   Repeat Problems 8-10 and 8-11 using the Ackerberg–Mossberg (AM) circuit shown in Figure 8.7(a).

8-13   Design LP filters using the AM circuit shown in Figure 8.7(a) for the following specifications using 741 OAs.

(a) $\omega_o = 3.4 \times 2\pi$ krad/s, $Q = 1$ and gain at dc $h_{lp} = 5$.

(b) $\omega_o = 60$ krad/s, $Q = 2.5$ and gain at dc $h_{lp} = 2$.

(c) $\omega_o = 200$ krad/s, $Q = \sqrt{2}$ and gain at dc $h_{lp} = 3$.

8-14   Using 741 OAs, design BP filters using the AM circuit shown in Figure 8.7(a) for the following specifications.

(a) $\omega_o = 1.59 \times 2\pi$ krad/s, $Q = 10$ and mid-band gain $h_{bp} = 2$.

(b) $\omega_o = 50$ krad/s, $Q = 5$ and mid-band gain $h_{bp} = 5$.

(c) $\omega_o = 250$ krad/s, $Q = 12$ and mid-band gain $h_{bp} = 1$.

8-15   Design and test a notch filter for the following transfer function using a two-integrator network and a summer configuration shown in Figures 8.8 and 8.9. The two-integrator loop filter needs to be a KHN type. Also find the sensitivity of the parameters ω_o and Q_o, and ω_z and Q_z with respect to the passive elements used.

$$T(s) = \frac{s^2 + 10^{10}}{s^2 + 5*10^4 s + 10^{10}}$$

8-16 Repeat Problem 8-15 employing a TT circuit in place of a AM circuit.

8-17 Repeat Problem 8-15 using the modified summation method shown in Figure 8.12, where AM biquad is employed.

8-18 Repeat Problem 8-15 using the GIC based biquadratic circuit shown in Figure 8.19.

8-19 Design and test a notch filter for the following transfer function using a two-integrator network and a summer configuration shown in Figures 8.8 and 8.9. The two-integrator loop filter needs to be a KHN type. Also find the sensitivity of the parameters ω_o and Q_o, and ω_z and Q_z with respect to the passive elements used.

$$H(s) = \frac{s^2 + 0.25 \times 10^8}{s^2 + 0.35 \times 10^4 s + 0.49 \times 10^8}$$

8-20 Repeat Problem 8-19 employing a TT circuit in place of a KHN circuit.

8-21 Repeat Problem 8-19 using the modified summation method shown in Figure 8.12, where AM biquad is employed.

8-22 Repeat Problem 8-19 using the GIC based biquadratic circuit shown in Figure 8.19.

8-23 Design and test a LP notch filter for the following transfer function using a two-integrator network and the summer configuration shown in Figures 8.8–8.9. The two-integrator loop filter needs to be a KHN type.

$$H(s) = \frac{s^2 + 1.44 \times 10^8}{s^2 + 0.35 \times 10^4 s + 0.49 \times 10^8}$$

8-24 Repeat Problem 8-23 using the modified summation method shown in Figure 8.12, where AM biquad is employed.

Chapter 9

Direct Form Synthesis: Element Substitution and Operational Simulation

9.1 Introduction

In the previous chapters, we studied realization of first-order and second-order filter sections. Though these filter sections are used as such, they are also used to generate higher-order filters employing different processes including the cascade process. However, a common alternate process for realizing second- or higher-order filter section is the *direct form* of synthesis. There are two broad categories in the direct form of synthesis: (i) element substitution method and (ii) operational simulation method. Though the filter realization procedures in the aforementioned categories differ, the starting point is the same. Initially, a passive structure with element values (mostly frequency and impedance normalized) is obtained. It is then converted into its active form. Although they have the same starting point, the construction and characteristics of the active circuit obtained through the direct form and that obtained through the cascade form differ on many counts, as shall be illustrated later.

The most common passive structure that is used to realize passive filters is the doubly terminated lossless ladder. A typical lossless ladder is shown in Figure 9.1 where R_{in} and R_L are the terminating resistors and the ladder contains only lossless elements, that is, inductors and capacitors; each series and shunt branch of the ladder can be any combination of inductors/ capacitors.

We will first discuss the *element substitution* type of direct form synthesis procedure, which is mainly the avoidance of the use of inductors. Therefore, simulation of inductors forms an important part of the chapter. Inductance simulation, configurations for inductance simulation and active filter realizations without using an inductor are discussed in Sections 9.2–9.5. Section 9.6 deals with the simulation of a floating inductance, mainly through using two circuit structures of grounded inductances. Another method in which the inductor can be eliminated from the general lossless ladder is through scaling of the structure by the complex frequency variable s. This method generates a new type of element called the *frequency dependent negative resistance* (FDNR). As simulation of inductors and FDNR requires impedance conversion configurations, it is important to study the basics of these concepts. The technique is included in Section 9.8.

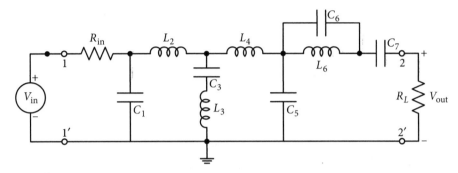

Figure 9.1   A typical doubly terminated lossless ladder structure with input resistance R_{in} and load resistance R_L.

An alternate method applied on ladders is based on the modeling of circuit equations and current–voltage relations of the circuit elements instead of direct element substitution. The electronic circuit is represented by a signal flow graph containing directional branches and nodes where branching takes place. It can also be represented in block diagram form with branches comprising blocks representing current–voltage relations of passive elements. Often, the employed blocks are integrators (or differentiators) interpreting inductors and capacitors. These blocks also incorporate summation of voltages and will be discussed in Section 9.9.

Once the principle of operational simulation is explained in detail, it is first utilized to get an LP (low pass) ladder and then for a BP (band pass) ladder structure in Sections 9.10 and 9.11, respectively. Since all networks may not be in as simple form as an LP or a BP, the scheme for realizing general ladders is studied in Section 9.12.

At first, the operational simulation method appears to be solving the problem in a roundabout manner compared to the element substitution method. In fact, the procedure is a bit lengthy, but it is observed that the method has certain advantages. In general, it employs a lesser number of active devices and deals well especially with floating inductors/FDNRs realizations as it does not require back-to-back matching circuits. As shall be shown later, a proper selection of integrators helps in considerable reduction in the non-ideal effect of OAs used, making the circuit useful in a comparatively larger frequency range.

Lossless Ladders: The lossless ladder structure is very popular among filter circuit designers because of its excellent property of very low sensitivity to component tolerances [9.1]. When such a ladder is converted to its active form, the property of low component tolerance sensitivity is transferred to it, which makes the structure attractive even at lower frequencies (audio or even up to a few hundred kHz; depending on the kind of active device used); whereas passive ladders continue to be used at higher frequency applications where active filtering is not suitable or where power supply is not available for active devices.

One important advantage in using lossless terminated ladders for active filters is that a large amount of literature is available in the form of filter structures, detailed description in terms of their transfer functions, pole locations and normalized element values from low to high-order filters [1.2]. Such available literature is of great help for active filter design. One of the main reasons for converting LC lossless ladders to their active forms is, as mentioned earlier, the non-availability of good quality inductors in most of the operating frequency range.

9.2   Gyrator and Inductance Simulation

As seen in Figure 9.1, inductances used in the ladder can have one terminal connected to the ground, which are known as grounded inductors (GIs), or none of their terminals connected to the ground, which are known as floating inductors (FIs). First, let us look at the simulation of a GI as shown in Figure 9.2(a). The method will be later extended to realize an FI as shown in Figure 9.2(b). For simulating a GI, it is required to find a circuit which contains only resistors, capacitors and some active device(s); moreover, its driving point impedance should appear as that of an inductance. Such a configuration is known as *impedance converter* or *gyrator*. Symbol of a simplified gyrator is shown within the dotted box in Figure 9.3, which is defined by the following equation [9.2].

$$I_1 = -(1/r)V_2 \text{ and } I_2 = -(1/r)V_1 \tag{9.1}$$

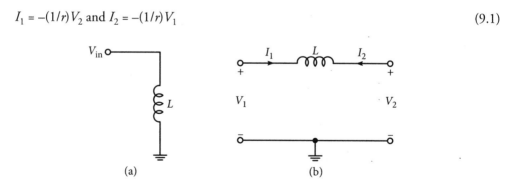

(a)   (b)

Figure 9.2   (a) Grounded inductance, and (b) a floating inductor representation.

The important parameter of a gyrator r is known as a *gyrator constant*; the constant has the units of ohm. From equation (9.1), the input impedance of the gyrator will be:

$$Z_{in}(s) = (V_1/I_1) = r^2/(1/sC) = sr^2C \tag{9.2}$$

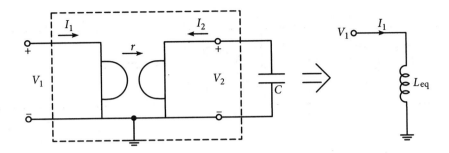

Figure 9.3   Grounded inductance simulation using a gyrator terminated in a grounded capacitor.

Therefore, a gyrator terminated in a capacitor C simulates an equivalent inductor L_{eq} given as:

$$L_{eq} = r^2 C \qquad (9.3)$$

Obviously, instead of using a capacitor C, termination can be done using any general load impedance $Z_L(s)$ and in that case, the simulated input impedance will become:

$$Z_{in}(s) = r^2 / \{Z_L(s)\} \qquad (9.4)$$

As mentioned earlier, the current–voltage relation of a *simplified* gyrator is presented in equation (9.1). In a general gyrator, the admittances in equation (9.1) are not necessarily equal, that is

$$(I_1/V_2) = y_{12} = (-1/r_2) = g_{m2} \text{ and } (I_2/V_1) = y_{21} = (1/r_1) = -g_{m1} \qquad (9.5)$$

The equivalent circuit of a general gyrator (Figure 9.4) can be obtained from equation (9.5). It is obvious that the practical realization of a gyrator is easy in terms of transconductance elements. Such a circuit, transconductance amplifier-based filter circuits, will be studied later in Chapter 15. Practical realization of gyrators using OAs will be discussed in the next section.

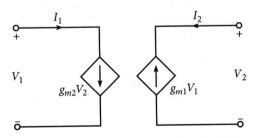

Figure 9.4   Small signal equivalent circuit of a general gyrator.

In a general ladder structure, it is also important to simulate FIs depending on the selected passive structure. Simulation of FIs is rather difficult and requires a cascade of two grounded gyrators and an embedded capacitor, as shown in Figure 9.5 [9.3]. Practical realization of an FI will be discussed after obtaining the practical realization of a GI.

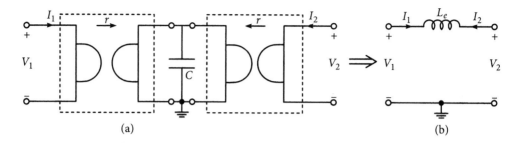

Figure 9.5   (a) Floating inductance simulation using back-to-back gyrators and (b) equivalent circuit.

9.3   Impedance Converters Using Operational Amplifiers

The terms 'gyrator' and 'impedance converter' can be used interchangeably as both convert the nature of impedance connected as termination. A general configuration which has been employed for the development of the impedance converter is shown in Figure 9.6. This configuration is found suitable for developing impedance converter circuits using OAs. In Figure 9.6, the schematic consists of R_1, a feedback resistor, and a *two-port network*, which has to be determined in order to make it an impedance converter. Assuming the two-port network has infinite input impedance and zero output impedance, it is desired to get the following voltage–current relation for simulating inductance at the input terminals.

$$(V_{in}/I_{in}) = Z_{in}(s) = sL \qquad (9.5)$$

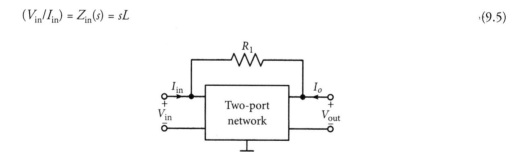

Figure 9.6   A schematic for impedance conversion.

With no current flowing into the two-port, $I_{in} = (V_{in} - V_{out})/R_1$, so substituting I_{in} in equation (9.5):

$$V_{in} = sL\frac{V_{in} - V_{out}}{R_1} \rightarrow \left(\frac{V_{out}}{V_{in}}\right) = (1 - R_1/sL) \qquad (9.6)$$

The required transfer function for the network, as shown in equation (9.6) is obtained by subtracting the gain of an integrator from a unity gain amplifier. Figures 9.7(a) and (b) show simple and known circuits for the non-inverting gain of 2 and the inverting ideal integrator with additional non-inverting input. These circuits are joined in Figure 9.7(c) for which

$V_2 = 2V_{in}$. With both the input terminals of OA2 being at the same potential V_{in}; the current–voltage relations for OA2 and at the input terminal gives the following relations:

$$(V_2 - V_{in})/R \rightarrow (V_{in}/R) = (V_{in} - V_{out})sC \tag{9.7}$$

$$I_{in} = (V_{in} - V_{out})/R_1 \tag{9.8}$$

Elimination of V_{out} in equations (9.7) and (9.8) gives:

$$Z_{in} = (V_{in}/R) = sCRR_1 \tag{9.9}$$

Hence, the circuit shown in Figure 9.7(c), which is known as the Riordan inductance simulation circuit [9.4] simulates an inductor with value $L = CRR_1 = CR^2$ for $R = R_1$.

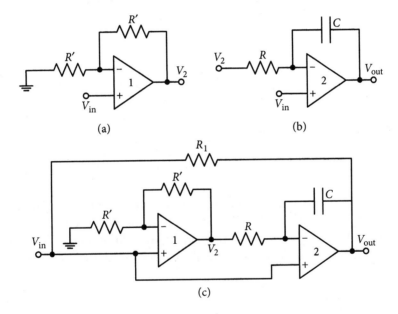

(a)   (b)

(c)

Figure 9.7   (a) Non-inverting amplifier with gain = 2, (b) Ideal non-inverting integrator and (c) Riordan inductance simulator circuit, performing impedance conversion through a combination of (a) and (b).

9.4   Antoniou's Inductance Realization

There are other possible configurations to realize the two-port network shown in Figure 9.6. A well-known configuration employs a non-inverting amplifier of Figure 9.8(a) and the inverting integrator in a differential input mode as shown in Figure 9.8(b). Here the feedback resistor of a difference amplifier is replaced by a capacitor C and the non-inverting terminal gets a potentially divided input $V_2 = V_1 R_5/(R_4 + R_5) = kV_1$. Applying KVL (Kirchhoff's voltage law) at the input terminal of the OA2 in Figure 9.8(b) and with the knowledge that the amplifier is an ideal OA with the inverting terminal voltage being equal to V_2, we get:

$$(V_1 - V_2)/R_3 = (V_2 - V_{out})sC_2 \tag{9.10}$$

or $(V_{out}/V_1) = k + \{(k-1)/(sC_2R_3)\} \tag{9.11}$

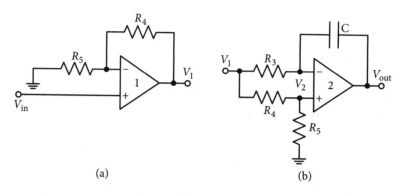

(a)   (b)

Figure 9.8   (a) A non-inverting amplifier with gain $(1 + R_4/R_5)$, and (b) inverting integrator in differential mode.

To make the circuit compatible with the relation of equations (9.6), k is to be eliminated in equation (9.11). It is done using a non-inverting amplifier with gain, $1 + (R_4/R_5) = (1/k)$ as shown in Figure 9.8(a) at the input terminal V_1 of Figure 9.8(b). However, connecting in this manner will result in a circuit containing five resistors (and one more resistor R_1 as feedback resistor). Instead, two resistors in the non-inverting amplifier can be saved by connecting the amplifier's inverting terminal directly to the inverting terminal of the OA2, whose voltage is also $V_2 = (kV_1)$ and $V_1 = (V_{in}/k)$. The resulting circuit, in addition to the feedback resistor R_1, is now shown in Figure 9.9(a). Analysis of the block inside the dotted line gives the following relation.

$$\frac{V_{out}}{V_{in}} = \left(1 - \frac{R_4}{sC_2R_3R_5}\right) \tag{9.12}$$

Comparing equation (9.12) with equation (9.6), the simulated inductance has the following expression

$$L_{eq} = (R_1R_3R_5C_2/R_4) \tag{9.13}$$

The circuit shown in Figure 9.9(a) is known as Antoniou's generalized impedance converter [9.5] of type I. Though presently it is shown to be simulating inductance, we will see later that its generalized form can be that of a generalized impedance converter (GIC). Type II GIC is obtained by simply interchanging R_4 with C_2, as shown in Figure 9.9(b), for which, the simulated inductance is given as:

$$L_{eq} = (R_1 R_3 R_5 C_4 / R_2) \tag{9.14}$$

If all the resistances used are selected equal in both types of GICs, the simulated inductance becomes $L_{eq} = R^2 C$; the same as obtained before in equation (9.9). It is to be noted that for the GIC, comparing the constant with the gyrator-based simulator, gyration constant $r = R$.

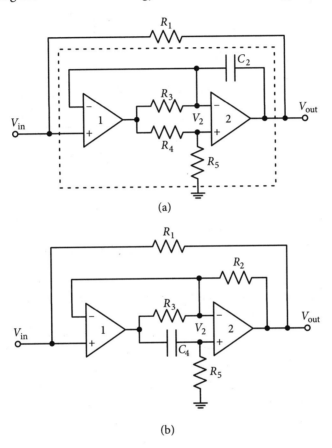

(a)

(b)

Figure 9.9   (a) Antoniou's general impedance convertor type I. (b) Antoniou's generalized impedance converter type II.

GICs have been used extensively for element simulations. Because of their importance, GIC needs to be studied carefully. Hence, instead of restricting the study to the GICs of Figure 9.9(a) and (b), the circuit configuration is redrawn in a form which is common in use and convenient for analysis. Figure 9.10 shows a general Antoniou's GIC in dotted rectangles; it was briefly discussed in Chapter 8 as well. Assuming OAs as ideal, voltage V_1, V_3, and V_5 shall be equal and the input impedance of the GIC is easily obtained as

$$Z_{in}(s) = \{Z_1(s)Z_3(s)Z_5(s)\}/\{Z_2(s)Z_4(s)\} \tag{9.15}$$

GICs of type I and II can easily be shown to be special cases of this general configuration. The general form is called the *generalized impedance converter* because of its ability to realize many other types of impedances depending on the kind of elements (or combination of elements) used for Z_i, i = 1 to 5. For example, selecting $Z_i = R_i$, i = 1, 2, 3 and 5 and $Z_4 = 1/sC$, the input impedance will be the same as for equation (9.14).

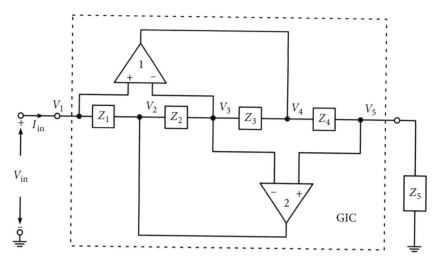

Figure 9.10 Commonly used structure for a generalized impedance converter.

9.5   Filter Realization Using Inductance Simulation

Once inductance simulation through an active RC circuit becomes available, active simulation of the LC ladder becomes simple enough. The obvious starting point is obtaining the passive filter structure and the values of the elements used. Inductances are then replaced by suitable active RC structures and the rest of the capacitors and resistors remain connected in the same position/location. Hence, the resulting overall circuit becomes an active RC structure.

Figure 9.11(a) shows the structure of a third-order HP passive filter section. This passive structure is suitable for the inductance simulation technique as it employs a GI. Hence, the structure shown in Figure 9.10 is easily used to simulate the inductor. Once the inductor is replaced, an active RC version of the third-order HP filter is conveniently obtained. It may be noted that the passive HPF is shown to have normalized terminating resistors R_1 and R_2. The element values of C_1, L_2 and C_3 are easily available from design tables [1,2]. Hence, for the given value of L_2, element values of the inductance simulator are evaluated using equation (9.14).

Example 9.1: Design a third-order HP active filter using the inductance simulation technique, having pass band ripples less than 1 dB and corner frequency of 200 krad/s. Compare its response with that of the passive filter.

Solution: For the given specifications, the circuit shown in Figure 9.11(a) will satisfy the requirements with the following normalized element values:

$$R_1 = R_2 = 1\Omega, \ C_1 = C_3 = 0.62645 \text{ F and } L_2 = 0.9118 \text{ H} \tag{9.16}$$

Using an impedance scaling factor of 10^4 and a frequency scaling factor of 200 krad/s, the denormalized element values for the passive filter from equation (9.16) will be:

$$R_1 = R_2 = 10^4 \ \Omega, \ C_1 = C_3 = 0.31322 \text{ nF and } L_2 = 0.04559 \text{ H} \tag{9.17}$$

For conversion of a passive filter to an active form, inductance L_2 from equation (9.17) is simulated using the circuit shown in Figure 9.10 and equation (9.14).

$$L_2 = 0.04559 \ H = CR^2 \rightarrow C = 0.4559 \text{ nF with } R = 10^4 \ \Omega \tag{9.18}$$

The GI shown in Figure 9.10, having element values as in equation (9.18), is substituted in Figure 9.11(a), resulting in the circuit shown in Figure 9.11(b). The simulated response is shown in Figure 9.11(c) with the following important observations.

Voltage gain at high frequencies = 0.4626, peak voltage gain = 0.498, ripple width = 0.88 dB and corner frequency = 28.63 kHz (179.96 krad/s). Obviously, there is a significant difference between simulated and design value of the corner frequency (−10.02%), and the high frequency gain is dropped by nearly 7.48%, though the shape of the characteristic remains intact. Deviations in parameters are due to the effect of the frequency-dependent gain

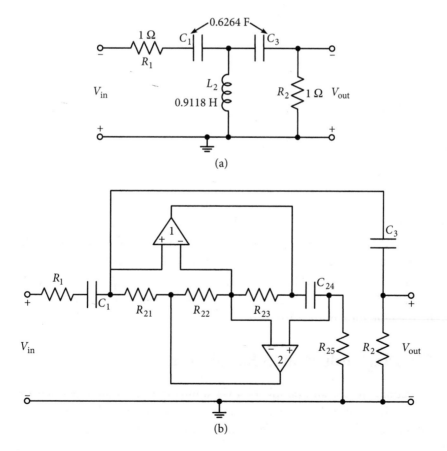

(a)

(b)

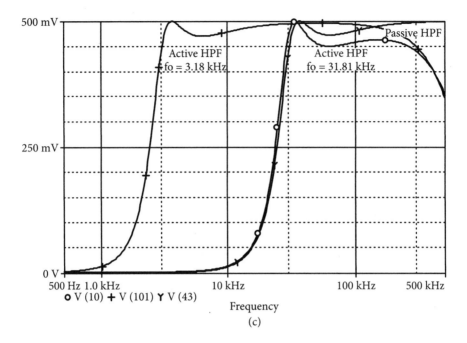

500 mV

Passive HPF

Active HPF
fo = 3.18 kHz

Active HPF
fo = 31.81 kHz

250 mV

0 V

500 Hz 1.0 kHz   10 kHz   100 kHz   500 kHz

○ V (10) ✚ V (101) Y V (43)

Frequency

(c)

Figure 9.11   (a) A third-order normalized passive high pass filter structure for Example 9.1. (b) The active RC version of Figure 9.11(a) through simulation of grounded inductance L_2 with inductance simulation using GIC (c) Response of the active third-order high pass filter from Figure 9.11(b) at lower and higher corner frequencies, and the response of the passive filter of Figure 9.11(a).

of the OA model. The same active filter, which is simulated for a lower corner frequency of 20 krad/s, had the following element values:

$$C_1 = C_2 = 3.132 \text{ nF}, \ L_2 = 0.4559 \text{ H} \rightarrow C_{24} = 4.554 \text{ nF} \qquad (9.19)$$

The simulated response is also shown in Figure 9.11(c) with the following observations:

Voltage gain at higher frequencies = 0.496, corner frequency = 3.15 kHz and ripple width = 0.5376 dB. Error in the corner frequency is now only –1% and gain deviates only by 0.8%.

Figure 9.11(c) also shows the PSpice simulated response of the passive filter which we can compare with that of the response of the active filter. The following are the observations.

Voltage gain = 0.5, ripple width = 0.5086 dB and corner frequency = 31.83 kHz (200.1 krad/s). The filter's voltage gain remains constant even at much higher frequencies, as it is not affected by the limitation of the OA.

9.6   Floating Inductance Simulation

In the last section, we saw an example of a GI simulation in a simple passive circuit. Quite often, a floating inductor (FI) also becomes a necessity and it has to be simulated as well. The FI shown in Figure 9.5(b) being a two-port structure is represented in terms of y parameters as:

$$[y] = \frac{1}{sL_{eq}} \begin{bmatrix} 1 & -1 \\ -1 & 1 \end{bmatrix} \qquad (9.20)$$

For the simulation of an FI, two gyrators in back-to-back form are to be joined as shown in Figure 9.5(a). In order to obtain a circuit realization of an FI using two back-to-back gyrators, the well-known technique of *lifting the element terminal from the ground* [9.6] is used on a GI circuit like that used in the circuit in Figure 9.10. The resulting configuration is shown in Figure 9.12. It is important to note that while using any circuit involving OAs, care has to be taken that a path for the flow of bias current remains available. Hence, for the circuit realization of the FI shown in Figure 9.12, a resistance each may be connected in parallel with the capacitors to enable the flow of biasing current. However, these extra resistors have to be of high value so that the parasitic inductance introduced due to these resistances is not significant.

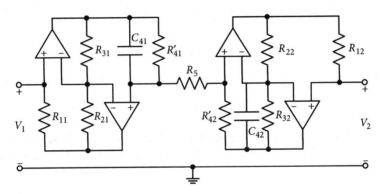

Figure 9.12 Floating inductance realization using back-to-back gyrator based grounded inductance simulators.

The major limitation of the process is that an FI simulator uses a large number of passive and active elements. Another significant issue crops up when the two gyrators are connected back-to-back. The gyrator constants need to be the same; otherwise, there will be a mismatch and the unity element in equation (9.20) will not be exactly unity, resulting in some parasitic elements. Obviously, it is not practically possible to exactly match the component values even in the IC form (mismatch can be minimized). This is a drawback in using such a configuration for FI. Hence, we need to look at other techniques of obtaining an FI circuit which do not require component matching. An alternative is to select a circuit which needs a lesser number of FIs. It is to be noted that L_{eq} in equation (9.20) will have the same expression as L_{eq} in equation (9.13) or (9.14) for GI.

Example 9.2: For the passive BPF shown in Figure 9.13(a), obtain an active RC filter using the inductance simulation method. Find the element values used with the center frequency of the filter as 20 kHz and bandwidth as 2 kHz.

Solution: The transfer function of the BPF of Figure 9.13(a) is obtained as:

$$(V_2/V_1) = (R/L)s/\{s^2 + (R/L)s + (1/LC)\} \qquad (9.21)$$

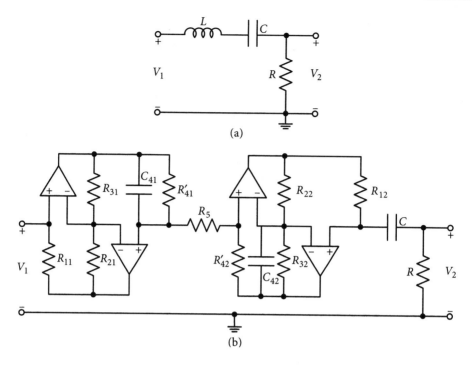

Figure 9.13 (a) Second-order prototype passive band pass filter, (b) its active RC version while simulating a floating inductor using the circuit shown in Figure 9.12. All resistances = 10 kΩ, but $R = 1$ kΩ and $R'_{41} = R'_{41} = 10$ MegΩ, $C_{41} = C_{42} = C = 0.795$ nF.

The important parameters of the filter are as follows:

$$\omega_o = \frac{1}{(LC)^{1/2}}, Q = \frac{1}{R}\left(\frac{L}{C}\right)^{1/2} \text{ and mid-band gain} = 1.0 \tag{9.22}$$

For $\omega_o = 1$ rad/s, the normalized element values from equation (9.22) are:

$$L = 1 \text{ H}, C = 1 \text{ F and as } Q = (20/2) = 10, R = 0.1 \text{ } \Omega \tag{9.23}$$

For the passive filter, an impedance scale factor of 10^4 and a frequency scaling factor of $20(2\pi)$ krad/s is used; this gives the following de-normalized element values:

$$R = 1 \text{ k}\Omega, C = 0.7954 \text{ nF and } L = 0.07954 \text{ H} \tag{9.24}$$

For the floating inductance, $L = 0.07954$ H, the circuit shown in Figure 9.12 is inserted in Figure 9.13(a) and application of equation (9.14) to find the element values for inductance gives an active filter circuit as shown in Figure 9.13(b). Resistances R_1, R_2 and R_3 in both the gyrators are 10 kΩ, and capacitances C_{41} and C_{42} are equal to 0.7954 nF. It is important to note that R_5 is 10^4 Ω and not the summation of two R_5s while connecting two back-to-back circuits, as one resistor acts as the terminating resistor for both sides. Bypass resistors R'_{41} and

R'_{42} are selected as 10 megΩ each, sufficient to allow the passage of bias current, and large enough so that their effect is minimal on the filter performance. The circuit is simulated and the response shown in Figure 9.14.

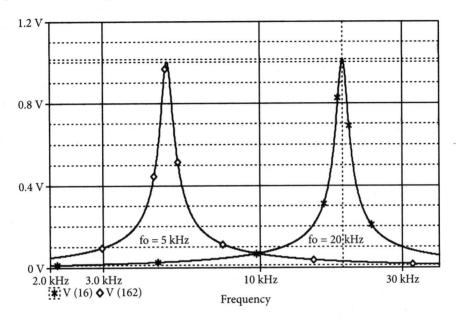

Figure 9.14 Magnitude response of the active band pass filter shown in Figure 9.13(b) at lower and higher center frequencies.

The simulated value of the mid-band gain is 1.017, center frequency is 18.889 kHz, upper and lower cut-off frequencies are 19.778 kHz and 18.027 kHz, respectively. Bandwidth being 1.756 kHz, Q becomes 10.75. Obviously, the main reason for deviation in the parameters is due to the frequency dependence of the OAs gain.

The same BPF was simulated for a lower center frequency of 5 kHz. All the calculated resistances remain the same but all the capacitances are now 3.1818 nF. The simulated magnitude response in this case is also shown in Figure 9.14. Mid-band gain was found to be 0.9987 at a center frequency of 4.927 kHz. The upper and lower cut-off frequencies of 5.177 kHz and 4.690 kHz, respectively gave $Q = 10.11$; this is now, a much smaller deviation in filter parameters, because of lower working frequency.

9.7   Generalized Inductance Simulation

The inductance simulation method is more useful for the circuit having inductances in grounded forms. Whenever FI is to be simulated, it involves a large number of components and their matching as well. It becomes a little confusing when a circuit contains both GIs and FIs. However, a technique known as the Gorski-Popiel (GP) embedding technique [9.7] is of great help in simulating a combination of GIs and FIs, which will require a lesser number of elements compared to the case of conventional direct simulation of GIs and FIs. In this technique, instead of simulating individual inductors, a complete sub-network comprising inductors (any number) is simulated through a GIC. This simulated network then simply replaces the original sub-network of the inductors.

To understand the generalized inductance simulation (GIS) technique, consider the GIC circuit shown in Figure 9.10 in the form of a two-port network, as shown in Figure 9.15(a). Assuming the OAs to be ideal means $V_2 = V_1$, and with $R_2 = R_3$ and Z_5 of Figure 9.10 replaced by (V_2/I_1), we get the driving point impedance at port 1 as

$$(V_2/I_1) = (sC_4R_1)(V_2/I_2) \tag{9.25}$$

Obviously, when the simplified block form of GIC shown in Figure 9.15(b) is terminated in a resistance R_1, the input will be an inductance as before, with the inductor expression as:

$$L_{eq} = (C_4R_1)R = L'R \tag{9.26}$$

The GIC in the block form, terminated in a resistance as shown in the dotted rectangle in Figure 9.15(b), yields the following from equation (9.25):

$$I_2 = (sC_4R_1)I_1 = (sL')I_1 \tag{9.27}$$

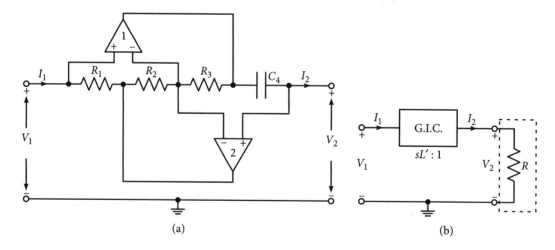

(a)   (b)

Figure 9.15 (a) Generalized impedance convertor as a two-port network and (b) representation of the GIC in a block form.

The important conclusion of the exercise is that if a GIC is placed at the input branch of a resistor, the resistor gets converted into an inductor. According to the generalized inductor simulation scheme, any number of branches of a network can have a GIC at its input which will convert all terminating resistors as inductors. In order to simulate an inductor L_{eq}, the terminating resistor value will be:

$$(L_{eq}/L') \rightarrow (L_{eq}/C_4 R_1) \tag{9.28}$$

From the point of view of IC fabrication of the active filter, it is preferable to use the same GICs with varying terminating resistors for the simulation of inductors with different values. However, if it results in a situation where the terminating resistor value becomes unsuitable for IC fabrication, a different GIC can be used.

Example 9.3: Realize a fifth-order Chebyshev LPF having a maximum of 1 dB ripples in the pass band. Let the corner frequency be 10 krad/s; use the GP technique.

Solution: Figure 9.16(a) shows the ladder structure and the normalized element values of the filter for the given specifications from Table 3.5. Cross points have also been shown in the figure for the application of the GP technique, where terminated GIC circuits will be inserted.

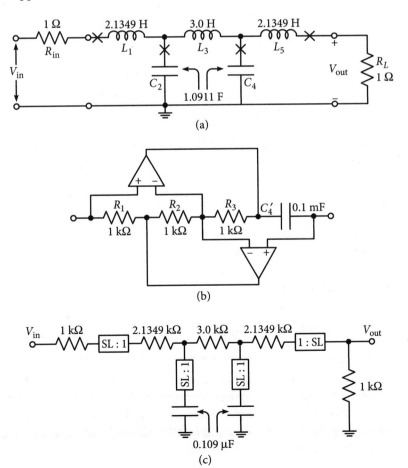

(a)

(b)

(c)

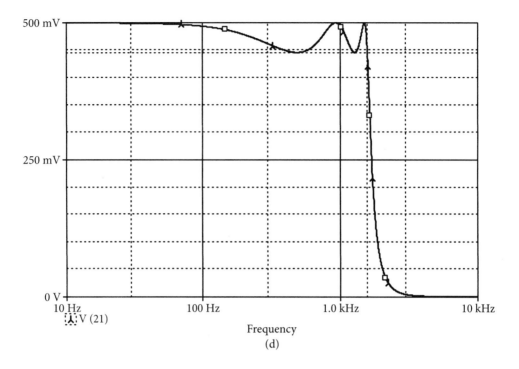

Figure 9.16 (a) Fifth-order passive Chebyshev filter for Example 9.3 (b) GIC circuit, with elements values, which is to be terminated and (c) the active version of the filter using the Gorski-Popiel ladder embedding technique. (d) Magnitude response of the filter in Figure 9.16(c).

Application of the frequency scaling factor of 10 krad/s and an impedance scaling factor of 10^3 results in the following element values:

$$R_{in} = R_L = 1 \text{ k}\Omega, L_1 = L_5 = 0.21349 \text{ H}, L_3 = 0.3 \text{ H}, \text{ and } C_2 = C_4 = 0.10911 \text{ µF} \qquad (9.29)$$

Since for the realization of inductances, terminated GICs are to be used at the cross point, Figure 9.16(b) shows a GIC in which resistances $R_1 = R_2 = R_3 = 1$ kΩ each. If C_4' is selected as 0.1 µF, the terminating resistances for the inductors L_1 and L_5 of equation (9.29) will be calculated using equation (9.14) as:

$$R_{L1} = R_{L5} = \frac{10^3 \times 0.21349}{10^3 \times 10^3 \times 10^{-7}} = 2.1349 \text{k}\Omega \text{ and } R_{L3} \text{ will be } 3.0 \text{ k}\Omega$$

Figure 9.16(c) shows the structure of the active ladder, where the GIC circuit of Figure 9.16(b) will be substituted at the four places as indicated; for the rightmost GIC, the inverted direction needs to be noted. The circuit was simulated and the magnitude response is shown in Figure 9.16(d), for which dc gain is 0.5, ripple width is 0.995 dB and pass band edge frequency are 1.584 kHz (9.956 krad/s); very close to the design values.

9.8 Filter Realization Using FDNR

Instead of replacing inductors through an active RC simulator, an alternate scheme was given by L. Bruton in 1969 [9.8]. In this scheme, all the elements are multiplied by a factor $(1/s)$. Such a transformation converts an inductive impedance (sL) to a resistor element of value L ohms. At the same time, it converts a resistor (R) to a capacitive element (R/s), that is, a capacitor with value $(1/R)$ farad. However, the capacitive impedance $(1/sC)$ gets converted to $(1/s^2C)$; not a conventional element. For $s = j\omega$, this converted impedance is:

$$Z'_c = (1/s^2C)\big|_{s=j\omega} = -1/\omega^2 C \tag{9.30}$$

Equation (9.30) shows that the impedance is negative, real and frequency dependent; hence, an appropriate name would be *frequency dependent negative resistance* (FDNR). Sometimes, it is also called a *super capacitor* as it is converted from a capacitor, and a usual symbol for it is three parallel lines. It is necessary to note that the transformation of element impedances by $1/s$ does not affect the transfer function of the network as it is a ratio of two polynomials in s.

After the transformation, RLC circuit now comprises resistors, capacitors and FDNRs (an RCD network). Obviously, to convert an RLD network to an active RC form, the FDNRs have to be simulated in the same way as the inductances were simulated, be it in the grounded or in the floating form. Fortunately, a large number of active RC circuits are available in literature for simulating FDNRs. One such circuit is obtained through the use of the GIC shown in Figure 9.10, for which the input impedance is given as

$$Z_{in}(s) = Z_1 Z_3 Z_5 / Z_2 Z_4 \tag{9.31}$$

Selecting $Z_1 = 1/sC_1$, $Z_5 = 1/sC_5$, $Z_3 = R_3$, $Z_2 = R_2$ and $Z_4 = R_4$, as shown in Figure 9.17(a), the circuit provides a grounded FDNR with the expression of its impedance as given here.

$$Z_{in}(s) = \frac{R_3}{s^2 C_1 C_5 R_2 R_4} = \frac{1}{s^2 D} \text{ with } D = \frac{C_1 C_5 R_2 R_4}{R_3} \tag{9.32}$$

Figure 9.17(b) shows the capacitor's symbolic representation of three parallel lines. Different values for the capacitors C_1 and C_5 and R_2 and R_3 can be chosen, but it does not give any specific advantage. Since equal value capacitors are desirable in integration, we prefer to select $C_1 = C_5 = C$. It has been shown that for a GIC, it is better to use $R_2 = R_3 = R$, hence, a simplified expression of input impedance from equation (9.32) will be as follows:

$$Z_{in}(s) = \frac{1}{s^2 C^2 R_4} \text{ or } D = C^2 R_4 \tag{9.33}$$

Application of the FDNR technique is obviously preferred for those networks which use more grounded capacitors, as such networks will have FDNRs in the grounded mode. At the same time, the floating inductance gets converted to resistors. For simulating FDNR in floating form, the method of realization and limitations are exactly the same as those in the case of FIs.

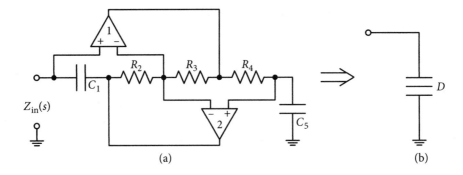

Figure 9.17 (a) Circuit diagram of a grounded FDNR obtained through GIC and (b) its symbolic representation.

While using the FDNR technique of converting passive RLC circuits to an active RC, there are certain practical glitches that need to be removed. For a doubly resistor-terminated ladder, the source resistor and the load resistor also get transformed to capacitors. The following example will help in designing an FDNR based filter and also illustrate the conversion of the two aforementioned resistors. The method of overcoming the practical glitches mentioned here will be discussed after the example.

Example 9.4: Obtain an active RC filter structure using FDNRs from a passive fifth-order Chebyshev LP filter having a cut-off frequency of 100 krad/s, a ripple width of 1 dB with source and load terminating resistors of 10 kΩ.

Solution: The structure and element values of a normalized fifth-order LP passive filter obtained from the standard design table or through the method described in Chapter 3 is shown in Figure 9.16(a). It is a minimum inductance configuration, and its normalized element values are already given in equation (9.29) and repeated here:

$$R_{in} = R_{out} = 1\ \Omega, L_1 = L_5 = 2.1349\ \text{H}, L_3 = 3\ \text{H and } C_2 = C_4 = 1.0911\ \text{F}$$

Application of $(1/s)$ transformation on the elements converts it to the circuit elements as shown in Figure 9.18(a).

If we use a normalizing frequency of 100 krad/s, the normalized pass band edge frequency will be at $\omega = 1$.

Once the passive filter structure and its element values are obtained/designed and $(1/s)$ transformation has been performed, the following are the next steps to design the converted FDNR(s).

Using equations (9.33), we get the element values for both FDNRs as:

$$D = 1.0911 \times R_4\ C^2 \rightarrow C = 1.04455\ \text{F for } R_4 = 1\ \Omega$$

Inductances converted as resistances will have normalized values as:

$$R_{L1} = R_{L2} = 2.135 \ \Omega \text{ and } R_{L1} = 3.0 \ \Omega.$$

Active FDNRs are then put in place of D_1, D_2 shown in Figure 9.18(a). The values of the resistor R_2 and R_3 are not selected so far; they are arbitrary and can also be selected after frequency and impedance de-normalization. The frequency de-normalization factor being 100 krad/s, we select an impedance normalization factor of 10 kΩ. The final circuitry of the active RC fourth-order LPF is shown in Figure 9.18(b) with the de-normalized value of the elements as follows:

$$C_{11} = C_{51} = C_{12} = C_{52} = 1.04455 \text{ nF}, R_{41} = R_{42} = 10 \text{ k}\Omega$$

$$R_{L1} = R_{L3} = 21.349 \text{ k}\Omega \text{ and } R_{L2} = 30 \text{ k}\Omega$$

Resistors R_2 and R_3 in each FDNR are selected as 10 kΩ, an arbitrary value; this equals resistor values already used in the circuit.

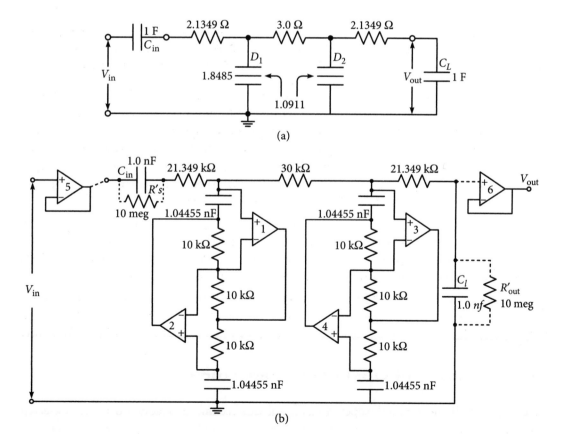

(a)

(b)

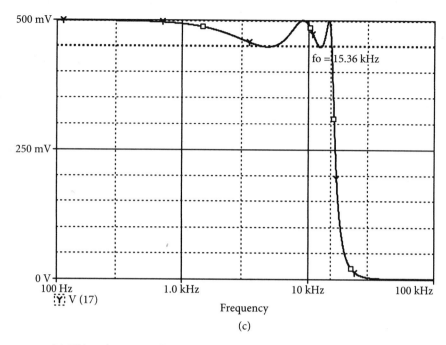

Figure 9.18 (a) Fifth-order passive doubly terminated low pass filter structure after (I /s) transformation and (b) active RC configuration using FDNRs after de-normalization of elements. (c) Response of the filter shown in Figure 9.18(b) while using grounded FDNRs.

Due to the conversion of terminating resistors as capacitors, it can be seen from Figure 9.18(b) that the input biasing current cannot flow in the non-inverting terminal of OA1 and 3 in the same way as in the case of inductance simulation. Using the same remedy in this case as well, the terminating capacitors C_{in} and C_L (1 nF each) are bypassed by large value resistors R'_{in} and R'_{out}, as shown linked through dotted lines in Figure 9.18(b). Obviously, the bypass resistors (which are equivalent to inductors in the original passive RLC circuit) have to be high enough, so as not to significantly affect the response of the filters.

Another issue to be resolved for the practical implementation of the filter is that the termination resistors which have been transformed as capacitors, have to be re-inserted in the circuit as these were not part of the lossless filter structure. The problem is solved through the use of non-inverting buffers at the input and output terminals as shown in Figure 9.18(b).

Figure 9.18(c) shows the PSpice simulated response of the active filter of Figure 9.18(b). Voltage gain at low frequencies is 0.4984, corner frequency is 15.368 kHz (96.6 krad/s) and ripple width is 0.947 dB, with gain at 40 kHz dropping by 62 dBs; a sufficiently good response.

9.9   Principle of Operational Simulation

To begin our discussion on the operational simulation technique, a ladder structure is selected for its simplicity and due to the fact that it is a common structure for filter realizations. Figure 9.19 shows a sixth-order ladder structure in block form along with the currents in the branches and voltage levels across branch immittances. The block diagram is valid for single or doubly terminated ladders. The circuit can be described in terms of the following currents and voltages.

$$V_1 = V_s - V_2, \; V_3 = V_2 - V_4, \; V_5 = V_4 - V_6 \tag{9.34}$$

$$I_2 = I_1 - I_3, \; I_4 = I_3 - I_5, \; I_6 = I_5 \text{ as } I_7 \text{ is zero} \tag{9.35}$$

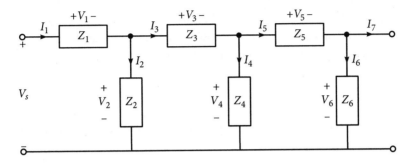

Figure 9.19 Block form representation of a sixth-order doubly terminated ladder. V_s is the source voltage and Z_1 contains the source resistance R_1. Z_6 contains the terminating resistor R_2.

In addition, the current–voltage relation for the series and shunt branches can be written as:

$$I_1 = (V_1/Z_1) = V_1 Y_1, \; I_3 = (V_3/Z_3) = V_3 Y_3, \; I_5 = (V_5/Z_5) = V_5 Y_5 \tag{9.36}$$

$$V_2 = I_2 Z_2 = (I_2/Y_2), \; V_4 = I_4 Z_4 = (I_4/Y_4), \; V_6 = I_6 Z_6 = (I_5/Y_6) \tag{9.37}$$

For the development of the procedure in which current–voltage relations of the branches can be simulated operationally, the aforementioned four equations can be combined in the following form.

$$I_1 = Y_1(V_s - V_2), \; V_2 = Z_2(I_1 - I_3) \tag{9.38a, b}$$

$$I_3 = Y_3(V_2 - V_4), \; V_4 = Z_4(I_3 - I_5) \tag{9.39a, b}$$

$$I_5 = Y_5(V_4 - V_6), \; V_6 = Z_6 I_5 \tag{9.40a, b}$$

Apart from the realization of the elements Z_i(or Y_i), which will be taken up later in the chapter, operational simulation faces the following two problems if the circuit is to be realized using OAs.

The first problem is that in equations (9.38) to (9.40), if the output side is in terms of voltage, the input is current or vice versa, whereas in the OAs, both input and output are in terms of voltages. The second problem is that summation of voltages is easier while using OAs, but differencing them is a bit involved. Both the problems are solved step by step.

The first of the two problems is solved by scaling equations (9.38) to (9.40) by a resistor R' and changing their way of representation. Hence, for equation (9.38a), we can write:

$$R'I_1 = R'Y_1 (V_s - V_2) \qquad (9.41a)$$

In equation (9.41a), $R'I_1$ becomes a voltage, which we will denote using a lower case voltage symbol with a subscript I as v_{I1}. Use of the lower case symbol is to identify that it was obtained after normalization through R' and subscript I denotes that, initially, this voltage was in the form of current. Another important point is that the term $R'Y_1$, which becomes dimensionless and is the ratio of two voltages as $v_{I1}/(v_s - v_2)$, will also become a transfer function h_{Y1}. Hence, in the modified form, equation (9.41a) is written as:

$$v_{I1} = h_{Y1}(v_s - v_2) \qquad (9.41b)$$

In equation (9.41b), subscript $y1$ on the transfer function indicates that it was obtained from admittance Y_1.

Following the same notation, equations (9.38) to (9.40) are written as follows, where both sides of the equation are in terms of voltages.

$$V_2 = (Z_2/R')(R'I_1 - R'I_3) \rightarrow v_2 = h_{z2}(v_{I1} - v_{I3}) \qquad (9.42)$$

$$R'I_3 = R'Y_3(V_2 - V_4) \rightarrow v_{I3} = h_{Y3}(v_2 - v_4) \qquad (9.43)$$

$$V_4 = (Z_4/R')(R'I_3 - R'I_5) \rightarrow v_4 = h_{z4}(v_{I3} - v_{I5}) \qquad (9.44)$$

$$R'I_5 = R'Y_5(R'V_4 - R'V_6) \rightarrow v_{I5} = h_{Y5}(v_4 - v_6) \qquad (9.45)$$

$$V_6 = (Z_6/R')(R'I_5) \rightarrow -v_6 = h_{z6}(-v_{I5}) \qquad (9.46)$$

In the transformed equations (9.41b) to (9.46), only v_{I1} is positive, v_6 has a negative sign and the rest of the voltages v_2, v_{I3}, v_4, and v_{I5} appear in both inverting and non-inverting form and in the voltage differencing form. Since differencing of the voltage is a bit involved, this differencing in voltages is to be avoided; in its place, voltage summation is used. In this case, at least four inverters are needed to get both inverting and non-inverting voltages v_2, v_{I3}, v_4, and v_{I5}. Use of these extra inverters will make the circuit complex and uneconomical. A better alternative is to make those transfer functions, which became available in these equations from the impedances Z_2, Z_4 and Z_6, inverting; h_{zi} is replaced by $-h_{zi}$. Under such a condition, the transformed equations (9.41b) to (9.46) will modify to the following:

$$v_{I1} = h_{y1}\{v_s + (-v_2)\}, \ (-v_2) = -h_{z2}\{v_{I1} + (-v_{I3})\} \qquad (9.47a, b)$$

$$-v_{I3} = h_{y3}\{(-v_2) + v_4\}, \ v_4 = -h_{z4}\{(-v_{I3}) + v_{I5}\} \qquad (9.47c, d)$$

$$v_{I5} = h_{y5}\{v_4 + (-v_6)\}, -v_6 = -h_{z6}(v_{I5}) \tag{9.47e, f}$$

Equation (9.47) can be implemented operationally using the symbolic notations shown in Figure 9.20. Now inverters are not required since v_2, v_{I3}, and v_6 are only negative, whereas voltages v_{I1}, v_4 and v_{I5} are only positive. The operational representation of equation (9.47) is shown in Figure 9.21, where only summers are needed. According to convention, signals originating due to current are placed on the upper line in the diagram and signals originating from voltages are placed in the bottom line. Figure 9.21 can be re-drawn as shown in Figure 9.22. It may be noted that every loop in both Figures 9.21 and 9.22 comprise one positive and one negative transfer function; this is important from the stability point of view.

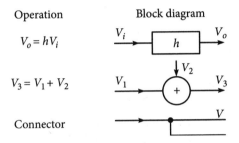

Figure 9.20 Block diagram symbols for the operational equations.

Use of alternate inverting and non-inverting transfer functions in the loop avoid the use of extra inverters, but it creates the possibility of the output being out of phase by 180°; this is not of much significance.

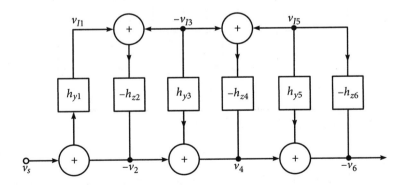

Figure 9.21 Operational representation of equation (9.47).

Avoidance of inverters is achieved by selecting the impedance-based transfer function as negative, resulting in equation (9.47) and its circuit representation in block form is shown in Figures 9.21 and 9.22. Alternatively, in the same way, admittance-based transfer functions

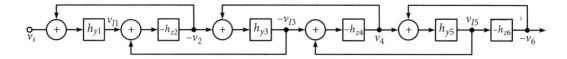

Figure 9.22 An alternate form of presentation of Figure 9.21.

may be made negative. Keeping in mind the negative sign, the transformed parts of equations (9.41) to (9.46) will modify as equations (9.48):

$$-v_{I1} = -h_{y1}\{v_s + (v_2)\}, \quad -v_2 = h_{z2}\{(-v_{I1}) + v_{I3}\} \tag{9.48a, b}$$

$$v_{I3} = -h_{y3}\{(-v_2) + v_4\}, \quad v_4 = h_{z4}\{v_{I3} + (-v_{I5})\} \tag{9.48c, d}$$

$$-v_{I5} = -h_{y5}\{v_4 + (-v_6)\}, \quad -v_6 = h_{z6}(-v_{I5}) \tag{9.48e, f}$$

Like in the previous case, equation (9.48) is represented in block form in Figures 9.23 and 9.24. Once the block form of the equations is available, each of the transfer functions is to be realized, which will depend on the element(s) used in a particular series and shunt branch.

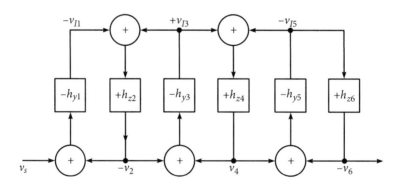

Figure 9.23 Operational block diagram or relationship for equation (9.48).

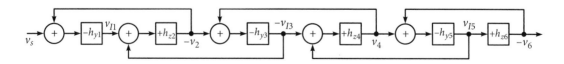

Figure 9.24 An alternate form of Figure 9.23.

The operational simulation method of particular types of ladders will now be discussed while applying the procedure just described.

9.10 Operational Simulation of a Low Pass Ladder

Even when a high pass (HP), band pass (BP) or band stop active or passive filter is to be realized, the procedure begins with a low pass (LP) structure. Later, it is transformed to the desired characteristics. Hence, the general block form structure shown in Figure 9.19 is now taken up to develop a procedure for an operationally simulated LP ladder. Figure 9.25 shows the structure of a sixth-order doubly terminated LPF, for which equations (9.36) and (9.37) will become the branch equations as shown here:

$$I_1 = \frac{V_1}{sL_1 + R_1}, \ I_3 = \frac{V_3}{sL_3}, \ I_5 = \frac{V_5}{sL_5} \tag{9.49}$$

$$V_2 = \frac{I_2}{sC_2}, \ V_4 = \frac{I_4}{sC_4}, \ V_6 = \frac{I_5}{sC_6 + G_2} \tag{9.50}$$

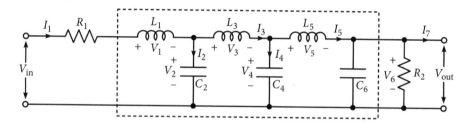

Figure 9.25 A sixth-order doubly terminated low pass ladder structure.

In order to convert the passive ladder into the form of operational representation shown in Figure 9.21 (or 9.22), we need to find the transfer functions h_{Y1}, h_{Y3}, h_{Y5}, h_{Z2}, h_{Z4} and h_{Z6}. To get these transfer functions, all branch immittances are scaled by a resistor R', as it was done in the previous section, and the transfer functions based on impedances, h_{Z2}, h_{Z4}, h_{Z6} are multiplied by (−1) in conformity with equation (9.47) and the block figures in Figure 9.21 or 9.22. Hence, we get:

$$h_{Y1} = R'Y_1 = \frac{R'}{(sL_1 + R_1)} = \frac{1}{\left(\dfrac{sL_1}{R'} + \dfrac{R_1}{R'}\right)} = \frac{1}{s\tau_1 + r_1'} \tag{9.51}$$

$$-h_{Z2} = -\frac{Z_2}{R'} = -\frac{1}{sC_2R'} = -\frac{1}{s\tau_2} \tag{9.52}$$

$$h_{Y3} = \frac{1}{sL_3 / R'} = \frac{1}{s\tau_3} \tag{9.53}$$

$$-h_{Z4} = -\frac{1}{sC_4R'} = -\frac{1}{s\tau_4} \tag{9.54}$$

$$h_{Y5} = \frac{1}{sL_5 / R'} = \frac{1}{s\tau_5} \tag{9.55}$$

$$-h_{Z6} = -\frac{1}{sC_6 R' + G_L R'} = -\frac{1}{(s\tau_6 + 1/r_L')} \tag{9.56}$$

In equations (9.51) to (9.56), τ_i is the time constant as $C_i R'$ or L_i/R', and $r_i' = R_i / R'$. It is observed that the realization requires a non-ideal non-inverting integrator for equation (9.51), two lossless inverting integrators for equations (9.52) and (9.54), two lossless non-inverting integrators for equations (9.53) and (9.55) and a finite gain inverting integrator for equation (9.56). It may be noted that had it been a case of fifth-order filter without C_6, an inverter would have sufficed for operationally simulating the resistor R_L for equation (9.56).

Hence, the problem boils down to the selection of proper, ideal and non-ideal, inverting integrators and non-inverting integrators. For the integrators, each integrator should have two inputs so that along with integration, it sums two voltages like v_s and $(-v_2)$ and v_{f1} and $(-v_{f3})$. In Chapter 8, a number of inverting and non-inverting integrators have been discussed. Figures 8.5 and 8.6 show inverting integrators, with and without active compensation, and Figure 8.7(a) shows a non-inverting integrator using an inverter. However, all the circuits have one input. We know from the circuit of an OA summer that addition of another resistor at the inverting input will do the job. Another important point is that all these integrators are lossless, and to make them non-ideal, a resistor (like QR) is to be connected in parallel with the feedback capacitor as in the Ackerberg–Mossberg biquadratic circuit shown in Figure 8.7(a). Based on this brief discussion, Figure 9.26(a) shows a two input non-ideal inverting integrator without any compensation and Figure 9.26(b) shows a two input inverting integrator with active compensation. Though any configuration can be used, the choice of this combination has the advantage that in any of the loops, the negative *quality* of the inverting integrator almost cancels the positive *quality* of the non-inverting integrator, making the combination better for an extended frequency range. Feedback resistance R_f has to be made open to realize ideal inverting and non-inverting integrators, as shown in Figures 9.26(a) and (b), respectively.

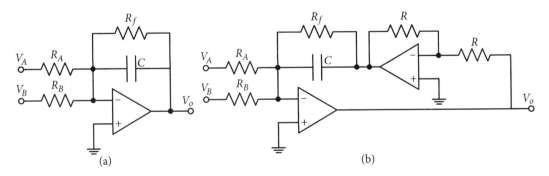

Figure 9.26 (a) A two input non-ideal inverting integrator without compensation and (b) a two input non-ideal non-inverting integrator with active compensation.

For Figure 9.26(a), with OA assumed as ideal:

$$V_o = -\frac{1}{sC+G_f}\left(\frac{V_A}{R_A}+\frac{R_B}{R_B}\right) \tag{9.57}$$

For the active integrator, a scaling factor R_s is used, which is independent of other branch constraints and values in order to provide flexibility in selecting proper component values. Hence, equation (9.57) modifies for the non-ideal integrator as:

$$V_o = -\frac{1}{sCR_s+G_fR_s}\left(\frac{R_s}{R_A}V_A+\frac{R_s}{R_B}V_B\right) = -\frac{1}{s\tau+r_s}(a_1V_A+a_2V_B) \tag{9.58}$$

With $G_f = 0$, equation (9.58) will reduce to the following for an ideal integrator:

$$V_o = -\frac{1}{sCR_s}\left(\frac{R_s}{R_A}V_A+\frac{R_s}{R_B}V_B\right) = -\frac{1}{s\tau}(a_1V_A+a_2V_B) \tag{9.59}$$

Analysis of the circuit in Figure 9.26(b), taking OAs as ideal gives:

$$V_o = +\frac{1}{sC+G_f}\left(\frac{V_A}{R_A}+\frac{V_B}{R_B}\right) \tag{9.60}$$

Once again scaling it by resistor R_s, equation (9.60) modifies as:

$$V_o = +\frac{1}{sCR_s+G_fR_s}\left(\frac{R_s}{R_A}V_A+\frac{R_s}{R_B}V_B\right) = +\frac{1}{s\tau+r_s}(a_1V_A+a_2V_B) \tag{9.61}$$

With $R_f = \infty$, expression for a two-input ideal non-inverting integrator will become:

$$V_o = +\frac{1}{s\tau}(a_1V_A+a_2V_B) \tag{9.62}$$

Now the results of the aforementioned equations for non-ideal and ideal cases can be applied on equations (9.51) to (9.56) along with equation (9.47). First, those three equations are taken which are based on admittance-based transfer functions, as these will be realized using the non-inverting integrator shown in Figure 9.26(b) with (or without feedback) resistor R_f.

$$v_{I1} = \frac{1}{\frac{sL_1}{R'}+\frac{R_{in}}{R'}}\{v_s+(-v_2)\} = \frac{v_s+(-v_2)}{s\tau_1+r'_{in}} \tag{9.63a}$$

$$-v_{I3} = \frac{1}{sL_3/R'}\{(-v_2)+v_4\} = \frac{(-v_2)+v_4}{s\tau_3} \tag{9.63b}$$

$$v_{I5} = \frac{1}{sL_5/R'}\left\{v_4 + (-v_6)\right\} = \frac{v_4 + (-v_6)}{s\tau_5} \tag{9.63c}$$

Next, two equations involving impedance-based transfer functions are taken up, which shall be realized using the inverting integrators shown in Figure 9.26(a) without R_f.

$$(-v_2) = -\frac{1}{sC_2R'}\left\{v_{I1} + (-v_{I3})\right\} = -\frac{v_{I1} + (-v_{I3})}{s\tau_2} \tag{9.64a}$$

$$v_4 = -\frac{1}{sC_4R'}\left\{v_{I1} + (-v_{I3})\right\} = -\frac{v_{I1} + (-v_{I3})}{s\tau_4} \tag{9.64b}$$

The last factor corresponds to equation (9.56), which will be realized using an inverting integrator.

$$(-v_6) = -\frac{1}{G_LR' + sC_6R'}\left\{v_{I5}\right\} = -v_{I5}/\left\{(1/r'L) + sC_6R'\right\} \tag{9.65a}$$

whereas, if C_6 is absent, it becomes a fifth-order LP filter. Then, the last factor corresponding to equation (9.56) will be realized using an inverter as:

$$(-v_6) = -\frac{1}{G_LR'}\left\{v_{I5}\right\} = -v_{I5}(r'_L) \tag{9.65b}$$

Combining the results of the aforementioned equations, the realized circuit for the fifth-order LP active filter through operational simulation is given in Figure 9.27. In this realization, three non-inverting, two inverting integrators and one inverter will be used; hence, a total of nine OAs were needed. If the block forms of Figure 9.23 or 9.24 are used, it will require three inverting and two non-inverting integrators and an inverter, which would need a total of eight OAs.

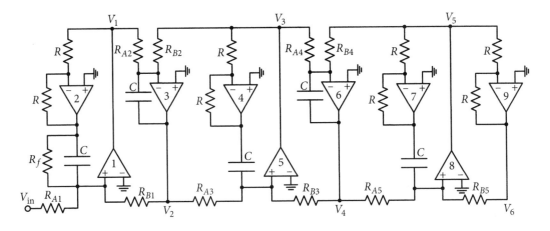

Figure 9.27 Realization of the fifth-order low pass ladder through operational simulation employing integrators of Figure 9.26.

In brief, operational simulation of an LPF can be completed in the following steps.

i.   After choosing an approximation method, a lossless LC ladder along with its element values and terminating resistances is obtained.

ii. All the branch elements are scaled by a resistor to convert the immittances to transfer functions.

iii. Either of the signal flow diagrams of Figure 9.22 or 9.24 can be selected; generally, the block diagram requiring lesser number of non-inverting integrators is chosen as it saves one OA.

iv. Blocks of integrating transfer functions are then replaced by active integrators, with each block having an independent impedance scaling factor for additional flexibility in selecting element values suitable for IC fabrication.

The following example will illustrate the procedure.

Example 9.5: For the given specifications, it was calculated that a fifth-order LP Chebyshev filter with pass band ripples of 0.5 dB will be suitable. Find an active filter using the operational simulation method which will have an LP cut-off frequency of $\omega_o = 10^5$ rad/s and a source resistance of 1 kΩ.

Solution: Figure 9.28 shows the structure and element values for a passive LP fifth-order Chebyshev filter having 0.5 dB ripples. The normalized element values, with cut-off frequency $\omega_o = 1$ rad/s and normalized input terminating resistance $R_{in} = 1$ Ω are as follows:

$L_1 = L_5 = 1.7058$ H, $C_2 = C_4 = 1.2296$ F, $L_3 = 2.5408$ H and $R_L = 1$ Ω

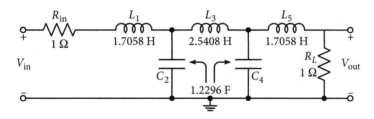

Figure 9.28 Fifth-order low pass filter with normalized element values having 0.5 dB ripples in the pass band.

Equations (9.51) to (9.56) can be written for these element values, wherein R' is the scaling resistance:

$$h_{Y1} = 1/\left(\frac{1.7058s}{R'} + \frac{1}{R'}\right), \quad -h_{Z2} = -1/1.2296sR' \qquad \text{(9.66a, b)}$$

$$h_{Y3} = 1/(2.5408s/R'), \quad -h_{Z4} = -1/1.2296sR' \qquad \text{(9.67a, b)}$$

$$h_{Y5} = 1/(1.7058s/R'), \quad -h_{Z6} = -1/R' \qquad \text{(9.68a, b)}$$

Transfer functions of the equations (9.66)–(9.68) are now compared with equations (9.63)–(9.65) and depending upon which transfer function is to be realized with an inverting or a non-inverting integrator or an inverter (or buffer), element values for each circuit will be obtained.

For the first transfer function h_{Y1} from equations (9.63a) and (9.66a):

$$v_{I1} = \frac{1}{\left(1.7058s / R'\right) + \left(1 / R'\right)}\left\{v_s + \left(-v_2\right)\right\} \tag{9.69}$$

When it is compared with equation (9.61), it gives the following relation for a non-inverting integrator:

$$v_{I1} = \frac{1}{\left(sCR_s + G_{f1}R_s\right)}\left\{\frac{R_s}{R_A}v_s + \frac{R_s}{R_B}\left(-v_2\right)\right\} \tag{9.70}$$

Since in the block diagram of Figure 9.21, v_s and $(-v_2)$ are added with equal weightage, the selection can be made such that $R_A = R_B = R_s$. Comparison of equations (9.69) and (9.70) gives:

$$1/(1.7058s/R') + (1/R') = 1/(sCR_s + G_{f1}R_s) \tag{9.71}$$

Individual terms on the right-hand side of equation (9.71) are compared with respective terms on the left-hand, and application of the block impedance scaling factor R_1 and the frequency normalization factor ω_o results in:

$$CR_s = \frac{1.7058(R_1 / \omega_o)}{R'} \quad \text{and} \quad \frac{R_s}{R_{f1}} = \frac{R_1}{R'} \tag{9.72}$$

In equation (9.72), there are two scaling factors, R' and R_1 which give enough flexibility in selecting practical values for the passive components. Selecting the value for capacitor $C = 1$ nF, and $R_1 = 2.5$ kΩ, equation (9.72) gives:

$$R_s R' = \frac{1.7058(R_1 / \omega_0)}{C} = \frac{1.7058 \times 2.5\,\text{k}\Omega}{1 \times 10^{-9} \times 10^5} = 42.645 \times 10^6 \Omega^2$$

$$R_{f1} = \frac{R_s R'}{R_1} = \frac{42.645 \times 10^6}{2.5 \times 10^3} = 17.058\,\text{k}\Omega$$

If we select $R_s = 5$ k$\Omega = R_{A1} = R_{B1}$, the aforementioned equations give $R' = 8.529$ kΩ and with this, all the components of the non-inverting integrator have been obtained.

For the next integrator, which is inverting and ideal, equation (9.64a) will be used.

$$\left(-v_2\right) = -\frac{1}{1.2296sR'}\left\{v_{I1} + \left(-v_{I3}\right)\right\} \tag{9.73a}$$

Equation (9.73a) is compared with equation (9.59) (with $G_f = 0$).

$$(-V_2) = -\frac{1}{sCR_s}\left\{\frac{R_s}{R_A}V_{I1} + \frac{R_s}{R_B}(-V_{I3})\right\}$$ (9.73b)

Since the scale factor of the two summing voltages V_{I1} and $-V_{I3}$ is unity, we select $R_s = R_A = R_B$. Then, the same impedance scaling and frequency de-normalization used in the first case is applied, which gives:

$$CR_s = 1.2296\, R'/R_1\omega_o$$

For a selected value of $C = 1$ nF as before:

$$\frac{R_s}{R'} = \frac{1.2296}{R_1 C\omega_o} = \frac{1.2296}{2.5\times10^3 \times 1\text{nF}\times 10^5} = 4.9184$$

It is important to note that since all integrators are independent of each other, the active scaling factor R_s in each case may be different, whereas the rest of the ladder will be scaled by the same factor $R' = 8.529$ kΩ which was obtained while designing the first integrator.

Hence, for the inverting integrator $R_s = R_{A2} = R_{B2} = (4.918 \times 8.529) = 41.94$ kΩ.

For the third integrator, from equation (9.63b):

$$(-v_{I3}) = \frac{1}{2.5408s / R'}\{(-v_2)+v_4\}.$$

Comparing this with equation (9.61), with $G_f = 0$:

$$(V_{I3}) = \frac{1}{sCR_s}\left\{\frac{R_s}{R_A}(-V_2) + \frac{R_s}{R_B}V_4\right\}$$ (9.74)

This provides the following relation after impedance and frequency scaling:

$$CR_s = \frac{2.5408(R_1 / \omega_o)}{R'}$$

If C is selected as 1 nF, $R_s = \frac{2.5408\times2.5\times10^3}{8.529\times10^3 \times 10^5 \times 10^{-9}} = 7.447\,\text{k}\Omega$.

Hence, having unity gain for $(-V_2)$ and (V_4), $R_s = R_{A3} = R_{B3} = 7.447$ kΩ.

For the fourth integrator from equation (9.64a), selecting the value of $C_2 = C_4$, the inverting integrator will be exactly the same as the second inverter.

For the fifth integrator, which is non-inverting, from equation (9.63c) $L_5 = 1.7058$ H, the same as L_1; hence, its circuit realization will be the same as that of the first non-inverting integrator; with R_f open since it needs a lossless integrator.

The last unit being an inverter, equation (9.61) reduces to the following:

$$(-v_6) = -\frac{1}{G_{f2}R_s}\left\{\frac{R_s}{R_A}(v_{I5})\right\}$$

(9.75)

Using equations (9.65b) and comparing $-v_6 = -\dfrac{R_1}{R'}(v_{I5})$ with equation (9.75):

For $R_{A6} = 5 \text{ k}\Omega$, $R_L = R_{in} = 2.5 \text{ k}\Omega$, $R_{f2} = 5 \text{ k} \times 2.5 \text{ k}/8.529 \text{ k} = 1.465 \text{ k}\Omega$

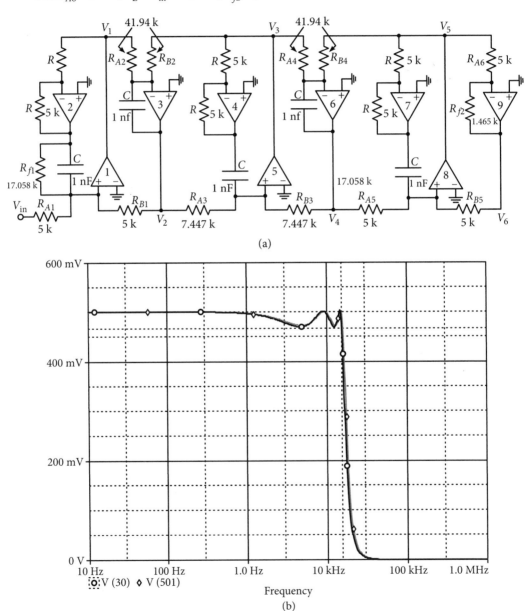

(a)

(b)

Figure 9.29 (a) Circuit realization for the low pass filter of Example 9.5. (b) Frequency response of the operationally simulated filter shown in Figure 9.29(a) and the passive filter shown in Figure 9.28.

Figure 9.27 is redrawn as Figure 9.29(a) with all calculated element values and Figure 9.29(b) shows the simulated magnitude response of the operationally simulated filter. Its corner frequency is 15.6 kHz (98.05 krad/s), voltage gain is 0.4998 and the ripple width is 0.555dB. At 200.8 krad/s (31.953 kHz), attenuation is 43.8 dBs. For comparison sake, the passive filter shown in Figure 9.28 was also simulated and its response is also shown in Figure 9.29(b). Its corner frequency is 15.902 kHz (99.95 krad/s), ripple width is 0.5061 dB and an attenuation of 42.53 dBs is at 31.888 kHz (200.4 krad/s). It can be easily observed that the operationally simulated response almost overlaps the passive response except that the ripple width is slightly more.

9.11 Operational Simulation of the Band Pass Ladder

In many cases, a BPF is designed using an initial design of a prototype LPF and then transforming it as discussed in Section 5.4 of Chapter 5. The effect of such a transformation is that the lossless inductance and capacitances get converted as series LC and parallel LC branches, respectively. Figure 9.30(a) shows a simple LP ladder and its converted BP form in Figure 9.30(b); the series resistor with the inductor gets converted into an RLC branch whose impedance is given as:

$$Z_1 = (R_{in} + SL_1) \rightarrow R_{in} + L_1 Q(s + 1/s) = R_{in} + sQL_1 + L_1\,Q/s$$

$$= R_{in} + sL_{1BP} + 1/sC_{1BP} \tag{9.76}$$

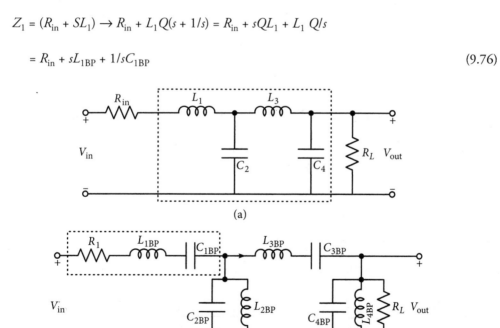

Figure 9.30 (a) Doubly terminated fourth-order lossless low pass ladder and (b) the band pass ladder obtained from (a) through low pass to band pass transformation.

The lossy capacitance at the output end gets converted into a shunt RLC branch having admittance as:

$$Y_4 = G_L + SC_4 \rightarrow G_L + C_4 Q(s + 1/s) = G_L + sC_4 Q + C_4 Q/s$$

$$= G_L + sC_{4BP} + 1/sL_{4BP} \tag{9.77}$$

whereas the lossless branches have been shown to be, respectively:

$$Y_2 = sC_{2BP} + 1/sL_{2BP} \text{ and } Z_3 = sL_{3BP} + 1/sC_{3BP} \tag{9.78}$$

In equations (9.76) and (9.77), elements of the normalized BP section are:

$$L_{1BP} = QL_1, \ C_{1BP} = 1/QL_1, \ L_{4BP} = 1/QC_4, \text{ and } C_{4BP} = QC_4 \tag{9.79}$$

Admittance of the series arm of equation (9.76) and the impedance of the shunt arm of equation (9.77) can be written as:

$$Y_{1BP} = \frac{(1/L_{1BP})s}{s^2 + \left(R_{in}/L_{1BP}\right)s + 1/(L_{1BP}C_{1BP})} \tag{9.80}$$

$$Z_{4BP} = \frac{\left(1/C_{4BP}\right)s}{s^2 + \left(G_L/C_{4BP}\right)s + (1/L_{4BP}C_{4BP})} \tag{9.81}$$

For the operational simulation of the BPF shown in Figure 9.30(b) in the form of the block diagram shown in Figure 9.23 or 9.24, equations (9.80) and (9.81) are scaled by R', in order to convert them into the form of the following voltage ratio transfer functions

$$-Y_{1BP}R' = -h_{Y1}(s) = -\frac{(R'/L_{1BP})s}{s^2 + \left(R_{in}/L_{1BP}\right)s + 1/(L_{1BP}C_{1BP})} \tag{9.82}$$

$$Z_{4BP}G' = h_{Z4}(s) = \frac{\left(G'/C_{4BP}\right)s}{s^2 + \left(G_L/C_{4BP}\right)s + (1/L_{4BP}C_{4BP})} \tag{9.83}$$

Expressions in equations (9.82) and (9.83) represent inverting and non-inverting BP functions with a finite value of pole-Q.

For the internal branches, which do not contain resistors, scaling by the resistor R' gives the respective transfer functions shown here, which are inverting and non-inverting BP functions with infinite pole-Q.

$$\frac{Z_{2BP}(s)}{R'} = h_{Z2}(s) = \frac{(G'/C_{2BP})s}{(s^2 + 1/L_{2BP}C_{2BP})} \tag{9.84}$$

$$-Y_{3BP}R' = -h_{Y3}(s) = -\frac{\left(R'/L_{3BP}\right)s}{(s^2 + 1/L_{3BP}C_{3BP})} \tag{9.85}$$

It is obvious from the aforementioned equations that a doubly terminated BPF obtained from an LPF through frequency transformation can be realized using only second-order filter sections which are able to realize finite as well as infinite Q. Such filter sections should be capable of adding two inputs, and these will be connected alternately inverting and non-inverting form; in exactly the same way as in the LPF case.

There are a number of active circuits for the realization of second-order BP functions employing one or more than one active device. Obviously, the advantages and limitations of the applied second-order section will be reflected in the overall functioning of the operationally simulated realization. Use of one OA BPF section will be economical but usually such realizations have high sensitivities, whereas multi amplifier sections may not be economical but are less sensitive. The following example will help in understanding the procedure for operationally simulating a BPF.

Example 9.6: A doubly terminated LP Butterworth approximated filter structure is shown in Figure 9.31. Using frequency transformation, convert the filter to a BPF having center frequency $\omega_o = 10^4$ rad/s, and $Q = 5$. Find a suitable active realization using the operational simulation method.

Solution: From equation (9.79), normalized value of the elements of the BPF are:

$$L_{1BP} = 5 \times 0.7645 = 3.8225 \text{ H}, \; C_{1BP} = 1/5 \times 0.7645 = 0.2616 \text{ F}$$

$$L_{2BP} = 1/5 \times 1.848 = 0.1082 \text{ H}, \; C_{2BP} = 5 \times 1.848 = 9.24 \text{ F}$$

$$L_{3BP} = 5 \times 1.848 = 9.24 \text{ H}, \; C_{3BP} = 1/5 \times 1.848 = 0.1082 \text{ F}$$

$$L_{4BP} = 1/5 \times 0.7645 = 0.2616 \text{ H}, \; C_{4BP} = 5 \times 0.7645 = 3.8225 \text{ F} \tag{9.86}$$

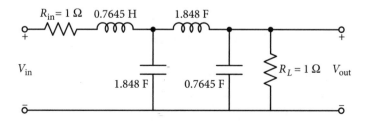

Figure 9.31 Fourth-order Butterworth filter structure with normalized elements values for Example 9.6.

After frequency normalization by $\omega_o = 10^4$ rad/s and impedance normalization by a factor of 10^3, element values become:

$$R_{in} = 1 \text{ k}\Omega, L_{1BP} = 382.25 \text{ mH}, C_{1BP} = 26.16 \text{ nF} \tag{9.87}$$

$$L_{2BP} = 10.82 \text{ mH}, C_{2BP} = 924 \text{ nF} \tag{9.88}$$

$$L_{3BP} = 924 \text{ mH}, C_{3BP} = 10.82 \text{ nF} \tag{9.89}$$

$$L_{4BP} = 26.16 \text{ mH}, C_{4BP} = 382.25 \text{ nF}, R_L = 1 \text{ k}\Omega \tag{9.90}$$

Figure 9.32 shows the transformed passive eighth-order BPF structure with de-normalized element values.

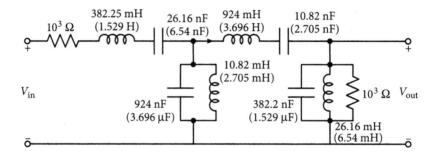

Figure 9.32 Eighth-order band pass filter with de-normalized element values from Figure 9.31 for Example 9.6.

Series and shunt branches comprising the elements in equations (9.87) and (9.90), respectively, are written like equations (9.80) and (9.81), and scaled by $R'(\text{k}\Omega)$ in order to be represented by transfer functions of equations (9.82) and (9.83). The resulting equations will be as follows:

$$-h_{y1}(s) = -\frac{\left(R'/0.38225\right)s}{s^2 + \left(10^3/0.38225\right)s + (0.38225 \times 26.16 \times 10^{-9})} \tag{9.91}$$

$$h_{z4}(s) = \frac{\left(G'/0.38225 \times 10^{-6}\right)s}{s^2 + \left(10^{-3}/0.38225 \times 10^{-6}\right)s + (0.02616 * 0.38225 \times 10^{-6})} \tag{9.92}$$

Taking the same steps, equations (9.88) and (9.89) will transform into the following equations:

$$h_{z2}(s) = \frac{\left(G'/924 \times 10^{-12}\right)s}{s^2 + 1/\left(0.1082 \times 924 \times 10^{-12}\right)} \tag{9.93}$$

$$h_{y3}(s) = \frac{\left(R'/0.924\right)s}{s^2 + 1/\left(0.924 \times 10.82 \times 10^{-12}\right)} \tag{9.94}$$

There are some choices available for the realization of the four transfer functions, equations (9.91)–(9.94). Separate configurations can be chosen for the inverting and non-inverting second-order BP sections, with finite and infinite Q. In the present example, a modified Tow–Thomas biquad with two inputs as shown in Figure 9.33 is used. The important feature of the modified circuit is that both inverting and non-inverting BP responses are available at the OA outputs and infinite Q is obtained simply by open circuiting the resistor R^*. Therefore, the design of all the four transfer functions becomes modular; an attractive feature for integration. Assuming ideal OAs, analysis of the circuit of Figure 9.33 gives its transfer function as:

$$V_{BP} = \left(\frac{V_A}{R_A} + \frac{V_B}{R_B}\right)\frac{(1/C_1)s}{s^2 + (1/C_1R^*)s + (1/C_1C_2R_1R_2)} \tag{9.95}$$

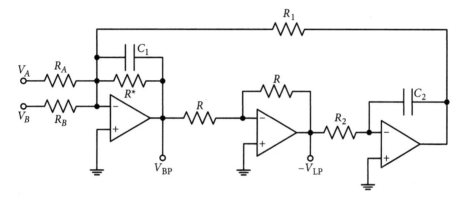

Figure 9.33 Modified Tow–Thomas biquad with two inputs.

From equation (9.95), parameters are

$$\omega_o = 1/(C_1\,C_2\,R_1\,R_2)^{0.5} \tag{9.96}$$

$$\frac{\omega_o}{Q} = \frac{1}{C_1R^*} \rightarrow Q = R^*\sqrt{\left(\frac{1}{R_1R_2}\frac{C_1}{C_2}\right)} \tag{9.97}$$

Normally, $C_1 = C_2 = C$, then the parameters are:

$$\omega_o = 1/C(R_1R_2)^{1/2} \text{ and } Q_{BP} = R^*\sqrt{\left(\frac{1}{R_1R_2}\right)} \tag{9.98}$$

To get infinite Q_{BP}, R^* is made open, which does not affect the expression for ω_o. Without affecting generality, $R_1 = R_2 = R$, and as the two voltages V_A and V_B is to be added with equal weightage, we select $R_A = R_B$. In the finite Q case, the simplified form of equation (9.95) becomes:

$$V_{BP} = \frac{(V_A + V_B)(1/CR_A)s}{s^2 + (1/CR^*)s + 1/(CR)^2} \tag{9.99}$$

Selecting $C = 10^{-7}$ F, in all the four branches, from equation (9.98), it gives:

$$R = 1/10^4 \times 10^{-7} = 1 \text{ k}\Omega \tag{9.100}$$

For evaluation of the rest of the elements of branch 1, comparing the parameters of equation (9.99) with equation (9.82), we get:

$$\frac{1}{CR^*} = \frac{R_{in}}{L_{1BP}} \text{ and } \frac{R'}{L_{1BP}} = \frac{1}{CR_A} \tag{9.101}$$

$$R^* = L_{1BP}/CR_{in} = 0.38225/10^{-7} \times 10^3 = 3.8225 \text{ k}\Omega \tag{9.102}$$

With $R' = 1$ kΩ, we get from equation (9.101):

$$R_{A1} = L_{1BP}/CR' = 3.8225 \text{ k}\Omega = R_{B1} \tag{9.103}$$

In equation (9.103), subscript 1 indicates elements for the first branch for which all elements have been calculated. For shunt arm 2, equation (9.99) (with R^* open) is compared with equation (9.84). It gives:

$$R_{A2} = C_{2BP}/G' C = 9.24 \text{ k}\Omega = R_{B2} \tag{9.104}$$

For the third (series) branch, R^* is open and comparison between equations (9.99) and (9.85) gives:

$$R_{A3} = L_{3BP}/CR' = 9.24 \text{ k}\Omega = R_{B3} \tag{9.105}$$

For the fourth (shunt) arm, comparison between equations (9.99) and (9.83), we get:

$$R^* = 3.8225 \text{ k}\Omega, R_{A4} = 3.8225 \text{ k}\Omega \tag{9.106}$$

Figure 9.34(a) shows the operationally simulated eighth-order BPF. Care is taken to keep alternate inverting and non-inverting blocks.

Figure 9.34(b) shows the simulated response of the active filter of Figure 9.34(a), having a center frequency of 1.5969 kHz (10.037 krad/s) with a mid-band gain of 0.547, instead of 0.5. With a bandwidth of 320.8 Hz, obtained $Q = 4.977$. It is observed that the pass band is also flat as it should be. Figure 9.34(b) also shows the simulated response at the same center frequency, with $Q = 5$, for the passive structure. With the mid-band gain as 0.5, the pass band is still flatter with $Q = 5.02$.

Figure 9.34(c) shows the simulated response at the same center frequency with $Q = 20$, for the operationally simulated as well as passive structure. De-normalized element values for the filter with $Q = 20$ are also shown in Figure 9.34(a) within brackets. The response is not flat and the mid-band gain is also increased to 0.713 instead of 0.5. The realized value of $Q = 19.6$.

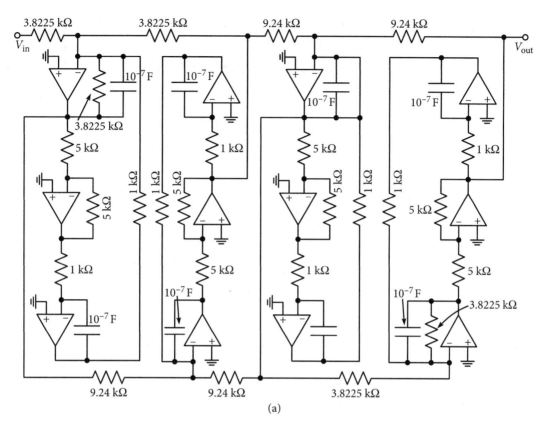

(a)

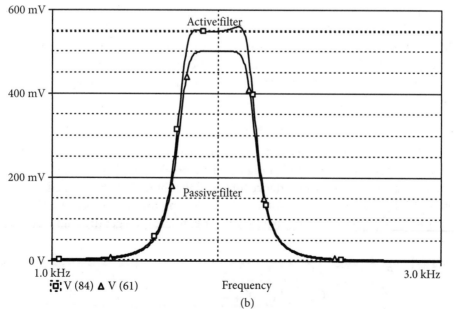

(b)

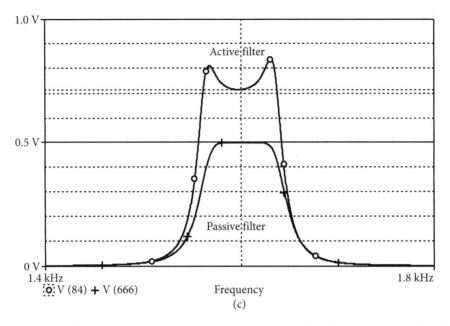

Figure 9.34 (a) Operationally simulated eight-order band pass filter for Example 9.6. (b) Simulated response of the filter with $Q = 5$, and passive filter (c) Response of the operationally simulated eighth-order band pass filter with $Q = 20$.

9.12 General Ladder Realization

LP and BP lossless ladders are very important from the point of view of filter design. From the discussion so far, it can be observed that the procedure for their realization through operation simulation is a bit long but follows a set pattern. Of course, the design depends on the kind of structure used for realizing a particular second-order transfer function for the inverting or non-inverting mode branch. However, there are many ladder configurations other than an all pole LP and BP in which the branches (series or shunt) are in a most general form. A branch can comprise inductors and capacitors in both series and parallel form and resistors as well. Figures 9.35(a) and (b) show one such general structure each in series and parallel form. Herein, in Figure 9.35(a), subscript s indicates an element in the series branch and in Figure 9.35(b), p indicates parallel (or shunt) branch elements. While operationally simulating, the series branches are expressed in admittance form and the shunt branch in impedance form, and both are scaled by the same value resistance R'. Following the same procedure and expressing them in continued expansion form, first for the series branch, we get the following expression.

$$R'Y(s) = \cfrac{1}{\cfrac{R_{s1}}{R'} + \cfrac{L_{s1}}{R'}s + \cfrac{1}{sC_{s1}R' + \cfrac{1}{sC_{s2}R' + \cfrac{1}{\cfrac{L_{s2}}{R'}s}}}} \qquad (9.107)$$

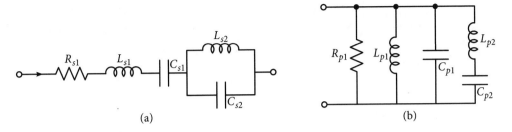

Figure 9.35 (a) Series and (b) shunt branch of a ladder with arbitrary combination of elements.

Since $R'Y(s)$ can be written as $h_Y(s)$ and the scaled elements can be written in lower case symbols, equation (9.107) modifies as:

$$h_Y(s) = \cfrac{1}{r_{s1} + sl_{s1} + \cfrac{1}{sC_{s1}} + \cfrac{1}{sC_{s2} + \cfrac{1}{sl_{s2}}}} \tag{9.108}$$

Similarly, for the shunt branch, the impedance function divided by R' and the resultant transfer function with negative sign are:

$$-\frac{Z(s)}{R'} = -\cfrac{1}{R'G_{p1} + sC_{p1}R' + \cfrac{1}{s\dfrac{L_{p1}}{R'}} + \cfrac{1}{s\dfrac{L_{p2}}{R'} + \cfrac{1}{sC_{p2}R'}}} \tag{9.109}$$

Since $Z(s)/R'$ can be written as $h_Z(s)$ and the scaled elements can be written in lower case symbols, equation (9.109) modifies as:

$$-h_Z(s) = -\cfrac{1}{g_{p1} + sc_{p1} + \cfrac{1}{sl_{p1}} + \cfrac{1}{sl_{p2} + \cfrac{1}{sc_{p2}}}} \tag{9.110}$$

Note that series branches result in a current, like $I_1 = Y_1\{V_s + (-V_2)\}$ in Figure 9.19, which gets converted to a voltage after resistance scaling, and the shunt branches result in voltages like $V_2 = Z_2(I_1 - I_3)$; scaled as well.

A suggested approach for the realization of the kind of transfer function of equations (9.108) and (9.110) uses a two-input summing inverter with an arbitrary transfer function $H(s) = (V_t/V_{out})$ and admittance $Y(s)$ in its feedback path as shown in Figure 9.36(a) and (b). Assuming OA as ideal, a simple analysis of Figure 9.36(a) shows that:

$$V_{out} = -\frac{1}{Y(s)H(s)}(V_A G_A + V_B G_B) = -\frac{G'}{Y(s)H(s)}\left\{\frac{G_A}{G'}V_A + \frac{G_B}{G;}V_B\right\} \tag{9.111}$$

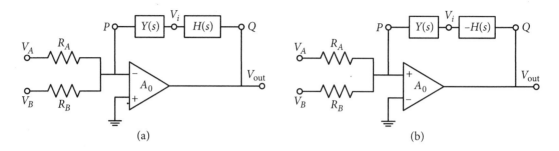

Figure 9.36 (a) A schematic realizing the transfer function for equation (9.111), with $H(s) = (V_i/V_{out})$ and (b) its dual for realization of the inverting transfer function.

Equation (9.111) shows the summation of two input voltages resulting in a voltage ratio transfer function which is inverting and proportional to the inverse of $Y(s) \times H(s)$, with $R' = 1/G'$ as the scaling resistor. Obviously, this arrangement is valid for the inverting transfer function blocks in operational simulation as represented by equation (9.110). In a dual scheme, shown in Figure 9.36(b), with two voltages connected to the non-inverting input of the OA and the inverting terminal grounded, and with transfer function $(-H(s))$, exactly the same output as in equation (9.111) results in a *non-inverting transfer function*, which is used for the non-inverting blocks in operational simulation as represented by equation (9.108). It is to be noted that, in each version, with the help of negative feedback, one inverting and one non-inverting combination of OA and $H(s)$ is maintained for the stability of the arrangement.

As mentioned earlier, $H(s)$ is an arbitrary function in s; hence, one can realize not only the transfer function of equation (9.108) and (9.110), but anything simpler or more complex than that. Presently, we will limit our discussion to the circuit realization of Figure 9.35, but it can be extended to other circuits or simplified to the LP and BP circuits, which are nothing but special simpler cases of the circuit in Figure 9.35.

9.12.1 Realization of shunt arm

To realize the inverting transfer function of equation (9.110), four feedback branches need to be connected between terminals P and Q of the circuit in Figure 9.36(a) corresponding to each branch of the circuit in Figure 9.35(b). For the realization of single elements, R_{p1}, L_{p1} and C_{p1}, the structures shown in Figure 9.37(a) will be connected between terminals P and Q, whereas, for realizing the parallel combination of L_{p2} and C_{p2}, the circuit shown in Figure 9.37(b) will be connected between terminals P and Q. Assuming both OAs as ideal, a routine analysis gives the following relations in Figures 9.37(a) and (b), respectively.

$$V_{c0} = H_0 \, V_{out}, \; V_{c1} = H_1 \, V_{out} \text{ and } V_{c2} = H_2 \, V_{out} \tag{9.112}$$

$$G_{31} \, V_{out} + \{Y_3 \, (-H_3) + Y_4 \, (-H_4)\} \, V_{c3} = 0 \text{ or } V_{c3} = G_{31} \, V_{out}/(Y_3 \, H_3 + Y_4 \, H_4) \tag{9.113}$$

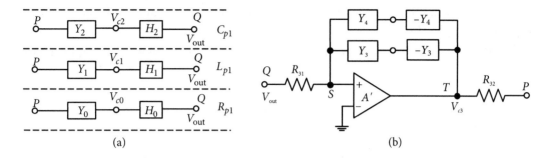

Figure 9.37 Expanded configuration for realizing (a) R_{p1}, L_{p1} and C_{p1} and (b) L_{p2} and C_{p2} of a ladder like that shown in Figure 9.35(b).

Substitution of Figures 9.37(a), (b) in Figure 9.36(a), yields the following relation:

$$G_A V_A + G_B V_B + V_{c0} Y_0 + V_{c1} Y_1 + V_{c2} Y_2 + V_{c3} G_{32} = 0 \tag{9.114}$$

Substituting V_{c0}, V_{c1}, V_{c2} and V_{c3} from equations (9.112) and (9.113) in equation (9.114) and scaling by resistor R'', the following output voltage is obtained.

$$V_{out} = -\left\{ \frac{R''}{R_A} V_A + \frac{R''}{R_B} V_B \right\} \frac{1}{Y_0 H_0 R'' + Y_1 H_1 R'' + Y_2 H_2 R'' + \dfrac{1}{\dfrac{Y_3 H_3}{G_{31} G_{32} R''} + \dfrac{Y_4 H_4}{G_{31} G_{32} R''}}} \tag{9.115}$$

Equation (9.115) shows the output voltage as a function of two voltages V_A and V_B, which are being added. If these voltages are to be added directly without any weightage, then R_A and R_B will be equal to R''. If the voltage levels are to be changed, which is sometimes required for changing the dynamic range of the filter section, then it is done by opting for a proper ratio between R'' and R_A and R_B. Components of the transfer function of equation (9.115) can be compared with the inverting transfer function of equation (9.110) and a procedure can be developed for finding the nature of the transfer function $H_i(s)$ and the value (expression) of elements to be used in it in terms of the elements of the circuit shown in Figure 9.35(b).

For the shunt arm of the ladder, to realize the element R_{p1}, we select $Y_0 = G_0$ and $H_0 = 1$, which means:

$$g_{p1} = Y_0 H_0 R'' \rightarrow G_{p1} R' = G_0 R'' \text{ or } R_0 = (R''/R') R_p = k_p R_p, \text{ where } k_p = R''/R' \tag{9.116}$$

For the realization of the capacitance C_{p1}, we select $Y_1 = sC_1$ and $H_1 = 1$, which results in:

$$sc_{p1} = Y_1 H_1 R'' \rightarrow sC_{p1} R' = sC_1 R'' \text{ or } C_1 = (R'/R'') C_{p1} = C_{p1}/k_p \tag{9.117}$$

It is better to fix the value of k_p from equation (9.117) and use equation (9.116) to get the value of R_0. If R_p is not present in the shunt branch of the original ladder, then R_0 shall be open circuited, and similarly, if C_{p1} is absent, C_1 shall will also be absent, and k_p can be selected using equation (9.116).

For the realization of inductor L_{p1}, we select $H_2 = (1/sC_2R_{21})$, a non-inverting integrator, and $Y_2 = G_{22}$, which means that:

$$\frac{1}{sl_{p1}} = Y_2 H_2 R'' \rightarrow \frac{1}{sL_{p1}/R'} = \frac{G_{22}}{sC_2R_{21}}R'' \rightarrow R_{21}R_{22} = \frac{L_{p1}}{C}k_p \text{ for } C_1 = C_2 = C \qquad (9.118)$$

Circuit structure for the realization of R_{p1}, L_{p1} and C_{p1}, corresponding to equations (9.116), (9.117) and (9.118) are shown in Figure 9.38(a).

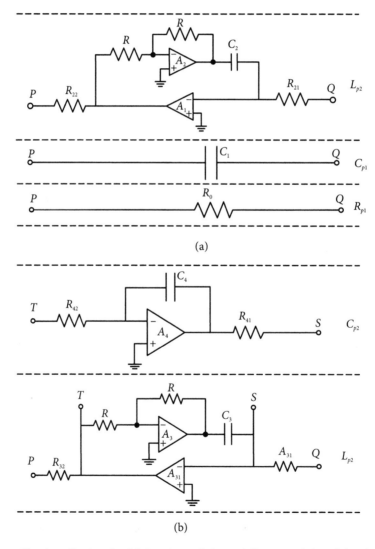

Figure 9.38 Circuit realization for (a) R_{p1}, L_{p1} and C_{p1} and (b) L_{p2} and C_{p2} of the shunt branch of a ladder like that shown in Figure 9.35(b).

For the branch containing a series combination of L_{p2} and C_{p2}, first to find the relations for L_{p2}, we select $H_3 = 1$ and $Y_3 = sC_3$; this gives the following relations:

$$sl_{p2} = \frac{Y_3 H_3}{G_{31} G_{32} R''} \rightarrow s\frac{L_{p2}}{R'} = s\frac{C_3 R_{31} R_{32}}{R''} \rightarrow R_{31} R_{32} = \frac{L_{p2}}{C_3}\frac{R''}{R'} = \frac{L_{p2}}{C} k_p \text{ for } C_3 = C \quad (9.119)$$

For the realization of the capacitor C_{p2}, we select, $Y_4 = G_{41}$ and $H_4 = (1/sC_4 R_{42})$ a non-inverting integrator, and we get:

$$\frac{1}{sc_{p2}} = \frac{Y_4 H_4}{G_{31} G_{32} R''} \rightarrow \frac{1}{sC_{p2} R'} = \frac{1}{s}\frac{G_{41} R_{31} R_{32}}{C_4 R_{42} R''}$$

Substituting for $R_{31} R_{32}$ from equation (9.119), we get:

$$R_{41} R_{42} = \frac{C_{p2}}{C_4}\frac{R'}{R''} \times \frac{L_{p2}}{C} k_p = \frac{C_{p2} L_{p2}}{C^2} \text{ for } C_4 = C \quad (9.120)$$

The circuit structures for the realization of L_{p2} and C_{p2}, corresponding to equations (9.119) and (9.120) are shown in Figure 9.38(b), which are to be combined with Figure 9.37(a).

Equations (9.116) to (9.120) serve as the design equations for the structure and the element values. Once k_p is fixed, and $C_1 = C_2 = C_3 = C_4 = C$ are selected, the products $R_{21} R_{22}$, $R_{31} R_{32}$ and $R_{41} R_{42}$, are found out. There is no constraint and we can select:

$$R_{i1} = R_{i2} \quad (9.121)$$

The remaining input resistances R_A and R_B are decided depending on the voltage gain given to the two voltages V_A and V_B. For the usual case of summing the voltages directly,

$$R_A = R_B = R'' \quad (9.122)$$

9.12.2 Realization of series arm

Admittance of a series arm, which is shown in Figure 9.35(a), is converted to a transfer function of equations (9.107) and (9.108). Its non-inverting expression is realized by similar structures as of Figure 9.37, shown in Figure 9.39, where input voltages are applied at the non-inverting terminal of OA, A_0 of Figure 9.36(a); the rest of the inputs are given to the inverting terminal. Analysis results in exactly the same output voltage expression as in equation (9.115) without the inverting sign as desired.

Using the same procedure as was adopted for the shunt branch, the following selections were made for the elements R_{s1}, L_{s1} and C_{s1}.

$$H_0 = H_1 = -1, \; -H_2 = -\frac{1}{sC_2 R_{21}} \quad (9.123a)$$

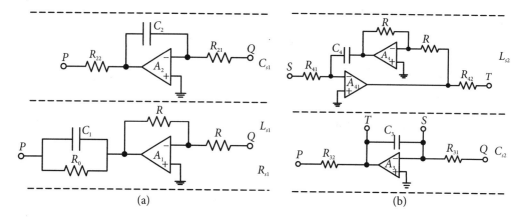

Figure 9.39 Circuit realization for (a) R_{s1}, L_{s1} and C_{s1} and (b) L_{s2} and C_{s2} of the series branch of a ladder like that of Figure 9.35(a).

Corresponding to equation (9.123a), the circuit structures to be connected between terminals P and Q of Figure 9.36(a) are shown in Figure 9.39(a). Selected components will be as:

$$Y_0 = G_0, \ Y_1 = sC_1, \ Y_2 = sC_2, \tag{9.123b}$$

The following selections were made for the elements L_{s2} and C_{s2}.

$$H_3 = 1, \ H_4 = \frac{1}{sC_4 R_{41}} \tag{9.124a}$$

Corresponding to equation (9.124a), the circuit structures to be connected between terminals P and Q of Figure 9.36(b) are shown in Figure 9.39(b). Selected components will be as:

$$Y_3 = sC_3, \ Y_4 = G_4 \tag{9.124b}$$

Component values are obtained as:

$$Y_0 H_0 R'' = R_{s1} / R' \rightarrow R_0 = (R'R'') / R_{s1} = (k_s^2) / R_{s1} \tag{9.125a}$$

$$Y_1 H_1 R'' = sL_{s1} / R' \rightarrow C_1 = L_{s1} / (R'R'') = L_{s1} / k_s^2 \tag{9.125b}$$

As before, after selecting a suitable value of $C_1 = C$, k_s is fixed from equations (9.125b); then, R_0 is obtained from equation (9.125a) and the value of k_s remains constant in the next steps. In case L_{s1} is not present, then $C_1 = 0$ and the value of k_s shall will be decided by equation (9.125a).

With C_2, C_3 and C_4 chosen equal to $C(C_1)$, the remaining resistances are obtained from the following:

$$R_{21} R_{22} = \frac{C_{s1}}{C} k_s^2, \ R_{31} R_{32} = \frac{C_{s2}}{C} k_s^2, \ R_{41} R_{42} = \frac{L_{s2}}{C} \frac{R_{31} R_{32}}{k_s^2} = \frac{L_{s2} C_{s2}}{C^2} \tag{9.126}$$

Once again without losing generality, we can select:

$$R_{i1} = R_{i2} = R_i \text{ and } R_{Ai} = R_{Bi} = R'' \text{ for unity gain.} \tag{9.127}$$

Example 9.7: Compare the responses of the passive circuit shown in Figure 9.40(a) and its active version as an illustration of the realization of shunt arm through operation simulation.

Solution: Impedance scaling factor R' of 1 kΩ and frequency scaling factor of 2 krad/s gives the element values as:

$$R_{p1} = 1 \text{ k}\Omega, \ L_{p1} = 0.2 \text{ H}, \ C_{p1} = 0.05 \text{ μF}, \ L_{p2} = 0.1 \text{ H and } C_{p2} = 0.1 \text{ μF} \tag{9.128}$$

Resistance $R^* = 1$ kΩ ($= R_{p1}$) has been connected in series to find the response in the passive case only.

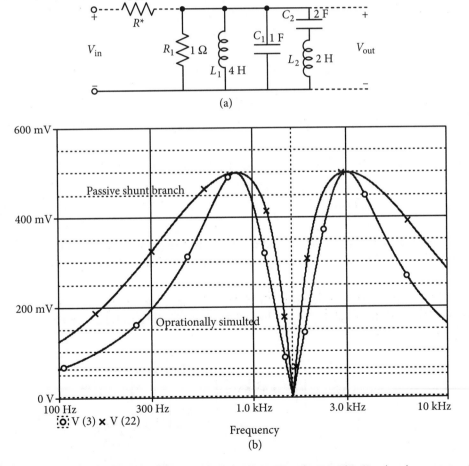

Figure 9.40 (a) Prototype normalized shunt arm for Example 9.7. (b) Simulated response of the passive and operationally simulated shunt arm of Figure 9.40(a).

Design of the shunt arm begins with the application of equation (9.117), and selection of scaling factor R'' as 1 kΩ. As impedance scaling factor R' was 1 kΩ, factor k_p = 1. It results in capacitor $C = C_{p1}$ = 0.05 μF. With k_p being unity, equation (9.116) gives R_0 = 1 kΩ.

While selecting $R_{i1} = R_{i2}$, equations (9.118)–(9.120) give $C_2 = C_3 = C_4$ = 0.05 μF, $R_{21} = R_{22}$ = 2 kΩ, $R_{31} = R_{32} = \sqrt{2}$ kΩ and $R_{41} = R_{42}$ = 2 kΩ.

Input scaling resistance R_A equals R'' when voltage gain of unity is desired. In the present case, for the passive circuit, gain magnitude is 0.5; hence, R_A is doubled to 2 kΩ.

Figure 9.40(b) shows the simulated frequency response of the de-normalized passive as well as its active version which match well with a notch at 10 krad/s and peak at the same frequencies of 19.92 krad/s and 0.823 krad/s.

Example 9.8: Compare responses of the passive circuit shown in Figure 9.41(a) and its active version as an illustration of the realization of series arm through operation simulation.

Solution: In the normalized passive circuit, resistance R_L = 1 kΩ is connected as a load for obtaining the circuit's transfer function. Impedance scaling factor of R' = 1 kΩ and a frequency scaling factor of 20 krad/s was employed, resulting in the value of elements as:

$$R_L = 1 \text{ k}\Omega, R_{s1} = 1 \text{ k}\Omega, L_{s1} = 0.2 \text{ H}, C_{s1} = 0.05 \text{ }\mu\text{F}, L_{s2} = 0.1 \text{ H}, C_{s2} = 0.1 \text{ }\mu\text{F} \qquad (9.129)$$

From equation (9.125b), selecting $C_1 = 0.05\mu\text{F}, k_s^2 = \left(L_{s1} / C_1^2\right) = \left(0.2 / 0.05 \times 10^{-6}\right) = 4 \times 10^6$.

Application of equations (9.125)–(9.127) and selecting $R_{i1} = R_{i2}$ gives the element values as:

$$R_0 = 4 \text{ k}\Omega, R_{21} = R_{22} = 2 \text{ k}\Omega, R_{31} = R_{32} = 2\sqrt{2} \text{ k}\Omega, R_{41} = R_{42} = 2 \text{ k}\Omega$$

$$C_2 = C_3 = C_4 = 0.05 \text{ }\mu\text{F} \qquad (9.130)$$

Since R' was selected as 1 kΩ, R'' = 4 kΩ. Hence, to bring the output voltage equal to half at the input (same as in the case of the passive circuit), the value of the resistance $R_A = 2R''$ = 8 kΩ.

Frequency responses of the de-normalized passive circuit and the active circuit are shown in Figure 9.41(b) with notch at 1.58 kHz and peaks at 1.123 kHz and 2.248 KHz.

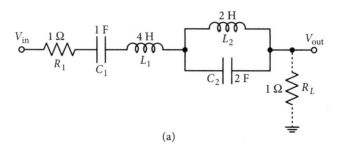

(a)

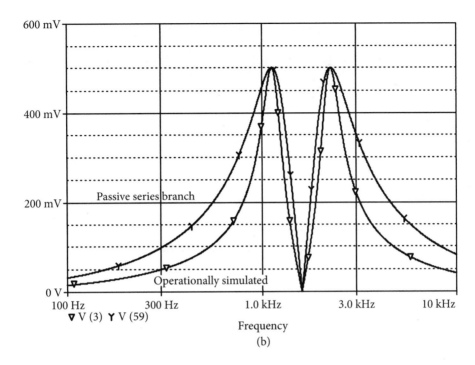

Figure 9.41 (a) Prototype normalized series arm for Example 9.8. (b) Simulated response of the passive and operationally simulated series arm of Figure 9.41(a).

Example 9.8: Redesign the eighth-order BPF of Example 9.6 using the general ladder realization approach.

Solution: Application of the frequency de-normalization by 10^4 rad/s and impedance scaling by a factor of 10^3, elements of the BPF were made available in equations (9.87)–(9.90). Now, its structure is shown in Figure 9.42(a), along with their values, which are calculated in following paragraphs. Design for the complete circuit proceeds in terms of individual branches and then they are properly interconnected.

Series branch number 1: Selecting scaling factor R'' equal to the original impedance scaling factor $R'(1\ \mathrm{k}\Omega)$ gives the coefficient $k_s^2 = 10^6$. Assuming all capacitors to be equal for the series branch, use of equations (9.125b), (9.125a), and (9.126), respectively, give the following values of the elements:

$$C_{11} = \frac{L_{s1}}{k_s^2} = \frac{1.529}{10^6} = 1.529\,\mu\mathrm{F} \tag{9.131a}$$

$$R_{01} = k_s^2 / R_s = 10^3\,\Omega \tag{9.131b}$$

$$R_{211} = R_{221} = k_s \sqrt{(C_{s1} / C_{21})} = 10^3 \sqrt{\left(\frac{6.54 * 10^{-9}}{1.529 * 10^{-6}}\right)} = 65.4\,\Omega \tag{9.131c}$$

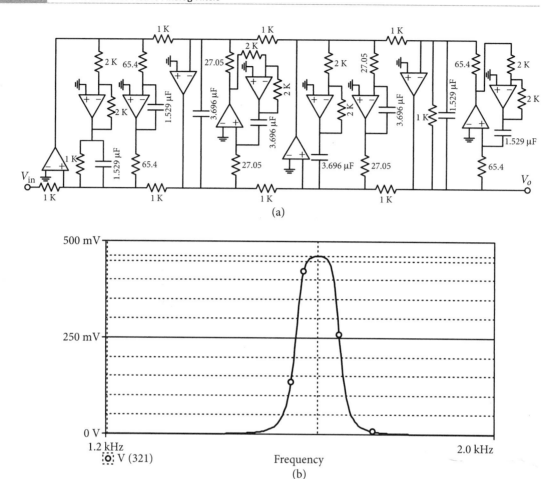

Figure 9.42 (a) Operational simulation circuit with element values for the eighth-order band pass filter of Example 9.9. (b) Response of the eighth-order band pass filter using general ladder realization approach for Example 9.9.

Assuming gain of the filter as unity, the input resistances are:

$$R_{A1} = R_{B1} = R'' = 1 \text{ k}\Omega \tag{9.132}$$

Shunt branch number 2: With all capacitors for the shunt branch taken as equal to C_{12} and $k_p = (R'/R'') = 1$, application of equations (9.117), (9.116), and (9.118), respectively, gives the following values:

$$C_{12} = C_{p1}/k_p = 3.696 \text{ μF, and } R_{02} = \infty \tag{9.133a}$$

$$R_{212} = R_{222} = k_p \sqrt{(L_{p2}/C_{12})} = \sqrt{\left(\frac{2.705 \times 10^{-3}}{3.696 \times 10^{-6}}\right)} = 27.05\,\Omega \tag{9.133b}$$

Input resistances $R_{A2} = R_{B2} = R'' = 1$ kΩ. In fact, for the rest of the two branches, input resistances will be the same.

Series branch number 3: Similar to series branch number 1, all capacitors in the branch remaining equal:

$$C_{13} = 3.696 \text{ }\mu\text{F, and } R_{03} = \infty \tag{9.134a}$$

$$R_{213} = R_{223} = k_s\sqrt{(C_{s3}/C_{23})} = 10^3 \sqrt{\left(\frac{2.705 \times 10^{-9}}{3.696 \times 10^{-6}}\right)} = 27.05\,\Omega \tag{9.134b}$$

Shunt branch number 4: Similar to the case of branch number 2, and capacitors remaining equal:

$$C_{14} = 1.529 \text{ }\mu\text{F and } R_{04} = k_p R_{p4} = 1 \text{ k}\Omega \tag{9.135a}$$

$$R_{214} = R_{224} = 10^3 \sqrt{\left(\frac{6.54 \times 10^{-9}}{1.529 \times 10^{-6}}\right)} = 65.4\,\Omega \tag{9.135b}$$

Figure 9.42(a) shows the combination of all the four branches connected in operational simulation form, along with the element values. The simulated response of the filter is shown in Figure 9.42(b). The simulated center frequency is 1.5878 kHz (9.945 krad/s) with a bandwidth of 78.99 Hz, resulting in $Q = 20.1$. Value of the mid-band gain is 0.462 instead of 0.5.

References

[9.1]   Orchard, H. J. 1966. 'Inductor less Filters,' *Electronics Letters* 2: 224–5.

[9.2]   Tellegen, B. D. H. 1948. 'The Gyrator, A New Electric Network Element,' *Philips Research Reports* 3: 81–101.

[9.3]   Holt, A. G. J., and J. R. Taylor. 1965. 'Method of Replacing Ungrounded Inductances by Grounded Gyrator,' *Electronic Letters* 1 (4): 105.

[9.4]   Riordan, R. 1967. 'Simulated Inductance using Differential Amplifiers,' *Electronic Letters* 3(2): 50–1.

[9.5]   Antoniou, A. 1969. 'Realization of Gyrators using Operational amplifiers and Their Use in Active Network Synthesis,' *IEE Proceedings* 16: 1838–50.

[9.6]   Dutta Roy, S. C. 1974. 'A Circuit for Floating Inductance Simulation,' *IEEE Proceedings* 62: 521–3.

[9.7]   Gorski-Popiel, G. 1967. 'RC-Active Synthesis using Passive Immittance Converter,' *Electronic Letters* 3: 381–2.

[9.8]   Bruton, L. T. 1969. 'Network Transfer Functions Using the Concept of Frequency Dependent Negative Resistance,' *IEEE Transactions on Circuit Theory* CT-16: 406–8.

Practice Problems

9-1   (a) Input impedance expression for the circuit in Figure 9.9(b) is obtained as $Z_{in} = Z_1 Z_3 Z_5 / Z_2 Z_4$ with OAs considered ideal. Resulting expression of the simulated inductance is given as $L_{eq} = R_1 R_3 R_5 C / R_2$. Show that, when OAs are modeled by their first-order roll-off model as $A_i = B_i / s$, following input impedance is derived for the working frequency range $\omega^2 \ll (B_1 B_2)$ as:

$$Z_{in}(s) \approx sLeq \frac{1+\left(\dfrac{R_2}{R_3}\dfrac{1}{A_2}+\dfrac{1}{A_1}\right)\left(1+\dfrac{1}{sC_4 R_5}\right)}{1+\left(\dfrac{1}{A_2}+\dfrac{R_3}{R_2}\dfrac{1}{A_1}\right)\left(1+sC_4 R_5\right)}$$

(b) With $s = j\omega$, simulated impedance is $Z_{in}\, j\omega = j\omega\,(L_{eq} + \Delta L) + R_s$. Find expressions for ΔL and R_s.

(c) What shall be the significance of selecting resistance $R_2 = R_3$.

9-2   Design a lossless ladder for the following specifications:

$$\omega_1 = 2\frac{krad}{s}, A_{max} = 1dB, \omega_2 = 4\frac{krad}{s} \text{ and } A_{min} = 20dBs$$

Use equal-ripple approximation with double termination ladder and realize the filter using GIC based inductor simulation. If possible, select a capacitor of 10 nF each.

9-3   Design an LC filter with Chebyshev approximation and minimum inductors and $R_{in} = R_{out} = 600\ \Omega$. Replace the inductors by an OA RC circuit; specifications are:

$f_1 = 12.5$ kHz, $A_{max} = 1$ dB, $f_2 = 56.25$ kHz, and $A_{min} = 35$ dBs

9-4   Obtain an active RC version of the ladder shown in Figure P9.1 using a gyrator, with $R_{in} = R_{out} = 1$ kΩ, $C_1 = C_3 = 20.94$ nF, $C_2 = 3.306$ nF and $L_2 = 0.8347$ H.

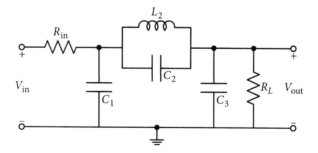

Figure P9.1

9-5   Design a GIC based third-order HP filter with Butterworth approximation having 3 dB frequency of 5 kHz. Use a doubly terminated ladder and capacitors need to be of 1 nF, if possible.

9-6   Realize the ladder shown in Figure P9.1 using Bruton's approach.

9-7   Design a fifth-order HP active RC filter using inductance simulation technique, having pass band ripples less than 1 dB and corner frequency of 100 krad/s. Compare its response with that of the passive prototype filter.

9-8   Realize the ladder shown in Figure P9.2 using Bruton's approach for a center frequency of 10 kHz and $Q = 10$.

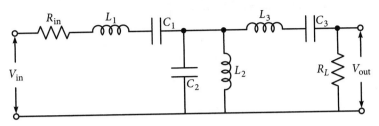

Figure P9.2

9-9   Design and test an LPF using operational simulation from a normalized third-order Butterworth filter. Cut-off frequency is 1 kHz and the terminating resistors are 2 kΩ each.

9-10   Repeat Problem 9-9 for a fifth-order Butterworth filter with the other specifications remaining the same.

9-11   Design and test an LPF using operational simulation from a normalized fifth-order Chebyshev filter. Pass band ridge frequency is 1 kHz and the terminating resistors are 2 kΩ each.

9-12   Specifications of a Butterworth LPF: range of pass band $0 \leq \omega \leq 8000$ rad/s and range of stop band 32000 rad/s $\leq \omega \leq \infty$ with $\alpha_{min} = 16$ dBs and $\alpha_{max} = 0.5$ dB. Design and test using the operational simulation approach.

Note: For Problems 9-13 to 9-16, employ the technique in Section 9.11.

9-13   A BPF is to be designed using a third-order LP Butterworth filter. Its center frequency is to be 2 kHz and quality factor $Q = 10$. Test the circuit using practical element values.

9-14   Repeat Problem 9-13 if its prototype is a fourth-order Butterworth filter.

9-15   Design a BPF with $\omega_o = 5$ krad/s and 3 dB bandwidth = 4500 rad/s. The prototype is to be a fourth-order maximally flat band pass.

9-16   An active BPF is to be constructed from a passive structure shown in Figure P.9.2 (Problem 9.16). Center frequency, lower and upper cut frequencies are 1265 Hz, 800 Hz and 2000 Hz, respectively. Employ operational simulation method to obtain the filter and test it.

9-17   In Figure 9.35(a), series resonance is to be at 4 krad/s and shunt resonance at 8 krad/s. Verify the frequency response with that of the passive structure. A resistance of 1 kohm is used in the de-normalized circuit.

9-18   Repeat Problem 9-17 for the circuit in Figure 9.35(b) with the shunt resistance in the de-normalized circuit being 1 kohm.

9-19   Use impedance scaling of 10^3 and frequency scaling of 10^5 for the series branch of Figure P9.3

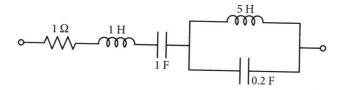

Figure P9.3

and compare frequency response with its de-normalized passive version.

9-20   Repeat Problem 9-19 for the circuit shown in Figure P9.4.

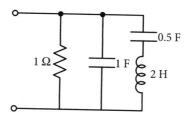

Figure P9.4

Chapter **10**

Cascade Approach: Optimization and Tuning

10.1 Introduction

In Chapter 2, a number of first-order active filters were realized along with the basic configuration of bilinear functions. Chapters 7 and 8 were devoted to the development of second-order filter sections using one or more than one OA (operational amplifier). However, as mentioned earlier, all filter specifications are not achievable only through second-order sections; higher-order filters become necessary. For the realization of higher-order filters, ladder simulation techniques through element (inductor/frequency dependent negative resistor: FDNR) substitution was discussed in Chapter 9. Ladder simulation using signal flow graph technique, which is better known as the *operationally simulated* method was also discussed in Chapter 9. The present chapter deals with another basic method of realizing higher-order filter sections known as the *cascade design method*.

Section 10.2 will discuss the basics of cascade design and the conditions to be satisfied by first- or second-order sections in order that these could be used for cascading and obtaining higher-order filters. After taking up some examples of cascade design, the importance of *cascade optimization* will be discussed in Section 10.3 through an example section. While cascading a number of second-order sections, a proper combination of poles and zeroes (Section 10.3.1), correct assignment of gain for each section (Section 10.3.2) and their proper order (Section 10.3.2) play a very crucial role. Hence, all the three aspects will be discussed in some details.

It is well-known that due to the tolerance associated with passive elements as well as with the parameters of the active elements, and their possible variation due to the change in the biasing voltage and operating temperature, filter parameters gets deviated. Therefore, it is imperative to provide on-chip tuning of the parameters, especially the pole frequency ω_o and pole-Q of the individual second-order sections used in the cascade. This chapter introduces the idea of filter parameter tuning. In Section 10.4, we get a better insight into the cascade design method of filter design.

10.2 Cascade Design Basics

To realize higher-order filters, cascading of second-order sections finds considerable favour in the eyes of a majority of filter designers; a first-order section is also cascaded in case of an odd-order filter. It will be shown later that such a technique has few advantages like the ability to tune filter parameters easily, and that it possess controlled tunability of the filter parameters with respect to the elements used.

Lower-order building blocks are connected as shown in Figure 10.1. Over all, the transfer function is simply the product of the transfer functions of the individual building blocks as shown here.

$$H(s) = H_1(s) \times H_2(s) \times \dots \dots \times H_n(s) \qquad (10.1)$$

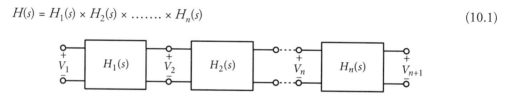

Figure 10.1 Cascading arrangement of n number of two-port networks.

Here, the transfer function $H_i(s) = V_{i+1}(s)/V_i(s)$ can be first-order; most often, second-order sections are used. However, this relation and the relation in equation (10.1) will be valid only when the individual sections are non-interactive, that is, the preceding sections do not load the previous section. From the basic knowledge of two-port networks, we know that perfect non-interactiveness will be achieved when the output impedance of each building block (Z_{oi}) is zero and their input impedances (Z_{ij}) are infinity. In practice, due to the finiteness of the input and output impedances of the individual sections, there will be some interaction between the two-port networks; hence, some deviation in the overall transfer function from the design is likely to occur. When two-port networks (second-order or first-order) are realized using OAs, it is always desirable to get its output at the output terminal of the amplifier. The output impedance of the OA being practically small, OAs are suitable for cascading. When passive two-port networks or an output terminal with higher impedance is to be cascaded, it is advisable to insert a buffer in between the networks as shown in Figure 10.2.

There are a number of advantages in cascade design method of higher-order filters. As the pole-Q and the critical frequency ω_o of each second-order section depends on one pole pair and the nature of the second-order filter depends on one zero (pair), it becomes easier to tune these parameters, which is very difficult in direct realization methods. Since individual second-order sections can be tuned easily, the overall tuning of response also becomes easy. It is to be noted that as only second-order sections are needed for higher-order filters, the best possible second-order section with all possible optimizations for that particular application can be used. For example, the selected section may have minimum possible sensitivity, and variability, with respect to the elements used.

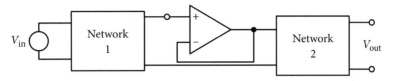

Figure 10.2 Insertion of buffer between two interactive two-port networks.

When a transfer function of order n, as given in equation (1.1) and repeated here, is to be realized, it is known that $n \geq m$. For n being even, there shall be $(n/2)$ pole pairs and $(n/2)$ second-order sections shall be connected in cascade.

$$H(s) = \frac{V_{out}}{V_{in}} = \frac{b_m s^m + b_{m-1} s^{m-1} + \ldots\ldots + b_1 s + b_0}{s^n + a_{n-1} s^{n-1} + a_{n-2} s^{n-2} + \ldots\ldots + a_1 + a_0} \tag{10.2}$$

If n is odd, $(n-1)/2$ second-order sections and one first-order section will be connected in cascade. In general, the transfer function of each second-order function will be expressed as:

$$H_i(s) = k_i \frac{\alpha_{2i} s^2 + \alpha_{1i} s + \alpha_{0i}}{s^2 + (\omega_{oi}/Q_i) s + \omega_{oi}^2} = k_i h_i(s) \tag{10.3}$$

Obviously, all second-order functions may not have the numerator coefficients α_2, α_1 and α_0 as finite. It will depend on the number of finite zeros and the type of response of that particular section.

Let us go through the basics of the cascade approach with the help of some simple examples.

Example 10.1: Realize a Butterworth LPF (low pass filter) which will satisfy the following specifications using the cascade form of synthesis.

α_{max} = 1 dB, α_{min} = 40 dBs, ω_1 = 2000 rad/s, ω_2 = 6000 rad/s

Solution: In Example 3.1, order of a filter in the form of a lossless passive ladder was found as 5 for the desired specification. For the fifth-order Butterworth filter, using the values of pole locations from Table 3.1 and with the cut-off frequency (de-normalization) given as $\omega_c = \omega_{CB} \times \omega_1 \cong 1.144 \times 2000 = 2288$ rad/s, the following forms of transfer function were obtained:

$$H(s) = \frac{1}{(s + 2288)(s^2 + 1414s + 2288^2)(s^2 + 3701.98s + 2288^2)} \tag{10.4}$$

$$= \frac{1}{s^5 + 7.408 \times 10^3 s^4 + 2.741 \times 10^7 s^3 + 6.2714 \times 10^{10} s^2 + 9.11478 \times 10^{13} s + 6.27018 \times 10^{16}} \quad (10.5)$$

Obviously, the next step is to find an active network topology containing suitable active devices, and values of the passive elements used. To realize higher-order filters using the cascade form, the transfer function of equation (10.4) needs to be broken to form a product of three transfer functions. Hence, a first-order function $H_1(s)$ and two second-order functions $H_2(s)$ and $H_3(s)$, with respective dc gains of k_1, k_2 and k_3 will be used. The resulting overall transfer function will be:

$$H(s) = H_1(s) \times H_2(s) \times H_3(s) \quad (10.6)$$

where expressions of the three transfer functions obtained from equation (10.4) will be:

$$H_1(s) = \frac{2288 k_1}{s + 2288} \quad (10.7)$$

$$H_2(s) = \frac{2288^2 k_2}{s^2 + 1414s + 2288^2} \quad (10.8a)$$

$$H_3(s) = \frac{2288^2 k_3}{s^2 + 3702s + 2288^2} \quad (10.8b)$$

Obviously, the dc gain of the product of the three transfer functions will become $(k_1 \times k_2 \times k_3)$, which should be equal to 1; the overall dc gain of the transfer function $H(s)$. Arbitrary values can be assigned to the individual dc gains in order to get their product as 1. However, one easy and convenient choice in the beginning is to make all three dc gains as unity (later, we shall see that the choice of dc gains for individual sections is not arbitrary for good designs).

The first-order transfer function of equation (10.6) with $k_1 = 1$ can be realized using an active section, along the lines followed in Section 2.3.1, as shown in Figure 10.3(a). Its transfer function is given as:

$$H_{11}(s) = \frac{1/CR_1}{s + 1/CR_2} \quad (10.9)$$

If the selected value of capacitor $C = 0.1 \ \mu F$, to get $k_1 = 1$, $R_1 = R_2 = 4.37 \ k\Omega$ in equation (10.7).

The remaining two second-order sections are realized using the circuit shown in Figure 2.15, redrawn in Figure 10.3(b), for which the transfer function is repeated here:

$$\frac{V_{out}}{V_{in}} = -\frac{(G_1 G_3 / C_1 C_2)}{s^2 + s\{(G_1 + G_2 + G_3)/C_1\} + (G_2 G_3 / C_1 C_2)} \quad (10.10)$$

It has the expressions for ω_o, dc gain and Q as:

$$\omega_o^2 = \left(\frac{G_2 G_3}{C_1 C_2}\right), \quad \text{dc gain} = \frac{G_1}{G_2}, \quad \text{and} \quad Q = \frac{C_1}{G_1 + G_2 + G_3}\left(\frac{G_2 G_3}{C_1 C_2}\right)^{1/2} \qquad (10.11)$$

For the transfer function $H_2(s)$ of equation (10.8a), it is compared with equation (10.10). Selecting $C_1 = 0.1\ \mu F$ and with $k_2 = 1$, use of equation (10.11) gives the value of $G_1 = G_2 = G_3 = 0.1244\ mA/V$, $C_2 = 0.0291\ \mu F$; so hence, $R_1 = R_2 = R_3 = 8.101\ k\Omega$.

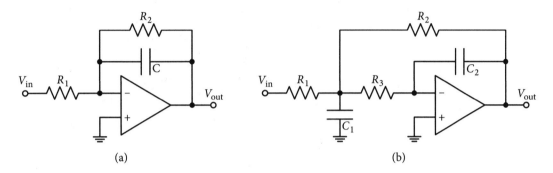

Figure 10.3 (a) An active first-order low pass circuit and (b) a second-order low pass filter circuit.

Similarly, for $H_3(s)$, element values are $C_1 = 0.1\ \mu F$, $C_2 = 4.252\ nF$, $R_1 = R_2 = R_3 = 21.196\ k\Omega$.

Three circuits are connected in cascade to get the overall transfer function. The complete circuit, having a sequence of sections as H_1 followed with H_3 and then H_2, along with the element values is shown in Figure 10.4(a). The simulated magnitude response is shown in Figure 10.4(b). It is observed that the output to input voltage ratio at 2000 rad/s is 0.889 or the attenuation is 1.012 dB, which is just above the design value of 1 dB. At 6000 rad/s, the output to input voltage ratio is 0.008106 which is equivalent of 41.2 dB attenuation; more than the design value of 40 dBs. However, there is a peak with a voltage of 1.28 volt at the output of the second stage. If the final output is limited to 10 volts, then the maximum input should not exceed 7.81 volts.

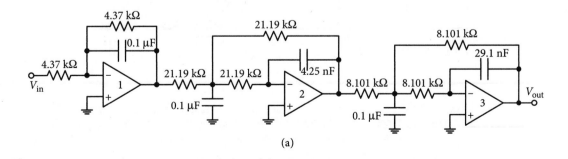

(a)

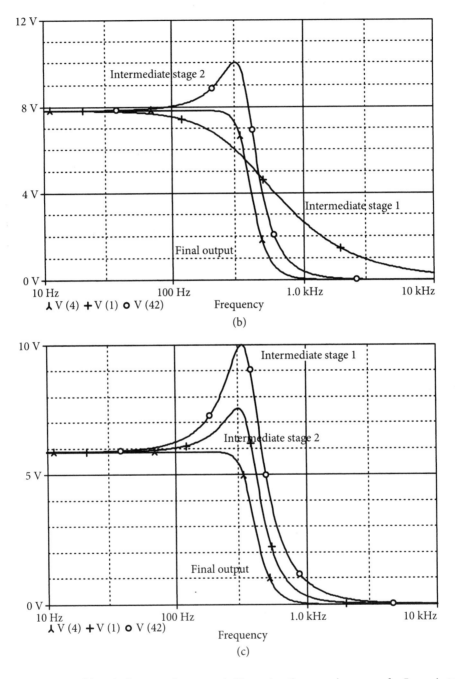

Figure 10.4 (a) Fifth-order low pass Butterworth filter using the cascade process for Example 10.1 and (b) the simulated response at intermediate and output nodes. (c) Response using cascade process with different sequence of sections for the circuit of Example 10.1.

If the sequence of sections is changed arbitrarily so that H_3 is followed by H_1 and then H_2, the simulated response is shown in Figure 10.4(c). The final output response is unaffected, but an intermediate peak has a voltage level of 1.7 volts, which means that with this sequence of sections, the allowable input reduces to 5.88 volts for a maximum output voltage of 10 volts. The effect of selection of another sequence (correct) will be taken up later.

Example 10.2: Realize an LPF with the following specifications using Chebyshev approximation. Obtain the frequency response while realizing it through the cascade process.

$$\alpha_{max} = 0.5 \text{ dB}, \ \alpha_{min} = 40 \text{ dBs}, \ \omega_1 = 2000 \text{ rad/s}, \ \omega_2 = 6000 \text{ rad/s} \tag{10.12}$$

Solution: In Example 3.3, the required filter order was obtained as 4 for the given specifications. The following was the normalized transfer function.

$$H(s) = \frac{0.3577}{\left(s^2 + 0.3508s + 1.0636\right)\left(s^2 + 0.8466s + 0.3563\right)} \tag{10.13}$$

Since for an even-order transfer function $H(0) = \alpha_{max} = 0.5$ dB or 0.944 (normalized), the value of the numerator in $H(s)$ becomes $(0.944 \times 1.0636 \times 0.3563) = 0.3577$ for maximum pass band gain of unity.

As frequency de-normalization is to be done by 2000 rad/s, the de-normalized transfer function is obtained as:

$$H(s) = \frac{0.3577 \times 2000^2}{\left(s^2 + 0.3508 \times 2000s + 1.0636 \times 2000^2\right)\left(s^2 + 0.8466 \times 2000s + 0.3563 \times 2000^2\right)} \tag{10.14}$$

As in Example 10.1, the transfer function of equation (10.14) is broken into the following two second-order LP functions:

$$H_1(s) = \frac{1.0636 \times 2000^2}{\left(s^2 + 0.3508 * 2000s + 1.0636 \times 2000^2\right)} \tag{10.15}$$

$$H_2(s) = \frac{0.3563 \times 0.944 \times 2000^2}{(s^2 + 0.8466 \times 2000s + 0.3563 \times 2000^2)} \tag{10.16}$$

Both the transfer functions $H_1(s)$ and $H_2(s)$ are realized using the circuit shown in Figure 10.3(b) whose transfer function and expressions of parameters are given by equations (10.10) and (10.11), respectively.

For transfer function $H_1(s)$, assuming $G_{21} = 10^{-4}$ mho (second subscript 1 corresponds to the first biquad) as the dc gain is 1.0, we get:

$$(G_{11}/G_{21}) = 1; \text{ hence, } R_{11} = R_{21} = 10 \text{ k}\Omega \tag{10.17}$$

Without losing generality, selecting R_{31} = 10 kΩ as well, with (ω_o/Q) = 0.3508 × 2000 and $\omega_o^2 = 1.0636 \times 2000^2$, application of equation (10.11) gives C_{11} = 0.4276 μF and C_{21} = 0.005498 μF.

For the transfer function $H_2(s)$ (with second subscript 2), assuming G_{22} = G_{32} = 10^{-4} mho, and with dc gain being 0.944, we get:

$$(G_{12}/G_{22}) = 0.944 \text{ or } R_{22} = 10 \text{ k}\Omega, R_{32} = 10 \text{ k}\Omega, R_{12} = 10.593 \text{ k}\Omega \tag{10.18}$$

Similar to the function $H_1(s)$, application of equation (10.11) for $H_2(s)$ gives C_{12} = 0.17387 μF and C_{22} = 0.04035 μF.

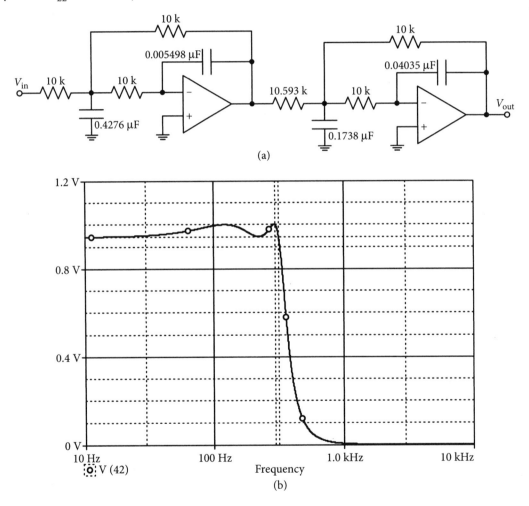

(a)

(b)

Figure 10.5 (a) Fourth-order low pass Chebyshev filter using cascade process for Example 10.2; (b) its simulated magnitude response.

Combining the circuit implementation of the transfer functions $H_1(s)$ and $H_2(s)$ in cascade, the overall circuit along with element values is shown in Figure 10.5(a). Its magnitude response through simulation is shown in Figure 10.5(b). The observed pass band edge frequency is 318.2 Hz or 2000.1 rad/s. Maximum output voltage at peaks being 1.002 V for input voltage of 1 volt and at the pass band edge frequency, voltage level is 944 mV, corresponding to an attenuation of 0.5 dB; the attenuation becomes 46.1 dBs at 6000 rad/s. DC gain of the filter is also 0.944, which corresponds to 0.5 dB. The circuit satisfies attenuation requirements at pass band and stop band edge frequencies with enough margin and shows maxima and minima as expected.

10.3 Optimization in Cascade Process

With the help of Examples 10.1 and 10.2, it is clear that process of cascading is simple. The main issue is to get an appropriate second-order section for each second-order function (and a first-order section, if needed). Unfortunately, actual implementation while cascading even a few second-order sections involves some serious considerations, which is important not only for the optimization of the performance, but also the correct functioning of the overall higher-order filter section.

For the simple fourth-order function of Example 10.2, two transfer functions $H_1(s)$ and $H_2(s)$ were shown to be cascaded. While forming these transfer functions, pole pairs and zeros were combined arbitrarily. Obviously, there is more than one possible combination that will result in the fourth-order filter. While cascading three sections, there are six possible combinations of pole pairs and zeros and many more for a larger number of sections to be cascaded. It will be shown in the next section that assignment of zeros with a pole pair should not be arbitrary as it affects the performance. A criterion has to be evolved for the proper combination of pole pairs and zeros.

Once the proper combination of poles and zeros is accomplished, there is more than one possibility in assigning the order in which the sections will be cascaded. Again, for the sixth-order function, there are the following six possible combinations in which the blocks may be cascaded.

$$H_1\,H_2\,H_3,\ H_1\,H_3\,H_2,\ H_2\,H_1\,H_3,\ H_2\,H_3\,H_1,\ H_3\,H_1\,H_2,\ H_3\,H_2\,H_1 \qquad (10.19)$$

Unless specified, one may wonder why the order of cascading is important. In fact, along with the third issue, which will be explained soon, proper ordering of second-order sections significantly affects the working of the overall filter and may deviate its performance drastically.

The third important issue is the assignment of gain for individual two-port blocks. Since the overall gain will be the product of the gain of the individual blocks, a designer might like to assign it arbitrarily to each block. However as mentioned earlier, the assignment of gain to individual blocks along with its ordering has to be done very carefully.

The reasons behind these considerations of pole–zero pairing, ordering of the sections and assignment of gain to the sections are two-fold. The first consideration is to check that the signal level at any internal or external node does not exceed or reach the saturation level of the active device decided by the level of the supply voltage or due to the constraint of the slew rate.

If this happens, harmonics will get generated and the resulting filter parameters will deviate from the design parameters. The second consideration is to maximize the dynamic range: the ratio of the undistorted signal to the noise level present in the system at each cascading stage. The basic principle used for maximizing the dynamic range is to *maximize the minimum level of the signal in the pass band* of each second-order filter section so that the effect of noise is minimum on the signal. This action produces a flatter type of response in which the ratio of maximum signal at any node to the minimum signal comes as close to unity as possible. In addition to the aforementioned considerations, the maxima of the output voltage for individual sections should also be made equal as will be shown later.

The idea explained here can be expressed mathematically as well, which forms the starting point of developing algorithms and computer programs for the purpose. Let $V_{o,\max}$ be the level of output voltage of the OAs used, which is the upper limit of the signal level set by the power supply level or the slew rate at all frequencies including both the pass and stop band. Then the maximum magnitude of the output signal for each individual biquad $|V_{oj}(j\omega)|$ should satisfy the following condition:

$$\max |V_{oj}(j\omega)| < V_{o,\max} \text{ for } 0 \leq \omega \leq \infty \tag{10.20}$$

The condition imposed by equation (10.20) needs to remain valid for all signal frequencies falling in the pass and stop band; the reason being that if any signal is overdriven even in the stop band, it may generate harmonics, which may interfere with the valid output.

On the other hand, it is necessary to check that the signal does not become so small at the intermediate stage or individual block output level that it gets corrupted by the circuit noise. If the signal to noise ratio becomes small, the signal becomes indistinguishable from the noise. Hence, another condition which is required to be fulfilled is that the smaller signals at all outputs of the biquad should be enlarged in the pass band as much as practically possible.

$$\min |V_{oj}(j\omega)| \rightarrow \text{maximize in the pass band} \tag{10.21}$$

The condition imposed by equation (10.21) needs to be valid only in the pass band; it is not required in the stop band as the smallness of signal there will not do any harm.

The brief discussion in this section and the conditions given in equations (10.20) and (10.21) imply that higher magnitude signals need to be pulled down to remain below $V_{o,\max}$ and smaller magnitude signals need to be amplified. This creates a kind of flatness in signals at the output of the individual second-order section as mentioned earlier. In other words, we need to make the ratio of maximum signal to minimum signal at all the intermediate levels and the final output as small as possible. Later this point will be taken up in mathematical terms.

10.3.1 Pole–zero pairing

In order to explain the idea of *flatness* of signals at all cascaded stages, let V_{ok} be the output voltage at the kth stage. Normally, the output at any stage is taken as the output of the OA. If the biquad uses only one OA, then V_{ok} shall will be its output; however, for a multi amplifier biquad, V_{ok} will be the largest output signal amongst all the amplifiers used. This implies that for the kth stage, there will be no signal greater than V_{ok}. Therefore, as discussed in the previous section, poles and zeros should be paired in such a way that the signal maximum $M_k = \max|V_{ok}(j\omega)|$ is minimized below $V_{o,\max}$ for all the input signal frequency range of pass and stop bands. At the same time, the signal minimum $m_k = \min|V_{ok}(j\omega)|$ is maximized in the pass band. Maximization of m_k and minimization of M_k means that $|h_k(j\omega)|$ in equation (10.3) should be as flat as possible in the frequency range of interest. For a mathematical treatment, *measure of flatness* of the signal can be expressed as:

$$d_k = (M_k/m_k) \rightarrow 1 \text{ with } k = 1, 2, \ldots\ldots, n \tag{10.22}$$

Obviously, the intention is to minimize d_k and the assignment of zeros to a pole pair in equation (10.3) should be such that it minimizes the maximum value of d_k, or:

$$d_{\max} = \max[d_k] \rightarrow \text{minimum for } k = 1, 2, \ldots\ldots, n \tag{10.23}$$

For comparatively smaller order filters say (4 or 5), evaluation of d_k, though laborious, can be done in a reasonable time period and then equation (10.23) satisfied. However, for higher-order filters, manual evaluation of d_k becomes highly time consuming and requires use of computer and an appropriate software. Fortunately, for quite a few cases, instead of taking recourse to computer programs, decisions based on intuition and experience became helpful. For example, when the transfer function has more than one zero at the origin, the type of filter section can be of different types for the same final transfer function. For example, for a numerator having a term s^2, the possible combinations can be s^2 and 1, or s and s, which will mean that realization can be in the form of a combination of a HP (high pass) and a LP (low pass) or a combination of two BP (band pass) sections, respectively. However, a much more important consideration comes in the form of a thumb rule of *forming pole–zero pair combinations which are closest to each other*. This thumb rule gets its idea from the fact that when the value of the combined pole and zero is close to each other, magnitude of the section will be a minimum; which is our aim.

Example 10.3: To design a sixth-order filter in cascade form, find the optimum pole–zero pair combination for the following transfer function:

$$H(s) = \frac{s(s^2+0.25)(s^2+2.25)}{(s^2+0.09s+0.83)(s^2+0.1s+1.18)(s^2+0.2s+1.01)} \tag{10.24a}$$

Solution: Transfer function zeros and poles are:

$$z_1 = 0, z_{2,3} = \pm j0.5 \text{ and } z_{4,5} = \pm j1.5 \tag{10.24b}$$

$$p_{1,2} = -0.045 \pm j0.9099, p_{3,4} = -0.05 \pm j1.0851 \text{ and } p_{5,6} = -0.1 \pm j1.0 \tag{10.24c}$$

Poles and zeros are shown in Figure 10.6, in the second quadrant of the complex frequency s plane. Applying the thumb rule of combining the nearest poles with zeros, we get the following combination of three second-order sections:

$$H_A(s) = \frac{(s^2 + 2.25)}{(s^2 + 0.1s + 1.18)}, \quad H_B(s) = \frac{(s^2 + 0.25)}{(s^2 + 0.09s + 0.083)}, \quad H_C(s) = \frac{s}{(s^2 + 0.2s + 1.01)} \quad (10.25)$$

Each transfer function in equation (10.25) will have to be assigned a gain k_a, k_b and k_c to get the required overall gain.

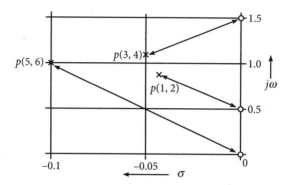

Figure 10.6 Poles and zeroes for the transfer function in Example 10.3.

It is significant to note that when performance of the fifth-order LP of Example 10.1 is to be optimized while using the cascade approach, the pole–zero pairing step is not required as all the zeros are at infinity. In fact, for all pole functions, the optimization process starts from the section ordering.

10.3.2 Section ordering

Once pole–zero assignment is made and n second-order sections are formed, it is required to decide their order in the chain of cascade. Once again, the aim is to get maximum dynamic range, and the procedure is very similar to that adopted for pole–zero pairing. We need to try and keep the variation of the signal for any individual section as flat as possible so that the ratio of the voltage at the output of the kth intermediate section to the input to the first section is as flat as possible.

As the mathematical treatment is similar for all cases, useful algorithms and computer programs have been developed. However, some simpler solutions, which do not require computer programs, are also available. It is generally desirable to keep either a LP or a BP section as the first in the cascade chain in order to suppress high frequency signal components which would have been generated due to the slew rate limitation. Similarly, a high or a BP section is put at the end of the cascade chain in order to suppress any low-frequency noise including dc offset and power supply ripples. However, a much more significant factor which decides the ordering of the sections is the value of the pole-Q of the biquads. Since a larger Q value means a larger peak gain, it is advised to keep the lowest Q section at the beginning of the cascade followed by sections with increasing values of the pole-Q. If Q_i is pole-Q of the ith section in an n section cascade, sections are selected with the following condition:

$$Q_1 < Q_2 < \ldots\ldots < Q_n \text{ for } i = 1, 2 \ldots\ldots, n \quad (10.26)$$

Example 10.4: For the transfer function of Example 10.3, find the optimum order of the second-order sections as derived in equation (10.25).

Solution: As suggested, a good choice is to put the sections governed by the condition given in equation (10.26). Hence, the quality factor of the three sections are calculated as:

$$Q_A = \frac{\sqrt{1.18}}{0.1} = 10.86, Q_B = \frac{\sqrt{0.83}}{0.09} = 10.12, Q_C = \frac{\sqrt{1.01}}{0.2} = 5.025 \tag{10.27}$$

This means that the most appropriate order of the sections given in equation (10.25) is H_C followed by H_B and then H_A.

For illustration purposes, the transfer function $H(s)$ of equation (10.24) is realized with two different orders of second-order sections; one depending on equation (10.27) and the other a different order. Presently, the gain assigned to each section is taken as unity.

Incidentally, the transfer functions H_C and H_B have already been realized in Chapter 7 as Examples 7.4 and 7.5, respectively, with filter circuits shown in Figures 7.14a and 7.15. In order to realize the overall transfer function, H_A is also designed selecting the *general differential input single OA biquad* of Section 7.4.2, which was used for H_C and H_B as well.

For the general configuration of Figure 7.12, selecting the auxiliary polynomial $Q(s) = s + 1$, we get:

$$\frac{N(s)}{Q(s)} = \frac{s^2 + 2.25}{s+1} = (s+2.25) - \frac{3.25s}{s+1} \tag{10.28}$$

This gives $y_a = s + 2.25$, $y_1 = \frac{3.25s}{s+1}$ (10.29)

$$\frac{D(s) - N(s)}{Q(s)} = \frac{0.1s - 1.07}{s+1} = -1.07 + \frac{1.17s}{s+1} \tag{10.30}$$

This gives $y_b = \frac{1.17s}{s+1}$, $y_2 = 1.07$ (10.31)

$$\frac{D(s)}{Q(s)} = \frac{(s^2 + 0.1s + 1.18)}{s+1} = s + 1.18 - \frac{2.08s}{s+1} \tag{10.32}$$

This gives

$$y_3 = s + 1.18, \; y_c = \frac{2.08s}{s+1} \tag{10.33}$$

Here, y_1, y_b and y_c are a series combination of resistance and capacitance, y_a and y_3 are a parallel combination of resistance and capacitance, whereas y_2 is only a resistor. Normalized element values are:

$R_a = 0.4444\ \Omega$, $C_a = 1$ F, $R_1 = 0.30769\ \Omega$, $C_1 = 3.25$ F, $R_3 = 0.8474\ \Omega$, $C_3 = 1$ F, $R_c = 0.4807\ \Omega$, $C_c = 2.08$ F, $R_b = 0.8547\ \Omega$, $C_b = 1.17$ F and $R_2 = 0.9345\ \Omega$

Frequency de-normalization with $1.18^{-0.5} \times 10^4$, to get its peak value at 10^4 rad/s, and impedance scaling of 10^4 gives the following element values, which are also shown in the filter circuit of Figure 10.7(a). Expected notch frequency is $1.18^{-0.5} \times 10^4 \times 2.25^{0.5}$ rad/s = (2197 Hz).

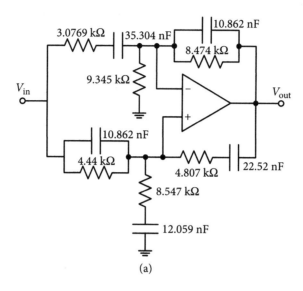

(a)

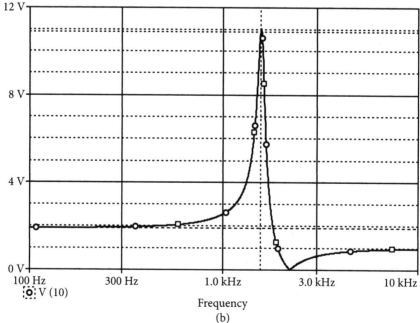

(b)

Figure 10.7 (a) Realization of the notch filter for Example 10.4. (b) Its magnitude response.

R_a = 4.444 kΩ, C_a = 10.862 nF, R_1 = 3.0769 kΩ, C_1 = 35.304 nF, R_3 = 8.474 kΩ, C_3 = 12.079 nF, R_c = 4.807 kΩ, C_c = 22.52 nF, R_b = 8.474 kΩ, C_b = 10.862 nF and R_2 = 9.345 kΩ

Element values are shown in the filter circuit of Figure 10.7(a). Magnitude response of the BR (band reject) filter is shown in Figure 10.7(b). Its dc gain is 1.91, high frequency gain is unity, peak gain of 11 occurs at 1.569 kHz and the notch frequency is at 2.204 kHz.

Example 10.5: Determine the correct order of the sections for the fifth-order LP Butterworth filter of Example 10.1 and find the allowable input voltage for a maximum output of 10 volts.

Solution: In Example 10.1, out of the three sections, $H_1(s)$ is a first-order section which shall be the first section in the cascade. For the functions $H_2(s)$ and $H_3(s)$, the respective pole-Q are Q = 1.618 and Q = 0.618. Hence, as per equation (10.26), $H_3(s)$ will follow the first-order section and $H_2(s)$ will be the last section.

With this sequence of sections $(H_1 H_3 H_2)$, the simulated response at all the outputs is shown in Figure 10.8. It is observed that all the three responses are monotonically decreasing with no peaking. The maximum voltage gain is unity at dc; hence, for the maximum output voltage of 10 volts, input can be 10 volts.

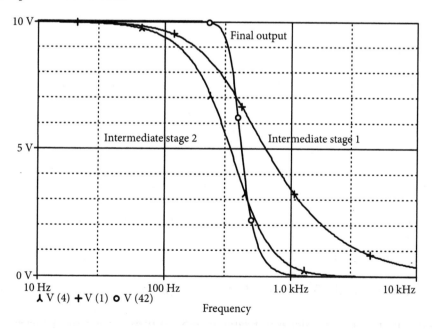

Figure 10.8 Fifth-order low pass active Butterwoth filter response using cascade process with correct sequence of sections; example 10.5.

10.3.3 Gain assignment

After completing the two steps discussed previously, we know the level of output voltage after every stage including that at the final output. Care was taken not to over-drive any output.

The next step is to assign gain to each section. Assignment of gain follows the principle that all internal output voltages become equal (as far as possible) in magnitude corresponding to the specified final output voltage. It may be noted that the absolute value of the overall gain is not to be obtained at all costs; a simple gain amplifier in cascade will be able to adjust the overall gain if needed.

For the transfer function $H(s)$ of equation (10.2) and using the notation of equation (10.3), we can define a constant M_i as:

$$M_i = \prod_{j=1}^{i} |h_j(s)| \text{ with } i = 1, 2, \ldots, (n-1) \tag{10.34}$$

The desired condition is that $M_i = M_n$, that is, the final output after n stages. Even after the first stage, that is,

$$i = 1, \, k_1 M_1 = \max |H_1(s)| = k_1 |\max h_1(s)| = K M_n \tag{10.35}$$

$$\text{where } K = \prod_{j=1}^{n} k_j \tag{10.36}$$

It gives the relation for k_1 as:

$$k_1 = K(M_n/M_1) \tag{10.37}$$

The relation given by equation (10.37) is true for intermediate stages as well; hence, for any consecutive stages,

$$k_i = (M_{j-1}/M_j), j = 2, \ldots, n \tag{10.38}$$

Equation (10.38) is used for assigning gain to each stage. Assignment of gain to each second-order section ensures near equal output at the final output, as well as at intermediate stages. It ensures maximum possible input signal that can be applied without the signals going beyond saturation in OAs. The statement is correct for a single OA second-order section, where output is taken at the OA output used. For a second-order section using more than one OA, care has to be taken as mentioned before.

Example 10.6: Realize the Butterworth filter of Example 10.5 with overall gain as 6 dBs.

Solution: With the correct order as in Example 10.5, unity gain was easily achieved with each section having a gain of one. If a gain of 6 dBs is desired, it can be done without assigning gains to individual sections as discussed in Section 10.3.3 for the all pole filters which were cascaded in correct ordering. Hence, simply having a gain of 2 in the first-order section will suffice; resistance R_1 in it is changed to 2.185 kΩ, and the rest of the circuit of Example 10.5 remains the same.

Example 10.7: Assign proper values of gain to the three second-order biquadratic sections of Example 10.3, when they are in the sequence suggested in Example 10.4, such that the overall gain is 40 and center frequency is 10 krad/s. Also find:

(a) Maximum allowable input signal if the maximum allowable output is 10 volts.

(b) What happens when the second-order sections are cascaded in correct order, but each section has unity gain?

Solution: (a) Equation (10.34) relates the maximum amplitude at each output node after the first, second and third section for given example. For individual sections, it can be found mathematically (or using a computer program). However, sections H_C and H_B have already been simulated, and from Figures 7.14(b) and 7.16, max $|H_C(j\omega)|$ = 5.0 and max $|H_B(j\omega)|$ = 7.14. Section H_A has been designed and simulated in Example 10.4, and its peak value is almost 11. Hence, using the notations given in Section 10.3.3:

$$M_1 = 5, M_2 = 5 \times 7.14 = 35.7, M_3 = 35.7 \times 11 = 392.7$$

With the overall gain to be 40, $KM_3 = 40$, so $K = 40/392.7$

Gain to be assigned to the respective sections are as follows:

$$k_1 = \frac{40}{5} = 8, \; k_2 = \frac{5}{35.7} = 0.14, \; k_3 = \frac{35.7}{392.7} = 0.0909 \tag{10.39}$$

With these gain values, the overall transfer function will be:

$$H(s) = k_1\,H_C(s) \times k_2\,H_B(s) \times k_3\,H_A(s)$$

$$= \frac{8s}{(s^2+0.2s+1.01)} \times \frac{0.14(s^2+0.25)}{(s^2+0.09s+0.83)} \times \frac{0.0909(s^2+2.25)}{(s^2+0.1s+1.18)} \tag{10.40}$$

Individual second-order sections are to be designed with the new assigned gain values. The same circuit structure and methodology, which was used so far (though, any other circuit can also be used), is used now.

For the general differential input single OA biquad, the auxiliary polynomial is again assumed to be $Q(s) = (s + 1)$; this gives the following relations for $k_1 H_C(s)$:

$$\frac{N(s)}{Q(s)} = \frac{8s}{(s+1)} \to y_a = \frac{1}{\frac{1}{8}+\frac{1}{8s}}, y_1 = 0 \tag{10.41}$$

$$\frac{D(s)}{Q(s)} = \frac{s^2+0.2s+1.01}{(s+1)} \to y_3 = (s+1.01), y_c = \frac{1.81s}{(s+1)} \tag{10.42}$$

$$\frac{D(s)-N(s)}{Q(s)} = \frac{s^2-7.8s+1.01}{(s+1)} \to y_b = (s+1.01), y_2 = \frac{9.81s}{(s+1)} \tag{10.43}$$

For $k_2 H_B(s)$, the corresponding relations are:

$$\frac{N(s)}{Q(s)} = \frac{0.14s^2 + 0.035}{(s+1)} \rightarrow y_a = (0.14s + 0.035), y_1 = \frac{0.175s}{(s+1)} \tag{10.44}$$

$$\frac{D(s)}{Q(s)} = \frac{s^2 + 0.09s + 0.83}{(s+1)} \rightarrow y_3 = (s + 0.83), y_c = \frac{1.74s}{(s+1)} \tag{10.45}$$

$$\frac{D(s) - N(s)}{Q(s)} = \frac{0.86s^2 + 0.09s + 0.795}{(s+1)} \rightarrow y_b = (0.86s + 0.795), y_2 = \frac{1.565s}{(s+1)} \tag{10.46}$$

For $k_3 H_A(s)$, the corresponding relations are:

$$\frac{N(s)}{Q(s)} = \frac{0.0909(s^2 + 0.225)}{(s+1)} \rightarrow y_a = (0.0909s + 0.2045), y_1 = \frac{0.2045s}{(s+1)} \tag{10.47}$$

$$\frac{D(s)}{Q(s)} = \frac{s^2 + 0.1s + 1.18}{(s+1)} \rightarrow y_3 = (s + 1.18), y_c = \frac{2.08s}{(s+1)} \tag{10.48}$$

$$\frac{D(s) - N(s)}{Q(s)} = \frac{0.9191s^2 + 0.1s + 0.8755}{(s+1)} \rightarrow y_b = (0.9191s + 0.8755), y_2 = \frac{1.6946s}{(s+1)} \tag{10.49}$$

For the three sections, with an impedance scaling factor of 10^4 and a frequency de-normalization factor of $(1.01)^{-\frac{1}{2}} \times 10^4$, $(0.83)^{-\frac{1}{2}} \times 10^4$ and $(1.18)^{-\frac{1}{2}} \times 10^4$ for the respective transfer functions, $k_1 H_C(s)$, $k_2 H_B(s)$ and $k_3 H_A(s)$ are used and the resulting structures are cascaded in the order as decided in Example 10.4. Figure 10.9 shows the complete filter with de-normalized element values.

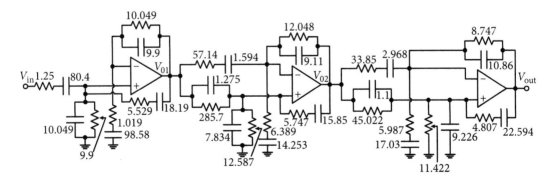

Figure 10.9 Cascade realization of the sixth-order BP filter for Example 10.7. All resistors are in kΩ and capacitors are in nF.

In Figure 10.10(a), the magnitude response of the sixth-order cascaded filter is shown with the output voltage on the y-axis on linear scale, while the same is shown in Figure 10.10(b) with the y-axis on log scale, depicting the notches clearly. Notch frequencies are 877.6 Hz and 2.314 kHz with 98.66 dBs and 59.3 dBs attenuation. Center frequency is 1.607 kHz (10.101 krad/s) with a gain of 40.65 against the design value of 40.

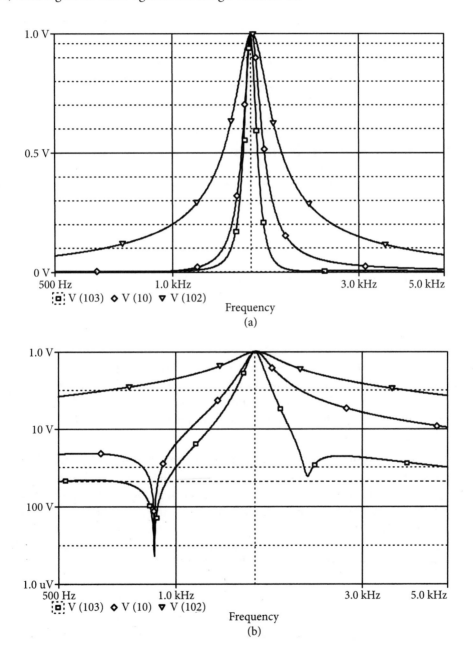

Frequency

(a)

Frequency

(b)

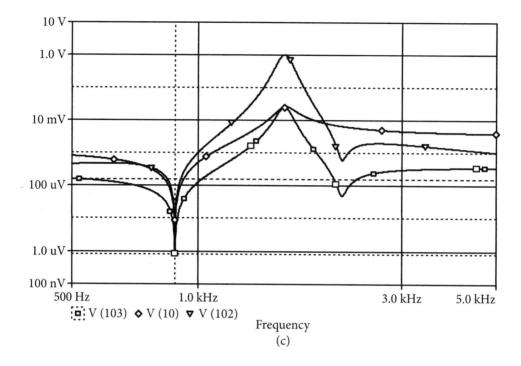

Figure 10.10 (a) Magnitude response of the sixth-order BP filter for Example 10.7. (b) Magnitude response of the sixth-order BP filter with the ordinate on log scale. (c) Magnitude response with order of the sections different from what is recommended.

(a) In order to make the level of the output voltage as 10 V, the input level can be as high as 250 mV (theoretically), and practically, 246 mV as simulated. If the order of sequence is changed with $k_2 H_B(s)$ at first, followed by $k_3 H_A(s)$ and then $k_1 H_C(s)$, the responses at the three outputs are shown in Figure 10.10(c). The final output level and the gain remains the same but the intermediate levels have gains of approximately unity. Notches do appear at 872.7 Hz and 2.31 kHz in the final output.

(b) If the section is formed with correct pole–zero combination and cascaded in correct sequence, but is assigned unity gain for each individual sections, instead of assigning gains as calculated, the simulated response is as shown in Figure 10.11. The final output amplitude becomes 10 V for an input voltage of 28.4 mV, compared to 250 mV for the same design but with correct gain assignment, as shown earlier. For this input level, the intermediate voltages are 142.53 mV and 1.015 V; obviously, all three voltage levels are not nearly the same and the overall gain becomes 350.14.

Cascade Approach: Optimization and Tuning   319

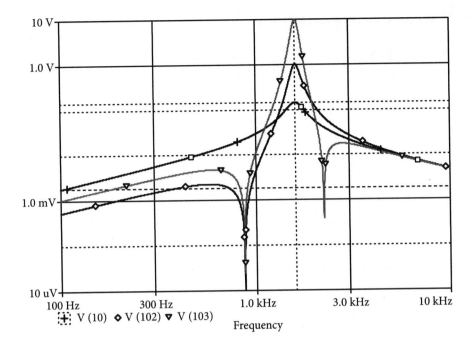

V (10) ◆ V (102) ▼ V (103)

Frequency

Figure 10.11 Magnitude response of the sixth-order BP filter for Example 10.7 with correct order of the sections but with unity gain assignment to the three sections.

Example 10.8: Discuss what happens with the response of the transfer function of equation (10.24) if the closest pole and zeros are not paired.

Solution: Let a different pole–zero pairing be done; the resulting three sections obtained by combining $z_{4,5}$ with $p_{3,4}$, $z_{2,3}$ with $p_{1,2}$ and z_1 with $p_{5,6}$ from the zero and pole locations given in equation (10.24b,c) are as follows:

$$H_1(s) = \frac{s}{\left(s^2 + 0.1s + 1.18\right)}, H_2(s) = \frac{\left(s^2 + 2.25\right)}{\left(s^2 + 0.09s + 0.083\right)}, H_3(s) = \frac{s^2 + 0.25}{\left(s^2 + 0.2s + 1.01\right)} \quad (10.50)$$

For the general differential input single OA biquad, the auxiliary polynomial is again assumed to be $Q(s) = (s + 1)$; this gives the following relations for $H_1(s)$:

$$\frac{N(s)}{Q(s)} = \frac{s}{(s+1)} \rightarrow y_a = s/(s+1) \quad y_1 = 0 \quad (10.51)$$

$$\frac{D(s)}{Q(s)} = \frac{s^2 + 0.1s + 1.18}{(s+1)} \rightarrow y_3 = (s+1.18), y_c = \frac{2.08s}{(s+1)} \quad (10.52)$$

$$\frac{D(s) - N(s)}{Q(s)} = \frac{s^2 - 0.9s + 1.18}{(s+1)} \rightarrow y_b = (s+1.18), y_2 = \frac{3.08s}{(s+1)} \quad (10.53)$$

For $H_2(s)$, the corresponding relations are:

$$\frac{N(s)}{Q(s)} = \frac{s^2 + 2.25}{(s+1)} \rightarrow y_a = (s+2.25), y_1 = \frac{3.25s}{(s+1)} \tag{10.54}$$

$$\frac{D(s)}{Q(s)} = \frac{s^2 + 0.09s + 0.83}{(s+1)} \rightarrow y_3 = (s+0.83), y_c = \frac{1.74s}{(s+1)} \tag{10.55}$$

$$\frac{D(s) - N(s)}{Q(s)} = \frac{0.09s - 1.42}{(s+1)} \rightarrow, y_b = \frac{1.33s}{(s+1)}, y_2 = 1.42 \tag{10.56}$$

For $H_3(s)$, the corresponding relations are:

$$\frac{N(s)}{Q(s)} = \frac{s^2 + 0.25}{(s+1)} \rightarrow y_a = (s+0.25), y_1 = \frac{1.25s}{(s+1)} \tag{10.57}$$

$$\frac{D(s)}{Q(s)} = \frac{s^2 + 0.2s + 1.01}{(s+1)} \rightarrow y_3 = (s+1.01), y_c = \frac{1.81s}{(s+1)} \tag{10.58}$$

$$\frac{D(s) - N(s)}{Q(s)} = \frac{0.2s + 0.76}{(s+1)} \rightarrow y_b = 0.76, y_2 = \frac{0.56s}{(s+1)} \tag{10.59}$$

For the three sections, an impedance scaling factor of 10^4 and a frequency de-normalization factor of $(1.18)^{-\frac{1}{2}} \times 10^4$, $(0.83)^{-\frac{1}{2}} \times 10^4$ and $(1.01)^{-\frac{1}{2}} \times 10^4$ are used for the transfer functions, $H_1(s)$, $H_2(s)$ and $H_3(s)$. The resulting structures, with element values calculated from the respective equations are shown in Figures 10.12(a), 10.13(a), and 10.14(a), respectively. Figure 10.12(b)

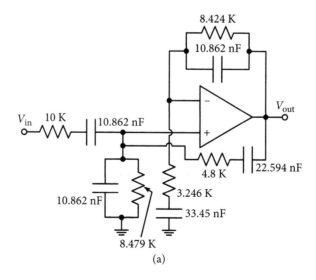

(a)

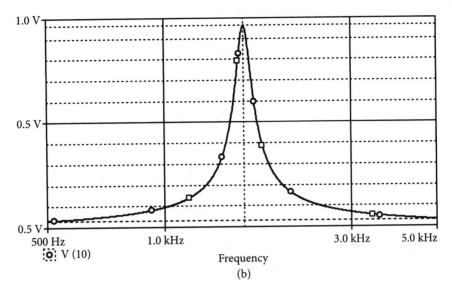

V (10)

Frequency
(b)

Figure 10.12 (a) Realization of the transfer function $H_1(s)$ for Example 10.8. (b) Magnitude response of the band pass function with unity gain.

shows the simulated response of $H_1(s)$, a BP response with center frequency of 1.6035 kHz, pole-Q of 10.46 and mid-band gain of 9.61. Figure 10.13(b) shows the simulated response of the function $H_2(s)$, BR characteristic with a notch at 2.639 kHz, dc gain being 2.71, high frequency gain being unity and peak gain of 30.79 at 1.607 kHz. Figure 10.14(b) shows a BR response for function $H_3(s)$ with a notch at 791.7 Hz, dc gain of 0.247, high frequency gain of unity and peak gain of 3.85 at 1.622 kHz.

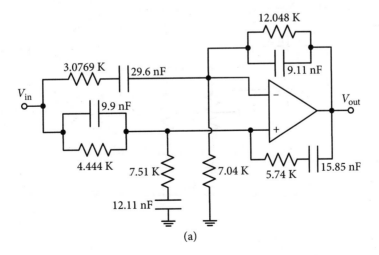

(a)

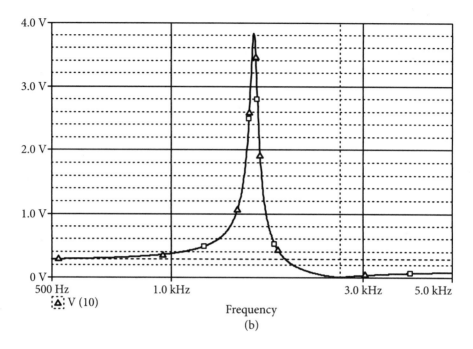

V (10)

Frequency

(b)

Figure 10.13 (a) Circuit realization of the transfer function $H_2(s)$ for Example 10.8 with unity gain. (b) Magnitude response of the transfer function $H_2(s)$ with unity gain.

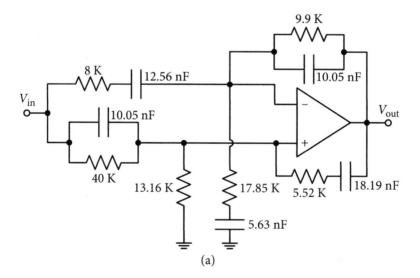

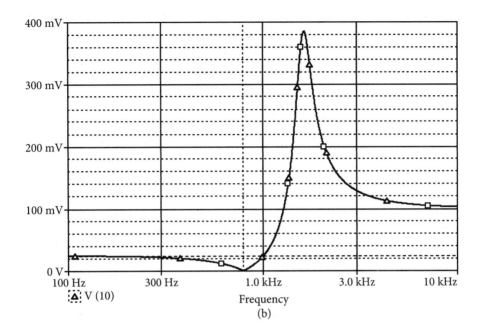

:▲: V (10)

Frequency
(b)

Figure 10.14 (a) Circuit realization of the transfer function $H_3(s)$ for Example 10.8. (b) Magnitude response of the transfer function $H_3(s)$ for Example 10.8 with unity gain.

The three sections are cascaded and Figures 10.15(a)–(b) show the response of the complete filter. Overall gain is 139.4; hence, to get 10 V output, the input has to be less than 71.7 mV.

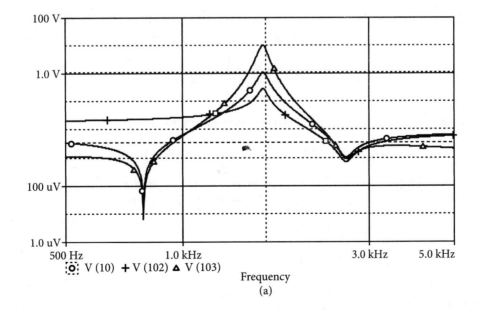

:○: V (10) + V (102) ▲ V (103)

Frequency
(a)

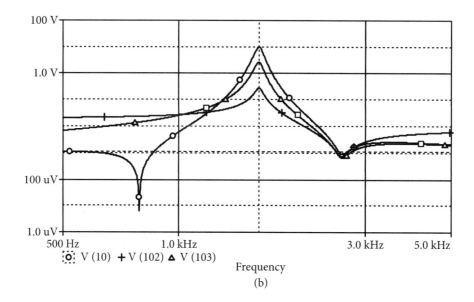

$$\begin{array}{c}\text{100 V}\\ \text{1.0 V}\\ \text{100 uV}\\ \text{1.0 uV}\end{array}$$

$\text{O: V (10)} \; + \text{V (102)} \; \triangle \text{V (103)}$

Frequency
(b)

Figure 10.15 (a) Response of the sixth-order filter for Example 10.8 with a pole–zero combination other than the case when the nearest poles and zeros are combined. (b) Response of the sixth-order filter while the sequence order is also different.

Example 10.9: Employing the cascade approach, realize an active RC filter for the following specifications, with dc gain as unity:

$$\alpha_{max} = 1 \text{ dB}, \; \alpha_{min} = 40 \text{ dBs}, \; \omega_1 = 2000 \text{ rad/s}, \; \omega_2 = 6000 \text{ rad/s}$$

Solution: In Example 3.4, it was found that a fourth-order inverse Chebyshev filter will be required for these specifications, for which the obtained transfer function is as given here:

$$H(s) = \frac{\{s^2 + 1.1757\}\{s^2 + 6.8283\}}{(s^2 + 0.3423s + 0.2559)(s^2 + 1.0091s + 0.31255)} \tag{10.60}$$

For the transfer function in equation (10.60), poles and zeros are:

$$z_{1,2} = \pm j1.0843, \; z_{3,4} = \pm j2.6131, \; p_{1,2} = -0.17117 \pm j0.47611, \; p_{3,4} = -0.50455 \pm j0.2408$$

Cascade optimization suggests a combination of $z_{1,2}$ with $p_{1,2}$ and $z_{3,4}$ with $p_{3,4}$ which results in the following two second-order functions from equation (10.60).

$$H_1(s) = \frac{\{s^2 + 1.1757\}}{(s^2 + 0.3423s + 0.2559)} \tag{10.61}$$

$$H_2(s) = \frac{\{s^2 + 6.8283\}}{(s^2 + 1.0091s + 0.31255)} \tag{10.62}$$

Parameters of the two second-order normalized functions are:

$\omega_{o1} = 0.50586$, $\omega_{o2} = 0.55906$, $Q_1 = 1.4778$ and $Q_2 = 0.554$

Both the transfer functions are LP notch, for which a number of circuits and approaches are available; we select the *modified summation method* of Section 8.6.1.

With the required overall dc gain being unity, use of equations (10.36) to (10.38) gives the value of gain coefficients as:

$k_1 = 0.2178$ and $k_2 = 0.04577$

With these values of gain coefficients, the transfer functions modify as:

$$H_1(s) = \frac{0.2178s^2 + 0.2559}{(s^2 + 0.3423s + 0.2559)} \tag{10.63}$$

$$H_2(s) = \frac{0.04577s^2 + 0.31255}{(s^2 + 1.0091s + 0.31255)} \tag{10.64}$$

To realize the aforementioned transfer functions, the Ackerberg–Mossberg biquadratic circuit is used in the modified summation approach; the respective design values of elements after de-normalization by 6000 rad/s and impedance scaling of 10 kΩ, are as follows:

$QR_1 = 14.778$ kΩ, $R_{31} = R_{41} = R_{51} = R_{61} = 10$ kΩ, $R_{y1} = 10$ kΩ, $C_{11} = C_{21} = 32.94$ nF and $C_{\alpha 1} = 7.1743$ nF

$QR_2 = 5.54$ kΩ, $R_{32} = R_{42} = R_{52} = R_{62} = 10$ kΩ, $R_{y2} = 10$ kΩ, $C_{12} = C_{22} = 29.811$ nF and $C_{\alpha 2} = 1.3644$ nF

The cascaded fourth-order filter is shown in Figure 10.16(a) and its simulated magnitude response is shown in Figure 10.16(b). The pass band is maximally flat, having unity gain at dc with an attenuation of 0.422 dB at 2000 rad/s and attenuation of 40.1 dBs at 6000 rad/s; this satisfies the specifications easily.

Example 10.10: Realize a maximally flat LPF in which it is desired that its dc gain remains unity and its gain drops by 1 dB at 20 krad/s by introducing transmission zeroes at 40 krad/s and 50 krad/s to increase rate of fall of attenuation.

Solution: For the given specifications, the desired transfer function was obtained in Example 3.7 of Chapter 3. The required sixth-order transfer function as obtained in Example 3.7 is as follows:

$$H_6(S) = \frac{3.119\left\{\left(\frac{S}{2}\right)^2 + 1\right\}\left\{\left(\frac{S}{2.5}\right)^2 + 1\right\}}{\left(S^2 + 2.511S + 1.744\right)\left(S^2 + 1.524S + 1.449\right)\left(S^2 + 0.475S + 1.234\right)} \tag{10.65}$$

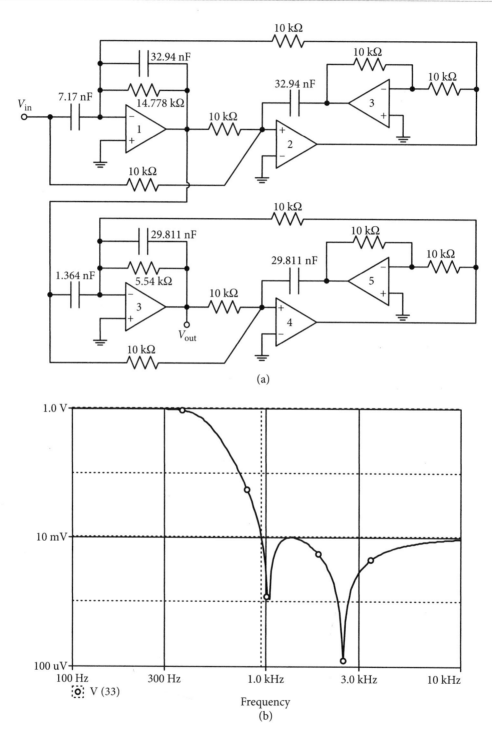

Figure 10.16 (a) Fourth-order inverse Chebyshev filter circuit for Example 10.9. (b) Its magnitude response.

Transfer function in equation (10.65) is to be broken into second-order sections. As a first step, the nearest poles and zeros are to be combined. Hence, the values of the poles and zeros are:

$$z_{1,2} = \pm j2, \, z_{3,4} = \pm j2.5, \, p_{1,2} = -1.256 \pm j0.41, \, p_{3,4} = -0.762 \pm j0.932, \text{ and } p_{4,5} = -0.238 \pm j1.085$$

Based on the closeness between poles and zeros, the following second-order sections are formed.

$$H_1(s) = \frac{\left\{(s/2)^2 + 1\right\}}{\left(s^2 + 2.511s + 1.744\right)} \tag{10.66}$$

$$H_2(s) = \frac{3.119}{\left(s^2 + 1.524s + 1.449\right)} \tag{10.67}$$

$$H_3(s) = \frac{\left\{(s/2.5)^2 + 1\right\}}{\left(s^2 + 0.475s + 1.234\right)} \tag{10.68}$$

The next step is the selection of the sequence of the sections for which the mentioned thumb rule in terms of pole-Q values can be applied. The respective values of the pole-Q for the transfer functions H_1, H_2 and H_3 are:

$$Q_1 = 0.5259, \, Q_2 = 0.7898 \text{ and } Q_3 = 2.3386$$

Hence, in the proposed cascade H_1 will be followed with H_2 and H_3 will be at the end. Otherwise, it will be preferred to have the LP section in the beginning.

Next, the assignment of gain is to be done for the individual sections so that dynamic range is maximized and the responses at all intermediate section nodes become as flat as possible. The voltage maxima at the output of each section are to be evaluated either by calculation or by inspection. As the function H_2 is an LP section with $Q_2 = 0.7899$, its maxima will occur at dc, which means $h_2 = 3.119/1.449 = 2.1525$; similarly, H_1 which functions as an LP notch, with $Q_1 = 0.5259$, has its maxima also at dc, which results in $h_1 = 1/1.744 = 0.5733$. As we are designing an LPF with gain unity at dc, for H_3, the value of h_3 was also taken at dc, which is $h_3 = 1/1.234 = 0.81$. From these values, we can write:

$$M_1 = |h_1|_{max} = 0.5733, \, M_2 = |h_1 \, h_2|_{max} = 2.1525 \times 0.5733 = 1.234 \text{ and } M_3 = |h_1 \, h_2 \, h_3|_{max} = 1.234 \times 0.81 = 1.0$$

As the final gain is to be unity, $KM_3 = 1$, and value of the gain coefficients of the three sections are evaluated as:

$$k_1 = 1/0.5733 = 1.744, \, k_2 = 0.5733/1.234 = 0.4646, \, k_3 = 1.234/1.0 = 1.234$$

Having obtained the gain coefficients, the biquadratic arrangement known as *modified summation method* of Section 8.6.1 was selected for the realization of the three transfer functions. For the sake of reference, the expression of the biquad is repeated here as equation (10.69).

$$\frac{V_{out}}{V_{in}} = -\frac{\alpha s^2 + s(k-\beta)/CR + \gamma/C^2R^2}{s^2 + \{s/(CRQ)\} + 1/C^2R^2} \qquad (10.69)$$

Comparing equation (10.66) with equation (10.69), $k = \beta = 0$, $\alpha = 0.25$ and $\gamma = \dfrac{1.744}{1.744} = 1$. If normalized $R = 1$, we get $C = 1/1.744^{0.5} = 0.7572$. As the frequency normalization was done by a factor of 20 krad/s in Example 3.7, the same is used for de-normalization, and with an impedance scaling factor of 10 kΩ, the values of the elements for $H_1(s)$ are obtained as:

$$C_1 = 3.7861 \text{ nF}, \ \alpha_1\,C_1 = 0.9465 \text{ nF}, \ R_1 = 10 \text{ k}\Omega, \ Q_1\,R_1 = 5.259 \text{ k}\Omega, \ R_1/\gamma_1 = 10 \text{ k}\Omega$$

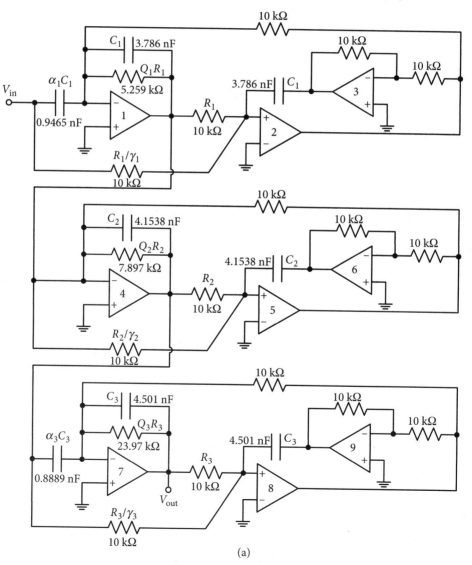

(a)

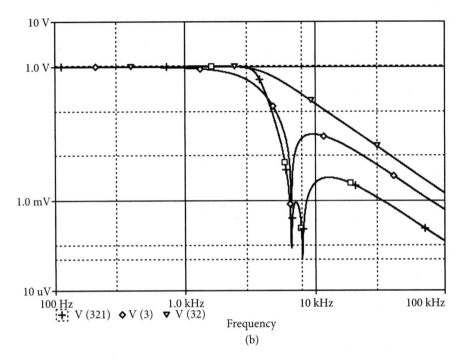

$$\text{V (321)} \quad \text{V (3)} \quad \text{V (32)}$$

Frequency
(b)

Figure 10.17 (a) Sixth-order cascaded low pass filter structure for Example 10.10. (b) Its magnitude response.

With the same circuit configuration and frequency and impedance scaling factors, element values for the remaining transfer functions H_2 and H_3, respectively, are as follows:

$$C_2 = 4.1538 \text{ nF}, \ \alpha_2 \, C_2 = 0, \ R_2 = 10 \text{ k}\Omega, \ Q_2 \, R_2 = 7.898 \text{ k}\Omega, \ R_2/\gamma_2 = 10 \text{ k}\Omega$$
$$C_3 = 4.501 \text{ nF}, \ \alpha_3 \, C_3 = 0.8889 \text{ nF}, \ R_3 = 10 \text{ k}\Omega, \ Q_3 \, R_3 = 23.386 \text{ k}\Omega, \ R_3/\gamma_3 = 10 \text{ k}\Omega$$

Figure 10.17(a) shows the complete sixth-order cascaded filter with element values and Figure 10.17(b) shows its simulated magnitude response. The filter's dc gain is unity, attenuation at 20 krad/s is 1.09 dB and zeros occur at 40.14 krad/s and 49.82 krad/s; this is very close to the design specifications.

10.4 Tuning of Filters

There has been a significant improvement in the fabrication processes and manufacturing methods of electronic components and devices and it has resulted in considerable advances in the performance capability of active filters. However, increasing complexities in various application fields such as communication, instrumentation and control has necessitated in more miniaturization. Fabrication processes of integrated circuits (ICs) have advanced from the thin and thick film technology to the hybrid and to the monolithic form. Unfortunately, even with tremendous advancements, both passive and active components suffer from manufacturing tolerances of varying degrees, affecting the performance parameters of the filters. Deviation in the filter parameters depend on the amount of tolerance of the components. Therefore, it becomes necessary to adopt correction methods such that the practically obtained performance is within the prescribed limits of the design, and the circuit and the system works as desired.

It has been mentioned a number of times that for OA based continuous-time filter circuits, parameters are set by RC products; this is barring OA-R and OA-C circuits where parameters depend on gain bandwidth product as well. In OTA based circuits, the parameters depend on the capacitance to the trans-conductance ratios. This means that not only the passive components but also the active components should be realized with maximum accuracy and their parameters should remain stable.

The presence of production tolerance in components requires that post-design adjustment is an essential step in meeting tight specifications. While pre-distortion is also an important step for compensating the effects of component imperfections, it is the post-fabrication parameter tuning which is almost essential. Of course, the specific tuning employed depends upon the function to be realized, the network configuration used and the technology of implementation. Over the years, *functional tuning* [10.1], *deterministic tuning*[10.2] and *automatic tunings* have been developed and used; these are briefly discussed here.

Functional and Deterministic Tuning Functional tuning is performed by adjusting the parameters. This is done by changing the components at a frequency of known phase shift while the circuit is in operational mode. Selection of phase, instead of magnitude is done because it was observed that change in phase is more pronounced than change in magnitude of filter near the region of the critical frequency, ω_o.

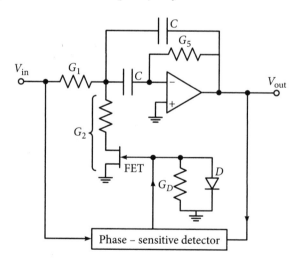

Figure 10.18 Self-tuned signal-tracking multi-loop feedback filter configuration.

Ideally, the process of tuning proceeds by controlling the parameters of interest in a non-interactive fashion through change in a single circuit element; which in almost all the cases, is a resistor. Obviously, complete non-interactiveness is practically difficult to achieve, which necessitates a tuning sequence. Without going into the details of the functional tuning, it can be said that the trial and error process for precision is slow and suitable only for simpler circuits. Moreover, the process of change in the controlling resistor through laser trimming is irreversible. Finally, this kind of tuning is possible in hybrid circuits and not in monolithic ICs.

In deterministic tuning, the time-consuming iterations of functional tuning are avoided by introducing a predictive step of performing initial circuit analysis. Exact design equations are formulated considering all known imperfections, like frequency dependent amplifier gain and capacitor losses. However, once the analysis is done without making the circuit inter-connections, final design equations become a set of non-linear expressions and demand a computer solution. At the end, laser trimming is performed, making this approach unsuitable for monolithic ICs.

10.4.1 Automatic tuning

In applications where the input signal frequency varies over a wide range, a wide bandwidth filter is not suitable for effective rejection of noise. One solution for the problem is to use a high Q BPF with its center frequency ω_o being continuously adjusted to a desired value. In such a case, it is also required that the bandwidth BW = ω_o/Q and mid-band gain H_o remain constant and unaffected by the change in ω_o. These conditions are achievable in OA-RC circuits in which tuning of ω_o is possible by a single resistor, and this resistor does not affect BW or H_o. There are quite a few multi-OA circuits like state variable and GIC based configurations satisfying the required conditions. The single OA multiple feedback circuit discussed in Section 7.3 comes in that category, for which equal capacitance C:

$$\omega_o = \sqrt{G_5(G_1+G_2)/C}, \ \mathrm{BW} = 2G_5/C \text{ and } H_o = -G_1/2G_5 \tag{10.70}$$

It is clear from equation (10.70) that G_2 is present in the expression of ω_o only but not in the expressions of BW and H_o. Automatic tuning is implemented by means of a phase detector as shown in Figure 10.18 [10.3]. The phase detector differentiates the input and output filter waveforms before their comparison and necessary steps, such as gating, smoothing and feeding to the FET acting as a voltage-variable resistor.

In practice, the arrangement of automatic tuning comprises the phase-locked loop (PLL), which tracks a given signal while passing signals only in a small bandwidth [10.4]. Such a system can be integrated in monolithic form. The method will be discussed later in connection with the realization of active R and active C filters in Chapter 17.

References

[10.1] Mossberg, K. 1969. 'Accurate Trimming of Active RC Filters by Means of Phase Measurements,' *Electronic Letters* 5 (21): 520–1.

[10.2] Lueder, E., and G. Malek. 1976. 'Measure-predict Tuning of Hybrid Thin-film Filters,' *IEEE Transactions* CAS-23 (7): 461–6.

[10.3] Deboo, G. J., and R. C. Hedlund. 1972. 'Automatically Tuned Filter Uses I.C. Operational Amplifier,' *Electrical Design News* 17: 38–41.

[10.4]   Grebene, A. B., and H. R. Cemenzind. 1969. 'Frequency-selective Integrated Circuit Using Phase-locked Techniques,' *IEEE Journal* SC-4 (4): 216–25.

Practice Problems

10-1   A LP filter has the following specifications:

$\omega_1 = 10$ krad/s, $\omega_2 = 36$ krad/s, $A_{max} = 1$ dB ripple and $A_{min} = 50$ dBs

Design an active RC filter to satisfy the specifications using single amplifier second-order sections in cascade approach and test it.

10-2   A HP filter has the pass band from 10^4 rad/s to infinity. The peak to peak ripple in the pass band has to be less than 2 dBs. For $\omega \leq 2$ krad/s, the loss must be greater than 50 dBs. Design and test an active RC filter to satisfy the specifications using the cascade approach.

10-3   Design and test an equal-ripple BP filter to satisfy the specifications: (a) the pass band extends from 2000 to 8000 rad/s. The ripple width in the pass band does not exceed 0.5 dB. (b) The magnitude is at least 30 dB down at 24 krad/s from its peak value in the pass band.

10-4   Realize a fifth-order LP Chebyshev filter using cascade approach. Optimize its dynamic range. Ripple width in the pass band is 2 dBs, and frequency is normalized by 25 krad/s. Use practical element values and test the filter.

10-5   Realize the following transfer functions by cascading second-order sections using single amplifier biquads while frequency de-normalization will be with 10 krad/s.

(a) $H(s) = \dfrac{4}{\left(s^2 + 0.77s + 1\right)\left(s^2 + 1.85s + 1\right)}$

(b) $H(s) = \dfrac{2}{\left(s^2 + 3s + 3\right)\left(s^2 + 1.414s + 1\right)}$

(c) $H(s) = \dfrac{s^2(s^2 + 1)}{\left(s^2 + 3s + 3\right)\left(s^2 + 1.414s + 1\right)}$

(d) $H(s) = \dfrac{1}{\left(s^2 + 0.77s + 1\right)\left(s^2 + 1.85s + 1\right)\left(s^2 + 1.414s + 1\right)}$

10-6   Repeat Problem 10-1 using multi amplifier biquads.

10-7   A fifth-order normalized LP filter can be realized by a cascade of first- and second-order functions as shown here:

$$H(s) = \frac{k_1 \times k_2 \times k_3}{(s+1)\left(s^2 + 1.61803s + 1\right)\left(s^2 + 0.61803s + 1\right)}$$

A possible design for the system is suggested in Figure P10.1. Prove its adequacy or provide an alternate design.

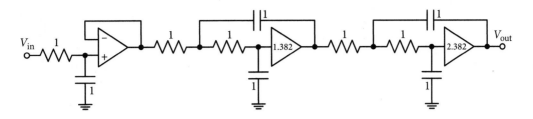

Figure P10.1

10-8 Specifications of an inverse Chebyshev function as shown in figure are:

$\alpha_{max} = 0.5$dB, $\alpha_{min} = 20$ dBs, $\omega_1 = 36$ krad/s and $\omega_2 = 80$ krad/s.

Determine the order of the filter and find transfer-function satisfying the specifications in terms of the product of second-order (a first-order also if needed) sections. Optimize dynamic range of cascaded filter.

10-9 Repeat problem 10-8 for the following specifications:

$\alpha_{max} = 0.5$ dB, $\alpha_{min} = 30$ dBs, $\omega_1 = 1$ krad/s, and $\omega_2 = 3.45$ krad/s.

Chapter **11**

Amplification and Filtering in Biomedical Applications

11.1 Introduction

There has been a lot of development in the use of biomedical equipment for diagnostic and treatment purposes in recent times. A good amount of literature has come from the interdisciplinary area of medical engineering, medical instrumentation, and so on, making medical electronics a specialized field of study. An important constituent of electronics engineering is the study of analog filters. Hence, the design and application of analog filters finds a place in the books and literature on medical engineering and medical instrumentation too. The design of analog filters especially the continuous-time types requires special attention. However, the available books on analog filter design do not adequately relate filter design with biomedical applications. A few simple examples included in this chapter will try to bridge the gap between theoretical design and the application of analog filters in this important field.

We have already studied the basics of continuous-time analog filter in the previous chapters. In this chapter, we will first connect the design of a filter to its utilization in the field of biomedical electronics.

It is a well-known fact that cells in the human body have different element (Na^+, K^+, Ca^{++}, Cl^-) ion concentrations inside and outside the membrane. This difference in ion concentration creates a small electric potential called *biopotential*. When there is a disturbance in a biopotential, an *action potential* is generated which is the result of depolarization and repolarization of the cells in the human body. It is the action potential at the location (nodes) on the body which is detectable and can be processed using biomedical circuits. When such signals generated by the heart are collected, it makes up the electrocardiogram (ECG). ECG detectors use electrodes to collect these signals, which are amplified, filtered and displayed for data analysis. ECG signals require filtering and amplification to produce high-quality signals. Not only are different stages of amplification used, specific signal processing is also required. Therefore, instead of a simple (DA) differential amplifier, instrumentation amplifiers (inst-amp) are used. Section 11.2 discusses the the necessity of converting DA into an inst-amp. Transformation of the inst-amp into a biopotential amplifier and as integrated circuit inst-amps especially suitable in medical instruments and devices are discussed in Section 11.3 along with a case study on a piezoelectric transducer.

A brief description of the characteristics of physiological signals is included in Section 11.4. Analog front-ends (AFEs) used in ECG detection is considered in Section 11.5, which shows that amplification and analog filtering are essentials. As an example, an AFE system using an on-chip filter is also discussed.

It is rightly assumed that signals from the heart are important. However, signals from other body parts also play an important role in diagnostics. The rest of the chapter discusses a few such cases. Utilization of inst-amp and filters for collecting electromyographic (EMG) signals is shown in Section 11.6. Filtering in brain–computer interface applications and chamber plethysmography is briefly discussed in Sections 11.8 and 11.9. The chapter concludes with the application of an eighth-order filter for simulating the frequency response of a human ear.

11.2 Instrumentation Amplifiers

Figure 11.1 shows a conventional simple differential amplifier (DA) using a single OA. In the circuit, the following resistance relation exists:

$$R_1 = R_3 \text{ and } R_2 = R_4 \tag{11.1}$$

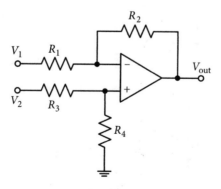

Figure 11.1 Commonly used differential amplifier circuit.

Then considering OA as ideal, the voltage gain of the DA is obtained as:

$$\frac{V_{out}}{V_{in}} = \frac{V_{out}}{V_2 - V_1} = \frac{R_2}{R_1} = \frac{R_4}{R_3} \tag{11.2}$$

The circuit is very useful and functions well as an inst-amp, which amplifies the differential signal $V_{in} = (V_2 - V_1)$ and rejects the common mode signal. However, as application requirements become strict, limitations of the circuit become obvious. For instance, input impedances at the inverting and non-inverting terminals are not very high; they are also unequal. Hence, if the signal is applied to only one terminal and the other terminal remains grounded, different currents flow in the two terminals; it degrades the common mode rejection ratio (CMRR).

In addition, the circuit requires a very close resistance ratio matching of equation (11.1). Even a small mismatch in the resistance ratios degrades the CMRR. This degraded value of CMRR may not be acceptable in many practical applications like precision instrumentation and especially in medical applications.

11.3 Differential Biopotential Amplifier

Figure 11.2 shows a simple differential biopotential amplifier. When bio-electrodes are placed on a body, electro-physical actions generate a potential difference across them. The first stage of the pre-amplifier rejects the changes in the common mode signals detected simultaneously by both the bio-electrodes. The pre-amplifier comprises OA1 as a DA, for which low frequency gain $\dfrac{R_3}{R_2} = \dfrac{R_4}{R_1}$ is generally the gain which is set at nearly 1000. However, with such a high gain, the pre-amplifier will work only when the dc offset is limited to 10 mV; otherwise, the OA will saturate. To use the circuit as part of the surface ECG amplifier, the design value of the gain is to be considerably reduced because offset may go even up to 300 mV [11.1].

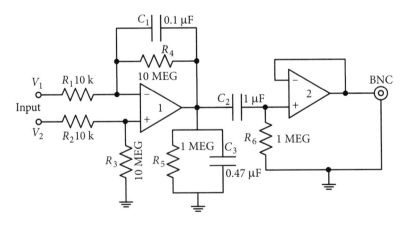

Figure 11.2 A simple differential bio-potential amplifier.

As capacitor C_1 offers low impedance at higher frequencies, resistance R_4 becomes ineffective and the first stage, using OA1, acts as a low pass filter (LPF). It also limits the bandwidth (3 dB frequency ≈ 160 Hz) of the amplifier eliminating high frequency noise. Capacitors C_1 and

C_3, which act as a bypass, clamp oscillation and reduce instabilities in the circuit. Capacitor C_2 along with R_6 acts as a high pass filter (HPF) (3 dB frequency ≈ 0.16 Hz), and the unity gain follower OA2 acts as signal buffer.

The circuit has also been found very useful in measuring the CMR and input impedance of a biopotential amplifier [11.1].

3 OA Instrumentation Amplifier- A simple and obvious method for improving the performance of the DA shown in Figure 11.1 is through the addition of high impedance buffer amplifiers. This arrangement works well up to a certain level of accuracy; hence, further improvements are made by operating the buffers with gain providing additional flexibility. Such an improved form of the circuit is shown in Figure 11.3; it is popularly known as an *instrumentation amplifier*.

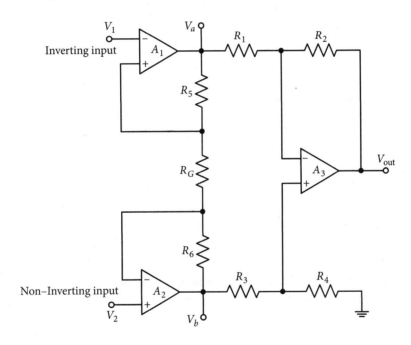

Figure 11.3 The classic 3OA instrumentation amplifier circuit.

Input OA1 and OA2 can be analyzed as non-inverting amplifiers. Therefore, applying super-position, output voltages V_a and V_b are found as:

$$V_a = \left(1 + \frac{R_5}{R_G}\right)V_1 - \left(\frac{R_5}{R_G}\right)V_2 \tag{11.3}$$

$$V_b = \left(1 + \frac{R_6}{R_G}\right)V_2 - \left(\frac{R_6}{R_G}\right)V_1 \tag{11.4}$$

With resistance equality of equation (11.1) satisfied, the gain of the DA OA3 is (R_2/R_1). Hence, the expression of the output voltage with $R_5 = R_6$ will become:

$$V_{out} = (V_b - V_a)(R_2/R_1) = (V_2 - V_1)(1 + 2R_5/R_G) \quad (11.5)$$

Therefore, the gain of the amplifier is $(1 + 2R_5/R_G)$ which is controllable with a single resistor R_G.

11.3.1 Integrated circuit monolithic instrumentation amplifier

In practice, bio-potential amplifiers are hardly built now in terms of individual OAs. Instead, all the components are integrated on a single chip to constitute an inst-amp, called an integrated circuit instrumentation amplifier (ICIA) or a monolithic instrumentation amplifier.

Monolithic IC inst-amp are mostly based on variations of 2OAs and 3OAs DA circuits and resistors are mostly laser-trimmed for improved accuracy. Additionally, both active and passive components are closely matched being on the same chip and placed very close to each other; which is an essential requirement of the circuit. It helps in getting high CMR. An additional advantage of a monolithic inst-amp is that they are easy to use as they are available in various packages like very small, very low-cost SOIC, MSOP or LFCSP (chip scales) packages. A good discussion on a large number of monolithic inst-amps is available in *Analog Devices* [11.2]. A large number of monolithic inst-amps are being manufactured for general and specific applications [11.3].

One of the commonly used ICIA is INA 102 from Burr Brown Corporation [11.4]. The INA 102 is a high-accuracy monolithic inst-amp designed for signal conditioning applications where low quiescent power is required. Its prominent features are: low quiescent current of 750 μA max, internal gains of 1, 10, 100 and 1000, high CMR of 90 dB min, low offset voltage of 100 μV max and high input impedance of $10^{10}\,\Omega$.

To illustrate the working of an inst-amp, a brief discussion and design steps of a specific case is presented here [11.5]. It is expected that the procedure can equally be applied to similar cases, being typical in nature.

A Case Study: A piezoelectric transducer produces signals which are conditioned and fed to a 250 kc/s, 12-bit ADC (analog to digital convertor) with an input voltage range of 0-5V. Maximum amplitude of the differential voltage from the sensor is expected to be 20 mV. Bandwidth of the sensor is not known but the presumed high frequency range needs to be band limited so as not to saturate the next stage inst-amp.

Based on the input signal magnitude, the frequency range, and the ADC to be employed, signal conditioning specifications are found and implemented using a chain of circuits. However, our focus being on analog filters, only that part will be highlighted here.

A pass band of 50 kHz was selected for an LPF, considering that almost all the relevant information from the transducer is expected to be contained in this frequency range. Permissible variation in amplitude was taken as 5%. This figure of 5% comes from the data sheet of the transducer, as it is calibrated within 5% of the actual value in the selected frequency pass band. A maximum of 5% non-linearity in phase was taken as a rule of thumb [11.6].

The signal conditioning arrangement is shown in Figure 11.4 in block form. The signal produced by the transducer is buffered and amplified by an inst-amp. It is then amplified to a level compatible with full ADC resolution. A LPF band limits the signal before it reaches the ADC. Data retrieved from memory is then un-inverted and further processed.

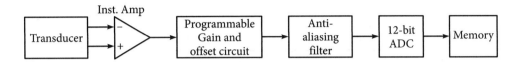

Figure 11.4 Signal conditioning sequence for a piezoelectric transducer signal [11.5].

A piezoresistive transducer is used with a Wheatstone's bridge built into the transducer packing. Output of the transducer circuit is a differential signal with an expected maximum amplitude of 20 mV.

A differential inst-amp is used to provide high-impedance buffering; it attenuates unwanted common-mode noise in the signal from the transducer. However, prior to the inst-amp, a first-order passive filter reduces undesired high-frequency signals. The first-order filter also forms the first pole of the seven poles of the anti-aliasing filter to be used before ADC.

In the present application, the inst-amp is required to have high DC precision, low noise, high CMRR and high input impedance. The AD8224B inst-amp shown in Figure 11.5(a) is selected because it can satisfy the aforementioned requirements. Its slew rate being 2 V/μs and with a maximum output of 100 mV (p-p for its gain of 5), full-power bandwidth (FPBW) is calculated using equation (11.6):

$$\text{FPBW} = \frac{\text{SR}}{2\pi V_{p-p}} = \frac{2 \times 10^6}{2\pi \times 0.1} = 3.18\,\text{MHz} \tag{11.6}$$

The calculated value of the FPBW shows that the slew rate limitation will not affect the inst-amp. Figure 11.5(b) shows the simulated response of the inst-ampl of Figure 11.5(a) with a voltage gain of 4.985 {(1 + 49.4/R_g) and R_g = 12.4 kOhm} and a 3-dB frequency of 59.1 kHz (1/150 pF × 2 × 8.87 k = 375.8 krad/s).

While using the inst-amp, it is important to consider the output load of the transducer, which is approximately (650±350) Ω. Obviously, this load will modify the 3-dB frequency of the passive filter.

Output of the inst-amp is passed on to a programmable gain circuit and voltage offset circuit. These circuits allow the system to bring the signal within the full scale range of the ADC [11.5].

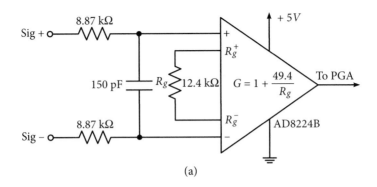

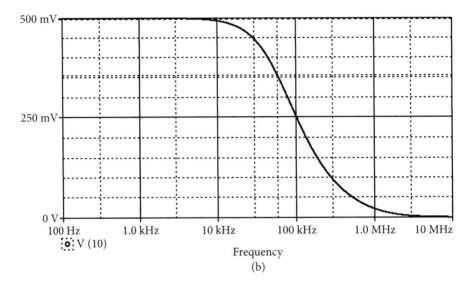

Figure 11.5 (a) The AD8224B instrumentation amplifier [11.5]. (b) Simulated magnitude response of the instrumentation amplifier shown in Figure 11.5(a).

Low Pass Anti-aliasing Filter: An ADC requires 6 dB of attenuation per bit at the Nyquist frequency [11.8]. Hence, a 12-bit ADC requires 72 dBs attenuation This means that if aliasing is allowed in the transition region (from 50 kHz to the sampling frequency of 250 kHz), 72 dBs attenuation is to be achieved at the sampling frequency (250 kHz) minus the highest non-aliased frequency (50 kHz), that is 200 kHz. For these specifications in terms of attenuation, the order of the filter can be calculated using the steps described in Chapter 3. Of course, the filter will depend on the type of the magnitude approximation used. In this case, Butterworth approximation was preferred and the following relation was used to find its order [11.8].

$$N \geq \frac{3 \times \text{Bits}}{10\log\left(f_s / f_c\right)} = \frac{3 \times 12}{10\log\left(250\,\text{kHz} / 50\,\text{kHz}\right)} \geq 6 \tag{11.7}$$

Here, $f_s = 250\text{kHz}$ is the sampling frequency and $f_c = 50\text{kHz}$ is the pass band frequency; the required order of the Butterworth filter is seven. One of the choices for realizing a seventh-order filter is to apply the cascade approach. Hence, from Table 3.1, location of the poles for the seventh-oder filter is obtained as:

$$s_{1,2} = -0.9009 \pm j0.4338, \ s_{3,4} = -0.2225 \pm j0.9649$$

$$s_{5,6} = -0.6234 \pm j0.7818, \ s_7 = -1.0 \tag{11.8}$$

Out of the seven poles in equation (11.8), the real pole s_7 was already realized prior to the inst-amp. The remaining six poles are to be realized using three second-order LP sections. Many choices are available for realizing second-order sections. However, after comparison between the Sallen and Key circuit (Section 7.7) and the multiple feedback (MF) topology (Section 7.3), the latter was selected in the report. The free Texas Instrument Filter Pro Desktop tool was used to design the three filter sections [11.7]. The sections were ordered in increasing value of pole-Q as described in Section 10.3.2. As input to the filter was limited to 4 V, the gain of the first two sections was unity and for the last section, gain was 1.25 to provide full 5 V to the ADC. The OA OP462 was employed as it has a slew rate of 10 V/μs for a maximum peak-to-peak voltage of 5 V.

Figure 11.6(a) shows the cascade arrangement in block form. The three second-order sections use the MF LP circuit of Figure 7.6 with the following respective element values; the corner frequency for the sections was 60 kHz.

$$R_{11} = 24 \text{ k}\Omega, \ R_{31} = 12 \text{ k}\Omega, \ R_{41} = 24 \text{ k}\Omega, \ C_{21} = 240 \text{ pF}, \ C_{51} = 100 \text{ pF}$$

$$R_{12} = 16.5 \text{ k}\Omega, \ R_{32} = 8.25 \text{ k}\Omega, \ R_{42} = 16.5 \text{ k}\Omega, \ C_{22} = 510 \text{ pF}, \ C_{52} = 100 \text{ pF}$$

$$R_{13} = 4.7 \text{ k}\Omega, \ R_{33} = 2.61 \text{ k}\Omega, \ R_{43} = 5.9 \text{ k}\Omega, \ C_{23} = 4.7 \text{ nF}, \ C_{53} = 100 \text{ pF} \tag{11.9}$$

In addition, resistances of 24 kΩ, 16.5 kΩ and 5.23 kΩ were connected at the non-inverting terminals of the respective three sections to equalize input bias currents in the OAs.

The simulated response of the three sections combined with the first-order section (and gain of 5 for the inst-amp) is shown in Figure 11.6(b). The overall gain is 6.274 with a 3-dB frequency of 59.59 kHz. At 200 kHz, attenuation from the pass band is 73.55 dBs, which satisfies the specifications.

11.4 Physiological Signals

Signals resulting from physiological activities can be considered as a communication system transmitting information from one part of the body to another. We can detect these signals from those parts of the body where many neurons activate simultaneously to create a single event. Accurate detection of these signals is very important as their interpretation can indicate whether a body part is healthy or not. Due to their low amplitude and the method of observing/ detecting these signals from the body parts, they are processed using *biopotential amplifiers*. These amplifiers have high gain in the range of 1000 or more and some specific characteristics that will be described here in brief.

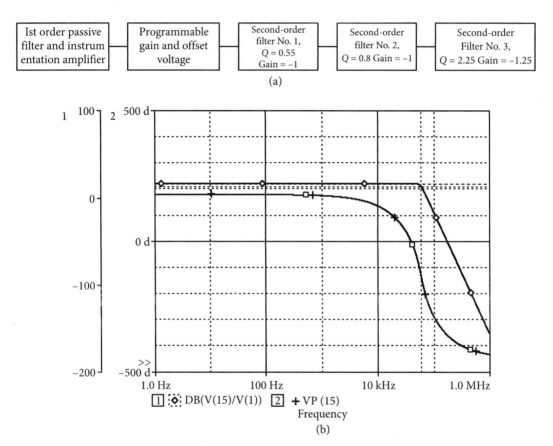

Figure 11.6 (a) Cascade arrangement in block form for filtering and amplification of signals from a piezoelectric transducer: a case study [11.5]. (b) Simulated magnitude and phase response of the seventh-order low pass filter for Figure 11.6(a), employing elements as given in equation (11.9).

Like any other amplifier, the bandwidth requirement of the biopotential amplifier depends on the frequencies present in the physiological signals of interest, which are sometimes in the range of a fraction of Hz to a few Hz or from a few Hz to a few kHz.

A number of noise signals corrupt the basic signal, like those due to magnetic induction in wires, equipment interconnections and imperfections. Noise signals are also generated due to the displacement current in the electrode leads and in the body. Another significant noise factor is due to the changing magnetic field of the AC power supply; this is at a rate of 50 Hz/60Hz and caused by induction. The changing magnetic field causes alternating current to flow to ground through the signal measurement system. All these noise signals are to be attenuated heavily.

A likely error may occur due to common-mode signals. Therefore, a very high value of common mode rejection ratio (CMRR) is required. Otherwise, even a small common-mode voltage gets amplified because of the large amplifier gain of the device used, which may saturate active components, causing great errors.

Physiological signals are collected using metal electrodes which are in contact with the body through an electrolyte. This process results in a voltage known as *half-cell potential*. This DC potential must be taken care of by the amplifier; otherwise, it will also saturate active components.

Most biopotential amplifiers are OA based. Single-ended OA base amplifiers were used in the past in the front-end stage. However, with the availability of low-cost integrated instrumentation amplifiers (IIA), single-ended and conventional DAs have been replaced. There are a number of IIAs available for general and specific applications from world class manufacturers [11.9].

11.5 Analog Front-end for ECG Signal

As mentioned in the previous section, electrical signals may be detected from a human body part. One of the most important signals comes from heart muscles. The measurement of these signals and their graphic representation are known as electrocardiogram (ECG or EKG). The basics of ECG measurements are almost the same for any type of application, for example, whether it is monitoring of heart rate that is required or diagnosis of any heart condition, or signals are obtained from any other part of the body. However, the electrical components and their specifications vary substantially depending upon application and usage.

All ECG signals (heart signals) collected through electrodes which are located at different parts of the body have amplitudes of a few mV. These signals are processed and finally displayed as a channel on the ECG print out. There is more than one channel for ECG monitoring; they are referred to as *leads*. For example, a 12-lead ECG device has 12 separate channels. The required electronic components for an ECG can be separated into the analog front-end (AFE) and the *rest of the system* [11.10]. Though AFEs are fundamentally similar, they differ in terms of the number of leads, fidelity of signal, nature and magnitude of noises, and so on. The primary function of the AFE is to digitize the heart signals for processing, which is then again converted as a function of time for display. The architecture of an AFE may have a *brute force* type which provides high fidelity over a wide frequency range or a *minimal AFE*, a consumer-grade ECG. Circuits in most ECG devices lie between the two extremes of brute force type or of minimal type.

When IC inst-amp (IIA) is used for amplification of small input signals, common-mode voltage rejection and elimination of noise also takes place. IIA also provides a buffer for the sampling capacitances of the ADC. Filters before and after the IIA remove noise and band limits the incoming signals. Without going into any further detail of the ECG system, let us take some practical examples of the application of analog filters in the medical field.

11.5.1 An AFE system with low power on-chip filter

Different types of ECG equipment include telemetry devices, Holter monitors, consumer ECGs and diagnostic ECGs. Diagnostic ECGs are used in hospitals and doctors' offices to perform high-quality ECG tests. For reasons of portability and durability of ECG equipment, an attempt is made to reduce the power consumption. In line with this requirement, the basic parts of a portable AFE system are shown in Figure 11.7(a). Here, the *acquisition board* converts the body signals to six leads. The acquisition board consists of an inst-amp, an HPF, a (50 Hz/60 Hz) notch filter and a common-level-adjuster [11.11]. Amplitude of the ECG signal is in the range of 100 μV to 4 mV [11.12]. With such a low level of signal, it is required that a signal-to-noise and distortion ratio (SNDR) must be at least 32 dBs (that is, 6 bits). The next important specification is that of the frequency range of useful ECG signals, which lies between 0.1 Hz and 250 Hz. Hence, the on-board analog LPF is designed for its 3-dB frequency of 250 Hz, and an HPF with a 3-dB frequency of 0.1 Hz is used to attenuate noise below 0.1 Hz. The active RC HPF structure shown in Figure 11.7(b) was used. The following element values are obtained employing design equations (7.69)–(7.70):

$$R_1 = 483 \text{ k}\Omega, R_2 = R_3 = R_4 = 100 \text{ k}\Omega, \text{ and } C_1 = C_2 = 15.9 \text{ μF}$$

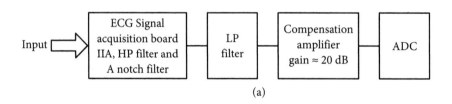

(a)

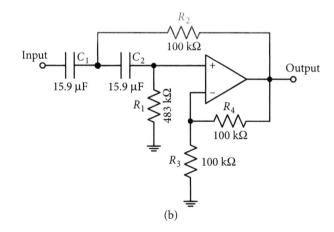

(b)

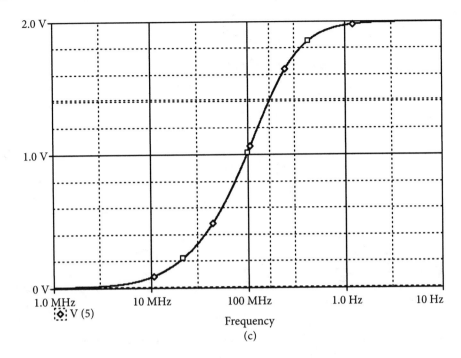

2.0 V

1.0 V

0 V

| 1.0 MHz | 10 MHz | 100 MHz | 1.0 Hz | 10 Hz |

◇ V (5)

Frequency

(c)

Figure 11.7 (a) An *analog front-end* system for portable EGC detection device [11.11]. (b) Second-order Sallen–Key high pass filter. (c) Simulated response of the high pass filter shown in Figure 11.7(b).

Using the 741 OA, its PSpice simulated response is shown in Figure 11.7(c). Its cut-off frequency is 0.167 Hz and high frequency gain is 2 as $R_3 = R_4$.

The last stage on the acquisition board is a notch filter using a twin-tee based active RC circuit shown in Figure 11.8(a) to attenuate the power supply noise frequency of 50 Hz/60 Hz. Its simulated response is shown in Figure 11.8(b). For a notch frequency of 60 Hz, $R = 530$ kΩ and $C = 5$ nF and for a notch frequency of 50 Hz, $R = 470$ kΩ and $C = 6.798$ nF.

Switched capacitor filters have been recommended for long-term physical detection and monitor systems [11.13]. However, power consumption in OAs increase due to leakage when the sampling frequency is low in the kHz range. It is for this reason that continuous-time operational transconductance amplifier (OTA) based filters are preferred for low frequency operation. For further reduction in power consumption, OTAs are to be used in sub-threshold regions with ultra-low transconductance [11.14].

As mentioned earlier, an anti-aliasing filter is to be used to attenuate the out-of-band interference before the ADC. A fifth-order ladder type Butterworth filter selected for its realization in OTA-C form is used. The advantage of a ladder structure have been discussed previously. OTA base fifth-order Butterworth filters with a designed 3-dB frequency of 250 Hz will be shown in Chapter 15.

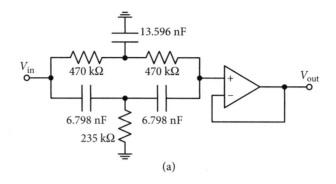

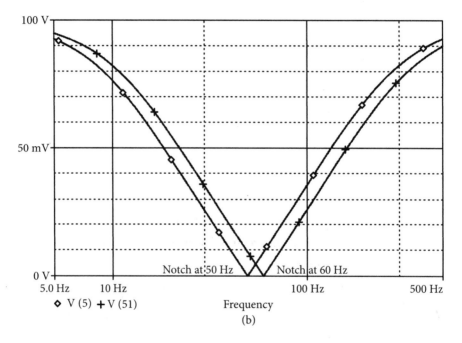

Figure 11.8 (a) Twin-T based notch filter circuit (notch at 50 Hz). (b) Response of the notch filter shown in Figure 11.8(b) at 50 and 60 Hz.

Why continuous-time analog filters, especially using OTAs are preferable over switched capacitor filters for low frequency medical applications is emphasized in another publication [11.15]. A 2.4 Hz sixth-order Bessel LP filter achieving 60-dB dynamic range was fabricated in 0.8 μm CMOS technology to illustrate the usefulness of continuous-time analog filters.

11.5.2 ECG signal acquisition system: an application example

A real-time ECG signal acquisition system has been introduced with DSP chip TMS320VC5509A at its core [11.16].

It has been mentioned before that ECG signals are bipolar, low frequency, low amplitude signals having an amplitude ranging from 10 µvolts to 4 milli volts. The frequency ranges in between 0.05 Hz and 100 Hz, but is mostly concentrated in 0.05–35 Hz. Because of the low level of the signals, they need to be analog filtered and amplified. In almost all cases, a preamplification is done using an inst-amp. After pre-amplification, a BPF (band pass filter) circuit employing OA OPA 2604 is used. It is a combination of an HPF and an LPF having corner frequencies of 0.05 Hz and 100 Hz, respectively. The circuit structure is shown in Figure 11.9(a). A combination of capacitors of 47 µF and 68 kΩ resistors work as HPF and a combination of 1 µF capacitors with 1.6 kΩ resistors provide LP response. The simulated response of the combined LP and the HP filters in Figure 11.9(b) shows its BP characteristics.

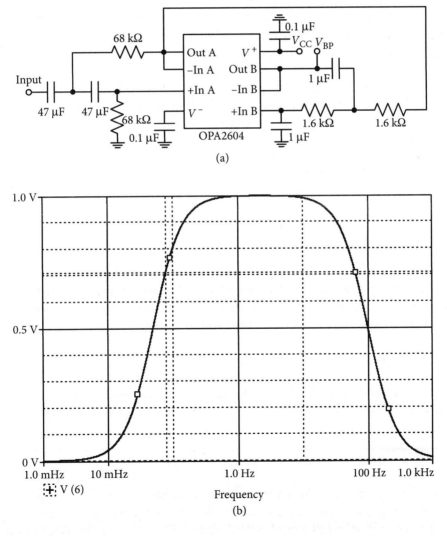

Figure 11.9 (a) Combination of high pass and low pass filters using OPA 2604 and [11.16]; {with permission from J Wang, et al} (b) its combined simulated response.

11.6 Filtering of Surface Electromyography Signals

Electromyography (EMG) relates to the study of electrical signals transmitted by body muscles, similar to the way nerves transmit signals. These electrical signals give rise to *muscle action potential*. Surface EMG is a method of collecting and recording the information present in these muscle action potentials.

The idea of having myoelectric control of the movement of artificial limbs, say hand, using EMG signals is sufficiently old. Great progress has occurred and collecting EMG signals from the human body is now a routine procedure, both in rehabilitation and in medical research. EMG signals are very weak, within a range of few μVolts to low milli volts (0-6 mV$_{pp}$), and they contain noise signals due to different reasons. The frequency range of the useful EMG signal is mostly in the 0-500 Hz range, whereas the dominant component lies in the 50–150 Hz range.

Because of the very small magnitude of the EMG, it needs amplification of the order of 1000 to 10,000, while keeping in mind that noise is to be minimized and not amplified during signal amplification. Figure 11.10 shows an arrangement in block form of amplification and noise elimination through filtering. Obviously, HP and LP filtering form important components of the process before the EMG signal is fed into an ADC [11.17].

Figure 11.10 Block diagram of amplification and filtering for electromyographic signals. {With permission from Wang et al.[11.17]}.

The preamplifier consists of an inst-amp INA 128 and an OA OPA 2604. In this case, the gain of the preamplifier was set at around 10 and a Sallen–Key second-order HPF having a corner frequency of 20 Hz was used. The circuit structure and element values are shown in Figure 11.11(a). Expressions for the corner frequency and pass band gain are:

$$f_c = 1/2\pi\sqrt{(R_1 R_2 C_1 C_2)}, \quad \text{Gain} = 1 + R_4/R_5 \tag{11.10}$$

For obtaining a fourth-order HPF with the same corner frequency, two sections of Figure 11.11(a) were connected in cascade. However, the element values for each of the two sections are given as follows for the corner frequency of 20 Hz:

$$R_1 = R_2 = 560 \text{ k}\Omega, \ R_3 = 51 \text{ k}\Omega, \ R_4 = 16 \text{ k}\Omega, \ C_1 = C_2 = 22 \text{ nF}$$

Every amplification stage is followed by a second-order Sallen–Key LPF as shown in Figure 7.22. Expression for the corner frequency and high frequency gain are same as that in equation (11.10). Two second-order LPFs are placed before and after the second amplification in Figure 11.10. For better operation, fourth-order LPFs can be used. It needs to be mentioned that selection of corner frequency depends on the sampling rate that is important in EMG signal acquisition.

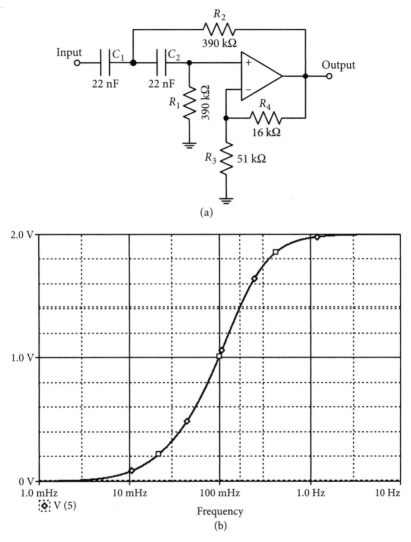

(a)

(b)

Figure 11.11 (a) Second-order Sallen–Key high pass filter for a 3 dB frequency of 20 Hz. {With permission from Wang et al. [11.17]}. (b) Magnitude response of the high pass filter shown in Figure 11.11(a).

Figure 11.12 shows the simulated responses of the second-order HPF with $f_c = 20$ Hz, fourth-order HPF with $f_c = 20$ Hz, second-order LPF with $f_c = 700$ Hz and $f_c = 1400$ Hz, fourth-order LPF with $f_c = 700$ Hz and $f_c = 1400$ Hz employed in this case study.

A similar study has been done in a conference publication [11.18].

Element values for the second-order LPF, for cut-off frequency of 709 Hz, are as follows:

$R_1 = 68$ kΩ, $R_2 = 68$ kΩ, $R_3 = 51$ kΩ, $R_4 = 24$ kΩ, $C_1 = 3.3$ nF, $C_2 = 3.3$ nF.

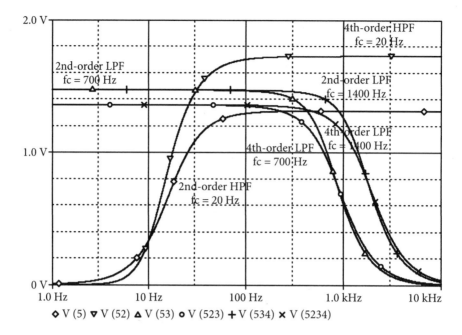

◇ V (5) ▽ V (52) ▲ V (53) ○ V (523) + V (534) ✕ V (5234)

Figure 11.12 Simulated responses of second-order high pass, fourth-order high pass, second-order low pass and fourth-order low pass filters for surface EMG signals.

11.7 Brain–Computer Interface Application

Electroencephalography (EEG) signals are commonly known for the waves representing the electrical activities of the brain. There are five major brain waves depending on the location of where the electrodes are placed. The five signals are different from each other on the basis of their frequency. In spite of the different frequency ranges of these wave types, all the relevant information is confined in the range 1–100 Hz.

Recently, the concept of controlling machines through brain activity, and not just by manual operation, has attracted attention. Being an interdisciplinary concept, researchers in the field of computer science, neuroscience and bio-engineering have joined to develop the prototype of a brain–computer interface (BCI) [11.19].

In general, systems used for the acquisition of EEG signals are similar in nature. However, there are differences on the basis of use of number of electrodes, degree of noise elimination by limiting signals to only the useful frequency range and amount of required amplification. The signals are then digitized and sent to an appropriate output device. Output signals can be used either for diagnostic purposes or other applications such as the BCI application. One of the objectives of developing BCI is to provide a communication channel for users who have lost their ability to communicate normally [11.20].

As mentioned earlier, differences in the practical EEG signal acquisition and the processing system depend on their process of detection and number of channels. However, significant difference also occurs due to differences in the technological advances in the tools used in signal processing and the required specification of the system itself. In one such example, an attempt was made to show the design of a battery-operated portable single-channel EEG signal acquisition system [11.21].

Magnitude of the EEG signals is in the range 5-500 μV. The signals are fed to an INA 128P inst-amp. Selection of the inst-amp is based on its suitability for battery-operated medical instrumentation having high CMRR, low offset voltage and low drift. A third-order Butterworth active HPF is used to eliminate those noise components which lie below 1 Hz. This kind of filter is shown in Figure 11.13(a). Element values shown in the figure realize a cut-off frequency of 1.0 Hz as given by relation $f_c = 1/2\pi RC$. Its simulated response is shown in Figure 11.13(b).

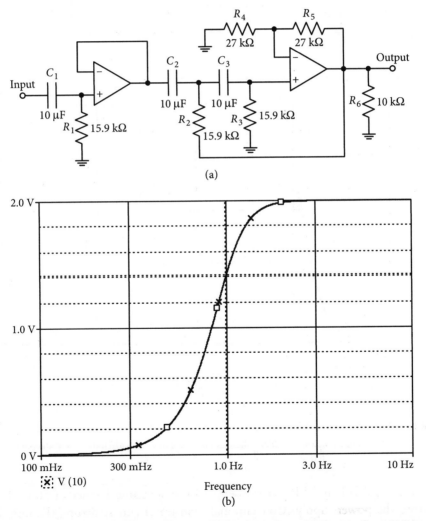

Figure 11.13 (a) Third-order high pass Butterworth filter for brain–computer interface. {With permission from Bhagwati and Chutia [11.21]} (b) Response of the high pass filter shown in Figure 11.13(a).

The next stage is a third-order Butterworth LPF with a 3-dB frequency of 100 Hz to attenuate noise signals above 100 Hz. Figure 11.14(a) shows the LPF with element values; Figure 11.14(b) shows its simulated response, along with that of the HPF.

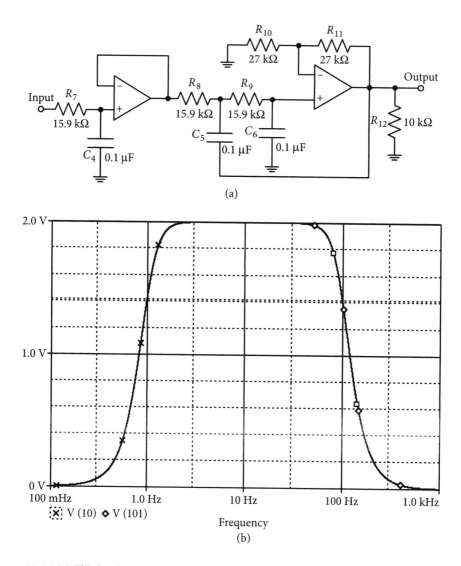

(a)

(b)

Figure 11.14 (a) Third-order low pass Butterworth filter used to attenuate signals above 100 Hz. (b) Responses of the low pass filter shown in Figure 11.14(a) and the high pass filter shown in Figure 11.13(a).

In addition to the HPF and LPF, it is essential to remove interference of 50/60 Hz due to leakage from the power supply. Conventional twin-tee circuit structure, the one shown in Figure 11.8(a), can be used and does not need any further consideration.

11.8 Chamber Plethysmography

Chamber plethysmography is used to accurately measure absolute changes in blood volume at the extremities, such as finger tips or ear lobes. The blood volume can be converted into blood flow rate. However, when only the pulse rate of the heart is to be found, only information about the relative volume is needed; the amplitude or shape of the signal is not required. In such a case, impedance plethysmography or photo-plethysmography (PPG) can be used [11.22].

In the present case study, PPG sensors have been used in transmission mode on an ear [11.23]. Light transmitted through the aural pinna is detected by a photo sensor. After the first step of optical detection, pulse filtering is done by a second-order BPF with the characteristic frequencies between 1.54 Hz and 2.34 Hz (3 dB frequencies need to be 1.0 Hz and 3.5 Hz) with a gain of 3700 at the peak. It is to be noted that there is specific reason behind selection of the mentioned frequency as it corresponds to wavelength near infrared, and the selected signals have strongest modulation due to the light absorption by the haemoglobin in the blood [11.23].

The circuit topology, shown in Figure 11.15, for realizing the BP characteristic consists of a passive first-order HPF and an active first-order LPF. The transfer function of the BPF is as follows:

$$H(s) = \frac{R_1 + R_2}{R_1} \frac{sC_0 R_0}{1 + sC_0 R_0} \frac{1 + s\dfrac{R_1 R_2}{R_1 + R_2}C_2}{1 + sC_2 R_2} \qquad (11.11)$$

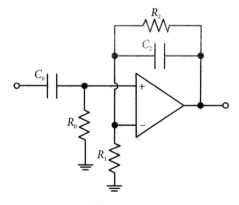

Figure 11.15 Combination of passive high pass and active low pass to realize a band pass function for chamber plethysmograph signal.

Critical frequencies of the LPF and HPF were so selected that the BPF peaks at 2 Hz. To increase the order of the filter, the structure shown in Figure 11.15 was implemented twice and a potentiometer (10 kΩ) was placed in between the two stages to vary the signal gain. Element values for the filter in Figure 11.15 are: $R_0 = 47$ kΩ, $C_0 = 2.2$ µF, $R_1 = 10$ kΩ, $R_2 = 1000$ kΩ, and $C_2 = 68$ nF. The complete circuit is shown in Figure 11.16(a).

PSpice simulated response of the BPF is show in Figure 11.16(b). Using op amp 741, a peak gain of 3300 (though theoretical value is 10200) occurs at 1.896 Hz while potentiometer branches were 1 kΩ and 9 kΩ.

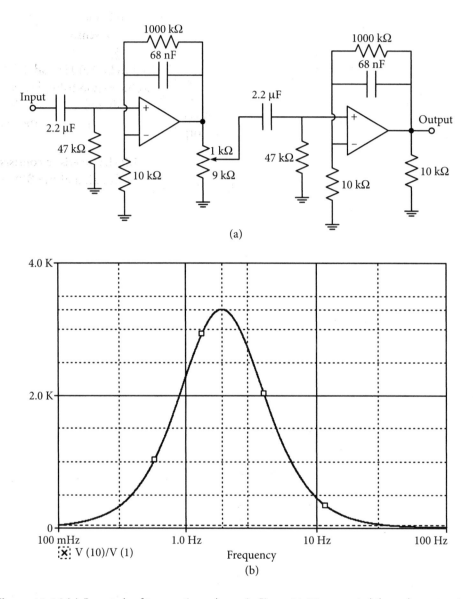

(a)

(b)

Figure 11.16 (a) A cascade of two sections shown in Figure 11.15 connected through a potentiometer. (b) Simulated response of the fourth-order band pass filter shown in Figure 11.16(a).

11.9 C-Message Weighting Function

A C-message weighting function simulates the frequency response of a human ear. It is a commonly specified test and a measurement filter for voice, audio and telecommunication applications in the US. C-message and similar filters are conventionally fabricated using a cascade of three BPFs and one LPF. The overall characteristics of the C-message filter depend on the parameters of the individual sections. A C-message filter specified by the IEEE standard 743-1984 is an all-pole filter having section parameters as shown in Table 11.1 [11.24].

Table 11.1 Parameters of the second-order sections of a C-message filter

Filter type	Pole values	Critical frequency (Hz)	Pole-Q
BP#1	$-1502 \pm j1267$	312.741	0.6540
BP#2	$-2437 \pm j5336$	933.761	1.2027
BP#3	$-4690 \pm j15267$	2541.886	1.7026
LP#1	$-4017 \pm j21575$	3492.728	2.7316

In actual practice, individual sections are fabricated with digitally programmable center frequency and pole-Q. Maxim Integrated in its application note 11 has realized such a C-message filter using switched capacitors. However, current examples have shown the same response with continuous-time second-order sections. For practical applications, these filters can also be made programmable.

The circuit shown in Figure 7.7 was used for the realization of the three BPFs, and the circuit shown in Figure 7.6 was used for the realization of LPF. It gives the following element values for the second-order sections.

BP#1 $R_1 = 49.28$ kΩ, $R_2 = 184$ kΩ, $R_3 = 66.6$ kΩ, $C_3 = 10$ nF, $C_5 = 10$ nF

BP#2 $R_1 = 16.01$ kΩ, $R_2 = 12.3$ kΩ, $R_3 = 40.9$ kΩ, $C_3 = 10$ nF, $C_5 = 10$ nF

BP#3 $R_1 = 2.82$ kΩ, $R_2 = 5.28$ kΩ, $R_3 = 21.3$ kΩ, $C_3 = 10$ nF, $C_5 = 10$ nF

LP#1 $R_1 = 5.4$ kΩ, $R_3 = 10$ kΩ, $R_4 = 10$ kΩ, $C_2 = 49.7$ nF, $C_5 = 0.4168$ nF

It needs to be mentioned that the voltage gain of the respective filter sections, as suggested in the application note 11 of Maxim Integrated is taken as 0.675, 0.685, 0.864, and 2.0, respectively. The overall eighth-order filter is shown in Figure 11.17(a) and its simulated response is shown in Figure 11.17(b) which matches with standard C-message.

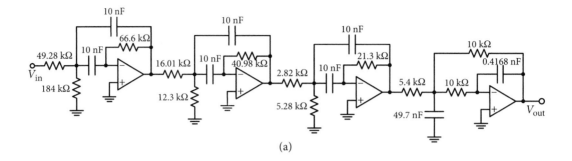

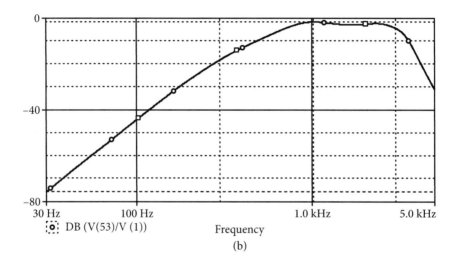

(b)

Figure 11.17 (a) Cascade of three band pass filters and a low pass filter for generating a C-message weighting function. (b) Simulated C-message weighting function for the circuit shown in Figure 11.17(a).

References

[11.1]   Prutchi, David and Michael Norris. 2005. *Design and Development of Electronic Instrumentation*. New Jersey: Wiley Interscience, John Wiley and Sons Inc., Publication.

[11.2]   Kitchin, Charles and Lew Counts. 2006. 'Monolithic Instrumentation Amplifier.' Chapter II and III A in *Designer's Guide to Instrumentation Amplifiers*. US: Analog Devices.

[11.3]   'USBPGF-S1. USB Programmable Single Channel Instrumentation Amplifier and Low Pass Filter'. Costa Mesa, CA, US: Alligator Technologies. https://alligatortech.com/downloads/USBPGF-S1_Data_Sheet.pdf.

[11.4]   Texas Instruments. 2019. *Low Power Instrumentation Amplifier INA 102 Data Sheet*. SBOSO27B. US: Burr Brown Corporation.

[11.5]   Larsen, Cory. A. 2010. *Signal Conditioning Circuitry Design for Instrumentation System*. SAND2011-9467. Cal. US: Sandia National Laboratories.

[11.6]   Walters, P. L. 2010. *Measurement System Engineering.* Short Course at Sandia National Laboratory. Cal. US: Sandia National Laboratories.

[11.7]   Reyerson, D. E. 1996. *Signal Conditioning Primer.* Cal. US: Sandia National Laboratories.

[11.8]   Texas Instruments. 2011. *Filter Pro Users' Guide.*

[11.9]   Texas Instruments. P O Box 655303, Dallas, Texas 75265 USA. Linear Technology Corporation 1630 McCarthy Blvd, Milpitas, C 95035-7417 (www.linear.com), Maxim Electronics Products and Analog Devices (www.maximintegrated.com/conact). ON Semiconductors P O Box 61312, Phoenix, Arizona85082-1312 USA. (http://onsemi.com). Microchip Technology Incorporated, 2355 West Chandler Blvd, Chandler AZ 85224-6199.

[11.10] Maxim Integrated Products. 2010. *Introduction to Electrocardiographs,* Tutorial 4693. US: Maxim Integrated Products, Inc.

[11.11] Lee, Shuenn-Yuh, Jia-Hua Hong, Jin-Ching Lee, and Qiang Fang. 2012. 'An Analog Front-End System with a Low-Power On-Chip Filter and ADC for Portable ECG Detection Devices.' In *Advances in Electro-Cardiogram-Methods and Analysis,* edited by Richard Miller., Germany: IntechOpen.

[11.12] Webster, G. J. 1995. *Design of Cardiac Pace Maker.* NJ: IEEE Press.

[11.13] Lasanen, K., and J. Lostamovaara. 2005. 'A 1V Analog Front-end for Detecting QRS Complexes in a Cardiac Signal,' *IEEE Transactions on Circuit Systems* 42 (10): 2161–8.

[11.14] Salthouse, C. D., and R. Sarpeshkar. 2003. 'A Practical Micropower Programmable Band Pass Filter used in Bionic Ears,' *IEEE Journal of Solid-State Circuits* 38 (1): 63–70.

[11.15] Solís-Bustos, Sergio, José Silva-Martínez, Franco Maloberti, and Edgar Sánchez-Sinencio. 2000. 'A 60-dB Dynamic Range CMOS Sixth-Order 2.4-Hz Low Pass Filter for Medical Applications,' *IEEE Transactions on Circuit and Systems-II, Analog and Digital Signal Processing* 47 (12): 1391–8.

[11.16] Wang, Kening, Shengqian Ma, Jing Feng, Weizhao Zhang, Manhong Fan, and Dan Zhao. 2012. 'Design of ECG Signal Acquisition System Based on DSP'. *SciVerse Science Direct, Procedia Engineering* 29: 3763–7.

[11.17] Wang, S., L. Tang, and J. E. Bronlund. 2013. 'Surface EMG Signal Amplification and Filtering,' *International Journal of Computer Applications* 82 (1): 0975–8887.

[11.18] Shobaki, Mohammed M., Noreha Abdul Malik, Sheroz Khan, Anis Nurashikin, Samnan Haider, Sofiane Larbani, Atika Arshad, and Rumana Tasnim. 2013. 'High Quality Acquisition of Surface Electromyography: Conditioning Circuit Design,' *IOP Conference Series: Material Science and Engineering* 53: 12027.

[11.19] Rani, M. S. A., and Wahida B Mansor. 2009. 'Detection of Eye Blinks from EEG Signals for Home Lighting Systems,' *Proc. ISMA09.* Sharjah, UAE: International Symposium on Mechatronics and Its Applications.

[11.20] Wolpaw, Jonathan R., Niels Birbaumer, William J. Heetderks, Dennis J. McFarland, P. Hunter Peckham, Gerwin Schalk, Emanuel Donchin, Louis A. Quatrano, Charles J. Robinson, and Theresa M. Vaughan. 2000. 'Brain-Computer Interface Technology: A Review of First International Meeting,' *IEEE Transactions on Rehabilitation Engineering* 8 (2): 164–173.

[11.21] Bhagwati, A. J., and R. Chutia. 2016. 'Design of Single Channel Portable EEG Signal Acquisition System for Brain Computer Interface Application,' *International Journal of Biomedical Engineering and Science* 3 (1): 37–44.

[11.22] Webster, John. G. (ed.). 1992. *Medical Instrumentation, Applications and Design*, second edition. Mass, US: Houghton Mifflin Co.

[11.23] Langereis, Geert. 2010. *Photo-plethysmography (PPG) System*, Version 2 (www.semanticscholar.org).

[11.24] Maxim Integrated. 1998. 'Programmable Universal Filter Implements C-Message Weighting Function,' Application Note 11. US: Maxim Integrated.

Practice Problems

11-1   Laplace representation is used for a three-level system of fluorescent spectroscopy by a simple cascade of filters as shown in Figure 2.7(b). Let life-time for fast relaxation be 10^{12} seconds and life-time for the slow relaxation is $0.5*10^4$ seconds life-time of the first-stage LPF for both the cases was set at 1.1 second. Design the filters and find the phase shift for the two sets at 10 kHz and 100 kHz.

11-2   Equivalent circuit for the transfer function with two life-time components is shown in Figure 2.8(a). Signal frequencies used for the slow and the fast transition rates in the circuit were 1 krad/s and 1 Mrad/s. Design the two filter sections and add their outputs while changing the weightage of the output of the slow transition state. What will be the phase shift at 1 kHz when the weightage of the slow state with respect to the fast state transition is 1, 0.75, 0.5 and 0.25.

11-3   Design an instrumentation amplifier using 741 type operational amplifiers (BW = $2\pi*10^6$ rad/s) as shown in Figure 11.3. Voltage gain value will be 1, 2, 5, 10, and 20 by varying a single resistance. Assume suitable values for the resistances and find the bandwidth of the instrumentation amplifier as a function of the gain value.

11-4   Instrumentation amplifier AD8224B is to be used to obtain an LPF for the following specifications. Output voltage should become 1 volt at dc for a differential input signal of 25 mV, and its 3-dB frequency is to be 25 kHz. Design and test the circuit.

11-5   Input to an 8-bit ADC working on sampling frequency of 100 kHz, receives non-aliased signal up to 30 kHz. The voltage level at the input of the ADC is to be 4 volts. Design a maximally flat LPF that provides sufficient attenuation for the proper working. The ADC employed the instrumentation amplifier to band limit the input signal and acted as a pre-amplifier with gain of 10. Input to the instrumentation amplifier was at an amplitude of 25 mV.

11-6   With other specifications remaining the same, repeat problem 11-5 for a 12-bit ADC.

11-7   Phase compensation is to be used to linearize the phase of the Butterworth response obtained in the problem 11-4. The compensation filter should have less than 10% phase non-linearity aver the pass band. Design analog phase compensating filter utilizing the circuit (or cascade) of Figure 4.6(b).

11-8   Amplitude of an ECG signal was 2 mV, which was pre-amplified by a factor of 100. The signal is to be band-limited up to 250 Hz using a third order Chebyshev approximation with 1 dB ripple such that its output voltage level becomes 1 volt. Design and test the filter.

11-9   Output of the LPF in problem 11-8 is fed to a Sallen–Key HPF structure to minimize low frequency noise. 3-dB frequency of the HPF is to be 0.1 Hz with its gain being 1.5. Design and test the filter.

11-10 Design a notch filter for suppressing the noise frequency of 50 Hz/ 60 Hz on the acquisition board of an 'analog front-end' of an ECG system. Employ the notch filter structure of Figure 8.10(a).

11-11 A BPF is to be designed for an ECG signal acquisition system employing the operational amplifier OPA 2604. The BPF is obtained as a combination of an LPF with cut-off frequency of 500 rad/s and an HPF with cut-off frequency of 0.3 rad/s.

At what frequency the output falls to half of its maximum magnitude?

11-12 Electromyography signal was pre-amplified and fed to an HPF. Design the HPF with gain of 10 using biquad structure of Figure 7.8(a) having 3-dB frequency of 20 Hz.

Cascade two second-order HPFs and re-design the filter with the same specifications.

11-13 An LPF is to be used in the chain of Electromyograph signal processing. Design a second-order filter using Sallen–Key structure and OPA 2604 operational amplifier, with cut-off frequency of 40 krad/s.

11-14 Combine the second-order HPF of problem 11-12 and a second-order LPF of problem 11-13 to yield a BPF. What are frequencies at which the gain falls to half of its mid-band gain?

11-15 ECG signal is to processed through a BPF which is a combination of a maximally flat third-order LPF and a third-order HPF. Both the filters are realized by cascading a first-order and a second-order filter section. In each case the second-order section employs general differential input single OA biquad configuration of Figure 7.13. Respective corner frequencies of the LPF and the HPF are 600 rad/s and 6 rad/s.

What is the frequency range of the pass band of the composite filter?

11-16 Circuit shown in Figure 11.15 is utilized to obtain a second-order BPF. Select component values such that dc gain of the LPF and the high frequency gain of the HPF is 100, and respective corner frequencies are 20 rad/s and 1.6 rad/s.

What is the peak gain of the BPF and at what frequency it occurs?

11-17 Critical frequencies of the filter components of the standard C-message weighting function are increased by 10%, and respective pole-Q are reduced by 5%. Obtain modified response and compare gain values between the two responses at 100 Hz, 1 kHz and 2 kHz.

Chapter 12

Audio Signal Processing and Anti-Aliasing Filters

12.1 Introduction

Two important areas of analog filter applications are included in this chapter: audio signal processing (ASP) and study of anti-aliasing/reconstruction (AA/ReF) filters.

In ASP, the frequency range of operation is from a few Hz to around 20 kHz. All the different types of filters, that is, LP (low pass), BP (band pass), HP (high pass), BE (band elimination) and AP (all pass) filters, can be used. Hence, in most applications, general purpose OAs (operational amplifiers) may be used while taking care of noise signals. Generally, order of the filters may not be very high; whereas, in AA/ReF, only LPFs are used for which the frequency range of operation depends on specific applications. Hence, the selection of OA is to be considered carefully. Order of the filter can also be high.

One of the major objectives in ASP is to improve audio signals employing different approaches. Section 12.2 discusses this aspect in terms of crossover networks and filters eliminating infrasonic and ultrasonic undesired signals. Section 12.3 discusses a specific case which addresses the unequal power output requirements of the satellite speaker and the subwoofer speaker system. Equalization using mid treble boost/cut and mid bass boost/cut, which are part of programmable equalizers are studied in Section 12.4. Section 12.5 takes up certain aspects of the Record Industry Association of America (RIAA) in brief. An application of phase approximated filters in ASP is explained in Section 12.6.

The essential nature of AA/ReF filters is described in Section 12.7 citing the example of a filter design for a 12-bit ADC. A detailed case is presented in Sections 12.8 and 12.9 showing the application of high frequency AA and ReF filters used in video signal processing.

12.2 Improvement in Audio Performance

Problems in the frequency response of audio systems and loudspeakers are resolved using active analog audio filters. Of course, the ways these filters are used depend on the required end result. For example, in some cases, filters are required to produce a response which is nearly the inverse of the response of the (say) loudspeaker or any other audio speaker. Hence, when the two responses are combined, it results in a nearly flat response. In some applications, undesirable signal peaks coming from music systems, cell phones, and so on, are to be reduced. Figure 12.1 illustrates such a compensation in a PDA (public digital assistance) speaker response [12.1]. Obviously, such compensations are used to improve the audio response; the compensations make the signals more pleasing and intelligible than the original.

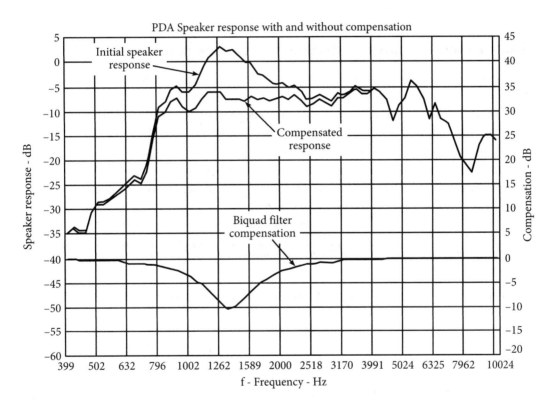

Figure 12.1 PDA response with and without compensation {Courtesy: Texas Instruments}.

Audio filters may be first-, second- or higher-order with a number of choices being available as shown in the earlier chapters, and vast literature is available elsewhere. Before considering some practical examples of audio filters, a brief introduction about sound measuring instruments is desirable [12.2].

In sound level meters (SLMs), the electrical signal from a transducer (say microphone) is fed to a pre-amplifier, and then to a weighted filter over a specified range of frequencies, if needed. After that, the filtered signal is further amplified and the output signal is either directly measured or it acts as input to some other device. It is to be noted that noise (sound) measuring instruments are widely used in the practice of occupational hygiene [12.2].

Frequency analyzers are used to determine the distribution of the overall signal level over a range of frequencies. For occupational hygiene, the most usual analysis for noise studies is done in the octave band. A number of frequency analyzers are available for use with SLMs. Quality analyzers can perform frequency analysis in all desired frequency bands simultaneously. Characteristics of the filters play an important role in quality analyzers. In most cases, filter attenuation is at the rate of 24 dB per doubling of frequency after the 3 dB level.

Improvement in audio performance is a subject of continuous study. The application report [12.3] is one such example. Along with a few other applications of LM 833 OA, the report contains some useful continuous-time filters.

12.2.1 Active crossover networks for loudspeakers

Loudspeaker systems generally consist of two or more transducers for different frequency bands of the audio frequency spectrum. Initially, passive filters were used as crossover filters. However, these filters required inductors and capacitors of large values. Such passive filters created noise and distortion; the system efficiency was also considerably poor.

A straightforward approach was to use low-level filters to divide the frequency spectrum. Separate power amplifiers then follow each driver or group of drivers. Though this arrangement appears to be obvious, it is very difficult to find a commercially feasible active dividing network. Fortunately, the objective could be achieved by employing a *constant voltage* crossover circuit as shown in Figure 12.2(a) [12.3]. The term *constant voltage crossover* means that once outputs of the LP and HP sections are added, the exact replica of the input signal will be produced. For the circuit in Figure 12.2(a), analysis gives the following relations:

$$V_{HP} = -(V_{in} + V_{LP}), \ V_1 = -(V_{in} + V_{LP})/sCR, \ V_2 = (V_1 + 4V_{HP})/sCR \qquad (12.1)$$

The resulting transfer functions for the LP and HP functions are as follows:

$$\frac{V_{LP}(s)}{V_{in}(s)} = \frac{a_1 s + 1}{a_3 s^3 + a_2 s^2 + a_1 s + 1} \qquad (12.2a)$$

$$\frac{V_{HP}(s)}{V_{in}(s)} = \frac{a_3 s^3 + a_2 s}{a_3 s^3 + a_2 s^2 + a_1 s + 1} \qquad (12.2b)$$

With the following selected values of coefficients, $a_3 = 1$, $a_2 = 4$ and $a_1 = 4$, denominator for both the mixed LP and HP filters become third-order as shown in equation (12.3).

$$D(s) = (s + 1)(s^2 + 3s + 1) \qquad (12.3)$$

For a second-order filter, expression for the critical frequency

$$f_c = (1/2\pi RC) \hfill (12.4)$$

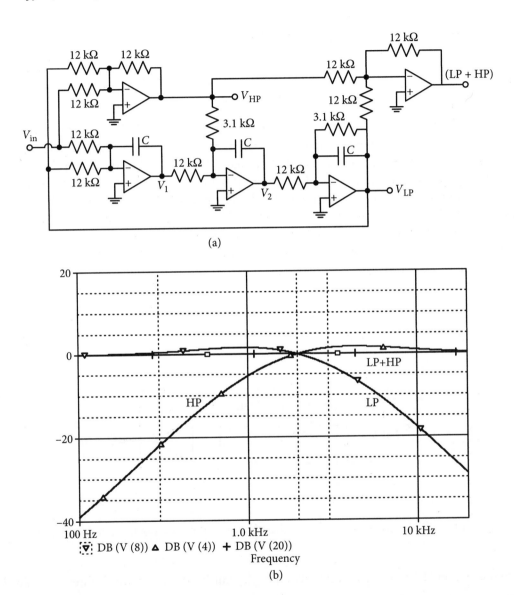

(a)

(b)

Figure 12.2 (a) Active crossover filter for loudspeakers {Courtesy: Texas Instruments}.
(b) Simulated response of the crossover filter shown in Figure 12.2(a).

Hence, for a selected crossover frequency of 2 kHz, with $R = 12$ kΩ we get $C = 6.628$ nF. To get the value of the coefficients a_1, a_2 and a_3, as mentioned earlier, two resistors will have their value as $0.25\ R = 3$ kΩ. If crossover frequency is to be changed, either the three capacitors are to be changed or all the resistors are to be changed simultaneously.

Figure 12.2(b) shows the simulated response of the LP and HP constant voltage crossover outputs and the summed output of the two responses. The final output is flat with a 0 dB gain as desired. The simulated crossover frequency is 1.9979 kHz.

12.2.2 Infrasonic and ultrasonic filters

There are a number of significant sources of noise signals above and below the limits of audibility. For example, the most significant source of noise in the low-frequency range is due to a phonograph arm/cartridge/disc combination. A low frequency, large amplitude signal at 0.556 Hz is also generated when the disc warps on $33\frac{1}{3}$ rpm records.

Under certain undesirable operating conditions, there are considerable noise signals beyond the audible frequency range as well. These ultrasonic noise signals affect the circuit elements in the audio band as well, and needs to be eliminated. Circuits shown in Figures 12.3 and 12.4 attenuate infrasonic and ultrasonic noise signals. The circuit in Figure 12.3 is a third-order Butterworth HPF with a cut-off frequency of 15 Hz. It is expected that the filter provides attenuation over 28 dBs at 5 Hz, while at 30 Hz, information is to be attenuated by only 0.1 dB. For the HPF shown in Figure 12.3, the following relations are obtained:

$$V_1(sC_1 + G_1 + sC_2) - V_2sC_2 - V_{in}sC_1 = 0 \tag{12.5a}$$

$$V_2(sC_2 + G_2 + sC_3) - V_1sC_2 - V_{out}sC_3 = 0 \tag{12.5b}$$

$$V_{out}(G_3 + sC_3) - V_2sC_3 = 0 \tag{12.5c}$$

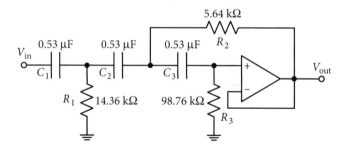

Figure 12.3 Third-order Butterworth HP filter for eliminating infrasonic noise {Courtesy: Texas Instruments}.

From equation (12.5), the transfer function of the third-order Butterworth filter is obtained as:

$$\frac{V_{out}}{V_{in}} = \frac{s^3}{s^3 + d_1s^2 + d_2s + d_3} \tag{12.6a}$$

$$\text{With } d_1 = \left(\frac{1}{C_1R_1} + \frac{1}{C_1R_3} + \frac{1}{C_2R_3} + \frac{1}{C_3R_3} \right) \tag{12.6b}$$

$$d_2 = \left(\frac{1}{C_1 R_1 C_2 R_3} + \frac{1}{C_1 R_1 C_3 R_3} + \frac{1}{C_1 R_2 C_3 R_3} + \frac{1}{C_3 R_3 C_2 R_3} \right) \tag{12.6c}$$

$$d_3 = \frac{1}{C_1 R_1 C_2 R_2 C_3 R_3} \tag{12.6d}$$

$$\frac{V_{out}}{V_{in}} = \frac{s^3}{s^3 + \left(\dfrac{1}{R_1} + \dfrac{3}{R_3} \right) s^2 + \left(\dfrac{1}{R_1} + \dfrac{1}{R_2} \right) s \dfrac{2}{CR_3} + \dfrac{1}{R_1 R_2 R_3 C^3}} \quad \text{for } C_1 = C_2 = C_3 = C \tag{12.6e}$$

Normalized transfer function of the third-order HP Butterworth filter is:

$$H(s) = \frac{s^3}{s^3 + 2s^2 + 2s + 1} \tag{12.7}$$

Using the coefficient matching approach with equations (12.6) and (12.7), the normalized element values are obtained as:

$$R_1 = 0.718, \; R_2 = 0.282, \; R_3 = 4.938 \text{ and } C = 1 \tag{12.8}$$

Frequency de-normalization is done by a factor of $15 \times 2\pi$ rad/s and by an impedance scaling factor of 20 kΩ. It resulted in the element values for the filter as:

$$R_1 = 14.36 \text{ k}\Omega, \; R_2 = 5.46 \text{ k}\Omega, \; R_3 = 98.76 \text{ k}\Omega \text{ and } C = 0.5303 \text{ }\mu\text{F} \tag{12.9}$$

The circuit in Figure12.4 is a fourth-order Bessel filter having negligible attenuation of the phase with an expected magnitude attenuation of 0.65 dB at 20 kHz because its design cut-off frequency is 40 kHz.

The transfer function of a fourth-order LP Bessel filter is obtained from Table 4.2 as:

$$H_B(s) = \frac{105}{(s^2 + 5.79242s + 9.14013)(s^2 + 4.20758 + 11.4878)} \tag{12.10}$$

The Bessel filter is realized as a cascade of two second-order filters. The normalized parameters of the two sections are evaluated from the equation (12.10) as:

$$w_{o1} = 3.023 \text{ rad/s}, \; Q_1 = 0.522, \; w_{o2} = 3.389 \text{ rad/s}, \; Q_2 = 0.806 \tag{12.11}$$

Else, parameters of the two sections are also available from table 4.3. The low frequency gain for both the sections will be unity. For the realization of the two sections, the circuit shown in Figure 7.22 is used with unity gain. Using equations (7.64) and (7.65), the normalized elements are calculated as:

$$G_{11} = 2, \; G_{12} = 7.204, \; C_{11} = 0.7518, \; C_{12} = 1 \tag{12.12a}$$

$$G_{12} = 5, \ G_{22} = 5.98, \ C_{12} = 1.465, \ C_{22} = 1 \tag{12.12b}$$

Frequency de-normalization by a factor of $2\pi \times 40$ krad/s and impedance scaling factor of 3.977 kΩ was used. Calculated values are shown in Figure 12.4.

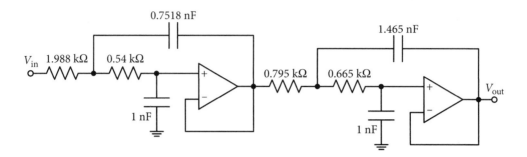

Figure 12.4 Fourth-order Bessel filter for eliminating ultrasonic noise {Courtesy: Texas Instruments}.

The filters in Figures 12.3 and 12.4 have been simulated using economical 741 OAs. However, to realize these filters with extremely low third-harmonic distortion, LM 833 is preferred [12.3]. The simulated amplitude response of the two filters in cascade, where LPF precedes HPF is shown in Figure 12.5. The simulated cut-off frequency of the HPF is 14.9 Hz, and at 5 Hz attenuation is 28.6 dBs; whereas at 30 Hz, attenuation is only 0.063 dB. For the LPF case, cut-off frequency is 46.62 kHz and attenuation is only 0.539 dB at 20 kHz.

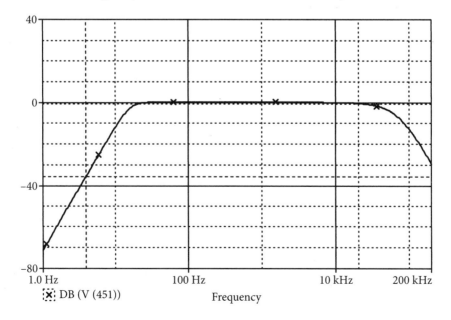

Figure 12.5 The combined response of the infrasonic and ultrasonic filters in Figures 12.3 and 12.4.

12.3 Satellite/Subwoofer Speaker Systems

We now address the unequal output power requirements of satellite speakers and the subwoofer. Limitations of the conventional solutions have been claimed to be overcome in an application note [12.4]. Solution of the unequal power requirement in audio systems is shown in Figure 12.6. Here, a MAXIM 9791 Window Vista®-complaint 2×2 W stereo amplifier with stereo head driver and the MAXIM 9737 is used, which is a 1×7 W mono class D amplifier.

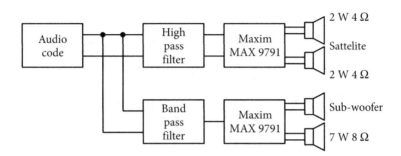

Figure 12.6 Utilization of analog filters in an audio system with unequal output power {Courtesy: Maximum Integrated}.

Presently, we are interested in the utilization of HP and BP filtering shown in the arrangement in Figure 12.6.

HPF with a 3-dB frequency of 500 Hz with a unity gain Butterworth response is used. It employs the filter structure of Figure 7.8(a). Utilization of equation (7.33) gives normalized component values:

$$C_1 = C_3 = C_4 = 1, R_2 = 0.4714 \text{ and } R_5 = 2.121 \tag{12.13}$$

For a cut-off frequency of 500 Hz, a frequency scaling factor of 1000π rad/s is applied, and an impedance scaling factor of 50 kΩ is selected. The resultant element values are given in equation (12.14a).

$$C_1 = C_3 = C_4 = 6.363 \text{ nF}, R_2 = 23.56 \text{ k}\Omega \text{ and } R_5 = 106.15 \text{ k}\Omega \tag{12.14a}$$

A resistance of 470 Ω is connected in series with the capacitance C_1, which is required to minimize the capacitive loading caused by the filter stage. Figure 12.7(a) shows the circuit structure and the simulated response in Figure 12.7(b) confirming the design. Two more responses with cut-off frequencies of 400 Hz and 600 Hz are also shown in the figure. Corresponding values of the elements for these responses are:

$$f_c = 400 \text{ Hz}, C_1 = C_3 = C_4 = 9.943 \text{ nF}, R_2 = 18.84 \text{ k}\Omega, R_5 = 84.92 \text{ k}\Omega \tag{12.14b}$$

$$f_c = 600 \text{Hz}, C_1 = C_3 = C_4 = 4.419 \text{ nF}, R_2 = 28.27 \text{ k}\Omega, R_5 = 127.38 \text{ k}\Omega \tag{12.14c}$$

Similarly, the BF filter which is used for the subwoofer is constructed by combining a second-order LPF and a second-order HPF. Cut-off frequencies of the HPF and the LPF range between 80 Hz and 100 Hz, and between 100 and 500 Hz, respectively. For the LPF, the second-order structure of Figure 7.6 is used. From equation (7.29), relations for the parameters are obtained as:

$$\omega_o^2 = \frac{G_3 G_4}{C_2 C_5}, \frac{\omega_o}{Q} = \frac{(G_1 + G_3 + G_4)}{C_2} \text{ and dc gain} = \frac{G_1}{G_4} \tag{12.15}$$

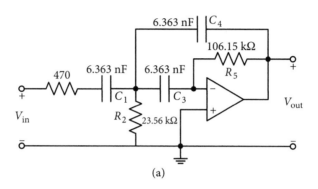

(a)

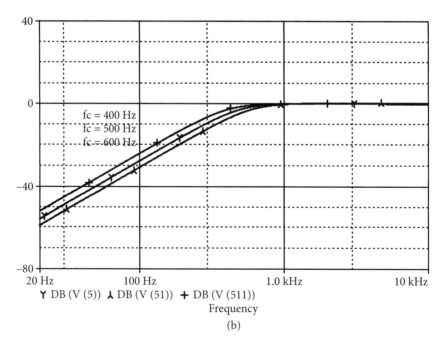

(b)

Figure 12.7 (a) High pass filter circuit structure of Figure 7.8(a), employed in Figure 12.6.
(b) Responses of the filter shown in Figure 12.7(a).

For unity gain LP Butterworth response, with a cut-off frequency of 2500 rad/s (397.7 Hz), application of equation (12.15) yields the component values as $R_1 = R_2 = R_3 = 21.218$ kΩ, $C_1 = 40$ nF, $C_2 = 8.888$ nF, and an input capacitance in series with R_1. For the HP section, the filter structure shown in Figure 7.8(a) is used. Application of equation (7.33), for the cut-off frequency of 500 rad/s (79.5 Hz) yields the component values as $C_3 = C_4 = C_5 = 200$ nF, $R_4 = 4.713$ kΩ, $R_5 = 21.218$ kΩ, and a resistance $R_{in} = 470$ Ω is connected in series with capacitance C_1. The combined circuit structure is shown in Figure 12.8(a). The simulated response of the combined LP and HP filters as a BPF is shown in Figure 12.8(b). Two more BP responses are shown in Figure 12.8(b), with the modified values of capacitors $C_3 = C_4 = C_5 = 250$ nF and 150 nF.

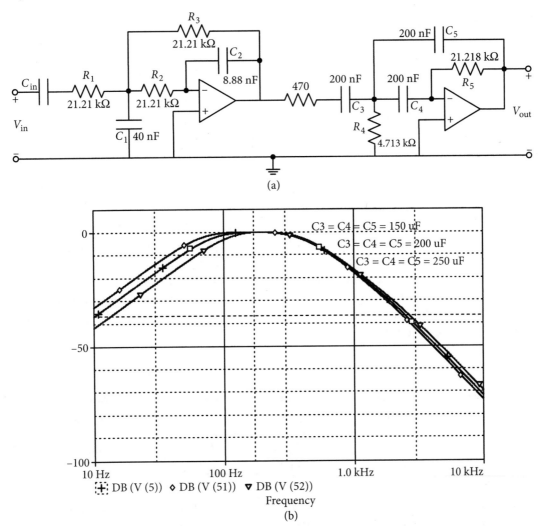

Figure 12.8 (a) Band pass filter circuit which is used with the sub-woofer in Figure 12.6. (b) Simulated magnitude responses of the band pass filter.

12.4 Four Band Programmable Equalizer

LE8 is a logic-controlled equalizer from Lectrosonics. The equalizer claims to provide an economical solution to the microphone equalization and notch filtering in an automatic modular audio processor system [12.5]. The heart of the LE8 equalizer is a microprocessor which controls all functions of the digitally programmed equalizer (EQ) and notch filters.

The EQ contains bass boost/cut, treble boost/cut filters as well as mid bass boost/ cut and mid-band treble boost/ cut band pass and inverted band pass types of filters. The bass/treble cut/boost filters can be realized with simple bilinear circuits and their description is included in Section 2.3.2 of Chapter 2.

The suggested center frequency of the mid bass is 300 Hz and pole-Q is 0.33; the corresponding values for the mid treble is 3 kHz and 0.33. Mid bass boost and mid treble boost being band pass type can be realized using second-order sections but for respective cut filters, it is preferable to cascade appropriate bass/treble boost/cut. Hence, for the sake of modularity, all the mid section filters are explained here with the same approach.

For mid bass boost realization, having a center frequency of 300 Hz, the treble cut circuit shown in Figure 12.9(a) is followed (cascaded) with a treble boost circuit shown in Figure 12.9(b). For the realization of the rest of the functions, namely, mid treble cut, mid treble boost and mid bass cut, the arrangement is given in Table 12.1, along with component values and cut-off frequency f_c of each section. The simulated response of the four types of filters is shown in Figure 12.10(a) and 12.10(b). As all the modules in Table 12.1 were designed for 10 dBs of boost or cut, their cascading results in a little less amount of boost or cut for the resulting mid frequency filter. Moreover, the center frequency of the filters will depend on the critical frequencies of the modules used; these have to be on either side of the 300 Hz or 3 kHz frequency band; for example, for the mid treble boost, calculated $f_c = 30$ kHz for the treble cut and $f_c = 1$ kHz for the treble boost. Now, all that remains is to find a rigorous mathematical relation between the cut-off frequencies of the two modules with the pole-Q and cut-off frequencies of the circuits forming the mid treble or mid bass cut/boost

Table 12.1 Component values and arrangement for the four mid frequency EQs

Components	Mid bass boost $f_o = 300$ Hz	Mid treble cut $f_o = 3$ kHz	Mid treble boost $f_o = 3$ kHz	Mid bass cut $f_o = 300$ Hz
	Treble cut $f_c = 900$ Hz	Bass boost $f_c = 300$ Hz	Treble cut $f_c = 9$ kHz	Bass boost $f_c = 100$ Hz
R_1 kΩ	200	63.2	200	63.2
R_2 kΩ	200	200	200	200
R_3 kΩ	92.3	92.7	63.2	92.7
C_1 nF	0.6047	1.813	0.067	16.32

Contd.

	Treble boost $f_o = 100$ Hz	Bass cut $f_c = 30$ kHz	Treble boost $f_o = 900$ Hz	Bass cut $f_o = 900$ Hz
R_4 kΩ	63.2	63.2	100	100
R_5 kΩ	136.8	136.8	216	216
R_6 kΩ	200	63.2	316	100
C_2 nF	36.78	0.1255	2.327	0.7758

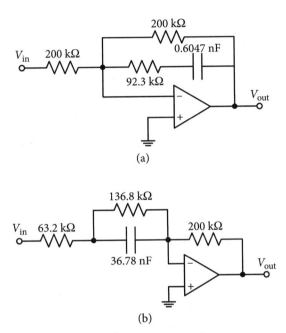

Figure 12.9 (a) A treble cut realization with cut-off frequency of 900 Hz. (b) A treble boost realization with cut-off frequency of 100 Hz.

Notch Filters: In general, specifications given for the design of notch filters include notch frequency and pole-Q. It is expected that attenuation at the notch will be as large as practically possible. However, in case of the equalizer LE8, it is to be noted that the required attenuation at the notch frequency is to be only 12 dBs so that it compensates for the effect of ringing, which has a magnitude of 12 dBs. The required three notch filters are to have programmable notch frequency, with possible variations in notch frequency from 50 Hz to 500 Hz (frequency resolution of 1.76 Hz) and from 500 Hz to 5 kHz (frequency resolution of 19.4 Hz); the notches are 1/7 of an octave wide.

Programming of the notch frequency is done by varying the components used; these components are controlled by the microprocessor and the circuit structure of the filter. The approach used in Section 8.6 (that of obtaining a multi output biquad using a summing amplifier) is used here to obtain a notch filter.

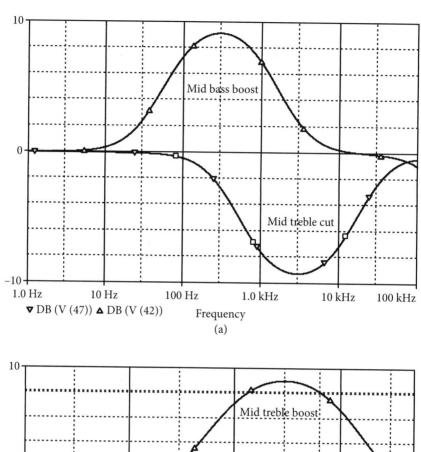

▼ DB (V (47)) ▲ DB (V (42))   Frequency
(a)

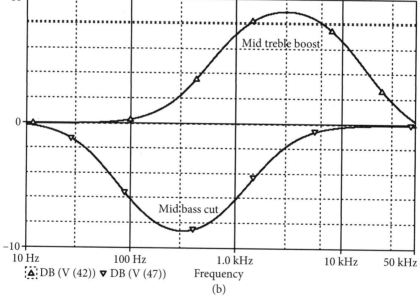

▲ DB (V (42)) ▼ DB (V (47))   Frequency
(b)

Figure 12.10 (a) Mid bass boost and mid treble cut responses using the cascade scheme. (b) Mid bass cut and mid treble boost responses using the cascade scheme.

In Example 8.3, the notch filter is realized with $Q = 5$ at a notch frequency of 3.18 kHz and 0.318 kHz. The filter obtains an attenuation of 40.17 dBs at a notch frequency of 3.18 kHz.

The notch becomes effective by making the coefficient of s in the numerator as zero in a near ideal case. Attenuation at the notch may be decreased while creating a non-ideal situation by making the coefficient of s in the numerator non-zero.

Using the structure shown in Figure 8.14(a) for the notch frequency of 3.18 kHz, the same values of components were used. Only those component values required change which changed the width of the notch to 1/7 of an octave. Hence, Q is taken as 3.5, which results in the resistor $QR = 17.5$ kΩ. To make the non-ideal attenuation condition, coefficient β was made 0.250 (instead of 1/3.5; equation (8.34)), which resulted in the resistor $R_\beta = 14$ kΩ (or 23.3 kΩ by trial). The simulated response of the notch is shown in Figure 12.11, with the required attenuation of 12 dBs and notch width of 849 Hz or $Q = 3.72$.

Variation in the notch frequency from 3.18 kHz to 0.318 kHz was easily done by changing capacitors C_1 and C_2 from 0.01 µF to 0.1 µF. Its simulated response is also shown in Figure 12.11. Attenuation remains unchanged at 12 dBs and $Q = 3.73$.

In all audio circuits, it is desirable to use low noise OAs.

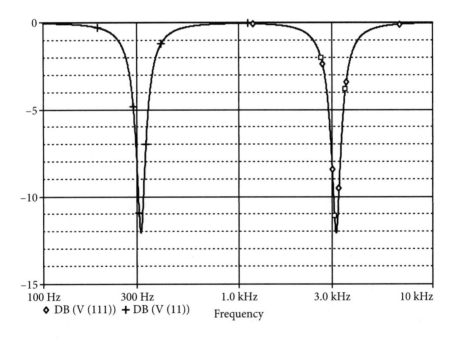

Figure 12.11 Notch filters with 12 dBs attenuation at different frequencies using the circuit structure of Figure 8.14(a).

12.5 Record Industry Association of America (RIAA) Equalization

For good quality hi-fi audio equipment, the phonographic record pickup has to be of magnetic type. However, when we play a disc, it generates a non-linear frequency response. In general, disc recording equipment does not give an exactly linear frequency response. Hence, to improve the performance, signals below 50 Hz and in the range of 500 to 2120 Hz, are recorded in a non-linear fashion. The nature and level of non-linearity is set by the Record Industry Association of America (RIAA). The RIAA curve, depicting the mentioned non-linearity is shown in Figure 12.12. The curve is described by 3 poles in time (frequency) terms as [12.6]:

$$T_1 = 3180 \ \mu s \ (50 \ Hz), \ T_2 = 318 \ \mu s \ (500 \ Hz), \ T_3 = 75 \ \mu s \ (2120 \ Hz)$$

These poles can be described as HP with T_1, LP with T_2 and again HP with T_3.

In 1992, the International Electrotechnical Commission (IEC) proposed a new superior version of the RIAA curve by adding another pole, a HP at 7950 μs (20 Hz). An additional pole was included to reduce the subsonic output of the amplifiers caused by the rumble of the turn table [12.6].

From the aforementioned discussion, it can be inferred that when a disc is played through a magnetic pickup system, its output must go to the power amplifier via a pre-amplifier; the frequency equalization curve of the pre-amplifier should be the exact inverse of the curve shown in Figure 12.12.

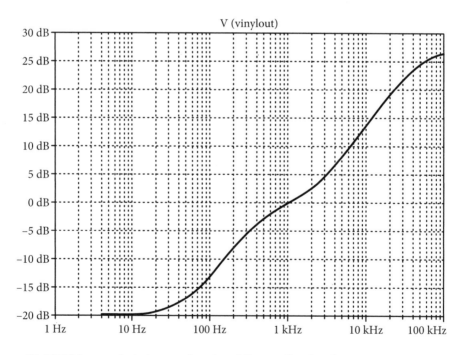

Figure 12.12 RIAA curve depicting non-linearity while recording [12.6].

12.5.1 RIAA phono pre-amplifier

Figure 12.13 shows an equalization schematic proposed by ON Semiconductors which consists of a pre-amplifier, an input buffer, a 5-band equalizer and a mixer using 5532 low noise OA [12.7]. The first stage is a pre-amplifier as the signal magnitude level in a magnetic pickup is very low. In the mid-band range, signal magnitude is only a few mV. Hence, a low noise pre-amplifier IC (rather than a simple OA) is necessary. Figure 12.14(a) shows one such configuration that uses a low noise OA, NE 5532A (any other low noise OA can also be used).

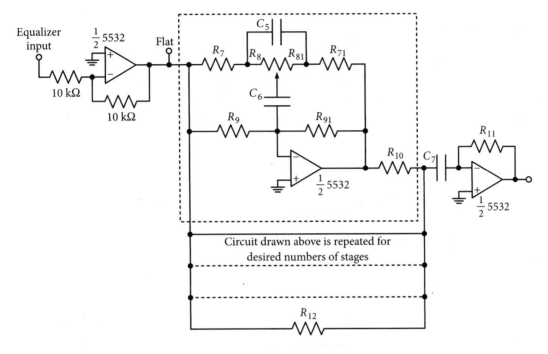

Figure 12.13 RIAA: Equalizer schematic [12.7]. {Used with permission from SCILLC dba ON Semiconductor}.

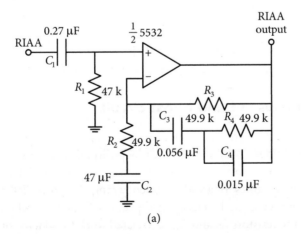

(a)

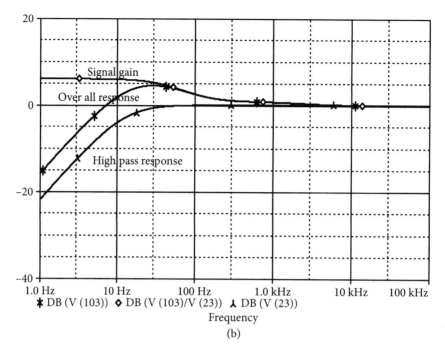

Figure 12.14 (a) RIAA phono pre-amplifier circuit [12.7]. (b) Magnitude response of the phono pre-amplifier and associated high pass filter. {Used with permission from SCILLC dba ON Semiconductor}.

The phono pre-amplifier is in effect a non-inverting amplifier with negative feedback applied through the potential divider formed by resistors R_2 and R_3. A combination of R_3, C_3, C_4, R_4 and R_2, C_2 network determines the signal gain. At high frequencies, capacitors, C_2, C_3 and C_4 have low impedance, so the gain is $\{1 + (R_3\|R_4)/R_2\}$, leading to nearly unity. At low frequencies, impedance of C_3 starts increasing and causes the ac gain to increase. At very low frequencies, the gain is limited to $(1 + R_3/R_2)$. The simulated response of the circuit is shown in Figure 12.14(b). Its low frequency gain is 6 dBs and the high frequency gain is unity, confirming the analysis. However, a combination of R_1 and C_1 acting as HPF with $f_c = (1/2\pi * 0.27\ \mu * 47\ k) = 12.5$ Hz changes the overall characteristics, reducing the gain at very low frequencies as shown in the simulated response; it also shows the HPF response.

12.5.2 Simultaneous cut and boost circuit

The main constituent of the equalizer in Figure 12.13 is an active band pass/notch filter which is shown within dotted lines. It can cut or boost by positioning the potentiometer to the right or left. Values of the resistors and capacitors for such circuits are available in literature. A table depicting the component values for a wide frequency range is also available in reference [12.7]. Frequency of the filter is changed by changing capacitors C_5 and C_6, where C_5 equals 10 times C_6 and the values of the resistors R_8 and R_{10} are related to R_9 by a factor of 10. The circuit was simulated for two center frequencies of 54 Hz and 541 Hz. Values of the elements which are common at both frequencies are as shown in equation (12.16a) [12.7]:

$$R_8 = 50 \text{ k}\Omega, \, R_7 = R_{71} = 5.1 \text{ k}\Omega, \, R_9 = R_{91} = 510 \text{ k}\Omega \tag{12.16a}$$

At 54 Hz, with $C_5 = 0.22 \, \mu\text{F}$ and $C_6 = 0.022 \, \mu\text{F}$, the respective values of the selected components for boost and cut case are:

$$R_8 = 2.5 \text{ k}\Omega, \, R_{81} = 47.5 \text{ k}\Omega, \text{ and } R_8 = 47.5 \text{ k}\Omega, \, R_8 = 2.5 \text{ k}\Omega \tag{12.16b}$$

At 541 Hz, with $C_5 = 0.022 \, \mu\text{F}$ and $C_6 = 0.0022 \, \mu\text{F}$, the respective values of the selected components for boost and cut case are:

$$R_8 = 15 \text{ k}\Omega, \, R_{81} = 35 \text{ k}\Omega, \text{ and } R_8 = 35 \text{ k}\Omega, \, R_{81} = 15 \text{ k}\Omega \tag{12.16c}$$

The simulated response of the cut and boost circuit is shown in Figure 12.15, reflecting the amount of cut and boost on the relative values of R_8 and R_{81}.

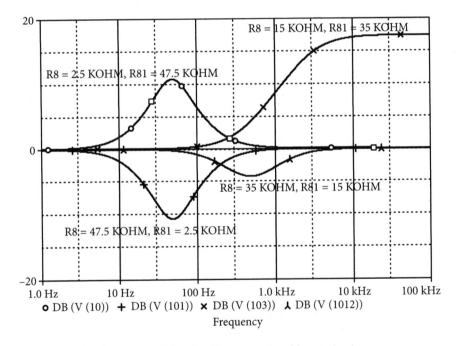

Figure 12.15 Simulated response of the simultaneous cut and boost circuits.

Tone Control Circuit: Tone control circuits are one of the most widely used types of variable frequency filter circuits. With the help of tone control circuits, one can change the frequency response of an audio system according to the requirement. A circuit similar in nature to the circuit of the previous section is also available; it boosts or cut signals at low or high frequencies.

One of the popular approaches to get an active tone control circuit employs a passive tone control network in the negative feedback loop of an OA. Such a system gives an overall signal gain rather than attenuation, when its controls are in a flat position. One such realization is shown in Figure 12.16(a); it provides 20 dBs of bass or treble boost or cut, set by the variable resistances $(R_2 + R_3)$ and $(R_7 + R_8)$. With capacitors C_2 to C_5 as 0.00625 μF its turn over frequency is 1 kHz as shown in the simulated response in Figure 12.16(b).

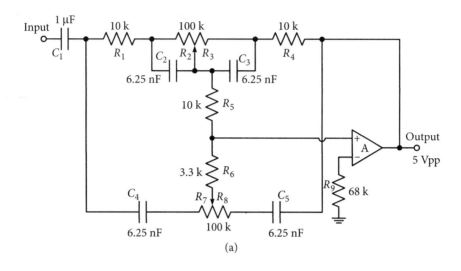

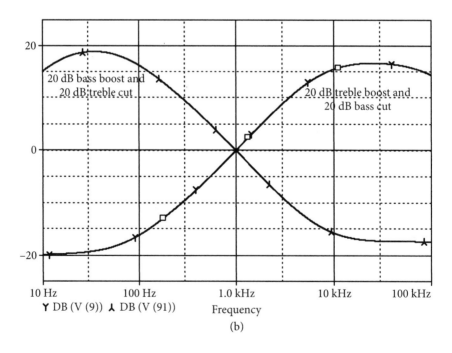

Figure 12.16 (a) Tone control circuit [12.7]. (b) Response of the tone control circuit shown in Figure 12.16(a). {Used with permission from SCILLC dba ON Semiconductor}.

12.6 Phase Delay Enhancement in 3D Audio

For the observance of sound in stereophonic form, speakers are usually placed at considerable distances. However, such an arrangement is not possible in cell phones or hand-held devices. In such cases, stereophonic sound effect is simulated by what is known as *trans aural cross talk* cancellation. In this scheme we employ wave interference in such a way that the left channel sound is canceled in the right ear and the right channel sound is canceled in the left ear.

Even a small difference in the arrival of sound waves at the left and right ears helps in determining the direction of the sound. This difference of time or time delay is known *as inter-aural time delay* (ITD). The ITD operates on the spectral properties associated with the construction of internal and external parts of the ears and determines the *head-related transfer function* (HRTF). The HRTF accounts for the audio source location in terms of the distance to listener's head, the separation distance between the two ears and frequencies of the sound [12.8].

The basic idea of simulating the 3D audio effect is by combining each source with the pair of HRTF corresponding to the direction of the source [12.9]. However, most multimedia products do not use the directional information for fully 3-D effects. They simply employ phase-delay circuits creating HRTF that produces an effect of receiving the sound as if it comes from a different direction. Therefore, the speakers which are placed closely are perceived as being placed at different locations.

Let us, in a typical hand-held device where the distance between speakers is less than 7 cm, assume that the listener's head width is 20 cm. It is shown mathematically that for a distance of 50 cm between the listener's ear and the device, a phase difference (or phase delay) in sound signals should be between $78.5°$ and $101.5°A$ [12.8]. Such phase differences are obtained using a simple AP filter section; this finds considerable application in audio electronics simulation. One of the most commonly used circuits for the purpose is the active first-order circuit shown in Figure 4.11(a). Currently, it is required to construct a phase shift filter with $f_o = 10$ kHz for the linear signal (or for proper channel) and $f_o = 1$ kHz for the quadrature signal. It is desired to have a phase shift of $90°$ between the quadrature signal over the frequency range of 1 kHz to 10 kHz.

Figure 12.17(a) shows a cascaded, first-order all pass filter that will achieve a phase shift of nearly $90°$ between L (left) and Q (right) outputs for the frequency range of 1 kHz to 10 kHz. Figure 12.17(b) shows the simulated phase response of the circuit with a reasonable approximation to $90°$ of phase shift.

To improve the 3D effect, more stages can be added and aligned appropriately to improve the frequency range and increase the closeness to the $90°$ phase shift between L and Q outputs. Audio ICs such as MAX 9775 incorporate such a phase delay circuit and audio amplifier on a single chip.

12.7 Anti-aliasing Filters

Precision analog to digital convertors (ADCs) constitute an essential part in a variety of applications, such as instrumentation and measurement, process control and motor control, where an analog signal is to be digitally processed. Currently used SAR (successive approximation register) ADCs go up to 18-bit or even higher resolution, while $\Sigma - \Delta$ ADCs can be 24- or 32-bit resolution at 100s of kc/s. Consequently, there are considerable design challenges and considerations associated with implementing analog and digital filters into the ADC signal processing chain to achieve optimum performance [12.10]. Figure 12.18 shows a typical data acquisition signal chain which employs analog and digital filtering technique or a combination of both. Presently, the interest is not to study the design challenges for a practical filter in detail, which can be seen in reference [12.10], but to show simple examples of analog filter used in data acquisition.

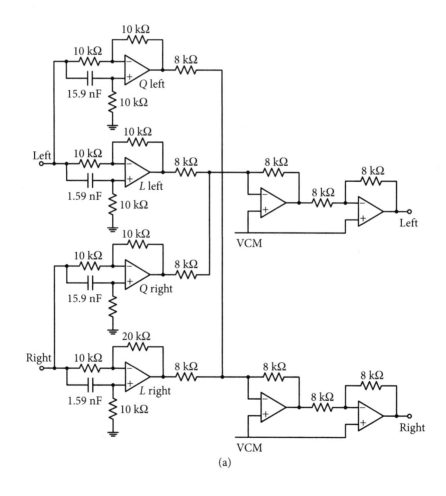

(a)

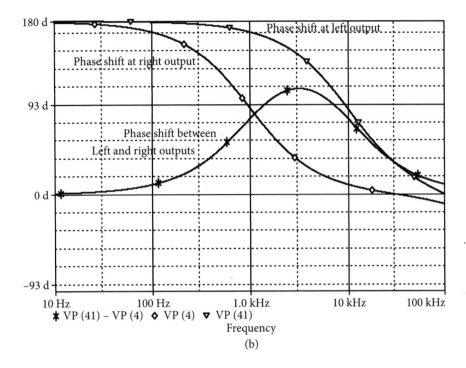

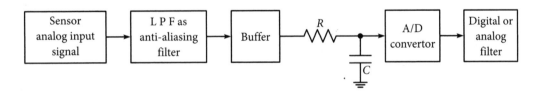

Figure 12.17 (a) Cascaded first-order active all pass circuits to achieve a phase shift of 90° between left and right signals {Courtesy: Maximum Integrated}. (b) Variation in phase shift for right and left signals and their phase difference.

Figure 12.18 A typical data acquisition system [12.10].

The analog filter is placed before the ADC in order to remove noise that is superimposed on the analog signal. Sometimes, there are noise peaks as well and it becomes necessary to eliminate these peaks; otherwise, noise peaks have the potential to saturate the analog modulator of the ADC. In addition, analog filtering is more suitable than digital filtering prior to ADC for higher speed systems (> 5 kHz). In these types of systems, the analog filter can reduce noise in the out-of-band frequency region, which reduces fold-back signals.

ADCs are usually operated with a constant sampling frequency. For sampling frequency f_c, typically called the Nyquist frequency, all input signals below $f_c/2$ folds back into the bandwidth of interest with the amplitude preserved. The frequency folding phenomena can be eliminated or attenuated by moving the corner frequency of the filter lower than $f_c/2$ or by increasing the order of the filter. In both the cases, the minimum gain of the filter at $f_c/2$ should be less than the signal-to-noise ratio (SNR) of the sampling system. For example, for a 12-bit ADC, the ideal SNR is 74 dB (> 6 dB per bit) less than the pass band gain. There are a number of papers available describing the utility and design of anti-aliasing filters. For example, Kyu et al.'s paper [12.11] discusses the design and implementation of a fourth-order Butterworth LPF using MATLAB and Circuit Maker for the data acquisition system. The following is also a typical example while having a comparison when LPF is implemented with different approximation approaches.

Anti-aliasing Filter Design: A Case Study[12.12]

In this example, the bandwidth of interest of the analog signal is from dc to 1 kHz; the output of the LPF will be fed to 12-bit SAR ADC. The sampling rate of the ADC will be 20 kHz. Hence, the design parameters for the LPF will be as follows:

i.   Cut-off frequency must be 1 kHz or higher.

ii.   Filter attenuates the signal by 74 dBs at 10 kHz

iii. Signal magnitude remains near constant in the pass band without any gain or attenuation.

(a) Implementation with Butterworth Approximation.

Using the design equations of the Butterworth approximation with $f_c = 1$kHz and a stop band frequency of approximately 5 kHz, we get the order of the filter as four. In the references note, a cascade of two Sallen and Key structures shown in Figure 7.22 was used (any other structure or design approach can also be used). Element values of the two second-order sections are:

First stage: $R_1 = 2.94$ kΩ, $R_2 = 26.1$ kΩ, $C_1 = 33$ nF and $C_2 = 10$ nF   (12.16)

Second stage: $R_1 = 2.37$ kΩ, $R_2 = 15.4$ kΩ, $C_1 = 100$ nF and $C_2 = 6.8$ nF   (12.17)

OA MCP 601 was used in both the stages. Simulated magnitude and phase responses of the filter are shown in Figure 12.19. The filter's gain at 10 kHz is 79.92 dBs against the theoretical value of 80 dBs; the gain remains flat in the pass band at unity.

(b) Implementation using Chebyshev Approximation

For the same specifications, but with 4 dBs of attenuation allowed in the pass band, the design equations give the required order of the filter as three. Hence, a cascade of a passive first-order and an active second-order (Sallen and Key) circuit was used; it is shown in Figure 12.20(a). The order of the filter is one less than in the Butterworth case; hence, it is economical. The simulated magnitude and phase responses are shown in Figure 12.20(b) which shows an attenuation of 74.12 dBs at 10 kHz against the theoretical value of 70 dBs. Along with the ripple present in the pass band, its phase response is worse than the Butterworth case.

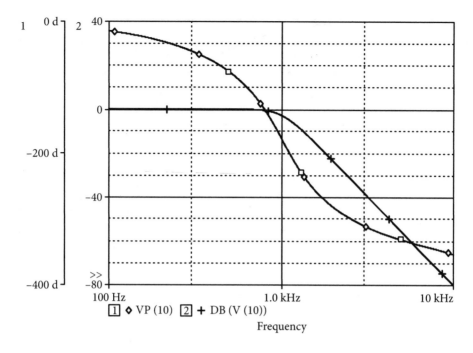

Figure 12.19 Response of fourth-order anti-aliasing filter using Butterworth approximation.

12.7.1 Digitally tuned anti-aliasing filter

LTC 1564 is a digitally controlled anti-aliasing/re-constructional filter. It is a high resolution eighth-order LPF that gives approximately 100 dBs of attenuation at 2.5 times of the corner frequency f_c. (f_c which is digitally controlled and ranges 10 kHz to 150 kHz in steps of 10 kHz). It also includes a digitally programmable gain amplifier (1V/V to 16 V/V in 1V/V steps). In addition to the mentioned attenuation, it has an SNR of 100 dBs. The filter has been shown to drive a 16-bit 500 ksps ADC for a complete 16-bit ADC interface [12.13].

The LTC 1564 is in a small 16-pin SSOP (Shrink Small Outline Package) and operates at a single or split supply range of 2.7 V to 10.5 V. Its pin connection diagram is shown in Figure 12.21.

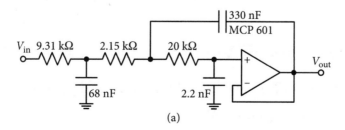

(a)

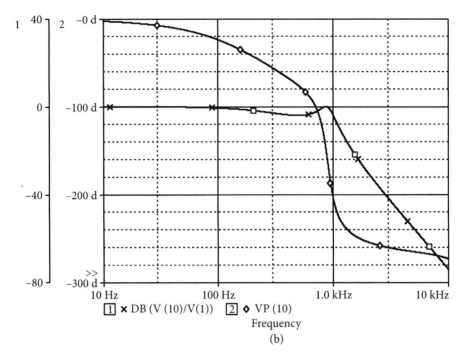

Figure 12.20 (a) Anti-aliasing filter using Chebyshev approximation. (b) Magnitude and phase response of the anti-aliasing filter shown in Figure 12.20(a).

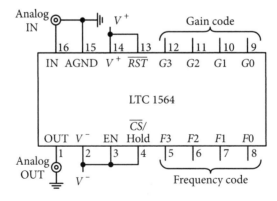

Figure 12.21 Pin connection diagram of an LTC 1564: a digital tuned anti-aliasing filter.

12.8 Active Filters for Video Signals: A Case study [12.14]

The following filters have been fabricated by Maxim Integrated using MAX 4450 OAs.

1. An ITU-601 anti-aliasing filter (AAF).
2. A 20 MHz AAF and reconstruction filter (ReF).
3. An HDTV ReF.

It is now well understood that AAF is used before an ADC to attenuate signals above the Nyquist frequency or half the rate of the sampling frequency of the ADC. In general, AAFs are designed with maximum roll-off in the stop band that is economically permissible. For ITU-601 and other similar applications, the objective is achieved by using analog filters combined with digital filters and an over-sampling ADC. In the case of ReFs (also called sin x/x or zero-order-hold correctors), they are placed after DACs to remove multiple images created by sampling. These filters do not require a steep roll-off in the stop band.

Both AAFs and ReFs have an LP characteristic to pass the video frame rate. Obviously, LPFs with best selectivity and lowest order are chosen. However, phase linearity requirement is also a serious consideration in video filter designs. It is for this reason that Butterworth approximation is most often used, though it may not be the most economical compared to Chebyshev and other approximations.

Group Delay Problems with Component Videos: In different formats and applications, the degree of sensitivity to group delay variations depends on the number of signals and their bandwidth. For example, a composite NTSC/PAL has only one signal with a group delay defined in ITU-470. Hence, the requirement is not difficult to meet; whereas for other component videos, each have multiple signals with equal or different bandwidths, requiring better group delay management.

Selection of OA to be Used: Since a video application operates in a wide bandwidth, and its typical magnitude is 2 $V_{p\text{-}p}$, gain bandwidth for an OA is an important consideration. In this case study, either Sallen and Key (Section 7.7), or multiple feedback or Rauch filter (Section 7.3) is used. Hence, the gain bandwidth of the OA is decided by the following relation of phase argument of a Rauch filter (effectively, it is the same for the Sallen and Key type filters).

$$\text{Arg}[H(j\omega)] = \pi - (\omega_c/GBW_{\text{rad}})(1 + R_f/R_i) \tag{12.18}$$

where R_f and R_i are the gain setting resistances and GBW_{rad} is the OA gain bandwidth product and ω_c is the cut-off frequency of the filter.

For AAFs, filter selectivity is determined by a template for ITU-601, as shown in Figure 12.22. It gives a specified bandwidth of 5.75 MHz ± 0.1 dB with an insertion loss of 12 dB at 6.75 MHz and 40 dB at 8 MHz. Variation in group delay is limited to ± 3 ns over a 0.1 dB bandwidth. This specification is difficult to meet easily using an analog filter alone; however, oversampling (×4) modifies the requirement to 12 dB at 27 MHz and 40 dB at 32 MHz; permitting the use of analog filters.

Using the procedure discussed in Chapter 3, it is easily observable that a 5-pole Butterworth filter with 3 dB cut-off frequency of 8.45 MHz will satisfy the aforementioned specifications. However, for satisfying the group delay specification, an important parameter of OA is the 0.1 dB, $2V_{p\text{-}p}$ bandwidth which is much lower [12.15]. Instead of designing the circuit, the circuits given in the application note 'Reference Schematic 31721' [12.14] of the Maxim Integrated is used. Figure 12.23(a) shows the schematic of the 5-pole, 5.75 MHz Butterworth filter for ITU-601 AAF, using a Rauch circuit with a delay equalizer employing MAX 4451

OAs. PSpice simulation of the magnitude and group delay is shown in Figure 12.23(b). In the simulated response, attenuation at 27 MHz and 40 MHz is 49.2 dBs and 67 dBs, respectively, and the overall group delay variation is of 0.1 dB up to 6.52 MHz.

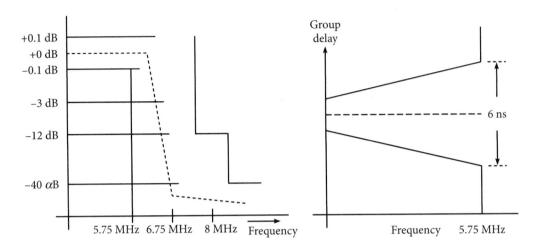

Figure 12.22 A template for determining the selectivity of an anti-aliasing filter {Courtesy Maxim Integrated}.

The next video filter example is that of a PC video, VESA. An XGA resolution (1024 × 758 at 85 Hz) has a sampling rate of 94.5 MHz. To have more than 35 dB attenuation at the Nyquist frequency of 47.75 MHz, a Rauch realization of a 20 MHz 4-pole Butterworth filter is used. Its circuit realization with element values employing MAX 4450 OA is shown in Figure 12.24(a). Figure 12.24(b) shows its simulated magnitude and group delay responses. The simulated response shows only 22 dBs of attenuation at 47.25 MHz.

12.9 Reconstruction Filter

An ReF removes all but the base band sample from the DAC output. Specification for NTSC/PAL ReF states that attenuation of more than 20 dB is required at 13.5 MHz and more than 40 dBs at 27 MHz; cut-off frequency depends on the applicable video standard. A 3-pole Butterworth with Sallen and Key configuration (gain value of 2) is used. Figure 12.25(a) and 12.25(c) shows filters circuits for NTSC and PAL designs, which are expected to satisfy gain and group delay specifications. Figures 12.25(b) and 12.25(d) show respective magnitude and group delay responses. In the simulated response of Figure 12.25(b), attenuation is 17.6 dBs at 13.5 MHz and 29.4 dBs at 27 MHz; somewhat less than the specifications. However, for the PAL case, the simulated response in the figure shows attenuation of 20.1 dBs at 13 MHz and 36.2 dBs at 27 MHz.

Another circuit for XGA is a 3-pole 20 MHz Butterworth filter. It uses the Sallen and Key configuration with gain 2, and it is shown in Figure 12.26(a). Complimenting the AAF of Figure 12.24, its simulated response is shown in Figure 12.26(b)

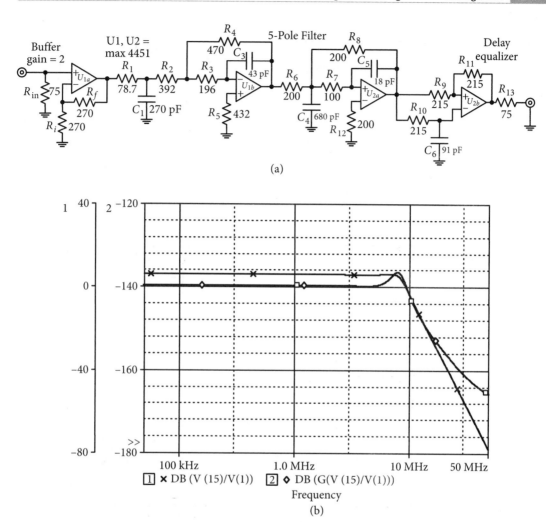

(a)

(b)

Figure 12.23 (a) A 5-pole 5.75 MHz Butterworth anti-aliasing filter {Courtesy: Maxim Integrated}. (b) Magnitude and phase response of the filter shown in Figure 12.23(a).

The last application in this case study is a reconstruction filter for HDTV. It is based on the template in the SMPTE-274 and 296M and its corner frequency is 29.7 MHz. A 5-pole Sallen and Key LP filter with a cut-off frequency of 30 MHz will have attenuation of more than 40 dBs at 74.25 MHz. Moreover, with the inclusion of group delay compensation, the filter will have a gain of +2 to drive a back-terminated 75 Ω co-axial cable. The complete circuit with element values is shown in Figure 12.27(a). Figure 12.27(b) shows the simulated magnitude and group delay responses. The filter's simulated low frequency gain is 2, cut-off frequency is 27.2 MHz; the attenuation of 41.6 dBs is achieved at 74.25 MHz.

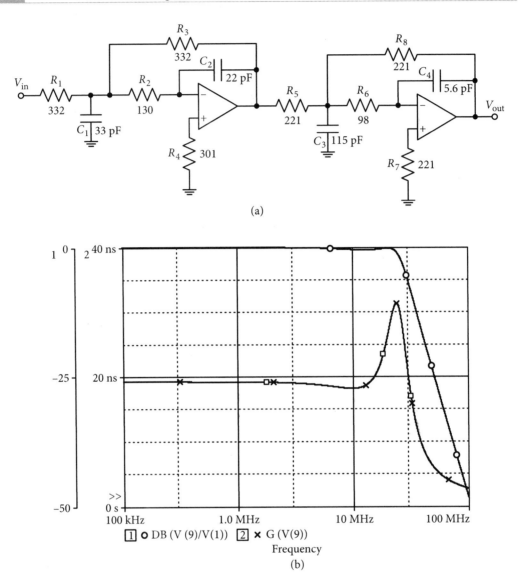

(a)

1 □ o DB (V (9)/V(1))   2 × G (V(9))
Frequency
(b)

Figure 12.24 (a) A 20 MHz, 4-pole Butterworth filter for a PC video {Courtesy: Maxim Integrated}. (b) Magnitude and phase response for the filter shown in Figure 12.24(a).

12.9.1 Video line driver for consumer video applications

A 5.25 MHz, third-order Butterworth filter having a gain of 2 V/V is realized using a MAX4390 OA. The proposed filter is capable of driving a 75 Ω back-terminated coaxial cable that is useful for video anti-aliasing and reconstruction filtering for composite (CVBS) or S-Video signals in standard definition digital TV applications. The third-order filter is to have an insertion loss of more than 20 dBs at 13.5 MHz and more than 40 dBs at 27 MHz [12.16].

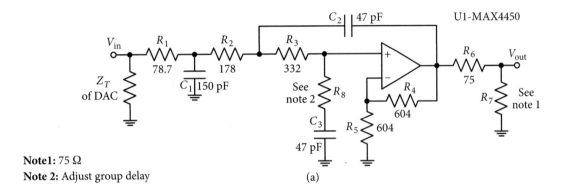

Note1: 75 Ω

Note 2: Adjust group delay

(a)

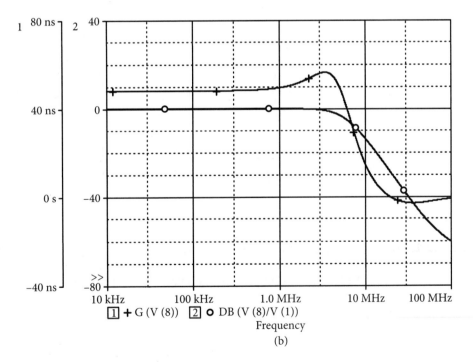

(b)

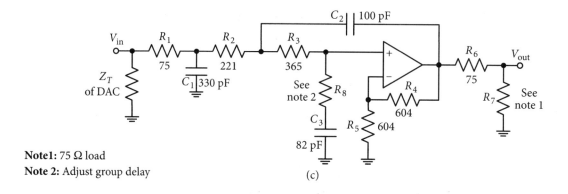

Note1: 75 Ω load

Note 2: Adjust group delay

(c)

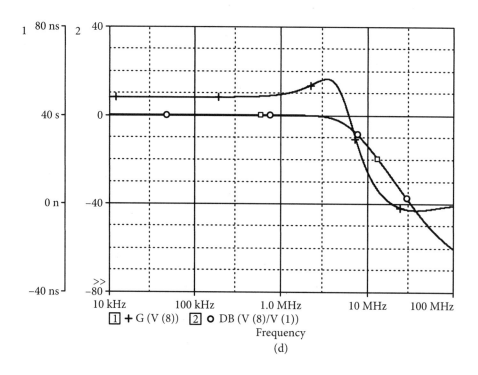

☐1 ✚ G (V (8))   ☐2 ○ DB (V (8)/V (1))

Frequency

(d)

Figure 12.25 (a) A reconstruction filter for PAL version {Courtesy: Maxim Integrated}. (b) Simulated response of the reconstruction filter shown in Figure 12.25(a). (c) A reconstruction filter for NTSC version {Courtesy: Maxim Integrated}. (d) Simulated response of the reconstruction filter shown in Figure 12.25(c).

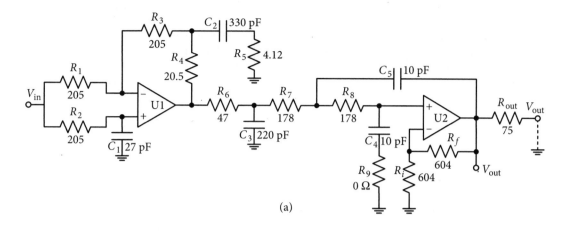

(a)

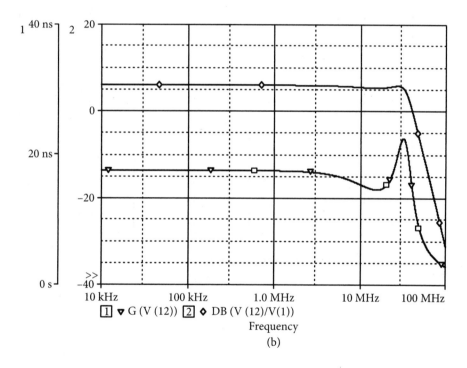

① ▼ G (V (12))   ② ◇ DB (V (12)/V(1))

Frequency

(b)

Figure 12.26 (a) A 20 MHz, 3-pole Butterworth reconstruction filter for XGA {Courtesy: Maxim Integrated}. (b) Simulated magnitude and phase responses of the filter shown in Figure 12.26(a).

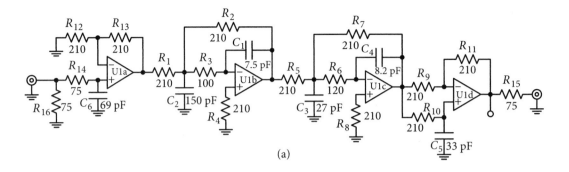

(a)

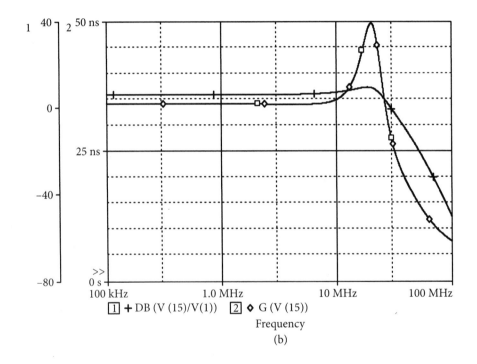

Figure 12.27 (a) A reconstruction filter for HDTV {Courtesy: Maxim Integrated}. (b) Simulated response of the filter shown in Figure 12.27(a).

For a standard quality video wave form reconstruction, group delay of the filter needs to be minimized and any group delay differential between it also needs to be minimized. The Sallen and Key filter, with a slight modification as shown in Figure 12.28(a) is used where addition of resistance R_8 in series with C_3 and R_3 creates a lag-lead network. The sum of R_3 and R_8 is kept constant equal to the original value that keeps the bandwidth of the dominant pole constant. When the value of R_3 is increased, it introduces a *lead* term, which lowers the group delay value. Figure 12.28(b) shows the simulated magnitude response with $R_3 + R_8 = 332\ \Omega$ and group delay variation over the pass band for three combination of resistances as shown in the simulation results.

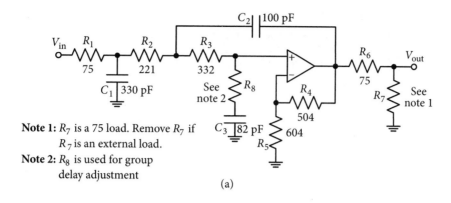

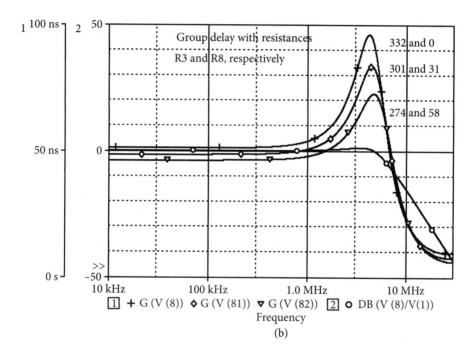

Figure 12.28 (a) Standard quality video wave form reconstruction filter {Courtesy: Maxim Integrated}. (b) Magnitude and group delay response of the 5 MHz video filter using modified Sallen and Key second-order filter.

The Ackerberg–Mossberg (AM) circuit shown in Figure 8.8 is used to provide a similar response using MAX 4390 OAs. A first-order LP section with a dc gain of 2 is cascaded to the second-order AM section as shown in Figure 12.29(a). Pole-location of the third-order Butterworth filter is as shown here:

$$s_1 = -0.5, s_{2,3} = -1 \pm j\, 0.866 \tag{12.19}$$

To realize the real pole s_1 for the first-order filter section in Figure 12.29(a), the design value of the elements is obtained using equations (1.25) and (1.26). The obtained element values for a dc gain of 2 are shown in Figure 12.29(a). For the AM second-order filter, relations in equation (8.23) and (8.25) are used; the obtained element values are as follows:

$R = 10$ kΩ, C $= 3.0303$ pF, $QR = 10$ kΩ, $(R/k) = 10$ kΩ, $R^* = 10$ kΩ.

Figure 12.29(b) shows the simulated response of the third-order filter with a 3 dB frequency of 5.045 MHz. The design value of pole-Q is unity in this case. Its attenuation is 21.7 dBs at 13.5 MHz and 40.5 dBs at 27 MHz; the realized attenuation is only a fraction better than the design. Figure 12.29(c) also shows the simulated variation in group delay response for $Q = 1.2$ and 1.4. For realizing $Q = 1.2$ and 1.4, all the component values remain the same except QR, which changes to 12 kΩ and 14 kΩ, respectively.

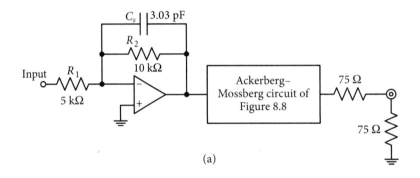

(a)

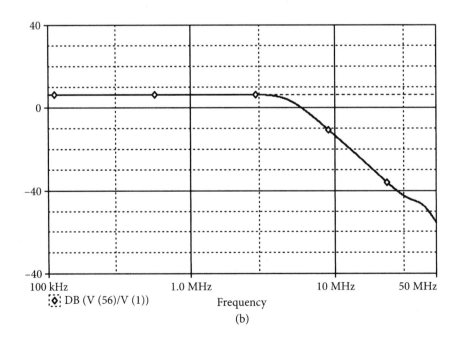

(b)

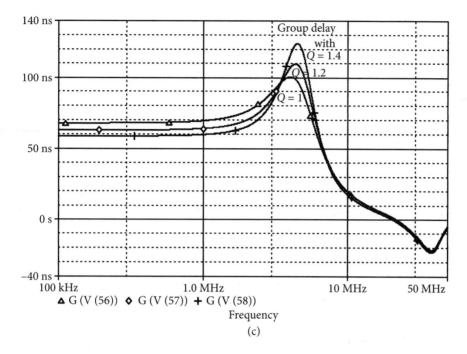

Figure 12.29 (a) Another scheme for reconstructing the standard quality video wave form using the Ackerberg–Mossberg filter configuration. (b) Simulated magnitude response of the 5 MHz video filter based on the AK biquad employed in Figure 12.29(a) and (c) variation in group delay.

References

[12.1]   Crump, Stephen. 2010. 'Analog Active Audio Filter,' *Application Report, SLOA 152*. US: Texas Instruments.

[12.2]   Malchaire, J. 2001. 'Sound Measuring Instruments,' *Occupational Exposure to Noise: Evaluation, Prevention and Control* 125–140. who.int/occupational health/publication/noise 6.pdf.

[12.3]   Texas Instruments. 2013. 'AN-346 Higher-Performance Audio Applications of the LM 833,' *Application Report, SNOA 586D*. US: Texas Instruments.

[12.4]   Maxim Integrated. 2009. 'Overview of 2.1 (Satellite/Subwoofer) Speaker Systems,' *Application Note 4046*. CA, US: Maxim Integrated.

[12.5]   LE8* Logic Controlled Equalizer Module-Operating Instructions, Patent No: 5,203,913, Lectrosonics, INC. Rio Rancho, NM 87174, USA 1996.

[12.6]   Marc_Escobeam. 2014. 'Professional RIAA Equalization with Analog Electronics'. https://www.instructables.com/id/RIAA-Equalization-with-analog-electronics/

[12.7]   ON Semiconductors. 2005. 'Audio Circuits Using the NE 5532/4,' *Application Note AND8177/D*. Arizona, US: ON Semiconductors.

[12.8]   Nicoletti, Robert. 2010. 'Phase Delay Enhances 3D Audio,' *Application Note 4632*. US: Audio Solution Group, Maxim Integrated Products Inc.

[12.9]   Casey, M., W. G. Gardener, and S. Basu. 1995. *Vision Steered Beam-Forming and Transaural Rendering for the Artificial Life Interactive Video Environment.* Cambridge, MA: MIT Media Laboratory.

[12.10]   Xie, Steven. 2016. 'Practical Filter Design Challenges and Considerations for Precision ADCs,' *Analog Dialog* 50–04.

[12.11]   Kyu, M. T., Z. M. Aung, and Z. M. Naing. 2009. 'Design and Implementation of Active Filter for Data Acquisition System,' *Proceedings of the International Multiconference of Engineers and Computer Scientists.* IAENG.

[12.12]   Baker, Bonnie. C. 1999. 'Anti-Aliasing, Analog Filters for data Acquisition System,' *AN 699.* Arizona, US: Microchip Technology.

[12.13]   Houser, M. W., and P. Karantzalls. 2002. 'A Digitally-Tuned Antialiasing/Reconstruction Filter Simplifies High Performance DSP Design,' *Design Note 276.* CA: Linear Technology Corporation.

[12.14]   Maxim Integrated Products. 2004. 'A Reference Schematic 3172,' *Application Note 3172.* CA, US: Maxim Integrated Products Inc.

[12.15]   Maxim Integrated Products. 2009. 'Ultra-Small, Low-Cost, 210MHz, Single-Supply Op Amps with Rail-to-Rail Outputs,' *MAX4450/MAX4451 Data sheet.* CA, US: Maxim Integrated Products Inc.

[12.16]   Maxim Integrated Products. 2002. '5 MHz, 3-Pole, Low-Pass Filter Plus Video Line Driver for Consumer Video Applications,' *Application Note 1799.* CA, US: Maxim Integrated Products Inc.

Practice Problems

12-1   Design and test active crossover network of Figure 12.2(a) for the crossover frequency of 10 krad/s.

12-2   (a) Design a third-order Butterworth filter for eliminating infrasonic noise signal having cut-off frequency of 200 rad/s; employ the structure shown in Figure 12.3.

(b) Design a fourth-order Bessel filter for eliminating ultrasonic noise signals having cut-off frequency of 250 krad/s.

(c) Obtain magnitude attenuation of the Bessel filter at 125 krad/s.

(d) Simulate combined magnitude response of the infrasonic and the ultrasonic filters. What are the frequencies at which gain becomes half of its maximum value?

12-3   To address the unequal output power requirement of the satellite speaker and the subwoofer, following filters are required. An HPF for a cut-off frequency of 3 krad/s and a BPF by combining a HPF and a LPF. Range of the respective cut-off frequencies of the HPF and LPF are from 500 rad/s to 600 rad/s and from 600 rad/s to 3 krad/s.

12-4   Repeat the problem 12-3 for the HPF to have the cut-off frequency of 4 krad/s and the composite BPF, for which range of frequencies for HPF is from 550 rad/s to 650 rad/s, and for the LPF frequency range is from 650 rad/s to 3.5 krad/s.

12-5   Obtain a mid bass boost filter circuit by combining a 12-dB treble cut having cut-off frequency of 800 Hz and a 12-dB treble boost having cut-off frequency of 200 Hz. What is the resultant peak mid bass boost and the frequency at which it occurs?

12-6   Obtain a mid treble cut filter circuit by combining a 12-dB bass boost having cut-off frequency of 200 Hz and a 12-dB bass cut having cut-off frequency of 20 kHz. What is the resultant peak mid treble cut and the frequency at which it occurs?

12-7   Obtain a mid treble boost filter circuit by combining a 12-dB treble cut having cut-off frequency of 8 kHz and a 12-dB treble boost having cut-off frequency of 800 Hz. What is the resultant peak mid treble boost and the frequency at which it occurs?

12-8   Obtain a mid bass cut filter circuit by combining a 12-dB bass boost having cut-off frequency of 200 Hz and a 12-dB bass cut having cut-off frequency of 800 Hz. What is the resultant peak mid bass cut and the frequency at which it occurs?

12-9   Design a notch filter, which has notch at 3 krad/s and width of the notch is 1/7 of an octave. Attenuation at the notch is to 6 dBs. Assume a suitable value of the quality factor.

12-10 Design a programmable notch filter which has the width of the notch as 1/7 of an octave and attenuation at the notch is to be 12 dBs to compensate the effect of ringing. Notch frequency of the filter are 2.5 kHz and 3.4 kHz. Assume suitable value of quality factor.

12-11 First-order APFs are cascaded to provide a phase shift of 90° between the two signals reaching left and right ears of a listener. Signal frequencies are in the range of 1500 rad/s and 15 krad/s. Design a suitable circuit and obtain the phase shift to the signal coming to the left and right ear, and the frequency range for which difference in the phase shift between the two outputs is 78° or more.

12-12 Analog signal with a maximum frequency of 1 krad/s is fed to a 10-bit SAR ADC. Sampling frequency rate of ADC will be 16 kHz. Design a maximally flat analog filter which would have sufficient attenuation (6 dB/bit) at the Nyquist frequency.

12-13 Repeat problem 12-12 if the analog signal frequency range extends up to 2 krad/s.

12-14 Repeat problem 12-12 employing Chebyshev approximation with maximum ripple width of 1 dBs in the pass band.

12-15 Repeat problem 12-13 employing Chebyshev approximation with maximum ripple width of 1 dB in the pass band.

12-16 Derive the transfer function for the circuit in Figure 12.25(a).

12-17 A third-order reconstruction filter is to be designed using Ackerberg–Mossberg configuration with following specifications. Insertion loss of more than 25 dBs at 15 MHz and more than 45 dBs at 30 MHz.

Test the circuit and find out simulated intrinsic loss at 15 MHz and 30 MHz.

12-18 What will be the value of group delay for the filter used in problem 12-17 with $Q = 1$, 1.2 and 1.5 at 1 MHz, 7.5 MHz and 15 MHz.

Chapter 13

Follow the Leader Feedback Filters

13.1 Introduction

One of the important issues in filter design is that parameter sensitivity has to be taken into consideration. A doubly terminated ladder is extensively used mainly because of the same reason, that is, parameters of the filter realized through it have low sensitivities. At the same time, an alternate synthesis method using the cascade approach has found favour because of its ability to tune specific pole–zeros in higher-order filters through the utilization of non-interactive second-order sections. However, in the cascade process, sensitivities increase, especially for high-Q filters. Since we know that negative feedback improves the performance of electronic circuits in a number of ways, the same have been applied to obtain what is known as multiple feedback (MF) topologies. An MF topology consists of a network with a single feed-forward path comprising unilateral (active structures are mostly unilateral unlike passive structures which are bilateral) second-order sections, having different kind of feedbacks. The nature of the feedback decides its final topology. One of the topologies is called *leap frog* and was discussed in Chapter 9. While performing operational simulation, the circuit had a topology shown in Figures 9.22 and 9.24, where it is easy to recognize the structure as a leap frog structure. Another topology under the broad area of MF topologies known as *follow the leader feedback* (FLF) is the subject of this chapter.

In Section 13.2, the basic FLF structure and the kind of transfer functions obtained is included. Also included in this section is the derivation of the structure's transfer function when either lossless integrators or lossy integrators are used in the feedback paths. Use of only a feedback block could provide all pole functions. Hence, feed-forward is also included, as will be discussed in Section 13.3, to improve the versatility of the scheme. A slightly modified feed-forward scheme, called the *shifted companion structure* is discussed in Section 13.4. It was observed that all these schemes were special cases of a general FLF structure that is given in Section 13.5. Without sacrificing the generality, synthesis of the filter becomes a bit easier if the feedback blocks have BP structures with equal value of quality factor, instead of different quality factors. Such a scheme, known as the *primary resonator block technique* is also discussed in this chapter.

13.2 Structure of the Follow the Leader Feedback Filters

Though the basics remain the same, there are slightly different FLF structures depending on the kind of basic blocks employed and whether the filter is an all-pole type or has finite zeros.

The basic structure of an FLF filter is shown in Figure 13.1, wherein the transfer function $H_i(i = 1, 2, \ldots, n)$ decides the nature of the final response. The transfer function of the blocks can be bilinear or biquadratic; it may be made of lossless or lossy integrators.

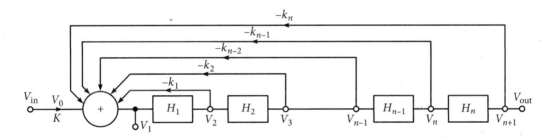

Figure 13.1 Basic structure of the follow the leader feedback filter scheme.

Without specifying the circuitry involved in providing the negative feedbacks $k_1, k_2, \ldots, k_n$ and the summation at the input, application of KVL (Kirchhoff's voltage law) at the input summing junction gives:

$$KV_0 - \{k_n V_{n+1} + k_{n-1} V_n + k_{n-2} V_{n-1} + \cdots\cdots + k_2 V_3 + k_1 V_2\} = V_1 \qquad (13.1)$$

As transfer functions of the blocks can be written as:

$$H_n = \frac{V_{n+1}}{V_n} = \frac{V_{out}}{V_n}, \; H_{n-1} = \frac{V_n}{V_{n-1}}, \ldots\ldots, H_2 = \frac{V_3}{V_2}, \; H_1 = \frac{V_2}{V_1} \qquad (13.2)$$

$$V_n = \frac{V_{n+1}}{H_n}, \; V_{n-1} = \frac{V_n}{H_{n-1}} = \frac{V_{n+1}}{H_{n-1} \times H_n}, \ldots\ldots,$$

$$V_2 = \frac{V_{n+1}}{H_2 \times \ldots \times H_n}, \; V_1 = \frac{V_{n+1}}{H_1 \times \ldots \times H_n} \qquad (13.3)$$

Equation (13.1) can be modified with $V_0 = V_{in}$ and $V_{n+1} = V_{out}$ as:

$$KV_{in} - \left\{ \begin{array}{l} k_n + \dfrac{k_{n-1}}{H_n} + \dfrac{k_{n-2}}{H_{n-1} \times H_n} + \ldots + \dfrac{k_1}{H_2 \times \ldots \times H_n} \\[3mm] + \dfrac{1}{H_1 \times \ldots \times H_n} \end{array} \right\} V_{out} = 0 \tag{13.4}$$

$$H(s) = \frac{V_{out}}{V_{in}} = \frac{K}{\left\{ k_n + \dfrac{k_{n-1}}{H_n} + \dfrac{k_{n-2}}{H_{n-1} \times H_n} + \ldots + \dfrac{k_1}{H_2 \times \ldots \times H_n} + \dfrac{1}{H_1 \times \ldots \times H_n} \right\}} \tag{13.5}$$

$$= \frac{K(H_1 \times H_2 \times \ldots \times H_n)}{\{ 1 + k_1 H_1 + k_2 H_1 H_2 + \ldots + k_{n-1} (H_1 \times \ldots \times H_{n-1}) + k_n (H_1 \times \ldots \times H_n) \}} \tag{13.6}$$

Equation (13.6) can realize a high-order filter in which the poles and zeros will depend on the nature of the transfer function of the blocks H_i and the feedback factors k_i. Zeros of the transfer function $H(s)$ will depend on the zeros of H_i and its denominator will be a polynomial in s. If H_i is a lossless integrator, the realized filter shall be an all-pole filter. However, if equation (13.6) is to be a general function with arbitrary zeros, H_i needs to a general biquad; this gives the reason for discussing different cases.

13.2.1 Use of lossless integrator blocks

One of the simplest cases is the one in which the transfer functions H_i blocks are only lossless integrators ($H_i = 1/s$). With the use of such blocks, equation (13.6) will modify as:

$$\frac{V_{out}}{V_{in}} = \frac{K \left(1/s^n \right)}{1 + \dfrac{k_1}{s} + \dfrac{k_2}{s^2} + \ldots + \dfrac{k_{n-1}}{s^{n-1}} + \dfrac{k_n}{s^n}} \tag{13.7}$$

$$= \frac{K}{s^n + k_1 s^{n-1} + k_2 s^{n-2} + \ldots + k_{n-1} s + k_n} \tag{13.8a}$$

It gives the following relation:

$$\frac{V_{in}}{1/K} - \left\{ s^n + \frac{s^{n-1}}{1/k_1} + \ldots + \frac{s}{1/k_{n-1}} + \frac{1}{1/k_n} \right\} V_{out} = 0 \tag{13.8b}$$

Equation (13.8) represents an LP filter section; though by taking the outputs at different points, HP and BP transfer functions can also be obtained.

In practice, inverting integrators are easier to realize than non-inverting ones. Hence, if all integrators are inverting lossless types, the feedback coefficients are to be multiplied by +1 or −1 to get all the feedbacks as negative. The following example will illustrate the procedure.

Example 13.1: Realize a fourth-order Chebyshev filter with a corner frequency of 20 krad/s and a ripple width of 1 dB.

Solution: The normalized transfer function of a fourth-order LP Chebyshev filter with a corner frequency of 1 rad/s is as follows

$$H(s) = \frac{V_{out}}{V_{in}} = \frac{K}{s^4 + b_3 s^3 + b_2 s^2 + b_1 s + b_0} \tag{13.9}$$

For a ripple width of 1 dB, the values of the coefficients in equation (13.9) are as follows:

$$b_0 = 0.2756, \ b_1 = 0.7426, \ b_2 = 1.4538, \ b_3 = 0.9528 \tag{13.10}$$

In order to get a gain of −1 dB at dc for the even order Chebyshev LP filter, value of K will be:

$$b_0 \times (-1 \text{ dB}) = 0.2756 \times 0.8912 = 0.2456.$$

Equation (13.9) can be modified in more than one way so as to be compatible with a circuit realizable in the form of Figure 13.1. The possible forms in which equation (13.9) can be written are as follows:

$$KV_{in} - (s^4 + b_2 s^2 + b_0) V_{out} - (b_3 s^3 + b_1 s) V_{out} = 0 \tag{13.11}$$

$$\frac{V_{in}}{1/K} - \left(s^4 + \frac{s^2}{1/b_2} + \frac{1}{1/b_0} \right) V_{out} + \left(-\frac{s^3}{1/b_3} - \frac{s}{1/b_1} \right) V_{out} = 0 \tag{13.12}$$

Equation (13.12) can be realized using four lossless integrators and two inverting summers as shown in Figure 13.2(a).

Comparing the coefficients in equation (13.12) with that in equation (13.8), we can write $k_1 = 1/b_3 = 1.0495$, $k_2 = 1/b_2 = 0.6878$, $k_3 = 1/b_1 = 1.346$, and $k_4 = 1/b_0 = 3.628$. Coefficients k_4, k_2 and k_0 are to be multiplied by (−1) and coefficients k_3 and k_1 are to multiplied by +1. The resulting circuit is shown in Figure 13.2(a), where frequency de-normalization is done by a factor of 20 krad/s and an impedance scaling factor of 10^3 was used to bring component values within a suitable range. The final element values are also shown in the figure.

The simulated magnitude response of the filter is shown in Figure 13.2(b). Its corner frequency is 3.1464 kHz (19.777 krad/s), maximum gain at ripple peaks is 0.9976 and dc gain is 0.8921, equivalent to 0.973 dB; all simulated parameters are found to be very close to the design values.

As mentioned earlier, equation (13.11) can be modified in other ways as well. In an alternate form, it is realized using three inverting lossless integrators, one inverting integrating summer and one inverting summer as shown in Figure 13.3(a). Its simulated response is shown in Figure 13.3(b); the pass band corner frequency is 3.118 kHz (19.6 krad/s) and ripple width is 1.008 dB.

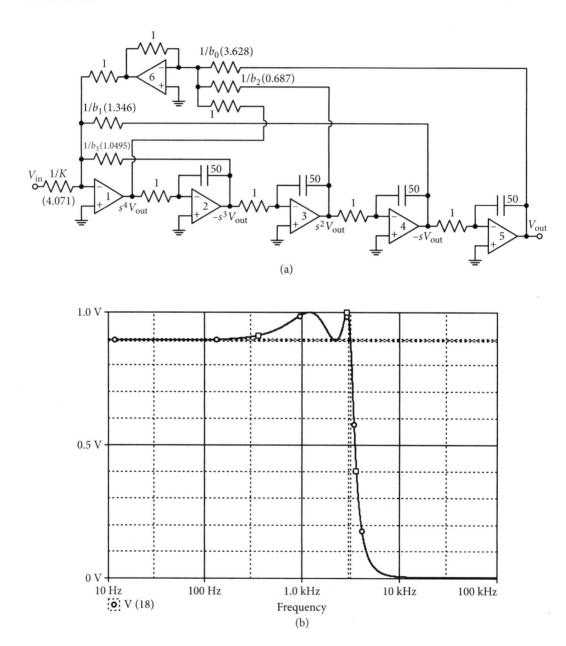

(a)

(b)

Figure 13.2 (a) Fourth-order normalized low pass filter realization for Example 13.1. All resistances are in kΩ and capacitors are in nF. (b) Magnitude response of the fourth-order FLF low pass Chebyshev filter using lossless integrator blocks for Example 13.1.

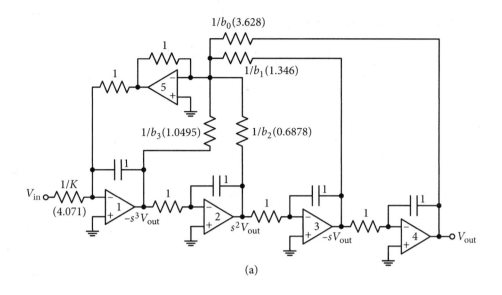

(a)

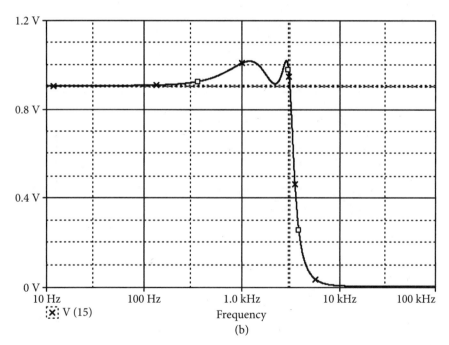

$\boxed{\times}$ V (15)

Frequency

(b)

Figure 13.3 (a) Alternate FLF structure for the fourth-order low pass filter shown in Figure 13.2(a). All resistances are in kΩ and capacitors are in nF. (b) Magnitude response of a fourth-order FLF low pass Chebyshev filter using lossless integrator blocks for Example 13.1: an alternate structure.

13.2.2 Use of lossy integrator blocks

A lossy inverting integrator shown in Figure 13.4(a) has the following form of transfer function:

$$\frac{V_2}{V_1} = -\frac{1}{s+a} \tag{13.13}$$

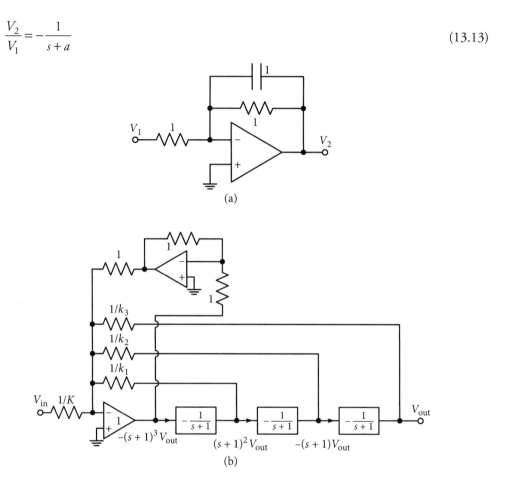

Figure 13.4 (a) Normalized lossy inverting integrator to be used in an FLF structure. (b) FLF structure for a third-order filter using the inverter shown in Figure 13.4(a) for realizing transfer function of equation (13.15).

If such lossy integrators are used to realize FLF structured filters, equation (13.6) will be written in the following form:

$$\frac{V_{out}}{V_{in}} = \frac{K}{(s+a)^n + k_1(s+a)^{n-1} + k_2(s+a)^{n-2} + \ldots\ldots + k_{n-1}(s+a) + k_n} \tag{13.14}$$

Equation (13.14) represents an all-pole LP filter of order n. If it is compared with the general all-pole LP response of equation (13.15) given in the following equation, coefficient matching can be done.

$$\frac{V_{out}}{V_{in}} = \frac{K}{s^n + b_{n-1}s^{n-1} + b_{n-2}s^{n-2} + \ldots\ldots + b_2 s^2 + b_1 s + b_0} \tag{13.15}$$

With coefficient matching between equations (13.14) and (13.15), the general relations between b_i and k_i can be written as:

$$b_{n-1} = na + k_1$$

$$b_{n-2} = \{n(n-1)/2\}a^2 + (n-1)ak_1 + k_2$$

$$b_0 = a^n + a^{n-1}k_1 + \ldots\ldots + a^2 k_{n-2} + ak_{n-1} + k_n \tag{13.16}$$

However, for comparatively lower-order filters, the relations will be simpler and can be used directly. For filter realization, a suitable value of a is selected and coefficients k_i are found in terms of b_i in equation (13.16). The transfer function of equation (13.15) is then realized through the basic structure shown in Figure 13.1 employing the lossy integrator shown in Figure 13.4(a) as the basic block. Care is taken to assign positive or negative sign for the coefficients k_i in the feedback in order to have all feedbacks negative. The process is illustrated here with an example.

Example 13.2: Design a third-order Chebyshev filter having a ripple width of 2 dBs and a pass band edge frequency of 10 krad/s using the lossy inverting integrator shown in Figure 13.4(a).

Solution: From Table 3.4, location of the poles for the desired filter is as follows:

$$s_{1-2} = -0.1845 \pm j0.9231, \; s_3 = -0.3689$$

Hence, the normalized transfer function of the all-pole LP filter having unity dc gain will be:

$$H(s) = \frac{V_{out}}{V_{in}} = \frac{0.3269}{s^3 + 0.7379s^2 + 1.02218s + 0.3269} \tag{13.17}$$

Putting $n = 3$ in equation (13.16), we get expressions for b_i and k_i as:

$$b_2 = 3a + k_1 \rightarrow k_1 = b_2 - 3a$$

$$b_1 = 3a^2 + 2ak_1 + k_2 \rightarrow k_2 = b_1 - 3a^2 - 2ak_1$$

$$b_0 = a^3 + a^2 k_1 + ak_2 + k_3 \rightarrow k_3 = b_0 - a^3 - a^2 k_1 - ak_2 \tag{13.18a}$$

For a selected value of the parameter, $a = 1$, and using values of the coefficients b_i from equation (13.17), equation (13.18a) gives the values of coefficients k_i as:

$$k_1 = -2.2621, \; k_2 = 2.5464, \; k_3 = -0.9574 \tag{13.18b}$$

With these values of coefficients k_i, for $n = 3$, equation (13.17) can be realized using the structure shown in Figure 13.4(b). For the design value of the pass band edge frequency, de-normalization is done by a factor of 10 krad/s and impedance scaling is done by a factor of 10^4, which makes the capacitance value in each integrator as 10 nF and the feedback resistor of 10 kΩ. The transfer function of equation (13.17) will become as follows.

$$\frac{V_{out}}{V_{in}} = -\frac{0.34727}{\left\{-(s+1)\right\}^3 + 2.2621\left\{-(s+1)\right\}^2 + (2.5464)\left\{-(s+1)\right\} + 0.9574} \tag{13.19}$$

$$\frac{V_{in}}{1/0.34727} + \left\{-(s+1)^3 + \frac{(s+1)^2}{1/2.2621} - \frac{(s+1)}{1/2.5464} + \frac{1}{1/0.9574}\right\}V_{out} = 0 \tag{13.20}$$

Figure 13.5 shows the circuit diagram of the filter with de-normalized element values. Figure 13.6(a) shows the phase response at the output of each OA, showing whether the outputs are in phase or out of phase with the input signal. Due to the negative sign with coefficients k_1 and k_3, output from integrators 1 and 3 are also applied directly to the summer OA0, with respective feedback resistance values of $(10^4/2.2621) = 4.4206$ kΩ and $(10^4/0.9574) = 10.444$ kΩ. The feedback resistance value from the output at integrator 2 is $(10^4/2.5464) = 3.927$ kΩ and input resistance is $(10^4/0.3269) = 30.59$ kΩ. Figure 13.6(b) shows the simulated magnitude response of the filter. Initially, dc gain is 0.945, with a ripple width of 1.96 dB. The ripple near the pass band does not reach its dc level, and the corner frequency is 1.532 kHz (962.9 rad/s). However, with small adjustments in the coefficient of $(s+1)^3$ from 1.0 to 0.98 by making the feedback resistor in the summing inverter-0 as 10.2 kΩ, the filter shows improvement in the shape of the response with ripple at the pass band reaching the dc level; corner frequency is now 1.55 kHz (974.3 rad/s), dc gain is unity and the ripple width is 1.85dBs.

13.3 Feed-forward Path Based FLF Structure

So far, the techniques used for the realization of FLF structure could give only all-pole filters. Researchers have shown that the required building blocks needs to be biquadratic sections for the realization of finite transmission zeros. However, some alternate methods which do not need biquadratic building blocks are also available. One such method which employs feed-forward paths will be shown here.

The structure shown in Figure 13.1 and implemented in the form of Figure 13.2(a), which employs lossless integrators as basic building blocks, realizes all-pole filters that have the transfer function as given in equation (13.5). In this basic structure, outputs are available in terms of s^n V_{out}, s^{n-1} V_{out}, ..., s V_{out} and V_{out}. These outputs can be used further in a feed-forward form to get a general nth order filter with arbitrary transmission zeros as shown in Figure 13.7. In the scheme, the original outputs from the lossless integrators are multiplied by coefficients a_i and then added together to obtain the final output.

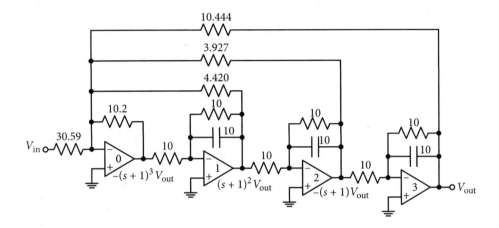

Figure 13.5 Circuit structure of the third-order filter for Example 13.2. All resistances are in kΩ and capacitors are in nF.

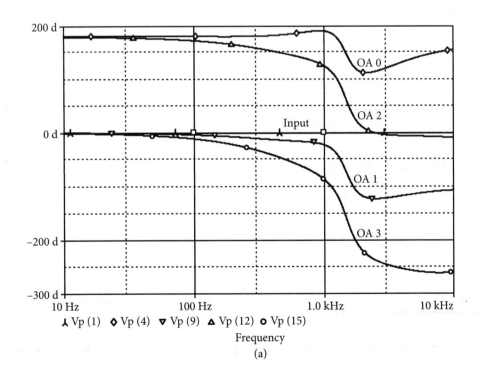

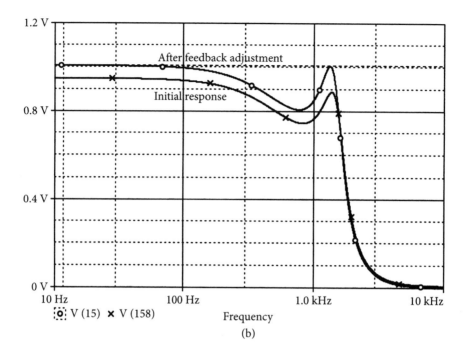

Figure 13.6 (a) Phase responses at the input, and outputs of the OAs in the third-order filter for Example 13.2. (b) Magnitude response of the third-order filter for Example 13.2.

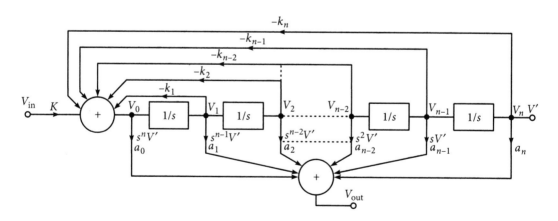

Figure 13.7 Realization of an nth order filter with finite zeros in the FLF structure employing lossless integrators and feed-forward paths.

As before, the ratio of the intermediate output V' to V_{in} in Figure 13.7 will be given as:

$$\frac{V'}{V_{in}} = H_1(s) = \frac{1}{s^n + k_1 s^{n-1} + k_2 s^{n-2} + \ldots + k_{n-1}s + k_n} = \frac{1}{D(s)} \qquad (13.21)$$

The ratio of the final output (V_{out}) with intermediate voltages V' will be

$$(V_{out}/V') = H_2(s) = a_0 s^n + a_1 s^{n-1} + \ldots + a_{n-2} s^2 + a_{n-1} s + a_n = N(s) \qquad (13.22)$$

Combining equations (13.21) and (13.22)

$$\frac{V'}{V_{in}} \times \frac{V_{out}}{V'} = H_1(s) H_2(s) \qquad (13.23)$$

$$\frac{V_{out}}{V_{in}} = H(s) = \frac{N(s)}{D(s)} = \frac{a_0 s^n + a_1 s^{n-1} + \ldots + a_{n-2} s^2 + a_{n-1} s + a_n}{s^n + k_1 s^{n-1} + k_2 s^{n-2} + \ldots \ldots + k_{n-1} s + k_n} \qquad (13.24)$$

Hence, a general transfer function of equation (13.24) can be realized using only lossless integrators. Once again, as inverting integrators are easy to use, appropriate positive or negative multipliers will be assigned to the coefficients a_j and k_j.

While doing final summation using the inverting summer, some of the received signals may have negative sign and some may have positive sign. Some coefficients may be zero as well; in these cases, feed-forward path will be simply open. The following example will illustrate the process.

Example 13.3: Realize an active LP filter in FLF mode with the feed-forward technique. Its pass band extends up to 20 krad/s with a maximum ripple width of 1.0 dB; the stop band is beyond 40 krad/s with a minimum attenuation of 34 dBs. The filter has unity gain in the pass band.

Solution: The design obtains the following normalized transfer function for the given specifications.

$$H(s) = \frac{V_{out}}{V_{in}} = K \frac{\left(s^2 + 5.1532\right)}{(s + 0.54)\left(s^2 + 0.434s + 1.01\right)} \qquad (13.25)$$

For the unity dc gain, $K = 0.10589$. Transfer function of equation (13.25) is expanded, and functions $H_1(s)$ and $H_2(s)$ corresponding to equations (13.21) and (13.22) are obtained from it as:

$$H_1(s) = \frac{V'}{V_{in}} = \frac{0.10589}{s^3 + 0.974s^2 + 1.2443s + 0.5454} \qquad (13.26)$$

$$H_2(s) = (V_{out}/V') = s^2 + 5.1532 \qquad (13.27)$$

Equation (13.26) can be written as:

$$\frac{V_{in}}{1/0.10589} - \left(\frac{s^2}{1/0.974} + \frac{1}{1/0.5454}\right)V' + \left(-\frac{s}{1/1.2443} - s^3\right)V' = 0 \tag{13.28}$$

Equation (13.28) is realized as the top portion of Figure 13.8(a) using inverting lossless integrators with the value of coefficient multipliers $k_1 = (1/0.974) = 1.0267$, $k_2 = (1/1.2443) = 0.8038$, and $k_3 = (1/0.5454) = 1.8335$. As $K = 0.10589$, V_{in} is to be divided by 9.4437.

While realizing $H_2(s)$ of equation (13.27), $a_0 = a_2 = 0$. Hence, it is written as:

$$V_{out} + \left(-s^2 - \frac{1}{1/5.1532}\right)V' = 0 \tag{13.29}$$

For the realization of the part played by equation (13.29), the lower half part of Figure 13.8(a) needs only one summing inverter with multipliers from V' being $(1/5.1532) = 0.194$ and unity from s^2V'.

Frequency de-normalization by a factor of 20 krad/s is done, and an impedance scaling factor of 10^4 is used to find suitable element values, which are shown in Figure 13.8(a).

Figure 13.8(b) shows the PSpice simulated magnitude response of the filter. Peak gain is 1.005 but the dc gain is 0.9862 and ripple width is 10.49 mV for an input of 100 mV, which corresponds to 0.962 dB. The corner frequency is 3.2168 kHz (20.219 krad/s) and gain at 40 krad/s (6.38 kHz) is 0.017427 (−35.17 dBs). A zero occurs at 7.24 kHz (from the numerator of equation (13.25)). All the simulated parameters are very close to the design parameters.

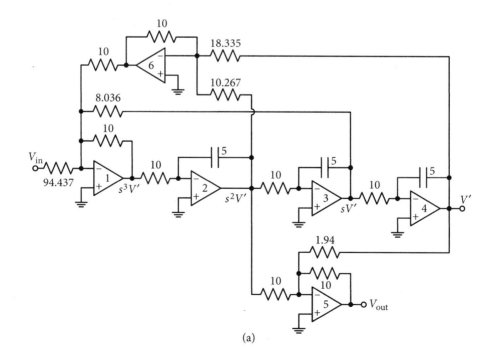

(a)

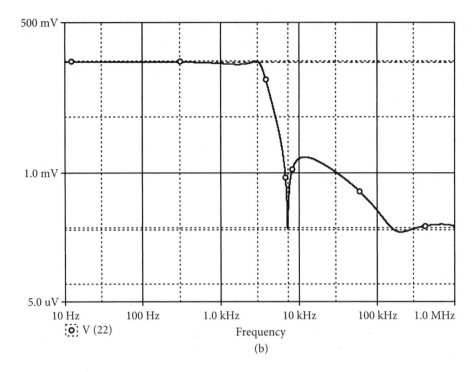

Figure 13.8 (a) Third-order filter section for Example 13.3, using feed-forward methodology. All resistances are in kΩ and capacitors are in nF. (b) Magnitude response of the third-order elliptic filter for Example 13.3.

13.4 Shifted Companion FLF Structure

In Section 13.2.2, lossy inverting integrators are used to realize all-pole filters where the first feedback coefficient, k_1 was derivable as $(b_{n-1} - na)$. The common practice is to select $a = 1$ for the sake of simplicity in calculations. However, if a is selected as $\left(\dfrac{b_{n-1}}{n}\right)$, k_1 will reduce to zero, which means that there will be no feedback after the first integrator (or any other building block used in place of the integrator). Such an arrangement is known as the *shifted companion feedback* (SCF) structure as shown in Figure 13.9(a)

Figure 13.9(b) shows the normalized inverting integrator when $a \neq 1$, which is to be used as the basic building block in the SCF scheme.

The SCF structure can also use feed-forward paths like those in Section 13.3 in order to realize the general transfer function having arbitrary zeros. The suggested process is also shown in Figure 13.9(a) with dotted lines joining a summer. An example will illustrate the procedure.

Example 13.4: Realize the third-order Chebyshev filter of Example 13.2 to have a pass band edge frequency of 5 krad/s using the SCF structure.

Solution: The normalized transfer function from Example 13.2 is repeated here for unity dc gain from equation (13.17):

$$H(s) = \frac{V_{out}}{V_{in}} = \frac{0.3269}{s^3 + 0.7379s^2 + 1.02218s + 0.3269} \tag{13.30}$$

Equation (13.30) can be written as:

$$\frac{V_{in}}{1/0.3269} - \left(\frac{1}{1/0.7379}s^2 + \frac{1}{1/0.3269}\right)V_{out} + \left(-\frac{s}{1/1.02218} - s^3\right)V_{out} = 0 \tag{13.31}$$

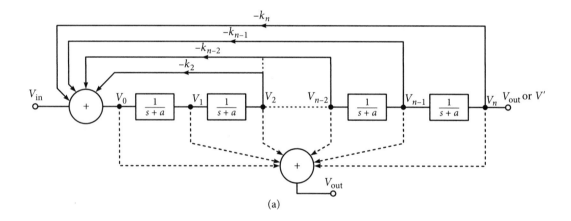

(a)

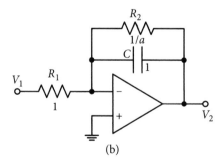

(b)

Figure 13.9 (a) Shifted companion feedback FLF structure (and feed-forward as well) (b) normalized inverting integrator to be used in part (a).

Equation (13.31) may be realized using the circuit of Figure 13.10(a). Calculation of the element values of the basic block, as well as values of the resistances providing feedback and feed-forward, are given here.

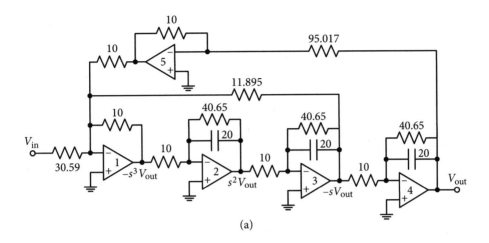

(a)

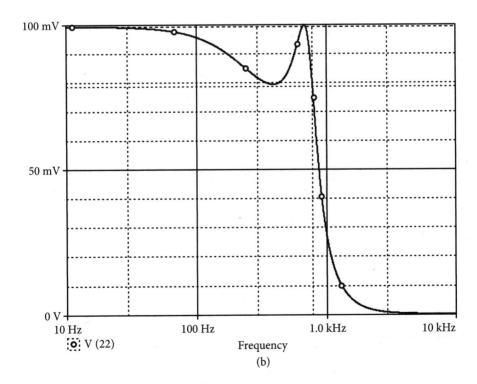

V (22)

Frequency

(b)

Figure 13.10 (a) Circuit realization of third-order Chebyshev filter for Example 13.4 using shifted companion feedback process. All resistors are in kΩ and capacitors in nF. (b) Response of the third-order filter.

Order of the filter n being three, the coefficient $k_1 = b_2 - 3a$. While making $k_1 = 0$:

$$a = (b_2/3) = 0.7379/3 = 0.24596$$

$$k_2 = b_1 - n(n-1) \times a^2/2 = 0.84069$$

$$k_3 = b_0 - ak_2 - a^3 = 0.10524 \tag{13.32}$$

A frequency de-normalizing factor of 5 krad/s and impedance scaling factor of 10 kΩ is used. For $a = 0.24596$, the selected capacitor in each inverting integrator shown in Figure 13.9(b) will be 20 nF; R_1 becomes 10 kΩ and $R_2 = 40.65$ kΩ. The rest of the resistors realizing the coefficients k_2 and k_3 are obtained after impedance scaling as ($1/0.84069 = 1.1895 \times 10^4\ \Omega$) and ($1/0.10524 = 9.5017 \times 10^4\ \Omega$), which are shown in Figure 13.10(a). To get the dc gain as unity, the input resistance after impedance scaling will be ($1/0.3269 = 3.059 \times 10^4\ \Omega$).

The simulated magnitude response of the circuit is shown in Figure 13.10(b). DC gain is found to be 0.995 and a ripple width of 19.96 mV for an input voltage of 100 mV is equivalent to 1.96 dB. Corner frequency is 791.5 Hz (4975 rad/s).

13.5 General FLF Structure

In Section 13.2, a basic structure of the FLF feedback filter was discussed. The structure was shown in Figure 13.1. Building blocks were initially lossless, but later on lossy integrators were used to realize all-pole filters. The idea was extended to include feed-forward paths to realize arbitrary transmission zeros while using lossless or lossy integrators. The shifted companion approach was introduced in Section 13.4.

All the aforementioned schemes can be shown to be special cases of a generalized FLF process. In this section, the FLF structure will be studied from a generalized view, which will not only help in revising the simpler versions studied so far in the chapter, but also leads to another useful special class of FLF filters, known as the *primary resonator block* (PRB) based filters.

For Figure 13.7, if the transfer functions of each block is taken as $H_j(s)$, we can write:

$$-V_0 = KV_{in} + \sum_{i=1}^{n} k_i V_i \tag{13.33a}$$

And the final output is written as:

$$V_{out} = a_0 V_0 + \sum_{i=1}^{n} a_i V_i \tag{13.33b}$$

Previous examples have shown that other than the basic blocks, the filter realization requires two summers using OAs; the realization of multipliers k_i and a_i is through resistors. A generalized form of FLF filters is easily derivable from Figure 13.7, which is shown in Figure 13.11. Here as:

$$k_i = \frac{R_{f0}}{R_{fi}}\Big|_{i=1\,to\,n} \text{ and } a_i = \frac{R_0'}{R_i}\Big|_{i=0\,to\,n} \tag{13.34}$$

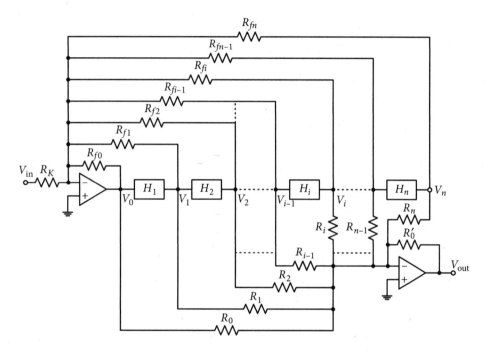

Figure 13.11 A general FLF structure where H_i could be first- or second-order sections, and summing amplifiers are assumed to be ideal.

Equations (13.33a) and (13.33b) can be written respectively as:

$$-V_0 = \frac{R_{f0}}{R_K}V_{in} + \sum_{i=1}^{n}\frac{R_{f0}}{R_{fi}}V_i \tag{13.35}$$

$$V_{out} = -\sum_{i=0}^{n}\frac{R_0'}{R_i}V_i \tag{13.36}$$

Internal voltages V_i are easily calculated from the following expression:

$$V_i = V_0\prod_{j=1}^{i}H_j(s) \quad i=1,2,\ldots\ldots,n \tag{13.37}$$

If V_i from equation (13.37) is substituted in equation (13.33a), then the expression of output voltage will be:

$$-V_0 = KV_{\text{in}} + \sum_{i=1}^{n} k_i \left\{ V_0 \prod_{j=1}^{i} H_j(s) \right\} \tag{13.38a}$$

$$= KV_{\text{in}} + k_1 V_0 H_1(s) + k_2 V_0 H_1(s) H_2(s) + \ldots + k_n V_0 H_1(s) H_2(s) \ldots H_n(s) \tag{13.38b}$$

Hence,
$$\frac{V_0}{V_{\text{in}}} = -\frac{K}{1 + \sum_{m=1}^{n} \left\{ k_m \prod_{j=1}^{m} H_j(s) \right\}} \tag{13.39}$$

In the expanded form, it can be modified as:

$$\frac{V_0}{V_{\text{in}}} = -\frac{K}{1 + k_1 H_1(s) + k_2 H_1(s) H_2(s) + \ldots + k_n H_1(s) H_2(s) \ldots H_n(s)} \tag{13.40}$$

Substituting V_0 from equation (13.39) in equation (13.37), transfer function at the ith stage is obtained as:

From
$$h_i = \left(\frac{V_i}{V_0} \right) \left(\frac{V_0}{V_{\text{in}}} \right)$$

$$h_i(s) = \frac{V_i}{V_{\text{in}}} = -\frac{K \prod_{j=1}^{i} H_j(s)}{1 + \sum_{m=1}^{n} \left\{ k_m \prod_{j=1}^{m} H_j(s) \right\}} \tag{13.41}$$

Since
$$h_i(s) = h_n(s) \prod_{j=i+1}^{n} 1/H_j(s) \tag{13.42}$$

Here, $h_n(s)$ is the transfer function for feedback network only and its expression is obtained from equations (13.42) and (13.41) as:

$$h_n(s) = \frac{h_i(s)}{\prod_{j=i+1}^{n} 1/H_j(s)} = \frac{K \prod_{j=1}^{n} H_j(s)}{1 + \sum_{m=1}^{n} \left\{ k_m \prod_{j=1}^{m} H_j(s) \right\}} \tag{13.43}$$

As discussed before, it is clear from equation (13.43) that for this transfer function, the zeros are decided by the zeros of $H_j(s)$. For realizing $h_n(s)$ with arbitrary transmission zeros, $H_j(s)$ should be a second-order function with finite transmission zeros, treatment of which is quite involved. Instead, if the basic blocks are second-order band pass (BP) functions, calculations become a little easier. Let the normalized second-order BP function for the building block in standard format be:

$$H_i(s) = h_{0i} \frac{(1/Q_i)s}{s^2 + (1/Q_i)s + 1} = h_{0i}h_i(s)$$

(13.44)

In equation (13.44), h_{0i} is the mid-band gain and Q_i is the quality factor of the ith stage.

Finally substituting V_i from equation (13.41) and V_0 from equation (13.39) in equation (13.33):

$$H(s) = \frac{V_{out}}{V_{in}} = -K \frac{a_0 + \sum_{m=1}^{n}\{a_m \prod_{j=1}^{m} H_j(s)\}}{1 + \sum_{m=1}^{n}\{k_m \prod_{j=1}^{m} H_j(s)\}}$$

(13.45)

Implementation of using a second-order BP section as $H_i(s)$ can be explained better with the help of a transfer function having a smaller value of n. Hence, for $n = 4$, $H(s)$ will be expanded as shown in equation (13.46) and will be studied further.

$$H(s) = \frac{V_{out}}{V_{in}} = K \frac{a_0 + a_1 H_1 + a_2 H_1 H_2 + a_3 H_1 H_2 H_3 + a_4 H_1 H_2 H_3 H_4}{1 + k_1 H_1 + k_2 H_1 H_2 + k_3 H_1 H_2 H_3 + k_4 H_1 H_2 H_3 H_4}$$

(13.46)

Calculations involving equation (13.46) become easier if the $H_i(s)$ are arithmetically asymmetric BPFs obtained from the low pass (LP) section using the transformation already discussed in Chapter 5, and repeated here.

$$S = \frac{\omega_o}{B} \frac{s^2 + \omega_o^2}{s} \rightarrow S = Q \frac{s^2 + 1}{s} \text{ for normalized } \omega_o$$

(13.47)

In equation (13.47), S is the normalized complex frequency variable of the LP prototype and $Q = (\omega_o/B)$ is the quality factor of the BP filter. The numerator and denominator of the BP transfer function of equation (13.44) is multiplied by (Q/s) to get the transfer function of the prototype LP.

$$H_i(s) = h_{0i} \frac{Q/Q_i}{Q\dfrac{s^2+1}{s} + \dfrac{Q}{Q_i}} \rightarrow h_{0i} \frac{Q/Q_i}{S + \dfrac{Q}{Q_i}}$$

(13.48)

Equation (13.48) is the first-order prototype LP transfer function, where h_{0i} becomes the dc gain of the LP filter $-H_{LP}(S)$ and if we write $(Q/Q_i) = q_i$, then

$$H_{LP}(S) = h_{LPi} \frac{q_i}{S + q_i}$$

(13.49)

To find the poles of the general FLF transfer function, if the BP to LP transformation is used, its order becomes $(n/2)$ and the poles are decided by the values of q_i and k_i. For the nth order

FLF, the transfer function expression of the $(n/2)$th order LP has been derived for further processing. However, as suggested earlier, we will consider the case of the eighth-order FLF filter for the sake of easier understanding.

Substituting the relation in equation (13.49) in the denominator of equation (13.46), we get:

$$
\begin{aligned}
= & (S+q_1)(S+q_2)(S+q_3)(S+q_4) + \\
& k_1\{h_{LP1}q_1(S+q_2)(S+q_3)(S+q_4)\} + \\
& k_2\{h_{LP1}q_1 h_{LP2}q_2(S+q_3)(S+q_4)\} + \\
& k_3\{h_{LP1}q_1 h_{LP2}q_2 h_{LP3}q_3(S+q_4)\} + \\
& k_4 h_{LP1}h_{LP2}h_{LP3}h_{LP4}q_1 q_2 q_3 q_4
\end{aligned}
\tag{13.50}
$$

When the relation in equation (13.49) is substituted in the numerator of equation (13.46), we get the following relation for finding the zeros of the LP filter.

$$
\begin{aligned}
= & a_0(S+q_1)(S+q_2)(S+q_3)(S+q_4) + \\
& a_1\{h_{LP1}q_1(S+q_2)(S+q_3)(S+q_4)\} + \\
& a_2\{h_{LP1}q_1 h_{LP2}q_2(S+q_3)(S+q_4)\} + \\
& a_3\{h_{LP1}q_1 h_{LP2}q_2 h_{LP3}q_3(S+q_4)\} + \\
& a_4 h_{LP1}h_{LP2}h_{LP3}h_{LP4}q_1 q_2 q_3 q_4
\end{aligned}
\tag{13.51}
$$

Coefficients of the denominator in equation (13.46) are found by comparing its expanded form in equation (13.50). However, we get four equations with $b_4 = 1$ (as in equation (13.9)), in terms of eight unknowns, k_1, k_2, k_3, k_4, q_1, q_2, q_3, q_4. One of the solutions which also simplifies the procedure is the one which assumes equal quality factors for all the second-order BP sections, that is, $q_1 = q_2 = q_3 = q_4 = q$. This means we need to use identical BP sections. Such a scheme is known as the *primary resonator blocks* (PRB) technique. Hence, we will continue the study of FLF filters in PRB form, which also includes the shifted companion approach.

13.5.1 Primary resonator block technique

For identical BP sections with the same value of quality factor q, if we compare equation (13.50) with the denominator of equation (13.46) and simplify, we get the following relations.

$$
k_1 = b_3 - 4qb_4
\tag{13.52a}
$$

$$
k_2 = b_2 - 6q^2 b_4 - 3qk_1
\tag{13.52b}
$$

$$
k_3 = b_1 - 4q^3 b_4 - 3q^2 k_1 - 2qk_2
\tag{13.52c}
$$

$$
k_4 = b_0 - q^4 b_4 - q^3 k_1 - q^2 k_2 - qk_3
\tag{13.52d}
$$

Here, there are four equations with five unknowns. It is used to advantage by making $k_1 = 0$, through selecting q given by the following relation from equation (13.52a):

$$k_1 = 0 = b_3 - 4qb_4 \rightarrow q = (b_3/4b_4) \tag{13.53}$$

It needs to be mentioned that for the nth order filter, equation (13.53) will become:

$$q = (b_{n-1}/nb_n) \tag{13.54}$$

The remaining feedback coefficients in equations (13.52) can be found recursively as has been shown in previous examples.

The process of finding coefficients while realizing transmission zeros is exactly the same as that for the case of poles; this has been explained in Section 13.3.

So far, the process of finding the feedback and feed-forward coefficients has been discussed when PRBs are used in FLF design. Obviously, finding the actual resistors realizing these coefficients should not be difficult. However, gain constants have not yet been specifically decided. In fact, these are free parameters and in a good design, they are selected to get maximum dynamic range without over-shooting of signals at any intermediate terminal, which would have distorted signals and thus might have given erroneous outputs. A process similar to the one discussed in the cascade process is to be followed for optimum design.

In case, one is interested in simulating a function without finite zeroes, no feed forward circuitry is required. In such case, numerator of the prototype LPF can be found out by substituting the relation of equation (13.49) in equation (13.43). Evaluation would then be replaced as shown below for a fourth order LP prototype:

$$N(S) = Ka_4 \, h_{LP1} \, h_{LP2} \, h_{LP3} \, h_{LP4} \, q_1 \, q_2 \, q_3 \, q_4 \tag{13.55}$$

Obviously, for identical prototyped expression will simplify as:

$$N(S) = Ka_4 \, (h_{LP} \, q)^4 \tag{13.56}$$

$K = 1$ in equation (13.56) for the dc gain of unity for the LP prototype.

Example 13.5: Design an eighth-order Butterworth BPF using the PRB structure having a center frequency of 2.5 kHz and quality factor of 20. Mid-band gain needs to be unity.

Solution: To realize the eighth-order Butterworth filter, we need to start with a fourth-order LP Butterworth filter, for which the normalized transfer function is given as:

$$H_{LP}(s) = \frac{V_{out}}{V_{in}} = \frac{1}{s^4 + 2.6131s^3 + 3.414s^2 + 2.6131s + 1} \tag{13.57}$$

To employ the shifted companion scheme and simplicity in FLF structure, equation (13.54) gives:

$$q = \frac{2.6131}{4} = 0.65328 \tag{13.58}$$

Applying equation (13.52), we get the coefficient values of equation (13.34) as:

$$k_2 = 0.8536,\ k_3 = 0.3838 \text{ and } k_4 = 0.2036 \tag{13.59}$$

As $q = Q_{BP}/Q_{LP} = 0.6532$ (from equation 13.48), for the given value of Q_{BP} as 20, $Q_{LP} = 30.616$. Therefore, expression of the normalized transfer function of the PRB is written as:

$$h_{PRB}(s) = \frac{h_0(1/Q_{LP})s}{s^2 + (1/Q_{LP})s + 1} \tag{13.60}$$

In equation (13.60), $h_0 = (1/q) = 1.5308$ in order to get unity mid-band gain for the PRB.

Since this example is for an all-pole filter, no feed-forward path is required.

The de-normalized transfer function of the PRB will be

$$\frac{V_2}{V_1} = \frac{1.5308(5000\pi/30.616)s}{s^2 + (5000\pi/30.616)s + (5000\pi)^2} \tag{13.61}$$

It is realized by a GIC based circuit shown in Figure 13.12(a), for which the expression of the transfer function, with $R_1 = R_3 = R$ and $C_2 = C$ is:

$$\frac{V_2}{V_1} = \frac{(1/CR_Q)\{1+(R_4/R_5)\}s}{s^2 + (1/CR_Q)s + (R_4/R_5)\{1/(RC)^2\}} \tag{13.62}$$

Comparison between equations (13.61) and (13.62) give the following component values.

With $R_4 = R_5 = R$, and $\omega_o = 2500(2\pi) = 1/RC$, for the selected value of $C = 20$ nF, we get $R = 3.1818$ kΩ and $(1/CR_Q) = 5000\pi/30.616 \rightarrow R_Q = 97.412$ kΩ. With $R_4 = R_5$, the realized mid-band gain becomes 2. Hence, to bring it down to 1.5308, a potential divider at the input is to be used. For the potential divider, the resistor ratio will be (R_Q/α) and $(R_Q/1-\alpha)$, where $\alpha = 1.5308/2 = 0.7654$; it results in the input resistors of 127.27 kΩ and 415.22 kΩ.

The calculated values of the components are shown in Figure 13.12(a). PSpice simulated response of the PRB section is shown in Figure 13.12(b). Its mid-band gain is 1.5296, center frequency is 2.488 kHz and with a bandwidth of 80.98 Hz, realized $Q_{LP} = 30.73$.

PRB of Figure 13.12(a) is used in the realization of the eighth-order Butterworth BPF shown in Figure 13.13(a). Feedback resistances are calculated to realize the coefficients k_3, k_2 and k_4 given in equation (13.59).

Selecting $R_{f0} = 10$ kΩ, we get $R_{f2} = 11.715$ kΩ, $R_{f3} = 26.155$ kΩ, $R_{f4} = 49.115$ kΩ, and for $K=1$, $R_{in} = 10$ kΩ. $\tag{13.63}$

The simulated response of the Butterworth BPF is shown in Figure 13.13(b). Cumulative effect of the gain (1.529) of the PRB results in the mid-band gain of unity. Center frequency is 2.492 kHz and obtained with a bandwidth of 123.44 kHz, Q_{BP} = 20.19 against the design value of 20. Rate of fall of the output corresponds to an eighth-order filter. Outputs at the other PRBs are also shown in Figure 13.13(b).

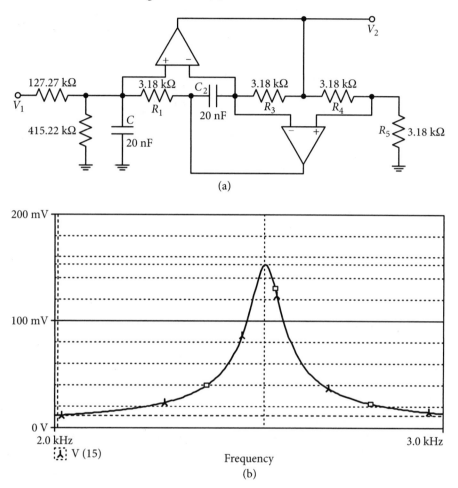

(a)

(b)

Figure 13.12 (a) Generalized impedance convertor based second-order PRB section for Example 13.5. (b) Its simulated magnitude response.

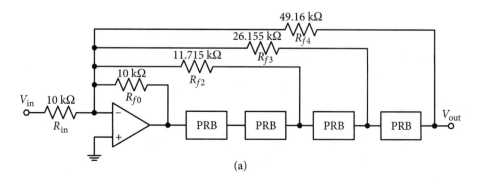

(a)

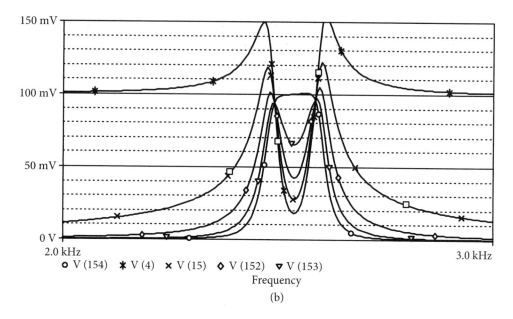

o V (154)   ✻ V (4)   × V (15)   ◇ V (152)   ▽ V (153)

Frequency

(b)

Figure 13.13 (a) Eighth-order band pass maximally flat filter using PRB of Figure 13.12(a) as the basic block for Example 13.5. (b) Simulated magnitude response.

Practice Problems

13-1   Design and test a fourth-order low pass filter with maximally flat characteristics having 1 dB attenuation at 1 kHz. Use the follow the leader feedback approach employing lossless integrators.

13-2   Repeat Problem 13-1 using integrators with loss. All capacitors should be below 10 nF.

13-3   Realize a fifth-order LP filter having Butterworth characteristics using the follow the leader feedback technique using lossless integrators. The filter is to have a cut-off frequency of 5 kHz.

13-4   Design the filter in Problem 13-3 using lossy integrators.

13-5   Realize a fifth-order LP filter having Chebyshev characteristics with ripple width of 2 dBs using the follow the leader feedback technique while employing lossless integrators. The filter is to have a cut-off frequency of 5 kHz.

13-6   Use shifted companion FLF structure to design a fourth order LP filter with maximally flat characteristic having 1 dB attenuation at 1 kHz.

13-7   Realize a notch filter in FLF mode with feed-forward technique for the transfer function:

$$H(s) = \frac{(s^2 + 0.25)}{(s^2 + 0.09s + 0.083)}$$

It is preferred that notch appears at 5 kHz. What is the peak value for the function and at what frequency it occurs?

13-8 Design a fourth-order Butterworth BPF using PRB technique. Its center frequency is 5 kHz and quality factor is 10. Use single amplifier biquad as the basic building block,

13-9 Repeat problem 13-8 using multi amplifier biquad as building block.

13-10 Apply a frequency de-normalization factor of 3.4 kHz for the following transfer function:

$$H(s) = \frac{0.274(s^2 + 2.41)}{(s + 0.635)(s^2 + 0.361s + 1.042)}$$

Design and test the realization using lossless or lossy integrators and feed-forward FLF structure approach.

Chapter 14

Switched Capacitor Circuits

14.1 Introduction

To meet the advances in technology, realization of monolithic electronic analog filters became an important issue. In the beginning, the problem of inductance realization was solved through its simulation using active RC circuits. However, integration needed the solution of a few more problems such that the realized filters perform as close to the design as possible. One of the main issues faced while integrating circuits was the amount of tolerance in passive elements; it was too large. This large tolerance resulted in large errors in the realized filter parameters. These errors could be reduced through electronic tuning of components or the tuning of filter parameters, with an obvious increase in circuit complexities. It was observed that the need of electronic tuning could be reduced through schemes in which filter parameters depend on the ratio of passive components, especially capacitors, as capacitor ratio tolerance is much better in integrated circuits. Switched capacitor filter (SCF) realization technique is one such scheme in which all filter parameters depend on capacitor ratios. Equally important is the fact that switched capacitor (SC) schemes do not use external resistors. This saved a large amount of chip area; a highly attractive feature for integrated circuits.

In this chapter, the basic concepts of SCF realization will be discussed. One important approach is the simulation of grounded and floating resistors using capacitors, switches, and non-overlapping clock signals. Such a realization and its simple applications including integrators are shown in Sections 14.2 and 14.3. As switches and clocks are the basic components of integrated circuits, these are further studied in Section 14.4. The rest of the chapter covers the topic of first- and second-order SCF sections.

14.2 Switched Capacitor Resistor

Let us examine a simple circuit as shown in Figure 14.1(a), containing two switches and a capacitor C_R. The port voltages are v_1 and v_2, and the two switches become ON and OFF continuously at a certain clock frequency f_c. This means that each switch can remain on for a maximum time duration $T_c = 1/2 f_c$. It is further assumed that either voltages v_1 or v_2 will remain constant during the small duration T_c when either of the switches is ON.

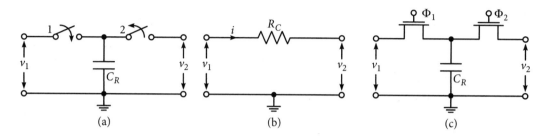

Figure 14.1 (a) Capacitor C_R switched alternately to voltages v_1 and v_2, (b) its equivalent resistance under certain conditions and (c) MOS based switches driven by a two-phase non-overlapping clock.

Initially, taking the switches as ideal, the capacitor C_R is first connected to port 1; as a result, on the kth switching cycle, the capacitor gets charged to a value $v_1(kT_c)$. Next, switch 1 is turned off and switch 2 is turned on for the half time period $T_c/2$. Now the capacitor C_R gets charged to $v_2(kT_c)$. Hence, the net charge transferred from port 1 to port 2 over the period T_c is given as:

$$\Delta Q = C_R\{v_1(kT_c) - v_2(kT_c)\} \tag{14.1}$$

This transfer of charge ΔQ takes place in a time period $\Delta T = T_c$. From the simple definition of current flow ($i = \Delta Q/\Delta T$), a current flows from port 1 to port 2 as:

$$i(kT_c) = \{C_R v_1(kT_c) - C_R v_2(kT_c)\}/T_c \tag{14.2}$$

If a resistance R_c is placed between port 1 and port 2 as shown in Figure 14.1(b), its current–voltage relation is simply:

$$i(kT_c) = \{v_1(kT_c) - v_2(kT_c)\}/R_c \tag{14.3}$$

Comparing equation (14.2) and (14.3), we get a very significant relation:

$$R_c = T_c/C_R = 1/f_c C_R \, \Omega \tag{14.4}$$

which implies that the switched capacitor circuit of Figure 14.1(a) behaves like an equivalent resistance depending on the value of the capacitance being switched OFF and ON, and the clock frequency f_c. It will be shown soon that the simulation of resistance using a capacitor and a clock has many advantages, but before that a few words of caution.

An important consideration is that Ohm's law applied in equation (14.3) is valid only when voltages v_1 and v_2 remain constant during the time interval T_c. One of its implication may be that f_c has to be so high that T_c is small enough and rate of variation in v_1 and v_2 is small as well. It is to be noted that, in practice, there are certain restrictions on f_c to be too large; though the conditions are not arbitrary as will be seen later.

Another important consideration is the non-idealness of the switches which are to be realized either using BJT (bipolar junction transistor), but mostly using MOS (metal–oxide semiconductors) transistors, as shown in Figure 14.1(c). The switches are shown to be operated by two non-overlapping clocks. The effect of the non-idealness of the switches on the performance of the switched capacitor circuits needs to be taken care of. Hence, a brief discussion on the working of a practical MOS switch is also important while realizing switched capacitor circuits. The reason for using two-phase non-overlapping clocks and the simple method of generating such a clock will also be discussed.

The active RC filter circuit's pole frequency ω_o depends on the RC product time constant. Because of larger tolerances in the absolute values of resistor and capacitor elements in integrated circuits, the practically obtained value of ω_o suffers from inaccuracy. If resistance is simulated as in Figure 14.1, time constant τ in an RC circuit will become:

$$\tau = R_c C = \frac{1}{f_c C_R} C = \frac{1}{f_c} \frac{C}{C_R} \tag{14.5}$$

This means that the time constant now depends on (i) a clock frequency, which can be realized very accurately, using a stable crystal-based oscillator and (ii) the ratio of capacitors which can be realized up to an accuracy better than 0.1%. As a result, in most non-critical situations, parameter tuning is not needed for ω_o.

Another significant advantage in SC circuits is the saving in chip area; resistance is not realized in the conventional form of metal/polysilicon strips, which may consume greater chip area. Moreover, capacitors simulating resistances are not realized in absolute value but in ratio form, as evident from equation (14.5). Hence, capacitors C and C_R can be as small as practically feasible resulting in saving in chip area.

14.3 Switched Capacitor Integrators

Figure 14.2(a) and (b) show an OA-RC integrator and its SC version with resistance being simulated using capacitor C_1. Figure 14.2(c) shows the wave forms of the clock signal for the two phases φ_1 and φ_2. For the circuit of Figure 14.2(a), the well-known voltage relation is:

$$V_{out}/V_{in} = -1/sC_2R_1 \tag{14.6}$$

For the SC version in Figure 14.2(b), application of equation (14.4) gives:

$$V_{out} = -\frac{I_2}{sC_2} = -\frac{1}{sC_2}\frac{V_{in}}{R_{C1}} \rightarrow \frac{V_{out}}{V_{in}} = -\frac{f_c C_1}{sC_2} \tag{14.7}$$

where f_c is the clock frequency and comparison of equation (14.7) with (14.6) shows that the switched capacitor C_1 represents an equivalent resistance R_{C1}.

Example 14.1: What shall be the output in the SC integrator of Figure 14.2(b) if R_1 = 500 kΩ and C_2 = 0.5 nF for input signal V_{in} = 5 sin $5\pi \times 10^3$.

Solution: If a clock frequency of 2×10^5 rad/s is used to simulate the resistance R_1 by switching a capacitance, equation (14.4) requires the value of the capacitance C_1 in Figure 14.2(b) to be:

$$C_1 = 1/2 \times 10^5 \times 5 \times 10^5 = 10 \text{pF}$$

If no parasitic capacitance is assumed in the switching act, use of equation (14.7) gives the peak-to-peak output voltage as:

$$V_{out}|_{pp} = -10 \frac{2 \times 10^5 \times 10^{-11}}{5\pi \times 10^3 \times 0.5 \times 10^{-9}} = -2.5454 \text{ volts}$$

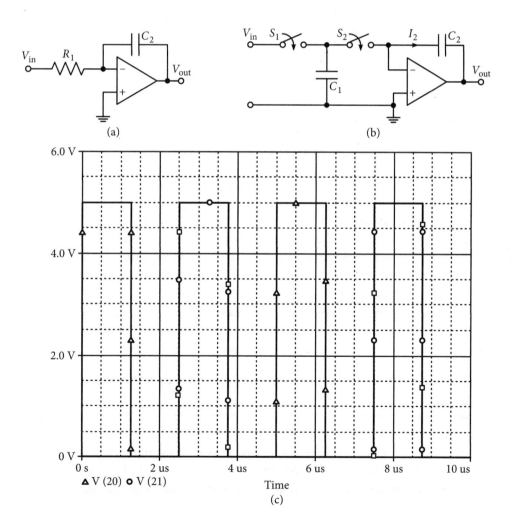

(a)

(b)

(c)

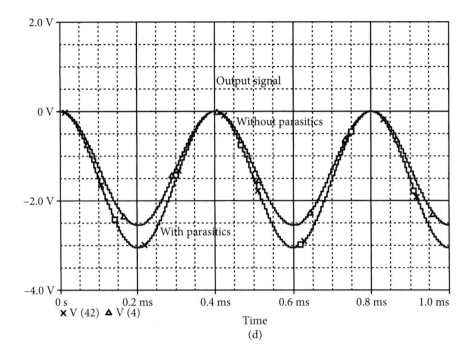

Figure 14.2 (a) Analog inverting integrator, (b) its switched capacitor version and (c) wave form of the non-overlapping clock signal applied to the switches. (d) Simulated output with and without the effect of parasitic capacitances for Example 14.1.

Figure 14.2(d) shows the PSpice simulated response, where peak-to-peak voltage of the output is 2.552 volts. It is common knowledge that there are parasitic capacitances when a MOS transistor is used for switching. The issue will be discussed later in the chapter; however, at present, if a parasitic capacitor of 2 nF is assumed to be present in parallel with C_1, the response will be affected and needs investigation. Figure 14.2(d) also shows that in such a case, peak-to-peak output voltage becomes 3.061 volts. It is an obvious effect of total switching capacitance becoming 12 nF. The effect of the parasitic capacitance and their elimination/minimization needs to be studied for accurate results in SC circuits.

The integrator circuit shown in Figure 14.2 can easily be modified to add and integrate two or more signals as shown in Figure 14.3(a)-(b). For Figure 14.3(a), an already known relation is:

$$V_{\text{out}} = -\frac{1}{sC_F}\left(G_1V_1 + G_2V_2 + G_3V_3\right) \tag{14.8}$$

For the SC version in Figure 14.3(b), the added currents of the inputs send more current in the form of a packet of charge to the capacitor C_F. In order to add the input currents, switches should become on and off at the same time. When these conditions are valid:

$$V_{out} = -\frac{I_F}{sC_F} \rightarrow V_{out} = -\frac{f_c}{sC_F}\left(C_1V_1 + C_2V_2 + C_3V_3\right) \tag{14.9}$$

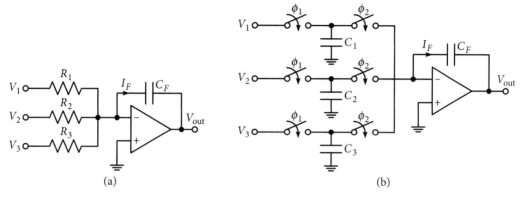

(a)   (b)

Figure 14.3 (a) Analog inverting summer, (b) its switched capacitor version.

For $C_1 = C_2 = C_3 = C_F$, input voltages will be added without any weightage during integration.

In an analog OA based circuit if a signal is to be subtracted, it needs to be inverted first. For example, in Figure 14.3(a) if V_1, V_3 were positive and V_2 was negative, the output would have been obviously:

$$V_{out} = -(G_1V_1 - G_2V_2 + G_3V_3)/sC_f \tag{14.10}$$

In the SC circuits, if an input signal is to be subtracted, it need not be inverted. Instead, it needs to be switched to the alternate phase compared to the other signals as shown in Figure 14.4. Obviously, now the current (charge) is sent to C_F in one direction from V_1 and V_2, but the current is sent in the reverse direction from V_3. Effectively, resistance R_1 and R_2 are simulated as positive resistances and R_3 as a negative resistance. It results in the following relation:

$$\left|V_{out}\right| = -\frac{f_c}{\omega}\left(\frac{V_1C_1}{C_F} + \frac{V_2C_2}{C_F} - \frac{V_3C_3}{C_F}\right) \tag{14.11}$$

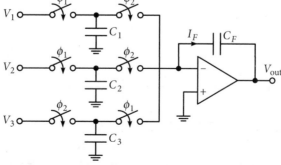

Figure 14.4 A signal (V_3) being subtracted while the other two are added in inverting summer.

Example 14.2: Using the SC inverting integrating summer of Figure 14.3(b), voltages $V_1 = 2\sin(5\pi \times 10^2\,t)$, $V_2 = 3\sin(5\pi \times 10^2\,t)$ and $V_3 = 1.0\sin(5\pi \times 10^2\,t)$ are to be added with respective multiplication factors of 1, 0.8 and 1.2. Find output voltage without any parasitic, and with a parasitic capacitance of 2 nF with each switch.

Solution: For input signals at 250 Hz, a clock frequency of 200 krad/s (arbitrary, but the same as in Example 14.1) is employed. With selected capacitance $C_F = 1$ nF, for the given multiplication factor of unity for the input signal V_1, the value of the capacitor C_1 is found using equation (14.9) as:

$$2 = \frac{2\times10^5}{5\pi \times 10^2 \times 10^{-9}}\left(C_1 \times 2\right) \rightarrow C_1 = 7.857\,\text{pF}$$

Similarly, for a gain of 0.8 for V_2 and a gain of 1.2 for V_3 values of the capacitors $C_2 = 6.285$ pF and $C_3 = 9.428$ pF. Using these elements, the simulated output is shown in Figure 14.5, where the peak-to-peak voltage is 11.34 volts; theoretically, $V\text{pp} = 11.19$ volts from equation (14.9).

If lump-sum parasitic capacitances (each of 2 pF) are also present at input, the effective input capacitances will increase and the output voltage will become:

$$V_{\text{pp}} = -\frac{2\times2\times10^5}{5\pi \times 10^2 \times 10^{-9}}\left(9.857 \times 2 + 8.285 \times 3 + 11.428 \times 1\right)\times10^{-12} = -14.25\,\text{volts}$$

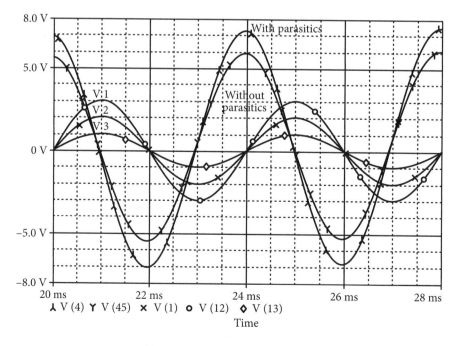

Figure 14.5 Simulated input and output voltages for the inverting summer in Example 14.2, with and without parasitic capacitors.

The simulated response in this case is also shown in Figure 14.5, where peak to peak output voltage is 14.31 volts confirming the predicted value.

14.4 MOS Switches

Some simple SC circuits were discussed in the previous sections assuming the switches to be ideal (and with lump sum arbitrarily connected parasitic). Before moving ahead to discuss first-order filter sections, it is important to briefly discuss some issues connected with the use of the MOS switch, its switching act and the effect of some of its important non-idealities.

14.4.1 Finite resistance

From its name *switched capacitor* circuits, it is evident that switches are the basic components along with capacitors. BJTs, MOS and CMOS transistors as transmission gates have been used widely as switches. Presently, we shall discuss only MOS transistor switches which are economical and also better suited for integration. Figure 14.6(a) shows a MOS transistor schematic and its conventional simple model.

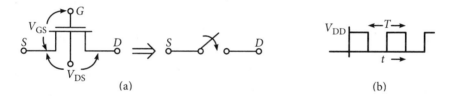

(a)   (b)

Figure 14.6 (a) MOS schematic and switch, (b) clock with 50% duty cycle.

For an ideal switch, its resistance should be zero when it is on and infinite when it is off. However, in practice, the resistance values depend on size and fabrication process of the transistor. In general, for a minimum size transistor, its ON resistance R_{on} ranges around 10^3 -10^4 Ω and when it is off, its resistance R_{off} ranges around 100–1000 MΩ.

14.4.2 Permissible clock frequency

As obvious from Figure 14.1, alternate charging and discharging of capacitors takes place in order to transfer net charge from one port to another. The equivalence of it is seen to be a measure of the total charge transfer per cycle. For practical implementation, it is essential to keep port voltages constant during respective half cycle. It represents the idea of *sample and hold* of the port voltages at a suitable clock frequency at which the capacitor (dis)charges. Frequency of the clock is important, as it sets the available half (or less; as shall be shown later) time period for charging of capacitors.

The practical time for charging and discharging depends on the on and off resistance of the switches and the involved capacitor. For example, if R_{on} = 5 kΩ, R_{off} = 500 MΩ and a capacitor C_R of 10nF is switched, then *on-time constant*, τ_{on} and *off-time constant*, τ_{off} will be respectively:

$$\tau_{on} = R_{on}\, C_R = 5 \times 10^3 \times 10 \times 10^{-12} = 12 \text{ ps} \tag{14.12a}$$

$$\tau_{off} = R_{off}\, C_R = 500 \times 10^6 \times 10 \times 10^{-12} = 5 \text{ ms} \tag{14.12b}$$

The ratio τ_{off}/τ_{on}, which is essentially R_{off}/R_{on} and generally known as the *discrimination factor* of the MOS switch is 10^5 or 100 dB in the aforementioned example. The larger the value of the discrimination factor, the closer it is to the ideal. The large discrimination factor enables the switched capacitor to acquire the charge through sampling, and holding it for relatively larger time, compared to the acquisition time of charge. Practically, it has been observed that in order to acquire a signal with sufficient accuracy during the half cycle of the clock period, the half cycle time duration should be greater than the time constant τ_{on}, that is, the following condition is required for the switch clock rate to get a sufficiently good output response.

$$0.5 T_c > 5\tau_{on} = 5R_{on}\, C_R \rightarrow T_c > 10\, R_{on}\, C_R \tag{14.13}$$

The Nyquist frequency definition states that in order to reconstruct a sampled signal, the sampling clock frequency has to be more than twice the frequency corresponding to that component of signal which has the maximum frequency, f_{max}. Only this condition provides a reconstructed signal without anti-aliasing. The condition, based on Nyquist criteria, helps in finding the lower limit of the switching frequency, whereas equation (14.13) mentions the condition of the upper limit of the clock frequency. Combining the two conditions gives the following relation governing clock frequency range.

$$2f_{max} \leq f_c \leq 1/5\, R_{on}\, C_R \tag{14.14}$$

14.4.3 Non-overlapping clock generation

It is essential that the clock pulses are non-overlapping with sufficient duration in which both the switches are off. This is considered as a *break-before-make* scheme in which switching in the next half-cycle starts only when it is ensured that capacitor C_R has been completely disconnected from the other port. If the two phases have 50% duty cycle, with no time when both the switches are off, it may create a serious problem. Even a small delay or advancement in one of the pulses creates specific time durations in which two switches become simultaneously on; this spoils the sequence of charging and (dis)charging of the capacitor.

Few circuits are available for generating non-overlapping two-phase clocks. One of these circuits is shown in Figure 14.7(a) employing two NOR, one inverter and a 50% duty cycle input clock φ_{in}. The input clock, its inverted output form delayed by one inverter delay φ_A and the final outputs φ_1 and φ_2 are shown in Figure 14.7(b). As each NOR has an input from the output of the other NOR, it ensures that output φ_1 and φ_2 will never be simultaneously on. In reality, the delay of the gates ensures a definite time gap between φ_1 becoming too high and φ_2 becoming too low, and vice versa.

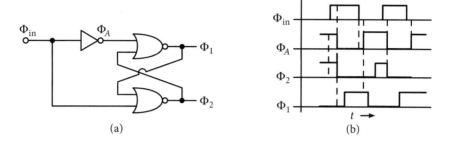

(a)   (b)

Figure 14.7   (a) Generation of a two-phase non-overlapping clock and (b) its waveforms.

14.4.4 Parasitic capacitances

One of the intended advantages of the SC circuits is the use of capacitors in ratio form with practically the smallest size capacitors, resulting in the consumption of as less chip area as possible. It is certainly a great advantage, but a potential threat to it comes from the fact that fabrication of capacitors itself generates quite a few number of parasitic capacitances. Hence, it is essential to not only know these parasitic elements, but account for them in such a way that their effect is either eliminated or minimized. Steps also have to be taken to assimilate the parasitic capacitances in the filter design.

There are many schemes of capacitor fabrication in integrated circuits; however double poly or poly-poly capacitors, which are realized using two polysilicon layers, with a thin silicon dioxide in between, are the most common approach. The intended capacitor to be realized is C_R which is fabricated between top and bottom poly layers and C_T and C_B are, top plate to substrate, and bottom plate to substrate parasitic capacitances, respectively, as shown in Figure 14.8(a). Here, C_B is generally 10–15% of the fabricated value of C_R and C_T is in the range of 1–2% of C_R. The behavior and effect of the parasitic capacitor depends on the type of fabrication, voltage level of the substrate depending on the kind of transistor used, NMOS (negative metal–oxide semiconductor) or PMOS (positive metal–oxide semiconductor), and whether C_R will be a floating, or a grounded capacitor in which the bottom plate terminal will be connected to ground. With V at ground level for NMOS, C_B becomes short circuited if C_R is a grounded capacitor; this is why grounded capacitances are preferred in active circuit designs.

There are other parasitic capacitances introduced due to switches and the input capacitance of OA used. Different techniques are used to counter the effect of all these parasitic capacitances either through the proper usage of switches or the formation of circuit structures as will be shown here and in the next section.

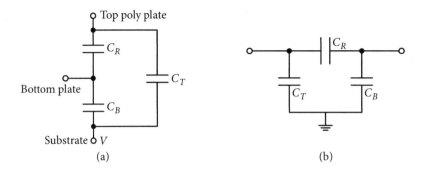

Figure 14.8 (a) Simplified view of important parasitic capacitances in a poly-poly capacitor; (b) its small signal equivalent C_R in floating form, and its parasitic.

For a good circuit design, effort is made such that charge flow in C_T and C_B becomes independent of the charge delivered by C_R. If C_R is to be used as a floating capacitor, Figure 14.8(a) will have to be modified to Figure 14.8(b) with the substrate at ground potential. To eliminate the effect of the parasitic capacitance, the symmetrically switched capacitor arrangement shown in Figure 14.9 can be used. The circuit uses four single-pole single-throw (SPST) switches with clock phases shown alongside. When φ_1 is on, φ_2 remains off, which makes C_R and C_T parallel, charging both of these to voltage v_1. At the same time, C_B gets short-circuited; hence, it is disconnected from C_R. In the next half cycle, φ_1 is off and φ_2 is on. Charge on C_T goes to ground, but C_R and C_B, which are in parallel get charged to v_2. Obviously, in the next half cycle, the charge on C_B goes to ground, making its presence also ineffective. The principle behind the operation is that the capacitor is switched to voltage sources in such a way that the parasitic conductances are parallel to the capacitor, thereby nullifying the effects of these parasitic conductances on filter performance.

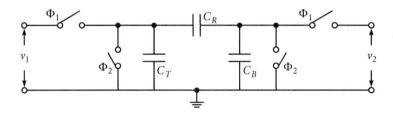

Figure 14.9 Symmetrically switched capacitor simulation of a resistor.

14.5 Parasitic Insensitive Integrators

Other than the parasitic capacitances of the switches, there are a few more parasitic capacitances and these are also to be accounted for. Figure 14.10(a) shows the conventional OA-RC integrator, in which the resistance is simulated using switched capacitor C_R. The circuit shows the parasitic capacitance of the capacitor C_R and the feedback capacitor C_F, parasitic capacitance of the switches C_S, and input capacitance of the OA, C_{OA}.

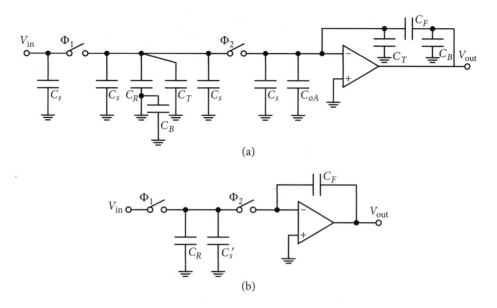

Figure 14.10 (a) Switched capacitor inverting integrator showing parasitic capacitances and (b) parasitic capacitances-accumulated simplified circuit.

As C_R is grounded, the corresponding C_B get short circuited. All parasitic capacitors connected to the inverting terminal of the OA also become ineffective as the terminal is at virtual ground. Switch capacitor C_s on the left of switch φ_1 and the bottom plate capacitance of C_F are driven by input and output voltages, respectively; they become ineffective. This means that the switched capacitor on the right of switch φ_1 and left of switch φ_2 and the top plate parasitic capacitance of C_R, combined together as one capacitance C_s' appear in parallel with C_R, as shown in Figure 14.10(b). This further means that the RC product of the integrator has changed and in comparison with equation (14.7), now the transfer function will become:

$$\frac{V_{out}}{V_{in}} = -f_c \frac{(C_R + C')}{sC_F} \tag{14.15}$$

If the value of C' was known correctly, it could have been included in the design. However, in addition to the value being probably incorrect, it is also quite variable. To eliminate the effect of C', the technique proposed in the previous section of applying symmetrically placed switches is used. Figure 14.11(a) and (b) show the application of symmetrical switching applied to the integrator of Figure 14.10, which results in a *parasitic insensitive* integrator in inverting and non-inverting mode, respectively. Practically, there is no difference in the operation of the circuit shown in Figure 14.11(a) with the circuit shown in Figure 14.11(b); both eliminate the parasitic C_1' and C_2' completely. Compared to the circuit shown in Figure 14.11 (a), in Figure 14.11(b), there is a difference in the ordering of the switching as shown. For φ_2 on and φ_1 off, C_R is charged by v_{in} but it is not connected with C_F, thus, no change in C_F takes place. Next for φ_2 off and φ_1 on, C_R is connected to C_F at the inverting terminal of the OA with its negative polarity; it gets discharged to ground. This negative discharging which equals $(-f_c C_R V_{in})$ flows into C_F and adds to its charge; this is opposite to what happens in the inverting integrator case. Hence, it can also be considered as though this kind of switching, simulated resistance becomes negative, $-R_R$. Therefore, the transfer function will be:

$$\frac{V_{out}}{V_{in}} = f_c \frac{C_R}{sC_F} \rightarrow -\frac{1}{sC_F(-R_R)} \qquad (14.16)$$

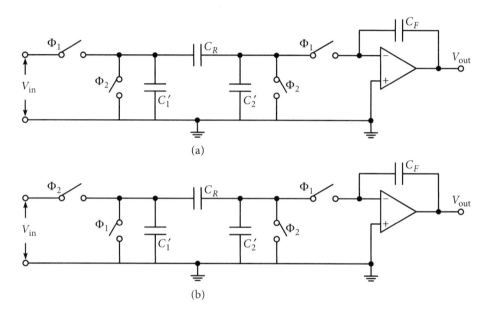

Figure 14.11 (a) Parasitic insensitive integrator using symmetrical switching in inverting mode and (b) in non-inverting mode.

Effectively, the transfer function in equation (14.16) realizes a non-inverting integrator. Hence, if there are more than one input signals, these can be added as in Figure 14.3. Moreover, subtraction can also be done by proper selection of switching. In Figure 14.3, if resistances R_1 and R_3 are simulated as positive resistances and R_2 is simulated as negative, it results in the following relation, with $s = j\omega$:

$$V_{out} = -\frac{f_c}{\omega}\left(\frac{V_1 C_1}{C_F} - \frac{V_2 C_2}{C_F} + \frac{V_3 C_3}{C_F}\right) \qquad (14.17)$$

In case of a simple subtraction without any weightage to the input voltages, $C_1 = C_2 = C_3$, and the output voltage expression becomes:

$$V_{\text{out}} = -\frac{f_c}{\omega} \frac{C_1}{C_F}\left(V_1 - V_2 + V_3\right) \tag{14.18}$$

Repeating Example 14.2 with V_2 becoming negative, the peak-to-peak output voltage from equation (14.17) becomes 1.61 volts. The circuit shown in Figure 14.12(a) was simulated, with intentionally introduced 2 nF parasitic capacitances as well at each switch location. Figure 14.12(b) shows the simulated response, with output as 1.55 volts p-p; the parasitic capacitances have no effect. It is to be noted that a high value resistance is connected in the feedback path; otherwise, the output voltage remains clamped to $+V_{\text{CC}}$.

14.6 Sampled Data Switched Capacitor Filters

While studying SC resistors in Section 14.2, it was assumed that the clock frequency is much greater than the signal frequency and the node voltages remain constant for a small duration (nearly half cycle) and then changes in small steps. This simply means that the signal is not continuous with time; therefore, its analysis as a purely analog circuit is not exactly correct. In reality, the correct procedure of analysis (and synthesis) is to use discrete-time equations rather than continuous ones. While using discrete-time equations, the frequency domain consideration is known as *sampled data characterization* and the frequency variable is also different. Instead of using Laplace transformation and complex frequency variable s, z-transform is used. A brief discussion about the relation between continuous-time domain and the sampled data characterization is now included with the help of the simple SC integrators discussed in Sections 14.3 and 14.5.

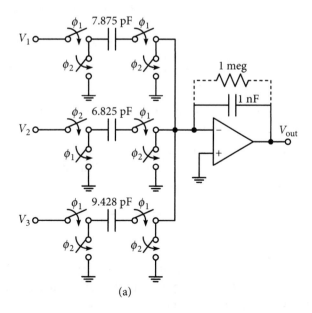

(a)

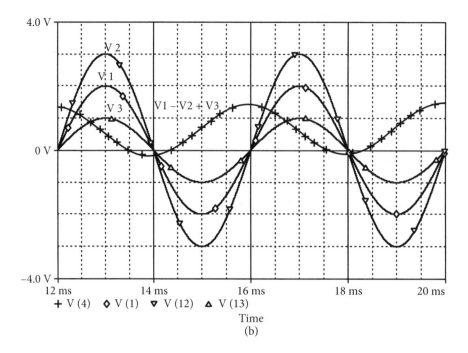

Figure 14.12 (a) Summing (and subtracting) switched capacitor integrator. (b) Simulated response in a parasitic insensitive subtractor using equation (14.18).

14.6.1 Sampled data integrators

The OA-RC inverting integrator and its SC version are shown in Figure 14.2(a) and (b) and the non-overlapping clock signal is shown in Figure 14.2(c). The clock signal φ_1 makes switches on at time instants $(n-1)T$, nT, $(n+1)T$, ... and the clock signal φ_2 makes switches on at time instants $\left(n-\frac{1}{2}\right), \left(n+\frac{1}{2}\right), \left(n+\frac{3}{2}\right)$, and so on. In contrast to the analog signal case, where the topology of the circuit remains unchanged with time, the topology of the circuit is distinct in two phases of the clock period in SC case. Figure 14.13(a) and (b) shows the topology of the inverting integrator during the two respective phases of the clock.

To convert the discrete-time domain relations into discrete frequency domain equations, the z-transform is to be applied to the two distinct sampling sequences, for phase φ_1 and for phase φ_2. One such transformations has been obtained as:

$$s \leftrightarrow \frac{1}{T}\frac{1-z^{-1}}{z^{-1/2}} \rightarrow s \leftrightarrow \frac{1}{T}(z^{1/2}-z^{-1/2}) \qquad (14.19)$$

Equation (14.19) is an important transformation relation between frequency variables s and z, and it is known as a *lossless digital integrator* (LDI) transformation. This transformation is obtained while considering phase 2 as the valid output. There are other transformations available based on the structure through which resistance is simulated in an SC circuit.

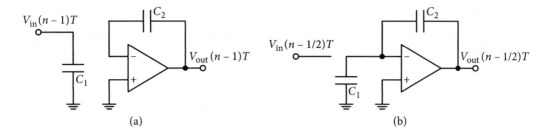

Figure 14.13 Inverting integrator equivalent circuits for the two respective non-over lapping clock inputs.

14.7 Bilinear Transformation: Warping Effect

As mentioned earlier, there are other *s to z* transformations. One of the most popular transformations is known as the *bilinear* transformation, which is given as:

$$s \leftrightarrow \frac{2}{T} \frac{1-z^{-1}}{1+z^{-1}} \tag{14.20}$$

The main reason behind the popularity of the transformation is that the entire analog frequency ranging from zero to infinity is transformed from zero to (π/T) in the sampled data frequency domain.

If we substitute $z = e^{j\Omega T}$ in equation (14.20), we get:

$$s \leftrightarrow \frac{2}{T} \frac{1-e^{-j\Omega T}}{1+z^{-j\Omega T}} \rightarrow \omega = \frac{2}{T} \tan \frac{T\Omega}{2} \tag{14.21}$$

For equation (14.21), it is observed that when $\omega = 0$, $\Omega = 0$ but when $\omega = \infty$, $\tan \dfrac{T\Omega}{2} = \infty$ implies $\Omega = (\pi/T)$. This also implies that when bilinear transformation is used while deriving $H(z)$ from $H(s)$, the response of $H(z)$ will get some what squeezed. This is known as the *warping effect* of the bilinear transformation.

The warping effect can be compensated by *pre-warping*. In an analog circuit, if ω_o is a critical frequency, like the center frequency of a BP filter, in the transfer function $H(s)$, then $H(s)$ is re-written with ω_o replaced by a pre-warped frequency ω_o^*, which is related as:

$$\omega_o^* = \frac{2}{T} \tan \frac{\omega_o T}{2} \tag{14.22}$$

Obviously, the *s* domain transfer function will change, which can be symbolized as $H(s^*)$. This operation results in the critical frequency Ω_o in $H(z)$ same as ω_o in $H(s)$:

$$\Omega_o = \frac{2}{T} \tan \frac{\omega_o^* T}{2} = \frac{2}{T} \tan^{-1} \left\{ \frac{T}{2} \frac{2}{T} \tan \frac{\omega_o T}{2} \right\} = \omega_o \tag{14.23}$$

The biquadratic transfer function expressions in the sampled data domain is given in Table 14.1 for the most commonly used analog transfer functions when the bilinear transfer function is applied with the following relations:

$$H(s) = N(s)/D(s), \quad D(s) = s^2 + (\omega_o/Q_o)s + \omega_o^2 \tag{14.24}$$

$$H(z) = h_D \frac{1 + a_{1N}z^{-1} + a_{2N}z^{-2}}{1 - a_{1D}z^{-1} + a_{2D}z^{-2}}, \quad D(z) = a^2 + (\omega_o^*/Q_o)a + \omega_o^{*2} \tag{14.25}$$

$$a = 2/T, \quad \omega_o^* = a\tan\left(\frac{\omega_o}{a}\right), \quad \omega_n^* = a\tan\left(\frac{\omega_n}{a}\right) \tag{14.26}$$

$$a_{1D} = 2(a^2 - \omega_o^{*2})/D(z), \quad a_{2D} = \left\{a^2 - (\omega_o^*/Q_o)a + \omega_o^{*2}\right\}/D(z) \tag{14.27}$$

Table 14.1 Sampled data transfer functions when bilinear transformation is applied to analog transfer functions

Filter type	Numerator $N(s)$ of the transfer function $H(s) = N(s)/D(s)$	$h_D \times D(z)$	a_{1N}	a_{2N}
LP	$H_{LP}\omega_o^2$	$H_{LP}\omega_o^{*2}$	2	1
HP	$H_{HP}s^2$	$H_{HP}a^2$	−2	1
BP	$H_{BP}(\omega_o/Q_o)s$	$H_{BP}(\omega_o^*/Q_o)a$	0	−1
AP	$H_{AP}\{1 - 2(\omega_o/Q_o)/D(s)\}$	$H_{AP}a_{2D}\,D(z)$	$-a_{1D}/a_{2D}$	$1/a_{2D}$
Notch	$H_N(s^2 + \omega_{on}^2)$	$H_N(a^2 + \omega_n^{*2})$	$-2\dfrac{(a^2 - \omega_n^{*2})}{(a^2 + \omega_n^{*2})}$	1

Example 14.3: Obtain the sample data transfer function for the following BP transfer function by applying bilinear transformation. Use a clock frequency of 16 kHz.

$$H(s) = \frac{1.5 \times 10^4 s}{s^2 + 1500s + 2.25 \times 10^8} \tag{14.28}$$

Solution: From equation (14.28), parameters are obtained as:

$$\omega_o = \sqrt{2.25} \times 10^4 = 1.5 \times 10^4 \,\text{rad}/\text{s}, \omega_o/Q_o = 1500 \rightarrow Q_o = 10, H_{BP} = 10$$

For clock frequency f_{CL} = 16 kHz, time period T = 62.5 µs. It gives pre-warped center frequency as:

$$\omega_o^* = 2f_{CL}\tan\left(\frac{\omega_o}{2f_{CL}}\right) = 2\times16*10^3\tan\left(1.5\times10^4\times180/32\times10^3\pi\right) = 16.19\,\text{krad}/\text{s}.$$

So, the pre-warped transfer function shall be:

$$H\left(s^*\right) = \frac{1.6197\times10^4 s^*}{s^{*2}+1.6197\times10^3 s^*+1.6197^2\times10^8} \tag{14.29}$$

From equation (14.24) to equation (14.27), parameters for the z-domain transfer function are evaluated as:

$a = 2/T = 32 * 10^3$

$D(z) = (32 * 10^3)^2 + 1.6197 * 10^3 * 32 * 10^3 + 1.6197 * 10^8 = 13.3817 * 10^8$

$a_{1D} = 2(32 * 10^6 - 2.6234 * 10^8)/D(z) = 1.13836$

$a_{2D} = \{(32 * 10^3)^2 - 1.6197 * 10^3 * 32 * 10^3 + 1.6197^2 * 10^8\}/D(z) = 0.9225$

$b_D = 10 * 1.6197 * 10^3 * 32 * 10^3/D(z) = 0.5183$

Therefore,

$$H\left(z\right) = \frac{0.5183(1-z^{-2})}{1-1.3836z^{-1}+0.9225z^{-2}} \tag{14.30}$$

It may be mentioned that $H(z)$ in equation (14.30) can also be obtained by applying the $s \leftrightarrow z$ transformation of equation (14.20) on the pre-warped analog transfer function of equation (14.29).

14.8 First-order Filters

Ideal integrators are first-order LP filters since the output starts dropping as soon as the frequency is finite. To have its cut-off frequency as finite, the integrators are made lossy through the simple addition of a resistance in the feedback path of the OA based integrator. Other first-order circuits are HP or bilinear ones. One of the simplest ways to obtain any of these first-order filters in SC form is to simulate the resistance, using switches and capacitors, in a corresponding OA-RC circuit. The parasitic insensitive approach is mostly used to eliminate errors. Hence, specific to first-order SCFs, no specific theory is involved. The following exercises will be helpful.

Example 14.4: Figure 14.14 shows an LP filter circuit. Design it for a 3-dB frequency of 3.4 kHz and dc gain of 5 dBs. Convert it as a first-order SC LPF while selecting a suitable clock frequency. Compare the output response of both the versions.

Solution: Both the resistors in Figure 14.14 are simulated as SC. The converted SC circuit is shown in Figure 14.15, for which the transfer function is obtained as:

$$H(s) = \frac{V_{out}}{V_{in}} = -\frac{1/C_F R_1}{s + 1/C_F R_2}$$

With

$$\omega_{3dB} = \frac{1}{C_F R_2}, \, \text{dc gain } H(0) = \frac{R_2}{R_1} \tag{14.31}$$

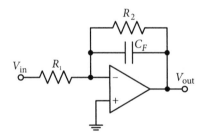

Figure 14.14   First-order OA-RC active low pass filter for Example 14.4.

For $\omega_{3dB} = 3.4(2\pi)$ krad/s, selecting $C_F = 10$ nF, the obtained value of $R_2 = 4.679$ kΩ and for the dc gain of 5 dB, $R_1 = 2.631$ kΩ. For these element values, its frequency response is now compared with that of the SC version. For converting the circuit to the SC form, capacitor values are preferably in the pF range. Hence, now the selection of $C_F = 20$ pF means the resistance values will become $R_2 = 2.3395$ MΩ and $R_1 = 1.3158$ MΩ. For the cut-off frequency of 3.4 kHz, we assume a maximum operating frequency of 34 kHz to get the whole range of the response spectrum. Since the clock frequency needs to be preferably 5 times the maximum operating frequency, the selected clock frequency $f_c = 180$ kHz. Now, if R_1 is simulated using capacitor C_1, its value will be $1/180 \times 10^3 \times 1.3158 \times 10^6 = 4.22$ pF and $C_2 = 2.374$ pF. For the SC version employing symmetrical switching shown in Figure 14.15, the time domain response is shown in Figure 14.16 for the OA-RC as well as the SC version. The respective voltage gains for both types of filters are shown in Table 14.2.

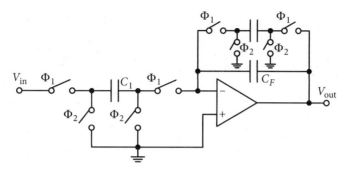

Figure 14.15   SC inverting integrator as first-order low pass filter from Figure 14.14.

Table 14.2   Voltage gains of the OA-RC and SC filters of Figures 14.14 and 14.15

Voltage gain at frequency (kHz)	0.20	3.4	5.0	10.0
OA-RC circuit	1.774	1.252	1.0	0.567
SC circuit	1.808	1.229	0.965	0.56

At 200 Hz, the respective gains are almost 5 dBs and 5.15 dBs, and at 3.4 kHz, the respective gains drop to 3.02 dBs and 3.35 dBs; this confirms their design cut-off frequency.

14.9 Simulation of SC Networks

Continuous-time active filters use PSpice and similar software for analysis. However, these simulation tools become too time-consuming for the analysis of SC networks when the MOS transistors are modeled as voltage-controlled switches. Therefore, one option is to simulate the SC network in time domain, as has been shown for Example 14.4.

In an alternate scheme, the s domain transfer function is first transformed into the z domain, and then the z domain transfer function is simulated in terms of delay lines and voltage-controlled voltage sources (VCVSs). The reason for the adoption of this technique lies in the following relation. From the z domain transfer function, the output signal $V_{out}(z)$ can be expressed as:

$$V_{out}(z) = k_1 V_i + k_2 V_i z^{-\left(\frac{1}{2}\right)} + k_3 V_i z^{-1} + \dots + A V_{o1} z^{-\left(\frac{1}{2}\right)} + B V_{o2} z^{-1} + C V_{o3} z^{-\left(\frac{3}{2}\right)} \quad (14.32)$$

Here, k_1, k_2, ...A, B, ... are constants, V_i, V_{o1}, V_{o2} ... are the z transformed input and output voltages, and $z^{-(1/2)}$, z^{-1} ... are delays by half-a-period, one full period, and so on. Obviously, the delay is represented by the transmission line with delay, and the constants are implemented either by VCVS or an amplification factor. The technique is illustrated by taking a specific case of Example 14.4.

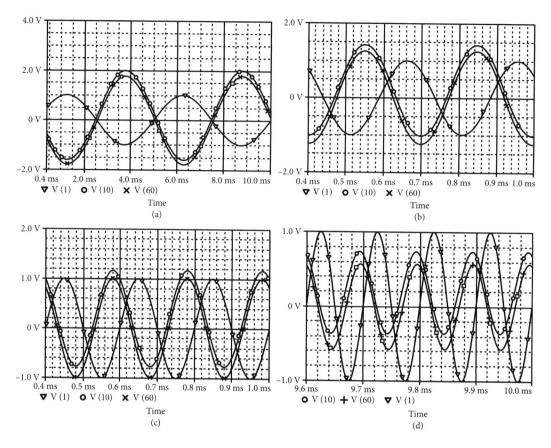

Figure 14.16 Time domain response of the OA-RC circuit of Figure 14.14 and the switched capacitor filter of Figure 14.15 at (a) 0.2 kHz, (b) 3.4 kHz, (c) 5 kHz and (d)10 kHz.

The transfer function of the first-order LPF given in equation (14.31) is re-written as:

$$H(s) = \frac{V_{out}}{V_{in}} = -\frac{1/C_F R_1}{s + 1/C_F R_2} = -\frac{a}{s + b} \tag{14.33}$$

Using bi-linear transformation equation (14.20), the z domain transfer function for $H(s)$ is obtained as:

$$H(z) = \frac{V_{out}(z)}{V_{in}(z)} = -\frac{a(1 + z^{-1})}{2f_s(1 - z^{-1}) + b(1 + z^{-1})} \tag{14.34}$$

It gives:

$$V_{out}(z) = -k_1 V_{in}(z) - k_3 V_{in}(z)z^{-1} + B V_{out}(z)z^{-1} \tag{14.35}$$

In equation (14.34), the clock frequency $f_s = 1/T$ and in equation (14.35):

$$k_1 = k_3 = \frac{a}{2f_s + b} \text{ and } B = \frac{2f_s - b}{2f_s + b} \tag{14.36}$$

Figure 14.17(a) shows the block diagram for the realization of equation (14.35) using basic delay, adder and multipliers. For simulating the block diagram of Figure 14.17(a), the expanded form is shown in Figure 14.17(b). For Figure 14.17 and for other simulations in the z domain, a systematic adopted procedure is outlined in the following.

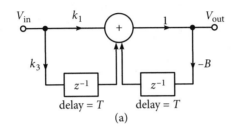

(a)

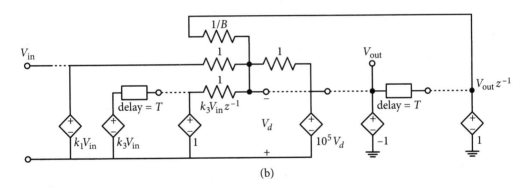

(b)

Figure 14.17 (a) Block diagram for realizing equation (14.35), employing inverting summer, delay lines and multipliers. (b) Expanded form of the block diagram shown in Figure 14.17(a).

The first step, which is already taken, is the formation of equation (14.35) for V_{out}, in which a feedback factor from the output will be a must. The next step is in the formation of basic units of half-a-delay $(T/2)$ isolated by a buffer (or inverter), because in some cases, factors of $(z^{-1/2})$ are also involved. Then, the delay blocks and the VCVSs for the implementation of buffers (or inverters) and gain (or attenuation) are appropriately joined together as suggested by basic equations like equation (14.35).

The above steps are completed for each OA sub-network in the original active RC network and proper connectivity is ensured in a network on the lines of Figure 14.17(b); the simulation is then performed.

Example 14.5: Simulate the first-order LPF of Example 14.4 using the delay line approach.

Solution: For the LPF of Example 14.4, the cut-off frequency was 3.4 kHz and the desired gain at dc was 5 dBs. Hence, comparison of equation (14.33) with equation (14.31) gives $a = 38$ krad/s and $b = 21.3714$ krad/s. Substitution of the values of a and b in equation (14.36) for the selected clock frequency $f_s = 32$ kHz, gives the following relation:

$$V_o = -0.4451V_{in} - 0.4451V_{in}z^{-1} + 0.4993V_oz^{-1} \tag{14.37}$$

For the implementation of equation (14.37) in the form of Figure 14.17(b), the netlist is shown in Figure 14.18. Since this realization does not require half the delay time, delay lines of only full delay time were used. The selected value of the clock frequency being 32 kHz, $T = 31.25$ μs. With $k_1 = k_3 = 0.4451$, and $B = 0.4993$, gain values of the amplifiers used and resistances employed are mentioned in Figure 14.18 and the PSpice simulated response is shown in Figure 14.19. This gives a dc gain of 1.778 and a cut-off frequency of 3.28 kHz (20.63 krad/s). A small difference between the simulated and the theoretical cut-off frequency is due to the warping effect of the BLT. To effectively overcome this problem, equation (14.26) is used to get the pre-warped values of constants a and b as:

$$b^* = 64 \times 10^3 \tan\left(\frac{3.4 \times 2\pi \times 10^3 \times 180}{64 \times 10^3 \times \pi}\right) = 22.193 \text{ krad / s and} \tag{14.38}$$

$$a^* = 1.778 \times b^* = 39.457 \text{ krad / s}$$

```
*FIRST-ORDER LOW PASS
          .SUBCKT DEL 1   2   4
     *CLOCK PERIOD     T=31.25US
            E11  3   0   1  2  1
    T1   3   0   4   0   ZO=50   TD=31.25US
            RT   4 0    50
             .ENDS   DEL
     *WITHOUT PRE-WARPING
        VS    1   0   AC   1
     E1    2   0   1   0   0.4451
     E2    10  0   1   0   0.4451
        X1    10   0   11   DEL
        E4    12   0   11   0 1
          R23    2   3   1K
          R133   12   3   1K
        E3    4   0   0   3   1E+5
          E5    6   0   4   0   -1
```

X3   6   0   7   DEL
R7   3   4   1K
E6   9   0   7   0   1
R93   9   3   2.0026K
.AC   DEC   100   10   100K
.PROBE
.END

Figure 14.18 Netlist for the first-order low pass filter of Example 14.5 without pre-warping.

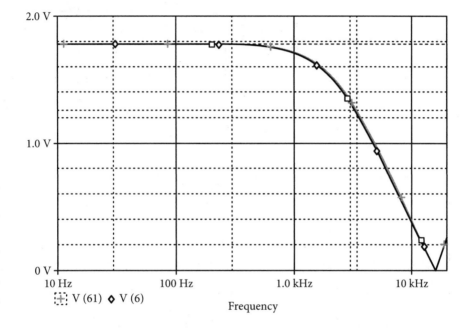

Figure 14.19 Simulated response of the first-order low pass filter for Example 14.5.

Use of the parameters in equation (14.38) gives new values of the coefficients as $k_1 = k_3 = 0.4581$ and $B = 0.4849$. PSpice simulation using these coefficients in the original netlist gives the simulated dc gain of 1.779 and cut-off frequency of 3.412 kHz, as designed is also shown in Figure 14.19.

Example 14.6: Simulate the second-order transfer function $H(z)$ in equation (14.30) using the delay line approach.

Solution: Transfer function $H(z)$ has been obtained for a BPF having the following specifications: $\omega_o = 15$ krad/s, $Q_o = 10$ and $H_{BP} = 10$. Equation (14.30) can be written as:

$$V_{out}(1 - 1.13836z^{-1} + 0.9225z^{-2}) = 0.5183(1 - z^{-2})V_{in} \qquad (14.39)$$

Comparing equation (14.39) with equation (14.36), it can be written as:

$$V_{out} = k_1 V_{in} - k_5 V_{in} z^{-2} + B V_{out} z^{-1} - D V_{out} z^{-2} \qquad (14.40)$$

In equation (14.40), $k_1 = -k_5 = 0.5183$, $B = 1.13836$ and $D = -0.9225$

Figure 14.20(a) shows a network implementing equation (14.40) for the realization of the BP function in terms of delay lines and amplifiers. The simulated response of the BPF is shown in Figure 14.20(b). The obtained center frequency is 2.343 kHz (14.72 krad/s), with a mid-band gain of 11.11 and a bandwidth of 243 Hz, resulting in pole-Q = 9.64.

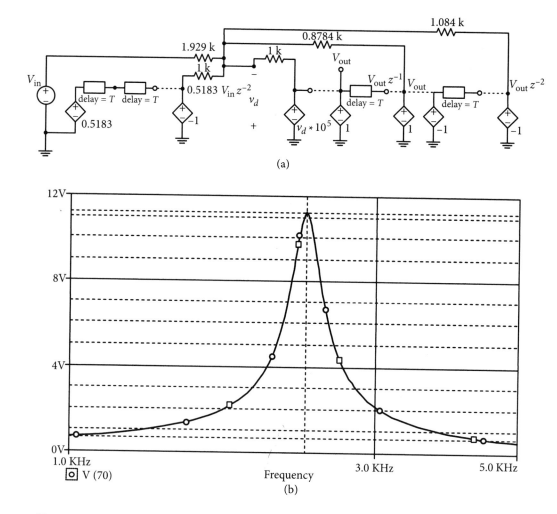

(a)

(b)

Figure 14.20 (a) Delay line-based simulation network for the second-order z domain transfer function in Example 14.6. (b) Simulated response in Example 14.6.

Example 14.7: Design and simulate a BPF with the following specifications:

$$\omega_o = 2\pi(3.4)\ \text{krad/s},\ Q = 5\ \text{and}\ h_{\text{obp}} = 10$$

Solution: The filter's transfer function in the s domain will be:

$$H_{BP}(s) = \frac{2\pi(6.8)10^3 s}{s^2 + 2\pi(0.68)10^3 s + \{2\pi(3.4)10^3\}^2} \tag{14.41}$$

Using equation (14.26), with the selected value of clock frequency $f_c = 16$ kHz, pre-warped center frequency will be:

$$\omega_o^* = 32000\tan\left(\frac{21371.4}{32000}\frac{180}{\pi}\right) = 25226.7\ \text{rad/s}. \tag{14.42}$$

With $Q = 5$, bandwidth will be $\omega_o^*/5 = 5045.3$ rad/s.

Using equations (14.24)–(14.27) and Table 14.1, z parameters are calculated as:

$$D(z) = \{32^2 + 32 \times 5.045 + 25.226^2\}10^6 = 1821.8 \times 10^6 \tag{14.43 a}$$

$$a_{1D} = \frac{2(32^2 - 25.226^2)10^6}{1821.8 \times 10^6} = 0.4255 \tag{14.43 b}$$

$$a_{2D} = \frac{1498.9 \times 10^6}{1821.8 \times 10^6} = 0.8227,\ h_D = \frac{1614.5 \times 10^6}{1821.8 \times 10^6} = 0.8862 \tag{14.43 c}$$

From parameter values in equation (14.43), z-domain transfer function becomes as:

$$\frac{V_{\text{out}}}{V_{\text{in}}} = \frac{0.8862(1 - z^{-2})}{1 - 0.4255z^{-1} + 0.8227z^{-2}} \tag{14.44}$$

Hence, the output voltage may be written as given below, where output voltage contains terms involving output voltage as well:

$$V_{\text{out}} = 0.8862(1 - z^{-1})V_{\text{in}} + 0.4255V_{\text{out}}z^{-1} - 0.8227V_{\text{out}}z^{-2} \tag{14.45}$$

Figure 14.21 (a) shows representation of equation (14.45) in terms of delay lines. PSpice simulated response is shown in Figure 14.21 (b), wherein center frequency is 3.324 kHz, $h_{\text{obp}} = 8.52$ and bandwidth is 568.8 Hz, resulting in $Q = 5.84$.

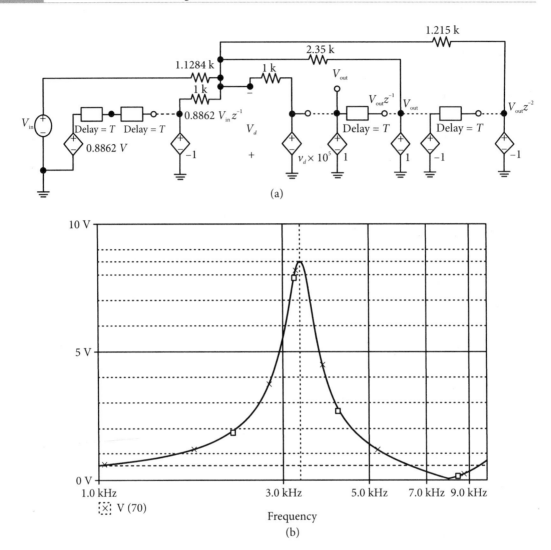

(a)

(b)

Figure 14.21 (a) Delay line-based simulation network for the second-order z-domain transfer function in Example 14.7, (b) simulated response of the network in Figure 14.21(a).

Practice Problems

14-1 An LPF is to be designed having a cut-off frequency of 250 Hz. Find the resistance values to be used, if the maximum value of the capacitor to be employed is 10 pF. Convert the active RC filter into a switched capacitor filter using a clock frequency of 256 kHz. What will be the size of the capacitor simulating the resistor?

14-2 Design a first-order ideal LP structure to get an output voltage of 1 V peak-to-peak, for a sine wave input having a peak value of 1 V at a frequency of 2.5 kHz. What will be the output voltage if a parasitic capacitance of 2 pF is present in parallel with the capacitor simulating the input resistance?

14-3   Design and test a switched capacitor inverting integrator for the given input voltages: $v_1 = 3 \sin(2\pi \times 10^3)$, $v_2 = 4 \sin(2\pi \times 10^3)$ and $v_3 = 2 \sin(2\pi \times 10^3)$. The respective weightage of the input voltages is (i) 1, 0.5 and 1.5 (ii) equal weightage.

14-4   Repeat Problem 14-3 if a 2pF parasitic capacitance is also present with each capacitor simulating the resistances.

14-5   Repeat Problem 14-3 for v_3 being subtracted instead of added to the inputs v_1 and v_2.

14-6   Repeat Problem 14-5 if a 2pF parasitic capacitance is also present with each capacitor simulating the resistances.

14-7   Design a first-order switched capacitor LPF to get the difference of two voltages v_1 and v_2. v_1 is to be multiplied by 3 dB and v_2 by 2 dB. The LPF should have a cutoff frequency of 1.59 kHz; assume clock frequency as 64 kHz.

14-8   Sketch the equivalent OA-RC circuit and find the transfer function for the circuit shown in Figure P-14.1. Element values in the OA-RC circuit are: $R_2 = R_3 = 50$ kΩ, $C_1 = 0$ and $C_4 = 2$ nF. Also find the transfer function of the equivalent discrete-time filter if the sampling frequency is 64 kHz.

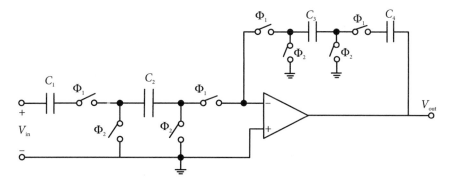

Figure P14.1

14-9   Repeat Problem 14-8 for $C_1 = 2$ nF.

14-10   Obtain the sample data transfer function for the following s domain function by applying bilinear transformation. Use a clock frequency of 32 kHz.

$H(s) = N(s)/D(s)$, $N(s) = 1.25 \times 10^4 s$, $D(s) = s^2 + 2500s + 6.25 \times 10^8$

14-11   Repeat Problem 14-10 for $N(s) = 1.25 s^2$.

14-12   Repeat Problem 14-10 for $N(s) = 1.25 \times 10^8$.

14-13   Repeat Problem 14-10 for $N(s) = 1.25(s^2 + 10^8)$.

14-14   Design the LPF of Figure 14.14 having a 3 dB frequency of 1.5 kHz and 6 dB gain at dc. Convert it to a switched capacitor form with a clock frequency of 128 kHz. Find the voltage gain of the OA-RC and the switched capacitor filters at 150 Hz, 1.25 kHz, 1.5 kHz, 1.75 kHz and 5 kHz.

14-15   Simulate the first-order switched capacitor LPF of Problem 14-14 using the delay line approach.

14-16 Simulate the first-order switched capacitor LPF of Problem 14-10 using the delay line approach.

14-17 Simulate the first-order switched capacitor LPF of Problem 14-11 using the delay line approach.

14-18 Simulate the first-order switched capacitor LPF of Problem 14-12 using the delay line approach.

14-19 Simulate the first-order switched capacitor LPF of Problem 14-13 using the delay line approach.

Chapter **15**

Operational Transconductance Amplifier-C Filters

15.1 Introduction

Operational transconductance amplifiers (OTAs) have emerged as a powerful alternative to OAs because of their current mode (CM) nature. Their CM nature allows OTAs to be used for much higher frequencies. Another major advantage of using OTAs is the electronic control it has over trans-conductance, which allows much easier tuning of filter parameters. The fact that only OTAs and capacitors are needed also makes the filter attractive for monolithic integrated circuit fabrication.

Since the development of OA-RC filter design is well established and extensively studied, it has been used to advantage while using OTAs. Filter design using OTAs follow a similar pattern and design procedures used in OA-RC synthesis; however, they are tailored to specific needs as will be shown in this chapter.

A brief review of the basic building blocks using OTAs is provided in Section 15.2. Simulation of grounded and floating resistors, inverting and non-inverting integrators, addition of voltages (and currents) and voltage amplifiers, for their stand-alone use or their application for realizing filter sections is included. First-order LPFs (low pass filter) and HPF (high pass filter) design is shown in Section 15.3 using integrators. Because of the suitability of OTA based circuits for differential outputs, the filter circuits discussed in Section 15.3 are then converted to differential mode. Similar to OA-RC synthesis, second-order filters are realized using the two-integrator loop method. Higher-order filters are realized using the *element substitution method* through the simulation of inductances and capacitances in Section 15.6 and 15.7. Filter realization using operational simulation is also discussed in Section 15.8.

15.2 Basic Building Blocks Using OTAs

OAs have been used for inverting and non-inverting amplification, summation (subtraction), integration and differentiation. In addition, while designing filters, OAs have been extensively used for simulating inductors and FDNRs (frequency dependant negative resistors). OTAs have also been used for all such entities; in addition, they are used for simulating grounded resistances (GRs) and floating resistances (FRs) as well. Hence, the resulting OTA-C circuits are not only realizable, but preferred as it makes them easily integratable in the monolithic form with more precise filter parameters.

To study and realize OTA based circuits, we will be using the circuit symbol, its ideal and first-order non-ideal model shown in Figure 15.1(a–c), where r_i, C_i and r_o, C_o are parasitic elements and g_m is trans-conductance.

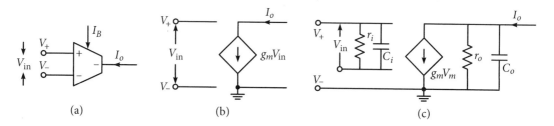

(a)   (b)   (c)

Figure 15.1 (a) Symbol of a single-ended OTA; (b) model of an ideal OTA; and (c) simple model of a non-ideal OTA.

15.2.1 Resistor simulation

The number of resistances used may not be large in a filter circuit, but these cannot be avoided and needs to be fabricated or simulated. It will be observed that the positive or negative resistor in the grounded or floating form can be easily simulated using a single OTA (or possibly two for a floating resistor) and the value of the resistor is controllable through the biasing current (or voltage) which changes trans-conductance (g_m) of the OTA. The external control of g_m is a great advantage in integrated circuits where it is not possible to make changes in elements after fabrication. Resistance realization through g_m saves considerable chip area as well compared to fabrication of large value resistors using diffusion or metal deposition. Another important advantage of such realizations is that they track with active devices.

In one of the simplest realizations, Figure 15.2 shows the simulation of a GR with one of its terminals connected to the ground. Assuming OTA as ideal for the circuit shown in Figure 15.2, the current–voltage relation is:

$$I_o = I_1 = (V_{in} - 0)g_m \rightarrow Z_{in} = (V_{in}/I_1) = 1/g_m \tag{15.1}$$

Meaning thereby that it realizes a resistor having a value $1/g_m$. If the input voltage is applied at the non-inverting terminal, the simulated resistance becomes negative, $-1/g_m$.

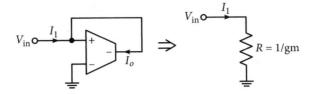

Figure 15.2 Grounded resistance simulation.

To realize an FR, a commonly used procedure of lifting a terminal off the ground in a GR can be used. The resulting circuit, using two such circuits in back-to-back form is shown in Figure 15.3. Once again, assuming ideal OTAs, currents are given as:

$$I_1 = I_{o1} = g_{m1}(V_1 - V_2), \quad I_2 = I_{o2} = g_{m2}(-V_1 + V_2) \tag{15.2}$$

Equation (15.2) yields the following admittance matrix, which represents a floating element.

$$[y] = \begin{bmatrix} g_{m1} & -g_{m1} \\ -g_{m2} & g_{m2} \end{bmatrix} = g_m \begin{bmatrix} 1 & -1 \\ -1 & 1 \end{bmatrix} \tag{15.3}$$

For $g_{m1} = g_{m2} = g_m$, the realized element is an FR having value $(1/g_m)$. An alternate scheme for simulating an FR using differential OTA will be taken up later.

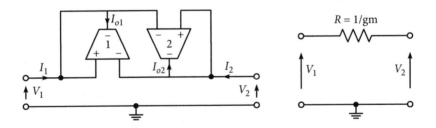

Figure 15.3 Simulation of floating resistance with two OTAs.

Simulation of GR and FR is highly advantageous as it converts active RC circuits to active C only and large value resistors are realized using only a small chip area.

15.2.2 Integrators

Integrators are required in both inverting and non-inverting mode; with or without loss. As the OTA output is current, it acts as an integrator when it terminates in a capacitor as shown in Figure 15.4(a). With OTA taken as ideal, the output voltage is easily obtained as:

$$V_{out} = -(I_o/sC) = -(V_+ - V_-)(g_m/sC) = -V_{in}(g_m/sC) \tag{15.4}$$

This means that the circuit integrates the input voltage, $V_{in} = (V_+ - V_-)$, without loss in the inverting mode. This basic integrator circuit leads very easily to a non-inverting integrator, with or without loss, and an inverting integrator with loss; a big advantage with circuits realized using OTAs. All the four structures are also available in differential output mode; an attractive and sometimes an essential feature as will be shown later.

If the input voltage V_{in} is applied at the inverting terminal with the non-inverting terminal grounded, the circuit realizes a lossless non-inverting integrator. In addition, either of the integrators become lossy if the terminating impedance contains a resistor in addition to the capacitor. Figure 15.4(b) shows such a configuration for a non-inverting integrator. The terminating resistor being simulated by OTA-2 has transconductance g_{m2}, and its transfer function is given as:

$$(V_{out}/V_{in}) = g_{m1}/(sC + g_{m2}) \tag{15.5}$$

Interchange between input terminals in either one of the OTAs will convert it to a lossy inverting integrator.

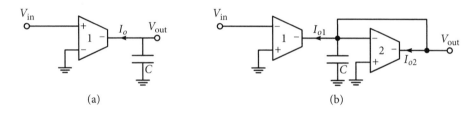

(a)   (b)

Figure 15.4 (a) Inverting ideal integrator and (b) non-ideal non-inverting integrator.

As of now, expressions have been derived without any parasitic element in the OTA model. Quite often it becomes essential to include the first-order parasitic because their value may not be negligible in comparison to the physical elements used. The commonly used OTA-3080, whose pin connection diagram is shown in Figure 15.5 and a first-order non-ideal model which was shown in Figure 15.1(c) has the following typical parameters when tested at bias current $I_B = 500$ μA and at full operating temperature: input resistance $r_i = 26$ kΩ, input capacitance $C_i = 3.6$ pF, output resistance $r_o = 1.5$ MΩ, and output capacitance $C_o = 5.6$ pF, forward transconductance $g_{mo} = 5400$ μS, and peak output current is 300 μA.

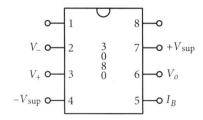

Figure 15.5 Pin connection diagram of the commonly available OTA-3080.

If the OTA is represented by its first-order model, the small signal equivalent circuit of the inverting integrator shown in Figure 15.4(a) becomes as shown in Figure 15.6; the input resistance is neglected as it is very high. The transfer function becomes:

$$\frac{V_o}{V_{in}} = -\frac{g_m}{s(C+C_o)+g_o} \tag{15.6}$$

Hence, the total effective load capacitance is increased by the parasitic output capacitance C_o and the output resistance ($r_o = 1/g_o$) makes it a bit lossy resulting in a 3 dB frequency $f_{3dB} = \{1/2\pi(C + C_o)r_o\}$, and a dc gain of $g_m r_o$.

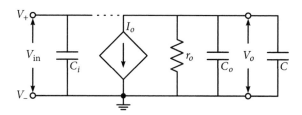

Figure 15.6 Small signal equivalent circuit of the OTA based integrator of Figure 15.4(a).

Figure 15.7(a) shows the small signal equivalent circuit of the lossy non-inverting integrator of Figure 15.4(b) with important parasitic elements. Here the current–voltage relations are:

$$I_{o1} = -g_{m1}\,V_{in},\; I_{o2} = g_{m2}\,V_o,\; V_o = -(I_{o1} + I_{o2})/(C + 2C_o + C_i)s + 2g_o \tag{15.7}$$

It gives the transfer function as:

$$\frac{V_o}{V_{in}} = \frac{g_{m1}}{(C+2C_o+C_i)s+g_{m2}+g_o} \tag{15.8}$$

Its 3 dB frequency and the dc gain expressions are obtained as:

$$f_o = (g_{m2} + g_o)/2\pi(C + 2C_o + C_i) \text{ and gain} = g_{m1}/(g_{m2} + 1/r_o) \tag{15.9}$$

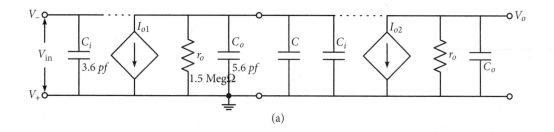

(a)

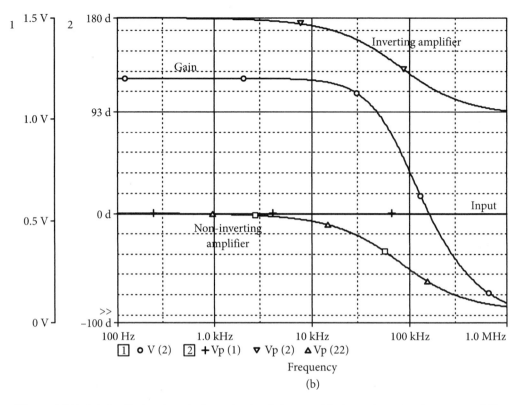

Figure 15.7 (a) Small signal equivalent circuit of the non-ideal non-inverting integrator of Figure 15.4(b). (b) Magnitude response of the non-inverting integrator of Example 15.1 and phase responses of the non-inverting and the inverting integrators, with input applied at the inverting and non-inverting terminals, respectively.

Example 15.1: Design a non-inverting integrator using the circuit shown in Figure 15.4(b), having a 3 dB frequency of 500 krad/s and a dc gain of 1.2.

Solution: Using the aforementioned parameters of the OTA and assuming g_{m2} = 4 mS, equation (15.9) representing equivalent circuit of Figure 14.4(a) with parasitics gives:

$$(C + 2C_o + C_i) = (g_{m2} + r_o)/2\pi f_o = (4 \times 10^{-3} + 1/1.5 \times 10^6)/(5 \times 10^5) = 8.0013 \text{ nF}$$

Substituting the values of the parasitic capacitances, the required value of the physical capacitance C = 7.9865 nF, and for the desired dc gain of 1.2, the required g_{m1} = 4.808 mS. Using these element values in the circuit shown in Figure 15.4(b), the magnitude response was simulated using PSpice.

Figure 15.7(b) shows the magnitude response as having a 3 dB frequency of 79.83 kHz (501.7 krad/s) and a dc gain is 1.2. Figure 15.7(b) also shows the phase response; there is no phase difference with the input at low frequencies, confirming that the circuit is a non-inverting integrator.

The non-inverting integrator can be easily converted into the inverting form by applying input at the non-inverting terminal. Its phase response, also shown in Figure 15.7(b), confirms the integrator as inverting, having a phase difference of 180° at low frequencies.

15.2.3 Current and voltage addition

As the output of the OTA is current, a number of output currents can be added (or subtracted) at a junction in a simple way. The same is the case with voltages; of course, through the means of output currents. For example, as shown in Figure 15.8(a), two voltages V_1 and V_3 can be added and V_2 subtracted therefrom, through the choice of input terminals. It is important to note that in this circuit, output current flows in a grounded resistor realized by OTA-4. Using the ideal model for the OTAs, we get the following relations.

$$I_{o1} = g_{m1}\ V_1,\ I_{o2} = -g_{m2}\ V_2,\ I_{o3} = g_{m3}\ V_3,\text{ and } I_{o4} = g_{m1}\ V_1 - g_{m2}\ V_2 + g_{m3}\ V_3 \qquad (15.10)$$

Addition (or subtraction) of currents is done easily, with weightage having unequal transconductance; without weightage, all transconductance are equal. With $I_{o4} = g_{m4} V_o$, we get:

$$V_o = \frac{g_{m1}}{g_{m4}} V_1 - \frac{g_{m2}}{g_{m4}} V_2 + \frac{g_{m3}}{g_{m4}} V_3 \qquad (15.11)$$

Equation (15.11) provides scaled algebraic addition of voltages with different weights. If all transconductance are made equal, the relation will simplify as:

$$V_o = V_1 - V_2 + V_3 \qquad (15.12)$$

It is obvious that the number of inputs can be increased easily.

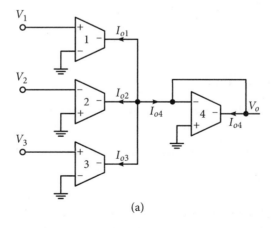

(a)

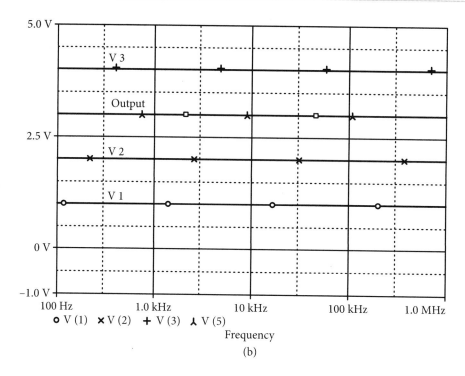

Figure 15.8 (a) Addition (and subtraction) of voltages through output currents. (b) Summing two voltages and subtracting the third voltage from it; input voltages have different weightages.

Example 15.2: Verify the summation circuit shown in Figure 15.8(a) for $V_1 = 1.0$ V, $V_2 = -2.0$ V and $V_3 = 4.0$ V, with a respective weightage of 1.0, 1.05 and 0.9.

Solution: If $g_{m4} = 4$ mS, for the given weightages, $g_{m1} = 4$ mS, $g_{m2} = 4.2$ mS and $g_{m3} = 3.6$ mS. With the respective voltages applied in the circuit shown in Figure 15.8(a), the output voltage becomes 2.5 V; confirmed in Figure 15.8(b). Performance can be repeated with equal weightage to input voltage, with all transconductance being equal. It can be observed that summation of voltages is valid at large frequencies as well.

15.2.4 Voltage amplifiers

If output current I_o of an OTA flows in a resistance, its output becomes a voltage amplifier. Figure 15.9 shows OTA-1 having transconductance g_{m1} loaded with OTA-2 forming a load resistor. The current–voltage relations are:

$$I_{o1} = g_{m1} V_{in}, I_{o2} = -I_{o1} \text{ and } V_{out} = (I_{o2}/g_{m2}), \text{ or } (V_{out}/V_{in}) = (-g_{m1}/g_{m2}) \qquad (15.13)$$

The circuit shown in Figure 15.9 realizes an inverting amplifier; the amplification is controllable by a single transconductance only. To get a non-inverting amplifier, only one of the input terminals needs to be swapped.

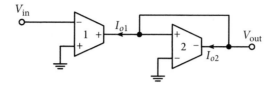

Figure 15.9 Inverting amplifier; easily convertible to non-inverting mode.

15.3 First-order Sections

The lossy integrator of Figure 15.4(b) acts as a first-order LP section. If the transconductance of the two OTAs is different, it provides the variable dc gain = g_{m1}/g_{m2}. If the variable gain is not essential, an alternate circuit shown in Figure 15.10(a) also serves as lossy integrator or a first-order LP section with the following relation:

$$V_{out} = -V_{in}\, g_m/(g_m + sC) \tag{15.14}$$

The non-inverting LP section is obtained by interchanging input and output terminals.

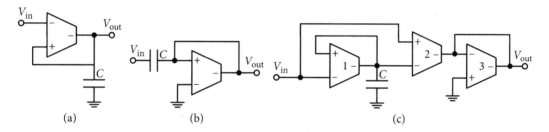

Figure 15.10 (a) First-order low pass section with unity dc gain. (b) First-order high pass section using a floating capacitor and (c) using a grounded capacitor.

The first-order HP section is obtained by either using a floating capacitor or a grounded capacitor as shown in Figure 15.10(b) and (c). Their respective transfer functions are:

$$\frac{V_{out}}{V_{in}} = \frac{sC}{(sC + g_m)} \tag{15.15a}$$

$$\frac{V_{out}}{V_{in}} = \frac{sC}{(sC + g_{m1})}\frac{g_{m2}}{g_{m3}} \tag{15.15b}$$

The first-order HP behaves like a lossy differentiator. A lossless differentiator cannot be realized directly as obvious from the expressions in equation (15.15). Hence, the technique shown in Figure 15.11 is applied, wherein a lossless integrator is connected in the feedback path of an amplifier. Its output voltage is obtained as:

$$V_o = -\frac{g_{m3}}{g_{m1}g_{m2}}sCV_{in} = -\left(\frac{sC}{g_m}\right)V_{in} \text{ for equal transconductance} \qquad (15.16)$$

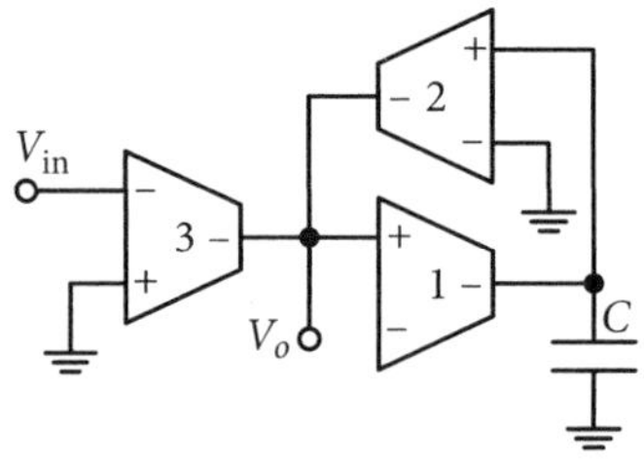

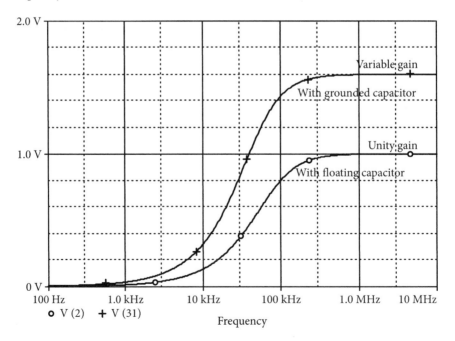

Figure 15.11 A lossless differentiator using feedback in an amplifier circuit.

Example 15.3: (a) Design an HP filter having a 3 dB frequency of 480 krad/s using the circuit shown in Figure 15.10(b).

(b) Design an HP filter using the circuit shown in Figure 15.10(c), with a 3 dB frequency of 50 kHz and a high frequency gain of 1.6.

Solution: (a) The circuit uses a floating capacitor and realizes a fixed gain of unity at high frequencies. From equation 15.15(a), if capacitor C is selected as 10 nF, the required $g_m = 4.8$ mS. Employing these elements, the simulated response of the circuit is shown in Figure 15.12; cut-off frequency is obtained as 76.45 kHz (480.54 krad/s).

Figure 15.12 Simulated response of first-order high pass filter with a high frequency gain of unity using a floating capacitor of Figure 15.10(c) and an OTA and a variable high frequency gain high pass filter using a grounded capacitor and three OTAs of Figure 15.11.

(b) From equation 15.15(b), for a selected value of capacitor C of 10 nF, required g_{m1} = 3.1428 mS, and for a high frequency gain of 1.6, with g_{m2} = 4 mS required, g_{m3} = 2.5 mS. Figure 15.12 shows the simulated response, having a cut-off frequency of 50.105 kHz and gain of 1.6; the response is not constrained to have only unity gain.

Example 15.4: Realize an APF (all pass filter) such that it has a phase shift of 90° at 200 krad/s.

Solution: A first-order APF section is shown in Figure 15.13(a), for which the transfer function is obtained as:

$$\frac{V_o}{V_{in}} = \frac{sC - g_{m1}}{sC + g_{m2}} \qquad (15.17)$$

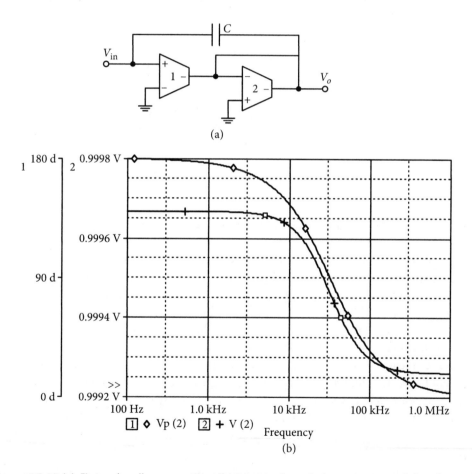

(a)

(b)

Figure 15.13 (a) First-order all pass section; (b) Magnitude and phase response of the all pass filter circuit.

Using non-ideal OTAs with the parasitic elements as mentioned in Section 15.2.2, an equal transconductance value of 4 mA/V is selected for realizing the APF. For the desired parameters required value of $C = 20$ nF. The circuit is simulated and its response in Figure 15.13(b) shows that the magnitude is almost unity, as its variation is only from 0.9997 V to 0.9993 V. Phase response is also shown in Figure 15.13(b), which starts dropping from 180° to zero and becomes 90° at 31.845 kHz (200.04 krad/s).

Figure 15.14 shows a general first-order section having the following transfer function with $C_1 = \alpha C$ and $C_2 = (1 - \alpha) \times C$.

$$\frac{V_{out}}{V_{in}} = \frac{s\alpha C + g_{m1}}{sC + g_{m2}} \qquad (15.18)$$

The previously discussed LP (low pass) section of Figure 15.4(b), the HP (high pass) section of Figure 15.11(a) and the AP (all pass) section of Figure 15.13 are easily shown as special cases of this general first-order section.

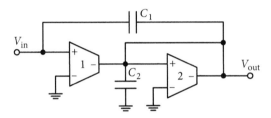

Figure 15.14 A general first-order section.

15.4 Single-ended to Differential Output Conversion

The advantages and requirements of using differential networks are well-known. These networks help in reducing common-mode noise considerably; though at the expense of more hardware. An added advantage in such circuits is the availability of the true output along with its inverted version; both the versions can possibly be used in the same network and elsewhere as well. Due to the desirability of fabricating linear filters on the same chip with digital circuits, the operating voltage of the filters had to be reduced which also requires reduction in the circuit noise. Hence, differential circuits became preferable and more common with OTAs. The reason being their inherent limited voltage–signal level capability, and more importantly, reduction in noise. Moreover, realization of a differential structure is simpler in OTA based networks compared to (say) OA based circuits.

For generating a differential output structure, a mirror-image circuit is combined with its original, in which each input and output sign is reversed; all non-essential circuitry is eliminated. Simultaneous change in the sign of the input and output does not affect the realized function. While combining the original and its mirror image, the input voltage is effectively doubled; this means that the output current is doubled. If a doubled output current flows in any capacitor, it doubles the voltage across it. To avoid this doubling of voltage, the capacitor value needs to be doubled. However, while adjustment of non-essential circuitry is done, a few capacitors can also be connected in differential mode; consequently, the total capacitance gets reduced. The following example will help illustrate the process.

Example 15.5: Develop a differential output, general first-order section from Figure 15.14.

Solution: Mirror image of the general first-order section of Figure 15.14 is drawn in Figure 15.15(a) with all input/output terminals interchanged; the input voltage becomes $-V_{in}$, and the output voltage becomes $-V_o$. They are combined with the original circuit of Figure 15.14, which results in Figure 15.15(b). The current–voltage relations of the circuit are:

$$I_{o1} = \{-V_{in} - (V_{in})\}\, g_{m1}, \quad I_{o2} = \{-V_o - (V_o)\} g_{m2} \tag{15.19}$$

As currents are doubled, charging of capacitors is also doubled; hence, doubling of capacitor values give:

$$I_{o1} + I_{o2} = 2sC_1\,(V_o - V_{in}) - 2sC_2\,V_o \tag{15.20}$$

Combining equations (15.19) and (15.20) with $C_1 = \alpha C$, $C_2 = (1 - \alpha)C$, we get:

$$\frac{V_o}{V_{in}} = \frac{sC\alpha + g_{m1}}{sC + g_{m2}} \tag{15.21}$$

As both the $2C_2$ capacitors are grounded, these can be connected in series, resulting in a single floating capacitor C_2 as shown in Figure 15.15(c). It gives an advantage of lesser overall capacitance from $4C_2$ to only C_2; but in floating mode.

With $\alpha = 1$, equation (15.21) gives an APF with unity gain. For the selected value of $g_{m1} = g_{m2} = 5$ mS and $C = 2$ nF, the circuit shown in Figure 15.15(c) is simulated, taking OTAs as non-ideal; $2C_1$ becomes 4 nF and C_2 is open circuited. Figure 15.15(d) shows that the magnitude is almost unity with variation from 0.9997 V to 0.9926 V only; the phase variations clearly depict the outputs in inverted and non-inverted forms.

An APF is to be designed with a low frequency gain of 1.1 and a high frequency gain of 0.5, with phase shift of 90° at 740 krad/s. For a high frequency gain of 0.5, $\alpha = 0.5$ from equation (15.21), and for a low frequency gain of 1.1, if $g_{m2} = 4$ mA/V, it will require $g_{m1} = 4.4$ mA/V. With these values of α and transconductance, $2C_1 = 4$ nF and floating capacitor $C_2 = 2$nF. The designed APF was simulated and its magnitude response shown in Figure 15.15(e) verifies the designed voltage gains at low and high frequencies. The simulated response in Figure 15.15(e) shows a phase variation of 90° for both the true and false outputs at 117.9 kHz (741.35 krad/s).

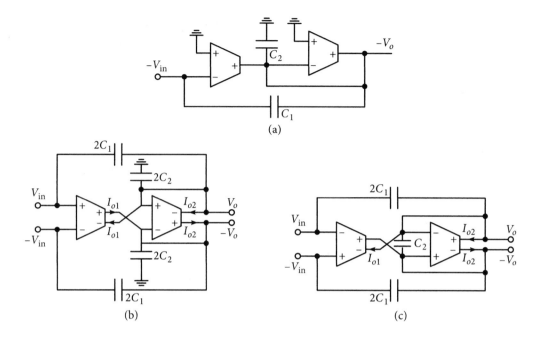

(a)

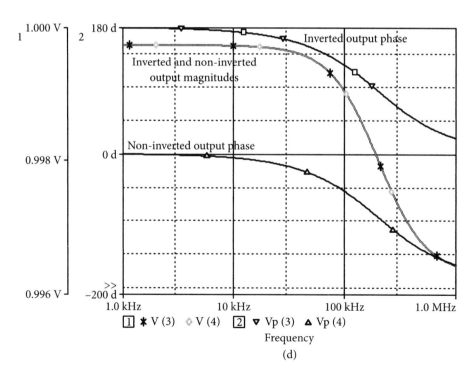

(b)

(c)

(d)

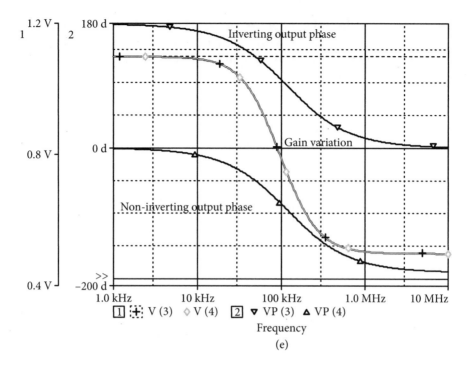

Figure 15.15 (a) Mirror image of the general first-order filter of Figure 15.14, (b) combination of Figures 15.14 and 15.15(a) with grounded capacitors $2C_2$, and (c) with floating capacitor C_2. (d) Magnitude and phase responses of the all pass filter using differential mode OTA in Figure 15.15(c) with C_2 open circuited. (e) Magnitude and phase responses of the all pass filter using the differential mode OTA shown in Figure 15.15(c).

15.4.1 Differential floating resistance and integrators

Floating resistance as simulated in Figure 15.3(c) using two single-output OTAs can also be realized using a single OTA with both outputs as shown in Figure 15.16(a). However, a differential resistance can be simulated as shown in Figure 15.16(b); which can become negative if the output terminals are interchanged.

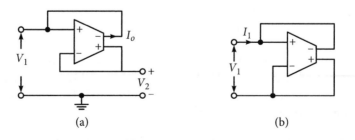

Figure 15.16 (a) Floating resistance simulation using single OTA and (b) a differential resistance simulation.

To obtain a differential inverting amplifier, we begin with a single-ended inverting amplifier as shown in Figure 15.9, where the output current of OTA-1 passes through a grounded resistor realized with OTA-2. Gain of the amplifier was obtained as

$$(V_o/V_{in}) = -(g_{m1}/g_{m2}) \tag{15.22}$$

The circuit shown in Figure 15.17 is obtained by combining a mirror image of the circuit in Figure 15.9(a) with the original, while taking care of the proper input sign; the generated circuit has the same gain as given in equation (15.22), but is differential in nature.

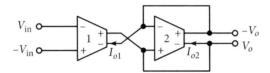

Figure 15.17 Inverting amplifier in differential mode.

In a similar way, the lossless integrator shown in Figure 15.4(a) will have a mirror image as shown in Figure 15.18(a), which can be combined with it. The resulting circuit will have either two grounded capacitors or one floating capacitor as shown in Figures 15.18(b) and 15.18(c), respectively. For Figure 15.18(a), the ratio of output to input voltage will be:

$$(-V_o/-V_{in}) = -(g_m/sC) \tag{15.23}$$

which is true for Figure 15.18(b) and (c) as well.

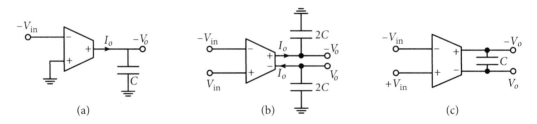

| (a) | (b) | (c) |

Figure 15.18 (a) Mirror image of the lossless integrator shown in Figure 15.4(a); (b) differential lossless integrator with two grounded capacitors and alternatively; (c) with one floating capacitor.

The lossy non-inverting integrator of Figure 15.4(b) can be converted into differential forms, with two grounded capacitors, and another with one floating capacitor, as shown in Figures 15.19(a) and (b), respectively. The corresponding transfer functions are the same as before in equation (15.5).

Example 15.6: Design a first-order LPF using the differential mode non-inverting amplifier shown in Figure 15.19(b) with the following specifications: cut-off frequency of 100 kHz with a dc gain of 1.6.

Solution: From equation (15.5), to get a dc gain of 1.6, if $g_{m2} = 2.5$ mA/V, we need $g_{m1} = 4$ mA/V, and for a cut-off frequency of 100 kHz, value of the capacitor C will be $= 2.5 \times 10^{-3}$ $/2 \times 2 \times \pi \times 10^5 = 1.9886$ nF (floating capacitance in the circuit shown in Figure 15.19(b) is halved). The circuit is simulated and the responses are shown in Figure 15.19(c). The simulated dc gain is 1.6 and the cut-off frequency is 98.945 kHz. The small deviation in the cut-off frequency is due to the parasitic components as effective value of capacitance increases due to C_i, C_o and other capacitances. Phase response in Figure 15.19(c) shows no phase shift in the output V_o at low frequency and a phase shift of 180° for $-V_o$.

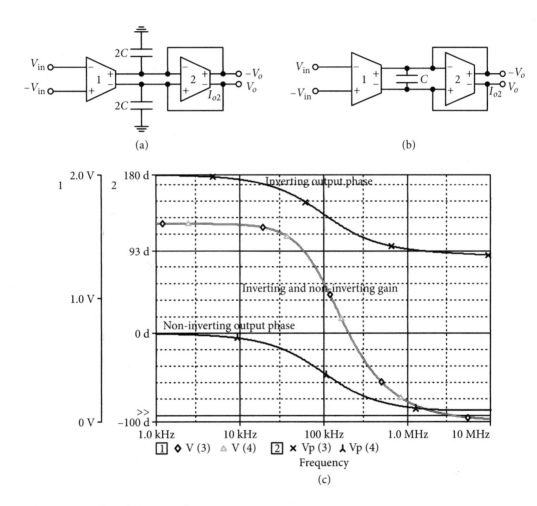

(a)

(b)

(c)

Figure 15.19 Combination of the non-ideal, non-inverting integrator shown in Figure 15.4(b) and its mirror image with (a) two grounded capacitors and (b) one floating capacitor. (c) Magnitude and phase responses of the non-inverting lossy integrator as a low pass filter from Figure 15.19(b).

15.5 Second-order OTA-C Filters

There are different approaches through which second-order OTA-C filters are obtained. Most of these are based on the same techniques as those in the active RC synthesis using OAs. Since alternatives are available for any given filter specifications, there are certain criterion on the basis of which a particular structure or approach is preferred. Apart from the sensitivity considerations, a few other important considerations are taken into account with respect to the spread and value of capacitors and OTAs' transconductance, economy in terms of OTAs used, flexibility in obtaining as many types of responses from a single circuit and ease in tuning or programmability of the filter parameters.

15.5.1 Two-integrator loop biquads

Out of the many second-order OA-RC sections, two-integrator loop sections are quite attractive because of their low sensitivity to passive elements. A circuit based on OA-RC Tow–Thomas biquad of Figure 8.2 is shown in Figure 15.20(a). Its analysis gives the following transfer function

$$\frac{V_{o1}}{V_{in}} = \frac{N(s)}{D(s)} = \frac{-(g_{m1}g_{m4}/C_1C_2)}{s^2 + (g_{m3}/C_1)s + (g_{m2}g_{m4}/C_1C_2)} \tag{15.24}$$

$$(V_{o2}/V_{in}) = -(g_{m1}/C_1)s/D(s) \tag{15.25}$$

The circuit provides LP and BP responses having parameters as:

$$\omega_o = \sqrt{g_{m2}g_{m4}/C_1C_2}, \ Q = (1/g_{m3})\sqrt{\{(g_{m2}g_{m4})(C_1/C_2)\}} \tag{15.26}$$

Mid-band gain of the BP section $= g_{m1}/g_{m3}$ \hfill (15.27a)

Low frequency gain of the LPF $= g_{m1}/g_{m2}$ \hfill (15.27b)

Hence, frequency ω_o can be tuned either by g_{m2} and/or g_{m4} and then Q can be changed through g_{m3} without effecting ω_o. The mid-band gain of the BP, or dc gain of the LP will be the last to be controlled through g_{m1}.

Example 15.7: Design a BPF, using the two-integrator loop circuit shown in Figure 15.20(a) having a center frequency of 400 krad/s, pole-Q of 2.5 and mid-band gain of 3.0.

Solution: Assuming a suitable value of 10 nF for both the capacitors C_1 and C_2, and with g_{m2} $= g_{m4}$, equation (15.26) gives their values as:

$$g_{m2} = g_{m4} = \left\{ C_1 C_2 \omega_o^2 \right\}^{0.5} = \left\{ \left(10^{-8} \right)^2 \times \left(4 \times 10^5 \right)^2 \right\}^{0.5} = 4 \, \text{mA} / \text{V} \qquad (15.28a)$$

$$g_{m3} = \left(g_{m2} g_{m4} \right)^{0.5} / Q = 4 \times 10^{-3} / 2.5 = 1.6 \, \text{mA} / \text{V} \qquad (15.28b)$$

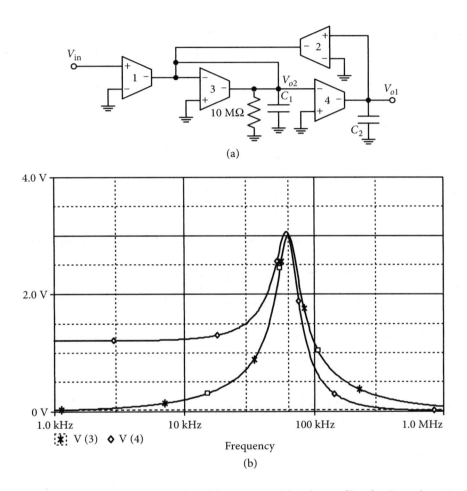

(a)

(b)

Figure 15.20 (a) Two-integrator loop-based low pass and band pass filter for Example 15.7; (b) its simulated magnitude response.

and from equation (15.27), for a mid-band gain of 3 and for the BP response

$$g_{m1} = 3 \times g_{m3} = 4.8 \, \text{mA/V} \qquad (15.28c)$$

The circuit shown in Figure 15.20(a) was simulated using the calculated elements and the magnitude response is shown in Figure 15.20(b). Center frequency of the BP is obtained as

63.407 kHz (398.56 krad/s) and mid-band gain is 3.0017. With a bandwidth of 25.331 kHz, pole-Q = 2.503. Gain of the LP at low frequencies is 1.2 and a peak gain of 3.065 occurs at f_{peak} of 61.108 kHz. From equation (2.46), center frequency is again obtainable as:

$$f_o = f_{peak}/\{1 - 1/2(Q_o)^2\}^{0.5} = 61.108/\{1 - 1/2 \times 6.25\}^{0.5} = 63.07 \text{ kHz}$$

KHN Biquadratic Structure: In addition to the LP and the BP responses, if it is desired to get HP and notch responses from the same circuit, OTA-C based KHN (Kerwin-Huelsman–Newcomb) biquad structures can be used which is shown in Figure 15.21(a); the circuit is derived from the one shown in Figure 8.1/8.2. Analysis shows the following relations.

$$\frac{V_o}{V_{in}} = \frac{1}{D(s)} \frac{g_{m1}\, g_{m5}\, g_{m6}\, (g_{m3} + g_{m4})}{g_{m4}(g_{m1} + g_{m2})C_1 C_2} \tag{15.29a}$$

$$D(s) = s^2 + \left\{\frac{g_{m2}\, g_{m5}\,(g_{m3} + g_{m4})}{g_{m4}(g_{m1} + g_{m2})C_1}\right\} s + \frac{g_{m3}\, g_{m5}\, g_{m6}}{g_{m4}C_1 C_2} \tag{15.29b}$$

Output voltage V_o gives LP response as shown in equation (15.29). Its parameters will be:

$$\omega_o = \sqrt{\left(\frac{g_{m3}\, g_{m5}\, g_{m6}}{g_{m4}C_1 C_2}\right)}, \; Q = \frac{(g_{m1} + g_{m2})}{g_{m2}(g_{m3} + g_{m4})} \sqrt{\left(\frac{g_{m3}\, g_{m4}\, g_{m6}C_1}{g_{m5}C_2}\right)} \tag{15.30}$$

With the same expressions for ω_o and Q, voltages V_2, V_4 and V_5 provide a notch, HP and BP responses respectively, as shown by the following expressions.

$$V_2 = \frac{1}{D(s)} \frac{g_{m1}}{(g_{m1} + g_{m2})}\left(s^2 + \frac{g_{m3}\, g_{m5}\, g_{m6}}{g_{m4}C_1 C_2}\right) \tag{15.31}$$

$$V_4 = \frac{1}{D(s)} \frac{g_{m1}(g_{m3} + g_{m4})}{g_{m4}(g_{m1} + g_{m2})} s^2 \tag{15.32}$$

$$V_5 = \frac{1}{D(s)} \frac{g_{m1}g_{m5}(g_{m3} + g_{m4})}{g_{m4}(g_{m1} + g_{m2})} \frac{s}{C_1} \tag{15.33}$$

In equation (15.31), poles and zeroes of the filter are the same for the notch response. If a general notch and AP responses are to be obtained, the process discussed in Chapter 8 can be followed. Output voltages V_o, V_2 and V_4 are to be added employing an adder circuit like that in Figure 8.8(a).

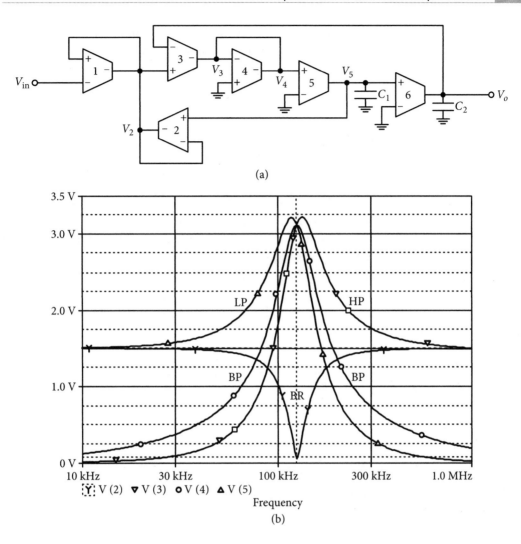

Figure 15.21 (a) KHN biquadratic configuration using six OTAs and (b) magnitude response of low pass, band pass, high pass and notch functions.

Example 15.8: Design the KHN based biquad shown in Figure 15.21(a) having the following specifications: center frequency = 800 krad/s, pole-Q = 2 and notch frequency will also be 800 krad/s.

Solution: As there are six OTAs and two capacitors, a few assumptions can be made in the beginning.

Let $g_{m3} = g_{m4} = g_{m5}$ and $C_1 = C_2 = 1$ nF, then from equation (15.30):

$$g_{m3} = \omega_o\, C_1 = 8 \times 10^5 \times 10^{-9} = 0.8\text{mA/V} = g_{m4} = g_{m5}$$

To get the notch frequency same as the center frequency, from equation (15.31), g_{m6} will also be 0.8 mA/V and value of the other transconductance will be obtained as:

$$Q = (1 + g_{m1}/g_{m2}) \times 0.5 \to g_{m2} = 1\text{mA/V}, \ g_{m1} = 3 \text{ mA/V}$$

With these elements, the circuit was simulated and the magnitude response is as shown in Figure 15.21(b); following are the observations:

The voltage V_o provides notch, which occurs at 126.005 kHz;

The voltage V_3 gives an LP response having a peak gain of 3.213 at 117.877 kHz;

The voltage V_4 gives a HP response which has a peak at 134.453 kHz with a peak gain of 3.221;

The voltage V_5 provides a BP response with a center frequency of 123.839 (778.4 krad/s) and a bandwidth of 60.289 kHz, resulting in $Q = 2.054$. Difference in the peaking of LP, HP or BP is due to $Q = 2$, in conformity with equation (2.46).

A two-integrator loop structure can be used in a different way, with more input voltages applied to some OTA terminals which were earlier grounded. For example, as shown in Figure 15.22(a), five input voltages are applied and outputs are taken as V_{o1} and V_{o2} having the following expressions.

$$V_{o1} \ D(s) \ C_1 \ C_2 = C_1 \ C_2 \ V_2 \ s^2 + (g_{m2} \ C_1 \ V_2 + g_{m3} \ C_2 \ V_3 + g_{m1} \ C_2 \ V_4 + g_{m1} \ C_2 \ V_1)s$$
$$+ (g_{m1} \ g_{m2} \ V_1 + g_{m2} \ g_{m3} \ V_2 + g_{m1} \ g_{m4} \ V_5) \tag{15.34a}$$

$$V_{o2} \ D(s) \ C_1 \ C_2 = C_1 \ C_2 \ V_4 \ s^2 + (g_{m4} \ C_1 \ V_5 + g_{m2} \ C_2 \ V_2)s + (g_{m1} \ g_{m2} \ V_1 + g_{m2} \ g_{m3} \ V_3) \tag{15.34b}$$

$$D(s) = s^2 + (g_{m2}/C_2)s + (g_{m1} \ g_{m2})/(C_1 \ C_2) \tag{15.34c}$$

It is obvious that a suitable choice of input voltages and transconductance will provide all general type of transfer functions.

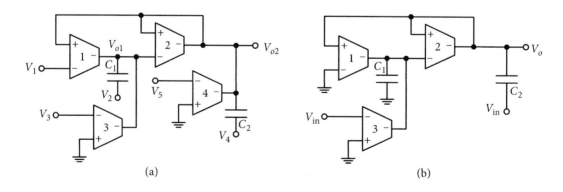

(a)   (b)

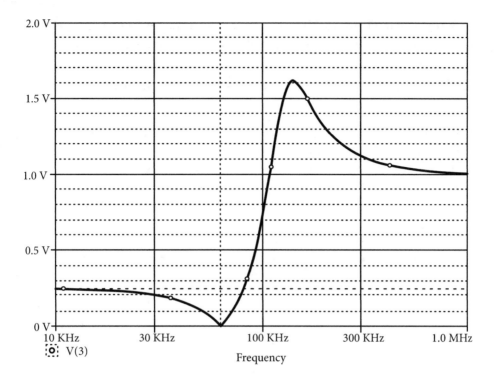

Frequency

Figure 15.22 (a) Two-integrator loop-based biquad with multiple inputs, (b) modified form to get a high pass notch response. (c) A high pass notch from a general two-integrators based biquadratic section with three OTAs for Example 15.9.

Example 15.9: Obtain an HP notch at 400 krad/s using the circuit shown in Figure 15.22(a) with pole frequency at 800 krad/s and pole-Q = 2.

Solution: To get a HP notch, from equation (15.34b), $V_4 = V_3$, $V_2 = V_5 = 0$ and V_1 can also be zero. For $g_{m1} = g_{m2}$ and selecting $C_1 \times C_2 = 4(\text{nF})^2$. It gives:

$$g_{m1} = 8 \times 10^5 \times 2 \times 10^{-9} = 1.6 \text{ m A/V} = g_{m2} \tag{15.35 a}$$

Equation (15.34c) is used to find values of capacitors C_1 and C_2 with Q = 2:

$$Q = \left(\frac{g_{m1}C_2}{g_{m2}C_1}\right)^{0.5} = 2 \rightarrow C_2 = 4C_1 \text{ or for } C_1 = 1 \text{ nF, } C_2 = 4 \text{ nF} \tag{15.35b}$$

Combining equations (15.34b) and (15.34c), for $\omega_z = \omega_o/2$, $g_{m3} = g_{m2}/4 = 0.4$ m A/V and V_3 as input, substitution of these element values and input voltages in the circuit shown in Figure 15.22(a) modifies it to Figure 15.22(b), which is simulated using PSpice. The simulated magnitude response in Figure 15.22(c) shows notch frequency at 63.61 kHz (399.83 krad/s), gain at low frequencies of 0.25 and gain at high frequencies reaching near unity.

15.5.2 OTA-C biquads derived from active RC circuits

Resistors can directly be replaced in OA-RC circuits using the circuits shown in Figure 15.2 and 15.3 for grounded and floating resistor, respectively, resulting in mixed OA-OTA-C circuits. In the case of floating resistance, transconductance g_{m1} and g_{m2} needs to be matched; the circuit shown in Figure 15.16(a) can also be used for floating resistance saving one OTA. Fortunately, many a times, the RC combination of integration can be replaced by the simple OTA-C integrator shown in Figure 15.18(a) resulting in an OTA-C circuit.

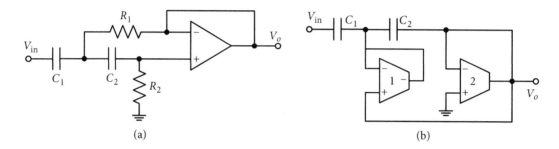

(a)   (b)

Figure 15.23 (a) Sallen and Key high pass filter section, (b) its OTA version.

Sometimes, the action of the OA may be easily substituted by a circuit using OTA; though it may not be practical economically in some cases, which makes it necessary to use OAs also in OTA-C circuits. Figures 15.23(a)–(b) and Figures 15.24(a)–(b) illustrate the point. OA in the HP Sallen–Key circuit and resistor R_1 in Figure 15.23(a) can be combined and replaced by OTA1; the resistance R_2 is realized using OTA2, resulting in all OTA HP filters, whereas in the LP case, only resistor R_1 is simulated; OA replacement would involve more components.

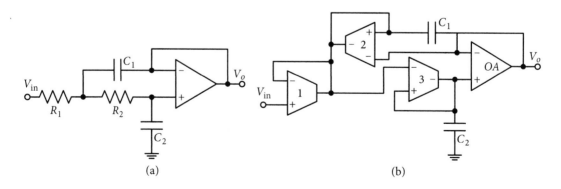

(a)   (b)

Figure 15.24   (a) OA-RC and (b) OTA with OA version of the Sallen and Key low pass filter.

15.6 OTA-C Filters Derived from LC ladders

Lossless doubly terminated ladders have been extensively studied and used because of their low sensitivities in passive as well as in active form. OTA-C filters based on these ladder structures also enjoy the same advantages and almost the same techniques are applied as in the OA-RC case. Hence, filter realization approaches making use of ladder structures will be discussed in this section.

15.6.1 Element substitution scheme

In a doubly terminated LC ladder, it is the inductance in the grounded and floating form which is replaced using OAs and R C elements. In the OTA-C filter realizations, in addition to the inductors, resistors are also to be simulated. It is also preferable to replace even floating capacitors by circuits using only grounded capacitors as will be described now.

Gyrators were found to be very successful in impedance conversion and hence, extensively used for inductance simulation. They essentially act as an interconnection of an inverting and a non-inverting voltage amplifier terminated in an impedance. The same idea is used here and it is found out that in the OTA-C case, it is much simpler compared to the OA-RC case, as shown in Figure 15.25(a), for which

$$I_{o1} = I_2 = g_{m1} V_1, I_{o2} = I_1 = g_{m2} V_2, \text{ and } V_2 = ZI_2 \tag{15.36a}$$

It gives $V_1/I_1 = Z_{\text{in}} = 1/\{g_{m1} g_{m2} (Z)\}$ \hfill (15.36b)

If $Z(s) = 1/sC$, the input behaves like an inductor having its expression as: $L = C/(g_{m1}g_{m2})$. For OTAs considered ideal, gyrator simulates the pure inductor. However, as simulation of inductors finds wide usage, effect of the non-ideality of the OTAs needs consideration. Figure 15.25(b) shows the small signal model of the gyrator with Z being an external capacitor. Analysis shows the input admittance as:

$$Y_{\text{in}}(s) = g_o + (C_i + C_o)s + 1 / \left\{ \left(\frac{C + C_i + C_o}{g_{m1} + g_{m2}} \right) s + \left(\frac{g_o}{g_{m1}g_{m2}} \right) \right\} \tag{15.37}$$

The circuit represented by equation (15.37) is shown in Figure 15.25(c), with parasitic resistances r_o and r_s, and capacitance C_p. The realized inductance value and effective increased terminating capacitance, respectively, are:

$$L_{\text{eff}} = (C_i + C_o + C)/g_{m1}g_{m2} \tag{15.38a}$$

$$C_{\text{eff}} = C_i + C_o + C \tag{15.38b}$$

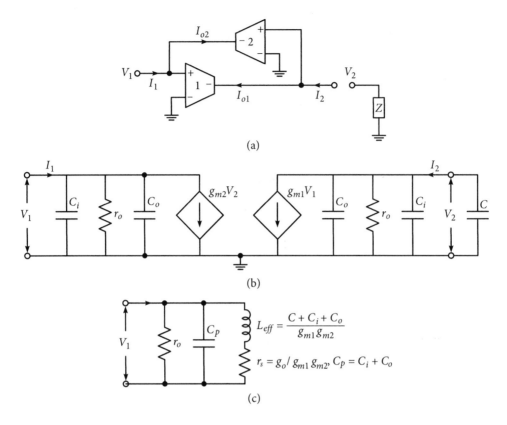

(a)

(b)

$$L_{eff} = \frac{C + C_i + C_o}{g_{m1} g_{m2}}$$

$$r_s = g_o / g_{m1} g_{m2}, \; C_p = C_i + C_o$$

(c)

Figure 15.25 (a) OTA-based gyrator terminating in impedance Z, (b) small-signal equivalent circuit of (a) showing parameters of OTA, and (c) the resulting non-ideal grounded inductor.

Differential GI (grounded inductors) can also be realized by employing the procedure discussed in Section 15.4. Figure 15.26(a) shows a mirror image of Figure 15.25(a), and as they are combined together, it results in the circuit shown in Figure 15.26(b).

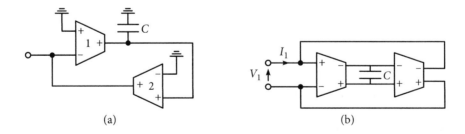

(a)   (b)

Figure 15.26 (a) Mirror image of the circuit in Figure 15.25(a) with Z as capacitor, (b) differential grounded inductance simulator.

Example 15.10: Verify that the circuit shown in Figure 15.25 simulates a GI by using it in a BPF with center frequency of 20 krad/s and $Q = 5$.

Solution: Figure 15.27(a) shows an RLC second-order BPF for which parameters are derived as:

$$\omega_o^2 = 1/LC_1, \quad Q = \tfrac{1}{2}R(C_1/L)^{0.5}, \quad \text{mid-band gain} = 0.5 \text{ for } R_L = R_{in} = R$$

For the given specifications, if C_1 is selected as 4 nF, the required value of inductance will be 1.0 mH. The inductance is realized using the circuit shown in Figure 15.25(a), with $g_{m1} = g_{m2}$ = 3 mA/V and the terminating capacitor C = 9 nF. For Q = 5, needed R = 5 kΩ.

Figure 15.27(b) shows the simulated magnitude response of the BPF. Center frequency is 79.49 kHz (499.6 krad/s), bandwidth is 15.88 kHz, giving Q = 5.02 and the mid-band gain is 0.498, verifying the simulation of inductance.

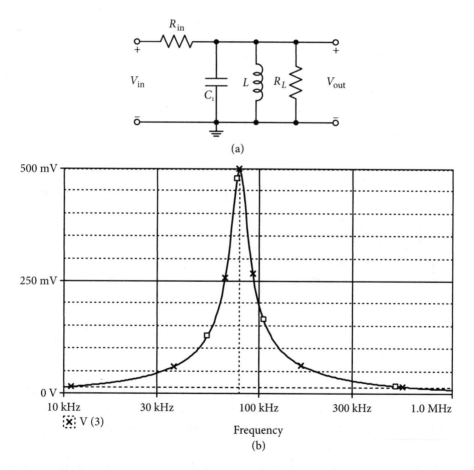

(a)

(b)

Figure 15.27 (a) Passive RLC band pass filter section for Example 15.10, (b) response of the band pass filter using the OTA simulated grounded inductor shown in Figure 15.25(a).

15.7 Single-ended and Differential Floating Inductance

Back-to-back gyrators terminating in a capacitor realize a FI (floating inductance) in the OTA-C case as well, as it is applicable in the OA-RC case. Figure 15.28(a) shows such a scheme and Figure 15.28(b) shows its simulated form. Analysis gives:

$$I_1 = I_{o2} = g_{m2}(-V'), \ I_2 = I'_{o2} = g'_{m2}V' \tag{15.39a}$$

$$I_{o1} = g_{m1}V_1, \ I'_{o1} = g'_{m1}(-V_2) \text{ and } V' = -(I_{o1} + I'_{o1})/sC \tag{15.39b}$$

For $g'_{m1} = g_{m1}$ and $g'_{m2} = g_{m2}$, equations (15.39a) and (15.39b) result in the following matrix:

$$\begin{bmatrix} I_1 \\ I_2 \end{bmatrix} = \frac{g_{m1}g_{m2}}{sC} \begin{bmatrix} 1 & -1 \\ -1 & 1 \end{bmatrix} \begin{bmatrix} V_1 \\ V_2 \end{bmatrix} \tag{15.40}$$

Matrix equation (15.40) represents the FI shown in Figure 15.28(b) with the expression of inductance as:

$$L = C/g_{m1}g_{m2} \tag{15.41}$$

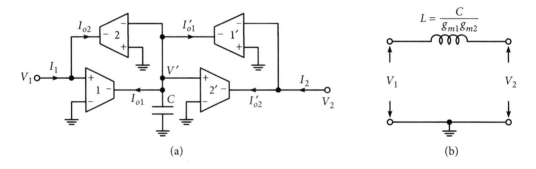

Figure 15.28 (a) Realization of floating inductance using back-to-back gyrators, (b) its equivalent circuit.

However, with non-idealities of the OTAs taken into consideration, the effective value of the simulated inductance will be the same as given by equation (15.18) for the GI, and the circuit representation will be as in Figure 15.25(c), but with both terminals ungrounded.

Example 15.11: Utilize the FI simulator shown in Figure 14.28(a) in a notch filter of Figure 15.29(a) having a notch frequency of 50 kHz.

Solution: Expression of the notch frequency will be $1/(LC)^{0.5}$. Hence, for the selected value of the capacitance $C = 10$ nF, required inductance is 1.0 mH. For the realization of FI, all the four transconductance are taken as 1 mA/V, which requires its terminating capacitor as 1.0 nF. Load resistance R decides the pole-Q. The simulated response shown in Figure 15.29(b) has a notch at 50.03 kHz.

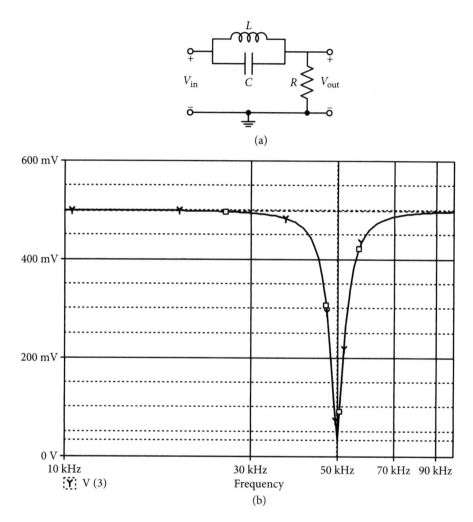

(a)

(b)

Figure 15.29 (a) Passive notch circuit for testing OTA-C floating inductor in part (a) of Figure 15.28(a), (b) magnitude response of the OTA-C notch filter.

An alternative FI using only three OTAs and a grounded capacitor is shown in Figure 15.30(a). It is obtained by combining OTA1 and OTA1' as OTA2. With OTAs considered ideal, the simulated inductance is the same as in equation (15.41) with $g_{m1} = g_{m3}$. Students can find the simulated equivalent circuit when OTAs are considered non-ideal.

Using the mirror image of Figure 15.28(a), and combining it with its original gives a floating differential inductor using four differential output OTAs as shown in Figure 15.30(b).

15.7.1 Floating capacitor simulation

From the circuit fabrication point of view, it is always better to use grounded capacitors, as parasitic capacitances are considerably reduced. That is why it is desirable to simulate the floating capacitor in terms of the grounded capacitor, wherever practically feasible. Tolerance is further reduced if the filter parameters are obtained in terms of the ratio of the grounded capacitors. A capacitor is simulated if the gyrator in Figure 15.25(a) terminates in a GI (or terminates in Figure 15.30(a) for FI). Using three OTA FIs of Figure 15.30(a), which terminates in a GI (using two OTAs and a grounded capacitor), a floating capacitor simulator is obtained as shown in Figure 15.31(a). With $g_{m1} = g_{m3}$, expression of the simulated capacitance is given as:

$$C = g_{m1}g_{m2}C'/g_{m4}g_{m5} \tag{15.42}$$

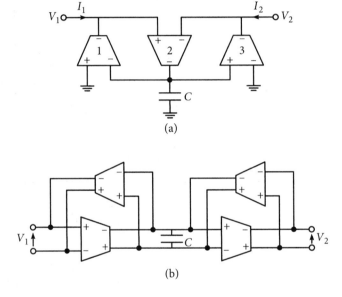

(a)

(b)

Figure 15.30 (a) Alternate floating inductance simulator using three OTAs. (b) Differential floating inductor from Figure 15.28(a).

It is obvious that the circuit not only simulates a floating capacitor, it also works as a capacitance multiplier with proper selection of transconductance.

Example 15.12: Figure 15.31(b) shows a passive BPF. Design and test the same using the OTA-C floating capacitor of Figure 15.31(a) for the filter to have center frequency of 500 krad/s and $Q = 10$ with a mid-band gain of 10.

Solution: For the circuit in Figure 15.31(b), expressions for center frequency and Q are:

$$\omega_o^2 = 1/LC, \text{ and } Q = (L/C)^{0.5}/R$$

Selecting $C = 5$ nF, the required inductance shall be 0.8 H and $R = 40$ kΩ. With $g_{m1} = g_{m2} = g_{m3} = g_{m4} = g_{m5} = 2$ mA/V grounded capacitance C' shall be 5 nF. The circuit was simulated and the response is shown in Figure 15.31(c). Simulated center frequency is 79. 49 kHz (499.65 krad/s), mid-band gain is 9.99 and with a bandwidth of 7.9489 k Hz, Q becomes 10; verifying the performance of the floating capacitor.

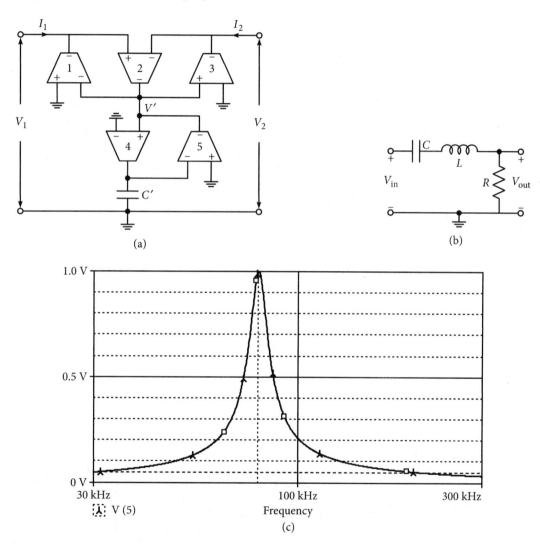

Figure 15.31 (a) Floating capacitor simulation using grounded capacitor (b) RLC band pass filter circuit and (c) simulation of the second-order band pass filter shown in part (b) while using the floating capacitor of part (a).

Example 15.13: Obtain a fifth-order Chebyshev filter employing the element substitution approach in the OTA-C form. The filter is to have a corner frequency of 800 krad/s and ripple width of 0.5 dB.

Solution: Figure 15.32(a) shows a fifth-order LP RLC ladder with normalized element values. The inductance simulator shown in Figure 15.30(a), and resistance simulators of Figures 15.2 and 15.3 can be used to obtain the OTA-C ladder shown in Figure 15.32(b). However, the circuit has been slightly modified by combining the adjacent OTAs 4 and 5, now shown as OTA45 in Figure 15.32(b). Note that the parasitic capacitances of the OTAs can be absorbed in the GCs. First, the elements are de-normalized using a frequency scaling of 8×10^5 rad/s and impedance scaling factor of 1.25 kΩ; the values become:

$$R_{in} = R_L = 1.25 \text{ k}\Omega, \ C_1 = C_5 = 1.7058 \text{ nF}, \ C_3 = 2.5408 \text{ nF}, \ L_2 = L_4 = 1.92125 \text{ mH}$$

For the realization of resistances R_{in} and R_L, use of equations (15.1) and (15.3) gives $g_{m0} = g_{m1} = g_{m8} = 1/1250 = 0.8$ mA/V. For simulating FIs, assuming $g_{m2} = g_{m3} = g_{m4/5} = g_{m6} = g_{m7} = 1.0$ mA/V, use of equation (15.41) gives $C_{L2} = C_{L4} = 1.92125$ nF. Using these element values in Figure 15.32(b), the circuit is simulated; the high value resistances (10 mega-ohm) are connected at some nodes where the path for bias current was needed.

Figure 15.32(c) shows the magnitude response having pass band edge frequency of 127.345 kHz (800.45 krad/s) and ripple width of 0.5058 dB and gain of 0.4998 at dc.

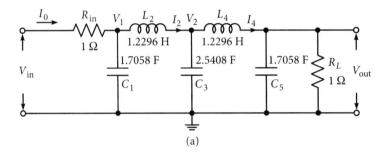

(a)

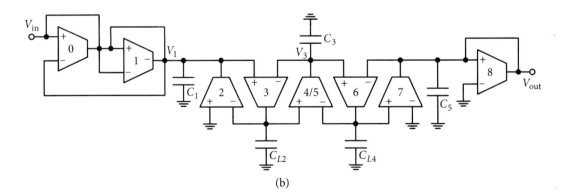

(b)

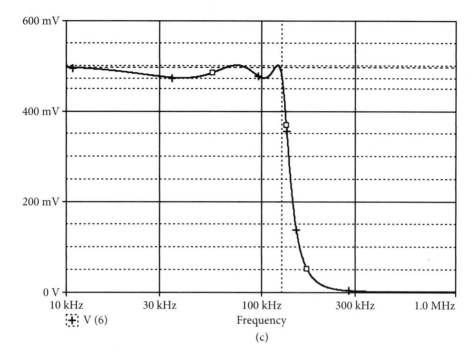

Figure 15.32 (a) Fifth-order LC ladder structure, (b) floating inductance and terminating resistances simulated OTA-C circuit and (c) simulated magnitude response of the fifth-order Chebyshev filter in (a)-example 15.13.

15.7.2 Anti-aliasing filter for an ECG detection device

A brief discussion on an analog front-end system for portable ECG (electrocardiograph) detection devices was included in Section 11.5.1 [15.1]. One of its important constituents is an anti-aliasing filter. Based on the system requirements, an OTA based filter is preferred at low frequency operation as transistors inside the OTA can be operated in the sub-threshold region to save power. To attenuate out of band interference before the ADC, a fifth-order ladder type Butterworth filter with cut-off frequency of 250 Hz was needed. As the aim was to show the utilization of OTAs in anti-aliasing, the present design of the filter is different from the filter mentioned in the reference on two counts. OTAs having transconductance of the order of mA/V are used instead of operating in the sub-threshold region, and element (floating inductance) substitution approach is used instead of operational simulation.

The structure of a fifth-order Butterworth passive filter is already shown in Figure 15.32(a). It can be used with different values of elements corresponding to the specification for this application. Normalized element values as obtained from Table 3.3 are:

$$L_2 = L_4 = 1.618 \text{ H}, \ C_1 = C_5 = 0.618 \text{ F}, \ C_3 = 2.0 \text{ F and } R_{\text{in}} = R_L = 1\Omega \qquad (15.43)$$

Applying frequency de-normalization by 250 Hz and using an impedance scaling factor of 10^3, element values modify as:

$$L_2 = L_4 = 1.029 \text{ H}, \ C_1 = C_5 = 0.3932 \ \mu\text{F}, \ C_3 = 1.272 \ \mu\text{F} \text{ and } R_{\text{in}} = R_L = 1 \text{ k}\Omega \qquad (15.44)$$

The inductance simulator of Figure 15.30(a), and resistance simulators of Figure 15.2 and 15.3 are used to obtain the OTA-C ladder as shown in Figure 15.32(b).

For the realization of resistances R_{in} and R_L, use of equations (15.1) and (15.3) gives $g_{m0} = g_{m1} = g_{m8} = 1/(1000 = 1.0 \text{ mA/V})$. For simulating FIs, assuming $g_{m2} = g_{m3} = g_{m4/5} = g_{m6} = g_{m7} = 1.0 \text{ mA/V}$, use of equation (15.41) gives $C_{L2} = C_{L4} = 0.9718 \text{ nF}$. Using these element values in Figure 15.32(b), the circuit is simulated, while connecting high value resistances (10 mega-ohm) at some nodes where path for bias current was needed.

Figure 15.33 shows the magnitude response having a pass band edge frequency of 258 Hz, and its attenuation of 46.5 dBs occurs at three times its cut-off frequency; this satisfies the requirements.

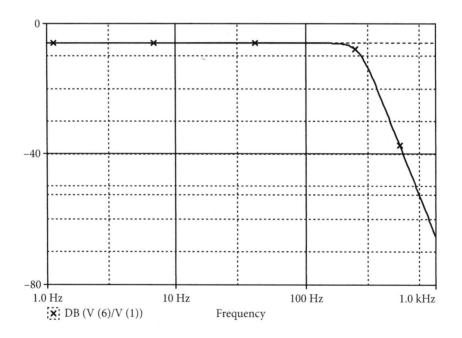

Figure 15.33 Response of a fifth-order anti-aliasing filter.

15.8 Operational Simulation Process

The process follows exactly the same steps as in OA based active RC circuits. Only inverting and non-inverting integrators are required for the RLC ladder. Lossless and lossy integrators of Figures 15.4(a) and (b), respectively are employed, while keeping in mind that alternate inverting/non-inverting modes are used. The RLC fifth-order low pass section of Figure 15.32(a) can be described by the following equations.

$$V_1 = \frac{I_o - I_2}{sC_1 + G_{in}}, \quad I_2 = \frac{V_1 - V_3}{sL_2}, \quad V_3 = \frac{I_2 - I_4}{sC_3} \tag{15.45a}$$

$$I_4 = \frac{V_3 - V_{out}}{sL_4}, \quad V_{out} = \frac{I_4}{sC_5 + G_L} \tag{15.45b}$$

Input series resistance R_i has been replaced by its equivalent shunt resistance in parallel with capacitor C_1 as depicted in equation (15.45a). All currents are to be converted as voltages through resistance scaling; hence, the following relations are obtained:

$$V_1 = \frac{RI_o - RI_2}{sRC_1 + RG_{in}} \rightarrow v_1 = \frac{v_{Iin} - v_{I2}}{sc_1 + g_{in}} \tag{15.46a}$$

Here, subscript I with v indicates a current signal converted to voltage through scaling. The remaining expressions are:

$$RI_2 = \frac{V_1 - V_3}{sL_2 / R} \rightarrow v_{I2} = \frac{v_1 - v_3}{sl_2} \tag{15.46b}$$

$$V_3 = \frac{RI_2 - RI_4}{sRC_3} \rightarrow v_3 = \frac{v_{I2} - v_{I4}}{sc_3} \tag{15.46c}$$

$$RI_4 = \frac{V_3 - V_{out}}{sL_4 / R} \rightarrow v_{I4} = \frac{v_3 - v_{out}}{sl_4} \tag{15.46d}$$

$$V_{out} = \frac{RI_4}{sRC_5 + RG_L} \rightarrow v_{out} = \frac{v_{I4}}{sc_5 + g_L} \tag{15.46e}$$

Using the lossless and lossy integrators of Figure 15.4, we obtain the OTA-C version in Figure 15.34(a). For $g_{m2} = g_{m1}$, $g_{m5} = g_{m4}$, $g_{m7} = g_{m6}$ and $g_{m9} = g_{m8}$, the current–voltage relations corresponding to equation (15.46) are:

$$v_1 = \frac{g_{m1}}{sc_1 + g_{m3}}(v_{in} - v_{I2}), \quad v_{I2} = \frac{g_{m4}}{sc_{L2}}(v_1 - v_3) \tag{15.47a, b}$$

$$v_3 = \frac{g_{m6}}{sc_3}(v_{I2} - v_{I4}), \quad v_{I4} = \frac{g_{m8}}{sc_{L4}}(v_3 - v_5), \quad v_{out} = \frac{g_{m10}}{sc_5 + g_{m11}}v_{I4} \tag{15.47c, d, e}$$

Equation (15.47) can be used as a design equation for finding the values of the capacitors and trans-conductance.

In case a floating capacitor is also present in the series branch in parallel with inductor L_2 or L_4, it needs no special consideration. The capacitor is to be placed between the same two terminals where the inductor is in the original RLC ladder, without any change. For example, if there is a capacitance C' in parallel with L_4, current through it will be $(V_3 - V_{out})sC'$. As these two node voltages V_3 and V_{out} are available in the OTA-C circuit and if C' is connected between them, it will carry the same current.

Example 15.14: Re-design the fifth-order ladder of Figure 15.32(a) using the operational simulation technique.

Solution: Figure 15.34(a) shows the operationally simulated OTA-C version of the filter ladder of Figure 15.32(a). Two non-ideal integrators simulating equations (15.46a–e) and three ideal integrators simulating equations (15.47b–d) have been joined as discussed in the text.

Application of the frequency scaling by 800 krad/s and impedance scaling of 1.25×10^3 yield the following element values for the ladder structure of Figure 15.32(a).

$$R_{in} = R_L = 1.25\ k,\ L_2 = L_4 = 1.92125\ \text{mH},\ C_1 = C_5 = 1.7058\ \text{nF and}\ C_3 = 2.5408\ \text{nF}$$

All integrators simulating equation (15.46) and (15.47) employ OTAs with transconductance $g_m = 1.0$ mA/V. It results in the value of capacitances $C_{L2} = C_{L4} = 1.9215$ nF. However, OTA1 and OTA11, which are simulating resistors have $g_m = 0.8$ mA/V.

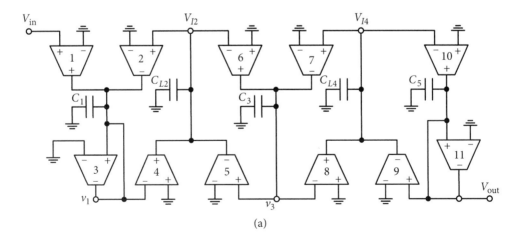

(a)

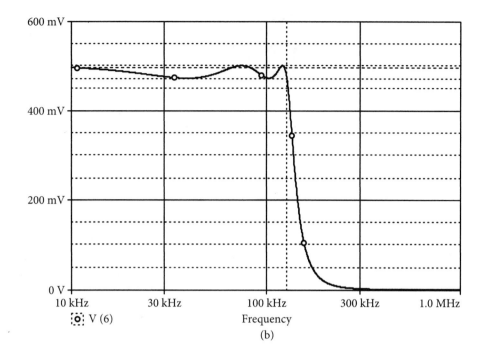

Figure 15.34 (a) Fifth-order Chebyshev filter through operational simulation technique. (b) Operationally simulated fifth-order Chebyshev filter response for Example 15.14.

The simulated response is shown in Figure 15.34(b), which is in close conformity with the design having a pass band edge frequency of 127.47 kHz (801.24 krad/s) and a ripple width of 0.495 dB.

References

[15.1]   Lee, Shuenn-Yuh, Jia-Hua Hong, Jin-Ching Lee, and Qiang Fang. 2012. 'An Analog Front-End System with a Low-Power On-Chip Filter and ADC for Portable ECG Detection Devices'. *Advances in Electro-Cardiogram-Methods and Analysis*. ISBN:9789533079233.

Practice Problems

15-1   Determine the output voltage in the circuit of Figure P15.1 at 1 MHz, with resistor being 10 k Ohm and applied voltages and trans-conductance are: $V_1 = 2V$, $V_2 = 3V$, $g_{m1} = 0.2\dfrac{mA}{V}$ and $g_{m2} = 0.3\dfrac{mA}{V}$.

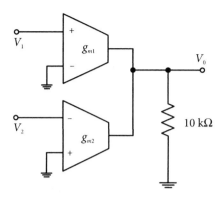

Figure P15.1

15-2 Derive the expression in equation (15.5). Design a non-inverting integrator having cut-off frequency of 10^6 rad/s and a dc gain of 2 using ideal OTAs. Find the percentage deviation in the cut-off frequency if the following values for the non-ideal model of the OTAs are used: input capacitance = 4 pF, output resistance = 1 MΩ and output capacitance = 10 pF.

15-3 Repeat Problem 15-2, for required dc gain of 20.

15-4 Determine the kind of function generated by the circuit in Figure P15.2

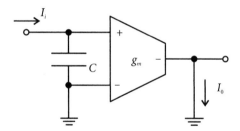

Figure P15.2

15-5 Design an inverting amplifier using OTAs for a gain of 10. Find its useful frequency range if the OTAs have parasitic input capacitance = 5 pF, output resistance = 1 MΩ and output capacitance = 8 pF.

15-6 In Figure 15.14, C = 5 nF, α = 0.2 and g_{m1} = 2 g_{m2} = 4mA/V. Sketch the voltage ratio transfer function equation (15.21) and give its dc gain and gain at high frequencies, if the OTAs are considered ideal.

15-7 Design and test an HP filter for a cut-off frequency of 300 krad/s using the circuit in Figure 15.10(b) when OTAs are considered ideal, and when OTAs have parasitic C_i = 3.6 pF, R_o = 1.5 MΩ and C_o = 5.6 pF.

15-8 Design and test an HP filter for a cut-off frequency of 400 krad/s and high frequency gain of 2, using the circuit in Figure 15.11 when OTAs are considered ideal, and when OTAs have parasitic C_i = 3.6 pF, R_o = 1.5 MΩ and C_o = 5.6 pF.

15-9 Realize an AP filter using the circuit of Figure 15.13(a), such that it has a phase shift of 90° at 500 krad/s. OTAs have the following non-idealities: C_i = 4 pF, R_o = 2 MΩ and C_o = 6 pF.

15-10   Test the circuit of Figure 15.15(c) as an AP filter such that its phase shift of 90° occurs at 100 kHz for both the outputs. How much variation takes place in magnitude when the signal frequency changes from dc to 100 kHz.

15-11   Design and test an LP filter using the differential mode inverting amplifier from Figure 15.19(b) for cut-off frequency of 120 kHz and dc gain of 2.

15-12   Design a BP filter using the two-integrator loop circuit in Figure 15.20(a) having a center frequency of 80 kHz and $Q_0 = 2$ and mid-band gain of 5. What is the peak gain of the simultaneously obtained LP response and the frequency at which the peak gain occurs?

15-13   Repeat Problem 15-12 for the following specifications:

$\omega_0 = 300$ krad/s, $Q = 5$ and mid-band gain = 10.

15-14   Derive the relation in equation (15.29a) and design an LP filter having the following specifications:

$\omega_0 = 250$ krad/s and $Q = 1.5$

15-15   Derive the relation in equation (15.32) and design a HP filter having the following specifications:

$\omega_0 = 200$ krad/s and $Q = 1.25$

15-16   Derive the relation in equation (15.33) and design a BP filter having the following specifications:

$\omega_0 = 300$ krad/s, $Q = 5$ mid-band gain = 10.

15-17   Design the KHN based biquad in Figure 15.21(a) for a notch frequency of 600 krad/s and $Q = 2$.

Note: For Problems 15-18 to 15-23, use circuit in Figure 15.22(a)

15-18   Obtain the signal V_{o1} as the LP response having a cut-off frequency of 500 krad/s, $Q = 2$ and dc gain of 5. What kind of response becomes available as V_{o2}?

15-19   Obtain the signal V_{o2} as the HP response having a cut-off frequency of 500 krad/s and $Q = 2$. What kind of response becomes available as V_{o1}?

15-20   Obtain the signal V_{o2} as the BP response having center frequency of 400 krad/s, $Q = 5$ and mid-band gain of 10. What kind of response becomes available as V_{o1}?

15-21   Obtain the signal V_{o1} as the BP response having center frequency of 500 krad/s, $Q = 5$ and mid-band gain of 5. What kind of response becomes available as V_{o2}?

15-22   Obtain the signal V_{o1} as an LP notch response having center frequency of 200 krad/s, and notch frequency of 500 krad/s.

15-23   Obtain the signal V_{o1} as an HP notch response having center frequency of 400 krad/s, and notch frequency of 200 krad/s.

15-24   Use the floating inductance of Figure 15.28(a) for the normalized fifth-order LP filter of Figure 15.32(a). Design and test it for pass band edge frequency of 100 kHz with values of the elements being:

$R_1 = R_L = 1\Omega$, $C_1 = C_5 = 1.1468$ F, $C_3 = 1.975$ F and $L_2 = L_4 = 1.3712$ H.

15-25 Obtain the expression of the floating capacitor of equation (15.42).

15-26 Design and test the notch filter of Figure 15.29(a) having a notch frequency of 60 kHz. Employ OTA based floating inductance as well as floating capacitance.

15-27 Repeat Problem 15-26 using operational simulation technique.

15-28 Design and test a fourth-order OTA based Butterworth filter having cut-off frequency of 80 kHz using operational simulation method.

Chapter 16

Current Conveyors and CDTA (Current Differencing Transconductance Amplifiers) Based Filters

16.1 Introduction

All conventional analog circuits are voltage mode (VM) circuits, where the performance of the circuit is determined in terms of voltage levels at all nodes including those at the input and output nodes; operational amplifiers (OAs), being the most commonly used active device, is a voltage controlled voltage source (VCVS). Unfortunately, VM circuits do suffer from some limitations: (i) voltages in the circuit cannot change very quickly when input voltage changes suddenly due to parasitic capacitances; (ii) the bandwidth of OA based circuits is usually limited to the audio frequency range unless high bandwidth OAs are used; and (iii) circuits generally do not have high voltage swings and require high supply voltages for better signal-to-noise ratio.

In the current mode (CM) approach, the circuit description is presented in terms of current. This implies that all signals, including those at the input and the output are taken in terms of current rather than voltage and the active devices used are preferably CM devices. Hence, CM signal processing techniques can be defined as the processing of current signals in an environment where voltage signals become irrelevant in determining circuit performance, although CM devices do generate VM circuits as well [16.1].

Advancements in IC (integrated circuits) technologies together with the demand for smaller and low power devices have necessitated in the development of monolithic IC filters, not only at audio frequency but at a much higher frequency range. The advent of sub-micron IC processing (0.5 μm and smaller) has facilitated the realization of filters even in the VHF frequency (30–300 MHz) band. Together with its high frequency operation, reduction in power consumption is another advantage of CM filter circuits. In view of this, attention is being paid toward signal processing in terms of current rather than voltage. This new type of signal processing is known as CM signal processing and is evolving into a better alternative to VM signal processing, especially in the high frequency region.

CM signal processing leads to a higher frequency range of operation because the signal current is delivered to a small (ideally short circuit) load resistance. Due to this small resistance, the parasitic pole frequency becomes very high. As a result, signals are processed without any appreciable deviation for much larger frequencies. In summary, the following are the main advantages of the CM approach: (i) much larger frequency range of operation, (ii) easy addition, subtraction and multiplication of signals, (iii) higher dynamic range, (iv) lower power consumption with reduced power supply voltage and (v) better suited to micro-miniaturization and simpler circuits.

Second-generation current conveyors (CCs) being the most versatile, its description and MOS (metal–oxide–semiconductor) implementation is discussed in Section 16.2. It is used to generate some basic building blocks (BBBs) like controlled signal sources, weighted current and voltage summers, lossless and lossy CM and VM integrators discussed in Section 16.3. First-order filters are studied in Section 16.4, and a variety of second-order filters are included in Sections 16.5 to 16.9. Some circuits of inductance simulation are shown in Section 16.10 for their use in direct form synthesis. Current differencing transconductance amplifiers (CDTA), its application for inductance simulation, and development of biquadratic filters is shown in Section 16.11.

16.2 Second Generation Current Conveyors (CCII±)

Current conveyors are very versatile elements and they have become available in many forms, like first generation current conveyors (CCIs), second generation current conveyors (CCIIs), differential voltage current conveyors (DVCCs), current controlled current conveyor (CCCIIs) and dual-X current conveyors (DXCCIIs).

As mentioned here, there are many versions of CCs, but more attention has been given to CCIIs in literature and in this chapter as well. The reason being that it is one of the most functionally flexible and versatile CC since its inception [16.2, 16.3].

A current conveyor (CC) is a 3-port device with four (even five) terminals, which has emerged as a powerful alternative to OAs for performing analog signal processing. The first CC introduced in 1968 by Smith and Sedra [16.2] and named as CCI is shown in block form in Figure 16.1(a), for which the current–voltage relationship is given as:

$$i_Y = i_X, \; v_X = v_Y, \text{ and } i_Z = \pm i_X \tag{16.1}$$

Here, current and voltage variables represent total instantaneous values. It is important to understand the meaning of the relations in equation (16.1) in conjunction with its block diagram in Figure 16.1(a). If an input current i_X flows in terminal X, it will force an equal current i_Y to flow in terminal Y. Similarly, if voltage v_Y is applied at terminal Y, an equal voltage v_X will appear at the terminal X. The same amount of current i_Z will go into or come out as i_X, but the terminal Z has a high impedance level and thus behaves like a current source.

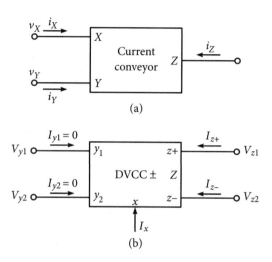

(a)

(b)

Figure 16.1   (a) A current conveyor in block form showing terminal voltages and currents. (b) Symbolic representation of a differential voltage CCII± with dual outputs (DVCCII).

Bipolar technology was used in the early stages of development of CCs in 1960s. However, fabrication of good quality pnp transistors, which was a necessity to get near ideal CCs, was difficult. Later, with the availability of CMOS (complementary MOS) technology, which could provide good quality matching MOS transistors, it became comparatively easy to fabricate quality CCs.

CCI was soon overtaken by its second version CCII in which current does not flow in terminal Y [16.3]. For the same block-form representation of CCII in Figure 16.1(b), the current–voltage relationship is given as:

$$
\begin{bmatrix} i_Y \\ v_X \\ i_Z \end{bmatrix} = \begin{bmatrix} 0 & 0 & 0 \\ 1 & 0 & 0 \\ 0 & \pm 1 & 0 \end{bmatrix} \begin{bmatrix} v_Y \\ i_X \\ v_Z \end{bmatrix}
$$

(16.2a)

The terminal Y has infinite input resistance. Terminal X follows the voltage at terminal Y, and ideally, it exhibits zero input impedance. Direction of current at terminal Z, which is at infinite output resistance ideally, defines the polarity of the CCII. As per convention, when current is going into the Z terminal, it is called a positive CCII (CCII+), and for outgoing current at Z terminal, it is called negative CCII (CCII-).

If CCs are compared to OAs, it is observed that CCs can be realized whose performance is very close to their ideal values. Practically, the impedance level at terminal X is close to zero, and at terminal Z, the level is close to infinity, making the device a near perfect current source.

Signal processing functions which were realized using OAs, can be designed in a comparatively simpler way using CCs. Additionally, there are some other advantages too. One of the major advantages in using CCs is that a higher voltage (or current) gain becomes possible over a larger signal bandwidth for small input signals, or large signal frequency conditions. This amounts to a much higher gain–bandwidth product than obtainable with OAs.

A number of structures are available for realizing CCs. One such CMOS implementation of a differential voltage input current conveyor (DVCCII) is available in reference [16.4] along with the size of the MOSFETs (MOS field effect transistors) used in it.

Equation (16.2a) represents the current–voltage relation of an ideal CC with no parasitics whereas practical CCs do have some parasitics. To study the non-ideal behavior, the relationship given in equation (16.1) can be modified as:

$$I_Y = 0, \ V_X = \beta V_Y, \ I_Z = \pm \alpha I_X \qquad (16.2b)$$

where the voltage transfer gain β from Y to X deviates from unity by an amount equal to the voltage transfer errors. Similarly, the current transfer gain α from X to Z represents a deviation from unity as current transfer error. The errors are expected to be quite low for integrated CCs for a large frequency range.

16.3 Some Basic Building Blocks

Current conveyors are able to realize almost all basic building blocks and complex networks which can be realized using OAs or OTAs (operational transconductance amplifiers). In fact, even at the very beginning of its usage, a large number of active network and signal processing basic blocks were given, showing the utility of CCs. Figures 16.2 and 16.3 show some of the basic building blocks using CCs, which can be utilized alone or for building more complex structures. Figures 16.2(a) and (b) show a voltage controlled current source (VCCS) and a current-controlled voltage source (CCVS), respectively.

For the VCCS:

$$I_1 = 0, \ I_x = (V_x/R) = (V_1/R) = I_z = I_2 \qquad (16.3)$$

Output current I_2 depends on input voltage V_1; and for the CCVS:

$$V_1 = V_{x1} = 0, \ I_{z1} = I_1, \ V_2 = V_{y2} = I_{z1}R = I_1 R \qquad (16.4)$$

Equation (16.4) shows that the output voltage V_2 depends on the input current I_1.

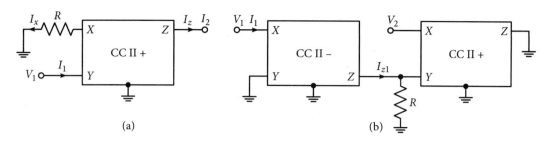

Figure 16.2 Current conveyor based (a) voltage controlled current source and (b) current controlled voltage source.

Some signal processing basic blocks like non-inverting current amplifiers, current integrators, and weighted current summers are shown in Figures 16.3(a), (b), and (c), respectively. Here, the relation for the current amplification for Figure 16.3(a) is given as:

$$V_y = I_{in}R_1 = V_x \rightarrow I_x = V_x/R_2 = I_{in}R_1/R_2 \rightarrow I_o/I_{in} = R_1/R_2 \tag{16.5a}$$

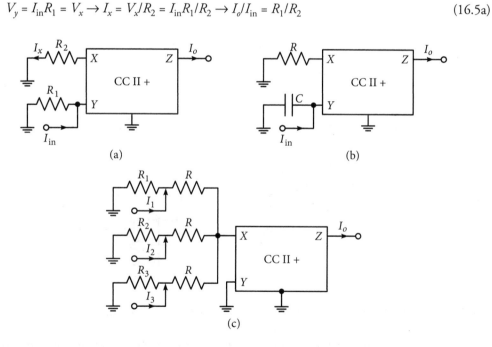

Figure 16.3 Current conveyor based (a) current amplifier, (b) current integrator and (c) current summation.

If R_1 is replaced by C, as shown in Figure 16.3(b), the CCII+ acts as an ideal current integrator for which expression of current gain is:

$$I_o/I_{in} = 1/sCR \tag{16.5b}$$

For the weighted current summer of Figure 16.3(c), the expression for the output current is given as:

$$I_o = I_1 \{R_1/(R + R_1)\} + I_2 \{R_2/(R + R_2)\} + I_3\{R_3/(R + R_3)\} \tag{16.6}$$

It is to be noted that the current amplification factor will be less than unity with this configuration and the weighted summer can become a simple adder when input currents are applied directly at the terminal X; the number of currents to be summed up can be arbitrarily large.

Figure 16.4(a) and (b) show lossy integrators. In Figure 16.4(a), the direction of currents I_x and I_z has been shown to be negative for CCII; hence:

$$V_x = V_y = V_{in}, \ I_x = (V_{in}/R_1), \ V_z = -I_z\{R_2/(1 + sCR_2)\} = V_{out} \tag{16.7}$$

$$(V_{out}/V_{in}) = -(R_2/R_1)/(1 + sCR_2) \tag{16.8}$$

For the inverting current integrator shown in Figure 16.4(b):

$$V_y = I_{in}(R_2/1 + sCR_2) = V_x, \ I_z = -I_x = -V_x/R_1$$

$$(I_{out}/I_{in}) = -(R_2/R_1)/(1 + sCR_2) \tag{16.9}$$

DC gain for both the VM and CM lossy integrators is (R_2/R_1) and half-power frequency is $1/CR_2$.

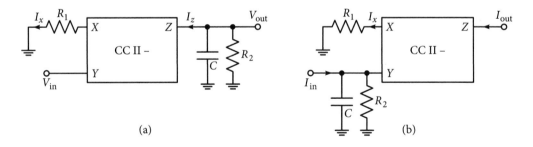

(a)   (b)

Figure 16.4 Lossy inverting integrator in (a) voltage mode and in (b) current mode.

If resistor R_2 is removed from Figures 16.4(a) or (b), it will result in respective lossless integrators.

16.4 Current Conveyor Based First-order Filters

Integrators of the circuits shown in Figure 16.4 behave as LPFs with the expression for 3 dB frequency as $1/2\pi RC$. However, there are a few simple circuits which act as all pass (AP) first-order filter structures, but can be simplified to act as an HPF (high pass filter) and LPF (low pass filter) as well.

As $V_x = V_y$ in Figure 16.5(a), impedances Z_1 and Z_2 are virtually parallel and current division in the two branches takes place accordingly; analysis give the following relations [16.5]:

$$V_x = V_y = Z_4 \left(\frac{Z_1 I_{in}}{Z_1 + Z_2} \right), \quad I_z = I_x = I_{in} \frac{Z_2}{Z_1 + Z_2} - \frac{V_x}{Z_3}$$

$$\frac{I_{out}}{I_{in}} = -\frac{I_z}{I_{in}} = -\frac{Z_2 Z_3 - Z_1 Z_4}{Z_3 (Z_1 + Z_2)} \tag{16.10}$$

For $Z_3 = Z_4 = R'$, $Z_1 = 1/sC$ and $Z_2 = R$

$$\frac{I_{out}}{I_{in}} = -\frac{sCR - 1}{sCR + 1} \tag{16.11}$$

It works as a current mode first-order APF; it also works the same way if $Z_1 = R$ and $Z_2 = (1/sC)$ with negative sign removed in equation (16.11).

For $Z_4 = 0$, $Z_3 = \infty$, $Z_2 = R$, and $Z_1 = 1/sC$, the circuit becomes a CM first-order HP section; whereas, it becomes an LP section with $Z_1 = R$ and $Z_2 = 1/sC$. It will be shown a little later that the circuit can also provide second-order sections with some other combination for impedances.

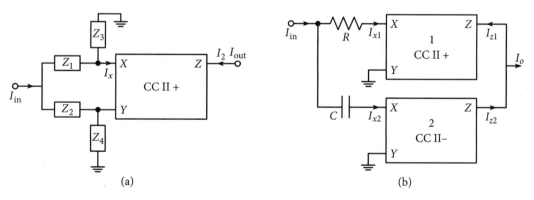

(a)   (b)

Figure 16.5 First-order current mode (a) general structure, and (b) alternate circuit using two CCs {With permission from Springer Nature}.

Another AP network is shown in Figure 16.5(b) using two CCs, a resistor and a capacitor. Resistor R and the capacitor C become practically parallel with both the Y terminals grounded. The circuit is a current input current output (CICO) configuration with the following relation:

$$V_{x1} = V_{x2} = V_{y1} = V_{y2} = 0, \quad I_{z1} = I_{x1} = \frac{1/sC}{R + 1/sC}, \quad I_{z2} = I_{x2} = \frac{R}{R + 1/sC}$$

$$I_o = I_{z2} - I_{z1} = \frac{sCR - 1}{sCR + 1} \tag{16.12}$$

Example 16.1: Obtain a CM HPF from the general structure shown in Figure16.5(a) having a 3 dB frequency of 1000 krad/s.

Solution: The HPF section is shown in Figure16.6(a) and derived from Figure 16.5(a). For selected value of C = 0.2 nF, R = 5 kΩ, its magnitude response is shown in Figure 16.6(b). High frequency gain is almost unity and the simulated 3 dB frequency is 159.46 kHz.

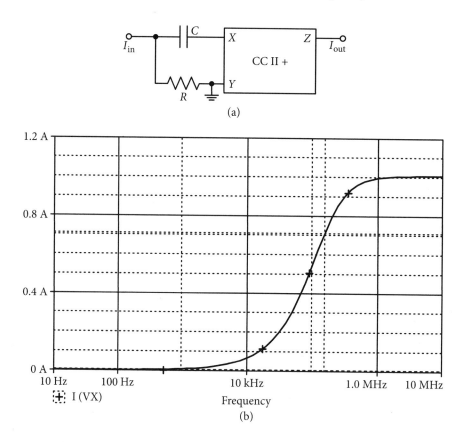

(a)

(b)

Figure 16.6 (a) Current mode HP filter from Figure 16.5(a) and (b) its simulated magnitude response.

16.5 Current Conveyor Based Second-order Filters

A large number of approaches have been used to obtain second-order filter structures using CCs. However, almost all of the approaches are like those employed in OA-RC or in OTA-RC cases. While most of the time CCIIs are used, sometimes CCIs are also used as will be shown later.

16.5.1 Wein bridge based structure

It was mentioned earlier that more options are available in Figure16.5(a) for the selection of branch impedances. For example, if we select Z_1 as a series combination of resistance R_1 and capacitance C_1, and Z_2 as a parallel RC combination as shown here:

$$Z_1 = (1 + sC_1R_1)/sC_1,\ Z_2 = R_2/(1 + sC_2R_2),\ Z_3 = \infty\ \text{and}\ Z_4 = 0 \tag{16.13}$$

a BP (band pass) response becomes available at the output, and equation (16.10) yields:

$$(I_{out}/I_{in}) = sC_1R_2/D(s) \tag{16.14}$$

$$D(s) = s^2 C_1 C_2 R_1 R_2 + s(C_1 R_1 + C_1 R_2 + C_2 R_2) + 1 \tag{16.15}$$

From equation (16.15), the second-order BPF parameters are:

$$\omega_o^2 = 1/\left(C_1 C_2 R_1 R_2\right) \tag{16.16}$$

$$Q = \frac{\sqrt{\left(C_1 C_2 R_1 R_2\right)}}{\left(C_1 R_1 + C_1 R_2 + C_2 R_2\right)} \tag{16.17}$$

$$\text{Mid-band gain } h_{mbg} = \frac{\left(C_1 R_2\right)}{\left(C_1 R_1 + C_1 R_2 + C_2 R_2\right)} \tag{16.18}$$

It is to be noted that the maximum obtainable value of Q with this configuration is less than 0.5.

Example 16.2: Design a second-order BPF with a 3 dB frequency of 400 krad/s and $Q = 0.4$ using the general configuration of Figure 16.5(a).

Solution: Assuming $C_1 = KC_2$ and $KR_1 = R_2$, and then using equations (16.16) and (16.17), the component values are:

$$4 \times 10^5 = 1/R_2 C_2 \rightarrow R_2 = 2.5\ \text{k}\Omega,\ C_2 = 1.0\ \text{nF} \tag{16.19a}$$

In addition, with the help of equation (16.17):

$$Q = 0.4 = \frac{\sqrt{KC_2 C_2 R_2 R_2 / K}}{C_2 R_2 + KC_2 R_2 + C_2 R_2} \rightarrow K = 0.5 \tag{16.19b}$$

So, $R_1 = 5\ \text{k}\Omega$, and $C_1 = 0.5\ \text{nF}$

The designed circuit with element values is shown in Figure 16.7(a) and its simulated response is shown in Figure 16.7(b). Center frequency is obtained as 63.096 kHz (396.6 krad/s) and with the bandwidth of 157.98 kHz, value of $Q = 0.399$; very close to the design.

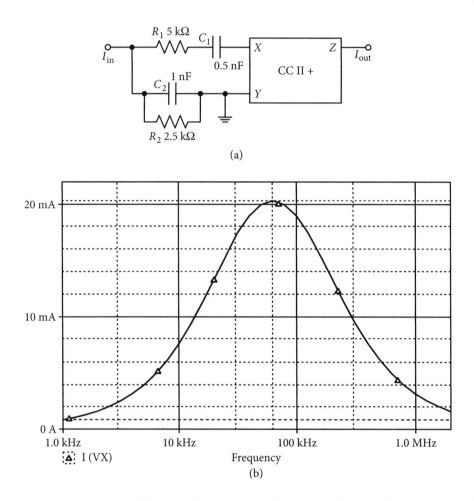

(a)

(b)

Figure 16.7 (a) Band pass filter from the general configuration of Figure 16.5(a); (b) its simulated response as a low-Q band pass CM filter.

In an alternate case, if we select the following combination of elements in Figure 16.5(a):

$$Z_1 = R_1 + (1/sC_1), Z_2 = R_2/(1 + sC_2R_2), Z_3 = R_3 \text{ and } Z_4 = R_4 \qquad (16.20)$$

The obtained CM transfer function is as follows:

$$\frac{I_{out}}{I_{in}} = \frac{R_4}{R_3} \frac{s^2 C_1 C_2 R_1 R_2 + s(C_1 R_1 + C_2 R_2 - C_1 R_2 R_3 / R_4) + 1}{s^2 C_1 C_2 R_1 R_2 + s(C_1 R_1 + C_2 R_2 + C_1 R_2) + 1} \qquad (16.21)$$

The circuit can now behave as an AP or a notch, under the following respective conditions:

$$\frac{R_3}{R_4} = \frac{2(C_1R_1+C_2R_2)}{C_1R_2}+1 \tag{16.22}$$

$$\frac{R_3}{R_4} = \frac{(C_1R_1+C_2R_2)}{C_1R_2} \tag{16.23}$$

The important parameters of the filters are obtained from equation (16.21) as:

$$\omega_o^2 = 1/C_1C_2R_1R_2 \tag{16.24}$$

$$\text{Gain at dc} = R_4/R_3 \tag{16.25}$$

It is to be noted that in the aforementioned configuration of Figure 16.5(a), only real poles are possible, that is why in Example 16.2, Q was taken less than 0.5. To obtain complex conjugate poles, one option is to replace CCII by CCI, because CCI can be considered as a combination of CCII with an additional current output feed back to the Y input [16.6]. Hence, with $Z_1 = 0$ and $Z_3 = \infty$, Figure 16.5(a) modifies to Figure 16.8(a) using CCI, which gives the following relation:

$$V_y = V_x = (I_{in} - I_x)Z_3 = (I_{in} - I_{out})Z_3, \; V_y = -\beta I_{out}Z_4$$

$$(I_{out}/I_{in}) = Z_3/(Z_3 - \beta Z_4) \tag{16.26}$$

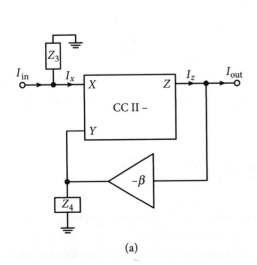

(a)

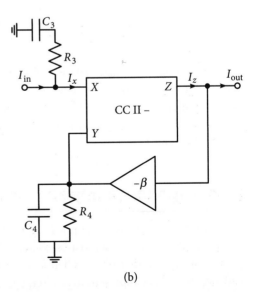

(b)

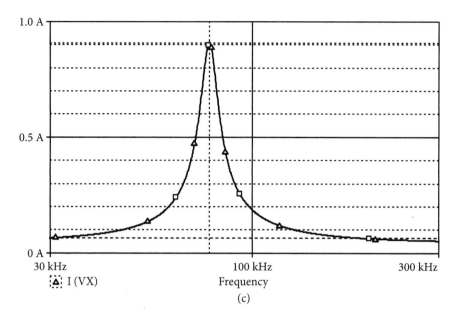

Figure 16.8 (a) First-order current mode filter which can realize complex poles {With permission from Springer Nature} (b) a CCI-RC version. (c) Response of the band pass filter for Figure 16.8(b) realizing high Q.

A CMBP (current mode band pass) transfer function can be obtained using equation (16.23), if we select $Z_3 = (R_3 + 1/sC_3)$ and $Z_4 = R_4/(1 + sC_4R_4)$. The resulting circuit is shown in Figure 16.8(b) and the transfer function is given as:

$$\frac{I_{out}}{I_{in}} = \frac{sC_3R_3/\beta}{s^2C_3C_4R_3R_4 + s(C_3R_3 + C_4R_4 - C_4R_3/\beta) + 1} \tag{16.27}$$

The BP response can have a large value of pole-Q on account of the difference term in the denominator. If the selected elements are $R_3 = 1$ kΩ, $R_4 = 2$ kΩ, $C_3 = 2$ nF and $C_4 = 1$ nF, center frequency will be 500 krad/s, and with feedback factor of 19/20, the expected value of $Q = 10$. The simulated circuit response is shown in Figure 16.8(c), where center frequency is 79.077 kHz and $Q = 9.65$.

16.6 High Input Impedance Biquads Using CCIIs

Filter realizations where the input is applied at terminal X do not have high input impedance because of the CC's characteristics. When it is important to have high input impedance of the filter, its input needs to be applied at terminal Y as $I_y = 0$. Quite a few circuits are available with input at Y. A representative circuit is shown in Figure 16.9(a)[16.7]. Its nodal current equations are:

$$I_x + (V_{in} - V_1)Y_1 = 0 \tag{16.28a}$$

$$V_1(Y_1 + Y_2 + Y_3) - V_{in}\,Y_1 - V_2 Y_3 = 0 \tag{16.28b}$$

$$V_2(Y_3 + Y_4) - V_1 Y_3 - I_z = 0 \text{ and } I_z = I_x \tag{16.28c}$$

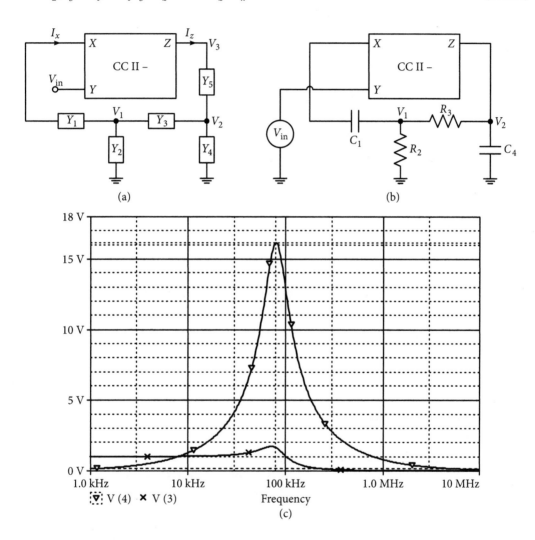

(a)   (b)

(c)

Figure 16.9 (a) High input impedance single CC biquad {With permission from Springer Nature}, and (b) circuit for providing high pass and band pass responses for Example 16.3. (c) Band pass and low pass responses from the circuit shown in Figure 16.9(b), when input is applied at the high impedance terminal Y.

Solving equations (16.28a–c) give the following transfer functions:

$$V_1/V_{in} = Y_1\,Y_4/D_1(s), \text{ and } V_2/V_{in} = Y_1\,Y_2/D_1(s) \tag{16.29a, b}$$

$$D_1(s) = Y_1 Y_4 + Y_2 Y_3 + Y_2 Y_4 + Y_3 Y_4 \tag{16.29c}$$

$$V_3/V_{in} = -Y_1 (Y_3 Y_4 + Y_2 Y_3 + Y_2 Y_4 + Y_2 Y_5)/Y_5 D_1(s) \tag{16.30}$$

The circuit provides LP and BP responses easily with $Y_1 = G_1$, $Y_2 = sC_2$, $Y_3 = sC_3$, $Y_4 = G_4$ and $Y_5 = G_5$ or zero. The important parameters of the second-order filter section will be:

$$\omega_o = (G_2 G_4 / C_2 C_3)^{\frac{1}{2}}, \; Q = (G_1 G_4 C_2 C_3)^{\frac{1}{2}} / (C_2 G_4 + C_3 G_4) \tag{16.31}$$

To get a combination of BP and HP responses, we need to select:

$$Y_1 = sC_1, \; Y_2 = G_2, \; Y_3 = G_3, \; Y_4 = sC_4 \text{ and } Y_5 = G_5 \text{ or zero}$$

It gives the denominator and the parameters as:

$$D_2(s) = C_1 C_4 s^2 + (G_2 + G_3) C_4 s + G_2 G_3 \tag{16.32}$$

$$\omega_o = (G_2 G_3 / C_1 C_4)^{\frac{1}{2}}, \; Q = (G_2 G_3 C_1)^{\frac{1}{2}} / (C_4)^{\frac{1}{2}} (G_2 + G_3) \tag{16.33}$$

Figure 16.9(b) shows the realized circuit with elements for which HP and BP responses will be:

$$V_1 D_2(s)/V_{in} = Y_1 Y_4 = s^2 C_1 C_4 \tag{16.34}$$

$$V_2 D_2(s)/V_{in} = Y_1 Y_2 = sC_1 R_2 \rightarrow \text{mid-band gain} = C_1/C_2 \tag{16.35}$$

Example 16.3: Design a BPF with a center frequency of 500 krad/s and pole-Q = 1.666. Verify the LP response as well.

Solution: Using equation (16.31), the calculated values of the elements are

$$R_1 = 1 \text{ k}\Omega, \; C_2 = 1 \text{ nF}, \; C_3 = 0.2 \text{ nF and } R_4 = 20 \text{ k}\Omega$$

Figure 16.9(c) shows the simulated response, where the center frequency of the BPF is 79.76 kHz (501.3 krad/s), which is close to the design value and the obtained Q = 1.637 is slightly less than the design value of 1.666. For the LP response, a peak occurs at 73.16 kHz with a peak gain of 1.72. The LP has unity gain at dc and its critical frequency is calculated as f_o = $73.16(1 - 1/2 \times 1.72^2)^{-0.5}$ = 80.25 kHz (504.4 krad/s).

16.7 Application of Voltage Following Property of CCs

We know that the X terminal follows the Y terminal voltage in a CC like the two input terminals of an ideal OA. This property can be utilized to convert some OA-RC circuits into CC based ones. For example, the well-known Sallen–Key circuit shown in Figure 16.10(a) is shown in Figure 16.10(b) as a CC based circuit. The circuit can provide VM as well as CM responses, using two CCs. Here, the output voltage is taken using the CCII+2 based voltage amplifier. In an OA-RC circuit, voltage at the inverting and non-inverting terminals is ideally equal. In the CC version, $V_{y1} = V_{x1}$ and if $R_3 = R_4$, voltage level at V_{y2} and hence, V_{x2} will also be equal to V_{y1}, making the CC version equivalent to the OA-RC circuit. It is significant to notice that the technique is applicable to other active RC filters to get either a unity or finite gain voltage amplifier. The following current–voltage relations are obtained for the CC based circuit:

$$V_{out} = V_{y1} = V_{x1}, \; I_{x1} = (V_{x1}/R_3) = (V_{out}/R_3) = I_{z1}, \; V_{y2} = I_{z1} \, R_4 = V_{x2} = V_{out}(R_4/R_3) = KV_{out},$$
$$K = (R_4/R_3) \tag{16.36a}$$

$$V_1 \, (G_1 + G_2 + sC_1) - V_{in} \, G_1 - V_{out} \, G_2 - KV_{out} \, sC_1 = 0 \tag{16.36b}$$

$$(V_1 - V_{out}) \, G_2 = sC_2 \, V_{out} \rightarrow V_1 = V_{out}(sC_2 + G_2)/G_2 \tag{16.36c}$$

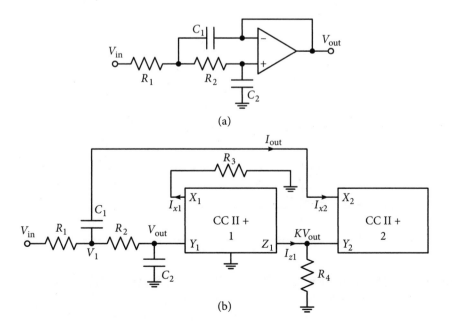

(a)

(b)

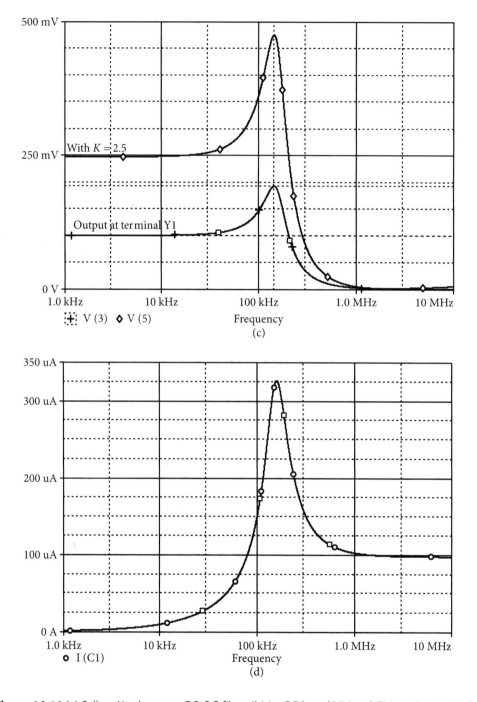

Figure 16.10 (a) Sallen–Key low pass OA-RC filter; (b) its CC based VM and CM version. (c) Voltage-mode low pass responses. (d) Current mode high pass response.

Voltage ratio transfer function of the LPF is obtained as:

$$(V_{out}/V_{in}) = (1/R_1R_2C_1C_2)/D(s) \tag{16.37}$$

$$D(s) = s^2 + s\left\{\frac{1}{R_1C_1} + \frac{1}{R_2C_2} + (1-K)\frac{1}{R_2C_2}\right\} + \frac{1}{R_1R_2C_1C_2} \tag{16.38}$$

The circuit also provides a mixed mode HP voltage–current transfer function:

$$(I_{out}/V_{in}) = (s^2/R_1)/D(s) \tag{16.39}$$

Second-order filter parameters are obtained from equation (16.38) as:

$$\omega_o^2 = \frac{1}{R_1R_2C_1C_2} \text{ and } Q = \frac{\left(R_1R_2C_1C_2\right)^{0.5}}{\left\{\dfrac{1}{R_1C_1} + \dfrac{1}{R_2C_2} + (1-K)\dfrac{1}{R_2C_2}\right\}} \tag{16.40}$$

Example 16.4: Obtain the VM LP response of equation (16.37) and CM HP response with pole frequency of 1000 krad/s and pole-$Q = 2$.

Solution: From equation (16.40), selecting equal value capacitors $C_1 = C_2 = 1.0$ nF, we get $R_1 = 1$ kΩ, $R_2 = 1$ kΩ, and with $Q = 2$, K will be 2.5; hence, selected $R_3 = 1$ kΩ and $R_4 = 2.5$ kΩ. Using these element values in Figure 16.10(b), the circuit is simulated and the responses are shown in Figures 16.10(c)–(d). With $Q = 2$, it is expected that a peak in LP response would occur at $1000/(1 - 1/2 \times 2^2)^{0.5}$ krad/s or 148.8 kHz. For the simulated LP response, low frequency gain is unity and a peak gain of 1.93 occurs at 144.35 kHz; close to the design value. The direct effect of K being 2.5 is reflected in the LP response at the Y input of CC1 with low frequency gain as 2.5. Incidentally, the circuit also gives a HP response in the form of current through the capacitor C_1. For the CM HP response, a simulated peak gain of 2.08 occurs at 167.74 kHz as shown in Figure 16.10(d); with $Q = 2$, theoretical value of the peak frequency is evaluated as $1000/(1 - 1/(2 \times 2^2)^{0.5} = 1069$ krad/s $= 168.9$ kHz, which is very close to the simulated value.

For the conversion of OA-RC networks into CM ones, the *adjoint network* technique has been found suitable. Adjoint circuits are those circuits where response and excitation are exchanged while replacing each branch (source) by its adjoint branch; there is no effect on the passive elements. As an example of converting an OA-RC circuit to a CM circuit, the circuit of Figure 16.10(a) is converted to a CM second-order section. Figure 16.11(a) shows the converted CM circuit with inputs and outputs interchanged. Since in this circuit the second CC is not used, $K = 1$. Input terminal of the OA are the controlling terminals of the controlled voltage source, which gets converted into output terminals Z and ground, with current I_Z through it.

Nodal current equations at nodes V_1, V_2 and X are as follows:

$$V_1(sC_1 + G_1 + G_2) - V_2G_2 = 0 \tag{16.41}$$

$$V_2(sC_2 + G_2) - V_1G_2 - I_x = 0 \tag{16.42}$$

$$V_1sC_1 - I_x + I_{in} = 0 \tag{16.43}$$

Solving equations (16.41)–(16.43), the following current ratio LP, BP and a combination of LP-BP transfer functions are obtained:

$$\frac{I(R_1)}{I_{in}} = \frac{-(G_1G_2)}{D(s)}, \frac{I(C_1)}{I_{in}} = \frac{-s(C_1G_2)}{D(s)}, \frac{I(R_2)}{I_{in}} = \frac{-s(C_1G_2)+(G_1G_2)}{D(s)}, \frac{I(C_2)}{I_{in}} = \frac{sC_2(sC_1+G_1+G_2)}{D(s)}$$
$$\tag{16.44}$$

$$D(s) = s^2C_1C_2 + sC_2(G_1 + G_2) + G_1 G_2 \tag{16.45}$$

Parameters of the second-order filters are:

$$\omega_o^2 = \frac{G_1G_2}{C_1C_2} \text{ and } Q = \frac{1}{(G_1+G_2)}\sqrt{\left(\frac{G_1G_2C_1}{C_2}\right)} \tag{16.46}$$

For selected values of the components as $R_1 = R_2 = 1$ kΩ, $C_1 = 4$ nF and $C_2 = 0.4$ nF, the design center frequency will be 125.7 kHz and Q will be 1.58.

Figure 16.11(b) shows the simulated LP response having a peak gain of 1.58 at 109.6 kHz, which gives a critical frequency of 122.5 kHz. Currents through other components simultaneously give HP and mixed BP responses. Center frequency of the BPF is 124.2 kHz with $Q = 1.52$. HP response shows a peak gain of 1.815 at a frequency of 135.03 kHz; this means an effective critical frequency of 124.3 kHz.

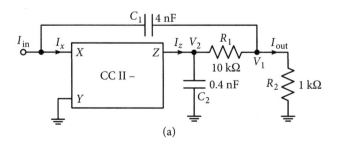

(a)

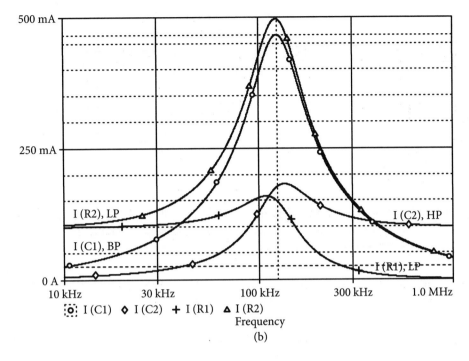

Figure 16.11 (a) CM second-order filter converted from Figure 16.10(a). (b) Simultaneous low pass, band pass and high pass, and mixed band pass responses from the circuit shown in Figure 16.11(a).

16.8 Biquads Using Single Current Conveyor

A general topology for obtaining biquads using single CC is shown in Figure 16.12(a). It shows three input currents I_1, I_2, and I_3. Different combinations of input currents and choice of elements have been given independently by several authors for such a configuration. For the general structure of Figure 16.12(a), the following transfer functions can be obtained while assuming the CC to be ideal.

$$I_{o1} = \{I_1(Y_1Y_2 + Y_1Y_3 + Y_2Y_3 + Y_3Y_4) + I_2Y_3\,Y_4 - I_3Y_1Y_4\}/D(s) \tag{16.47}$$

$$V_1 = \{I_2(Y_2 + Y_3) + I_3(Y_2 + Y_4)\}/D(s) \tag{16.48}$$

$$V_2 = \{I_3(Y_1 + Y_2 + Y_4) + I_2Y_2\}/D(s) \tag{16.49}$$

$$D(s) = Y_1Y_2 + Y_1Y_3 + Y_2Y_3 + Y_3Y_4 \tag{16.50}$$

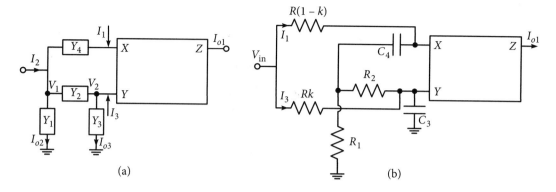

Figure 16.12 (a) A general C I C O biquad topology and (b) its implementation based on the feed-forward concept used in active filters for realizing notch and all pass response {With permission from Springer Nature}.

With reference to the choice of input currents by authors, Abuelmaatti [16.8] used only the input current I_2 with $I_1 = I_3 = 0$. For this choice, it is given that:

$$(I_{o1}/I_2) = \{Y_3 \, Y_4/D_1(s)\} \text{ and } (V_2/I_2) = \{Y_2/D_1(s)\} \tag{16.51}$$

Selection of the elements as mentioned in Figure 16.12(b) gives an HP response at I_{o1}, and V_2 provides a VM LP response; it can be converted to a CM LP response using another CC utilizing the equipotential property between X and Y terminals. Additionally, the current through $Y_3(I_{o3} = V_2Y_3)$ yields a BP response with the following selection of elements.

$$Y_1 = G_1, \, Y_2 = G_2, \, Y_3 = sC_3, \text{ and } Y_4 = sC_4 \tag{16.52}$$

However, it may be noted that current through Y_3 (or Y_1) will be taken after un-grounding Y_3 (or Y_1) and using an extra CC with its Y terminal grounded.

Other choices in terms of selection of passive elements are also possible to get these responses.

In equation (16.47), with $I_2 = 0$, the negative term associated with I_3 provides notch and AP responses with elements as selected earlier in equation (16.52). Notch and AP responses are obtained with the following respective conditions in equation (16.53) and (16.54):

$$I_1 \left\{ \frac{C_3}{R_1} + \frac{C_3}{R_2} \right\} = I_3 \frac{C_4}{R_1} \tag{16.53}$$

$$\left\{ \frac{C_3}{R_1} + \frac{C_3}{R_2} \right\} - \frac{C_4}{R_1} \frac{I_3}{I_1} = -\left\{ \frac{C_3}{R_1} + \frac{C_3}{R_2} \right\} \tag{16.54}$$

Obviously, the suggested configuration requires two input currents. Property of the equal voltage between X and Y terminals has been cleverly utilized by Anand Mohan [16.9] to circumvent it and uses only one input current. As shown in the circuit of Figure 16.12(b), input current division is obtained by two resistors having values of $R \times (1 - k)$ and Rk; here, $I_1 = kI_{in}$ and $I_3 = (1 - k) \times I_{in}$ while utilizing this technique. The transfer function for the circuit is obtained as:

$$\frac{I_{out}}{I_{in}} = k \frac{s^2 C_3 C_4 + s\left\{G_1\left(C_3 + C_4\right) - G_2\left(1 - k\right)/k\right\} + G_1 G_2}{s^2 C_3 C_4 + s\left\{G_1\left(C_3 + C_4\right)\right\} + G_1 G_2} \tag{16.55a}$$

The transfer function in equation (16.55a) can realize symmetrical notch and AP responses. Parameters of the notch or APF are as follows:

$$\omega_o^2 = \frac{G_1 G_2}{C_3 C_4} \text{ and } Q = \frac{1}{\left(C_3 + C_4\right)} \sqrt{\left(\frac{C_3 C_4 G_2}{G_1}\right)} \tag{16.55b}$$

Example 16.5: Design a notch filter using the structure shown in Figure 16.12(b), having one CC at the notch frequency of 350 krad/s (55.68 kHz). Convert it to an APF.

Solution: First, k is selected to be 0.5, for which current dividing resistors Rk and $R \times (1 - k)$ each are taken as 2 kΩ. With $I_3 = I_1$, and selecting $R_1 = R_2$, equation (16.53) gives $C_3 = 0.5C_4$. Finally, from equation (16.55b), the selected elements values for the notch frequency of 350 krad/s become:

$$R_1 = R_2 = 2 \text{ k}\Omega, C_4 = 2 \text{ nF}, C_3 = 1 \text{ nF}$$

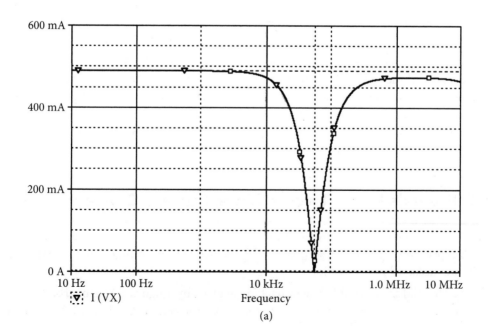

(a)

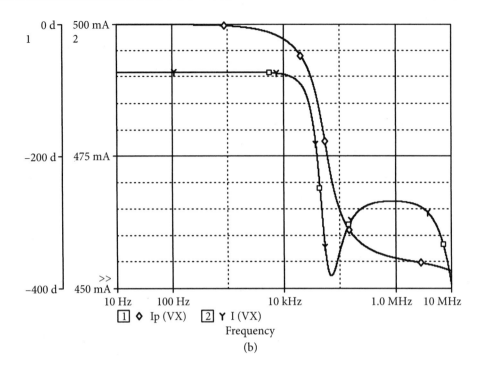

(b)

Figure 16.13 (a) CM notch filter response from Figure 16.12(b). (b) Magnitude and phase response of the CM all pass filter section of Figure 16.12(b).

The magnitude response of the simulated notch filter is shown in Figure 16.13(a), with notch at 56.08 kHz (352.5 krad/s) and attenuation of 55.1 dBs. Its gain at low frequency is 0.489 and gain at high frequencies is 0.474; k being 0.5.

To convert it to an APF, assuming the same current division with the same resistance values and the center frequency, equation (16.54) requires $C_4 = 4C_3$. For $C_3 = 1$ nF, the required values of the rest of the elements are $C_4 = 4$ nF and $R_1 = R_2 = 1.414$ kΩ. Using these elements, the simulated magnitude response of the APF is shown in Figure 16.13(b); variation in magnitude gain is from 0.491 to 0.452 only, when frequency was varied from 10 Hz to 10 MHz and the phase becomes 180° at a frequency of 55.47 kHz (348.67 krad/s) as shown in Figure 16.13(b).

16.9 Biquads Using Two or More Current Conveyors

A large number of biquadratic circuits using two or more CCs have been made available. Most of these depend on the tested ideas used in OA-RC active filter configurations. Some of these circuits are discussed in this section.

Figure 16.14(a) shows a multifunctional biquad given by Singh and Senani [16.10] which is based on the idea of well-known two-integrator loops, discussed in Section 8.2. CC1 and CC2 realize a lossless and lossy integrator respectively, using components R_1, C_1 and R_2, R_3 and C_2. LP and BP responses are available as V_{o1} and V_{o2}, respectively, and a notch or AP responses are obtained by subtracting the band pass output V_{o2} from the input V_{in} through CC3 and resistances R_4, R_5. In Figure 16.14(a), current–voltage relations are:

$$V_{x1} = V_{y1} = V_{o2} = I_{x1}R_3 \text{ and } I_{z1} = I_{x1} = V_{o1}sC_2 \qquad (16.56a)$$

$$V_{x2} = V_{y2} = V_{o1} = I_{x2}R_2 \text{ and } I_{z2} = I_{x2} \qquad (16.56b)$$

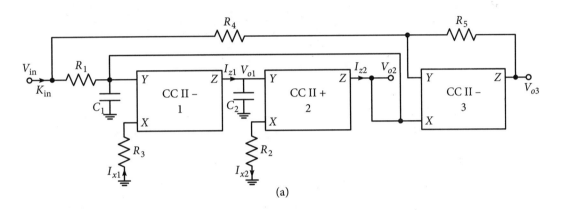

(a)

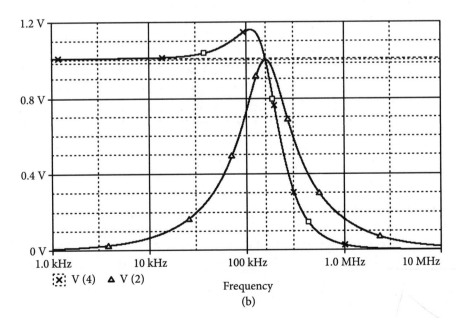

(b)

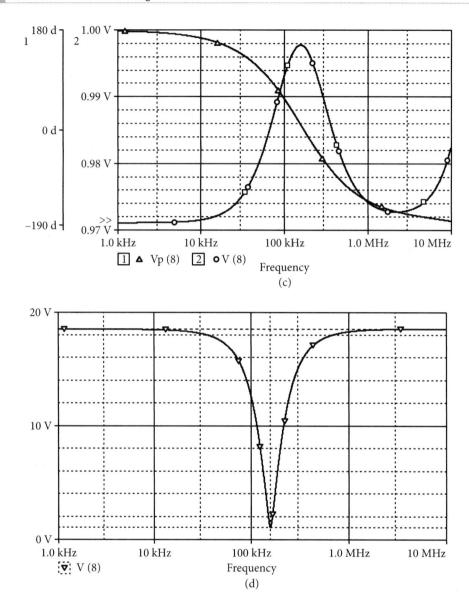

Figure 16.14 (a) Multi-function bi quad given by Singh and Senani {With permission from Springer Nature}. (b) Low pass and band pass responses using two-integrator loops and a summer for Example 16.6. (c) Magnitude and phase response of the all pass filter of Figure 16.14(a). (d) Notch filter response from the general biquad of Figure 16.14(a).

Nodal equation at terminal Y of CCII-1 will be:

$$V_{o2} (G_1 + sC_1) - I_{z2} - V_{in} G_1 = 0 \qquad (16.57)$$

Combining equations (16.56) and (16.57) gives the following LP and BP transfer functions:

$$(V_{o1}/V_{in}) = (1/R_2R_3C_1C_2)/D(s) \tag{16.58}$$

$$(V_{o2}/V_{in}) = (s/R_1C_1)/D(s) \tag{16.59}$$

$$D(s) = s^2 + \frac{s}{R_1C_1} + \frac{1}{R_2R_3C_1C_2} \tag{16.60}$$

From equation (16.60), expressions of the pole frequency and pole-Q are:

$$\omega_o^2 = \frac{1}{R_2R_3C_1C_2} \text{ and } Q = R_1\left(\frac{C_1}{R_2R_3C_2}\right)^{0.5} \tag{16.61}$$

To find the expression for output V_{o3}, the concerned relations are:

$$V_{x3} = V_{y3},\ I_{x3} = Iz_3 \text{ and } (V_{o2} - V_{in})\,G_4 + (V_{o2} - V_{o3})\,G_5 = 0 \tag{16.62}$$

Substituting V_{o2} from equation (16.59) in equation (16.62) gives:

$$\frac{V_{o3}}{V_{in}} = \frac{R_5}{R_4}\frac{s^2 - s\dfrac{R_4}{R_1R_5C_1} + \dfrac{1}{R_2R_3C_1C_2}}{D(s)} \tag{16.63}$$

AP and notch filter responses are available from equation (16.63) with the following respective relations:

$$R_5 = R_4 \text{ and } R_5 \gg R_4 \tag{16.64}$$

It is easily observed that while getting a notch with the condition of equation (16.64), the overall gain will become high.

Example 16.6: Design a BPF using the circuit of Figure 16.14(a) for center frequency of 1000 krad/s (159.09 kHz) and pole-Q of unity. Extend the same to realize AP and notch responses as well.

Solution: Selecting $C_1 = C_2 = 1$ nF, equation (16.61) gives the value of $R_2 = R_3 = 1\ \mathrm{k}\Omega$ and for $Q = 1$, $R_1 = 1\ \mathrm{k}\Omega$. Using these components in Figure 16.14(a), the simulated filter response is shown in Figure 16.14(b). Simulated center frequency and bandwidth are158.5 kHz and 158.7 kHz, resulting in $Q = 0.998$. For $Q = 1$, the obtained LP response shows a peak gain of 1.158 at a frequency of $f_{peak} = 110.37$ kHz. From the f_{peak} value, its 3 dB frequency will be $110.37/(1 - 1/2)^{0.5} = 156.11$ kHz; very close to the design value.

To get AP response, for a selected value of $R_4 = 1\ \mathrm{k}\Omega$, from equation (16.64), $R_5 = 1\ \mathrm{k}\Omega$. Figure 16.14(c) shows magnitude and phase responses of the APF. Variation in magnitude is only from 0.9977 V to 0.9711 V, and phase variation is very symmetrical from 180° to −180° with 0° of phase shift at 157.19 kHz.

With R_4 = 1 kΩ and R_5 = 20 kΩ, notch appears at 157.46 kHz. However, at low and high frequencies, the gain magnitude is nearly 19.5, and at notch, its value is almost unity (0.976). In some cases, these gain levels may not be entirely desirable. Fortunately, there are other choices available; one such scheme is given here.

Figure 16.15(a) shows modified connections for CC3 of Figure 16.14(a) which replaces CC3 in Figure 16.14 (a). With R_4 = R_5 = 1 kΩ, the notch occurs at 223.56 kHz with attenuation of over 42 dBs. Rise in the value of notch frequency (and pole frequency of BP and LP as well) occurs even though other elements do not change. This means a certain feedback, which is confirmed by changing the values of R_4 or R_5.

As the new notch frequency is $\sqrt{2}$ times the design frequency of 1000 krad/s, capacitances C_1 and C_2 are modified to $\sqrt{2}$ nF. Simulated response is shown in Figure 16.15(b) with notch and center frequency of BPF as 159.2 kHz; though bandwidth of the BPF of the configuration with three CCs is still 223.2 kHz

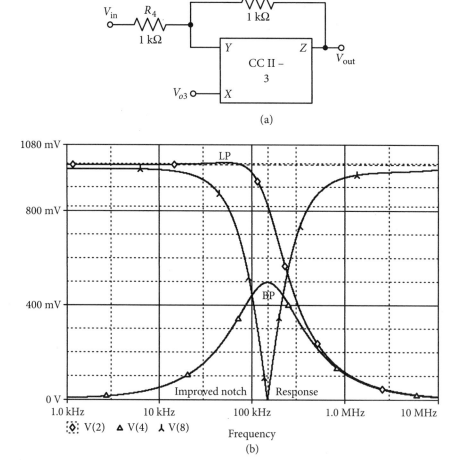

(a)

(b)

Figure 16.15 (a) Modification in the circuit of Figure 16.14(a) for getting better notch, and (b) its responses.

Different type of filter responses has been made available by Chang [16.11], who has given five biquadratic filter circuits employing four CCs and only grounded capacitors. Development of these circuits involve either of the following schemes: (i) connecting a grounded capacitor with a simulated lossless FI (floating inductance), (ii) using the idea of a KHN (Kerwin–Huelsman–Newcomb) biquad in which two integrator circuits are used to obtain LP and BP responses and (iii) use of a simulated GI (grounded inductance) shunted by a grounded capacitor multiplier. While simulating inductance, a practiced scheme is to first convert voltage to current and then re-convert the current to voltage. Figure 16.16(a) shows one of these circuits which is based on simulating FI using the same idea. The circuit can realize LP, BP and a notch response through simulating FI between terminals A and B.

Voltage V_{x1} being equal to V_{y1} (or at terminal A) and voltage V_{x3} being equal to V_{y3} (or V_{z4} at terminal B) means that $(V_A - V_B)$ is the voltage across the resistor R_1. Current through R_1 gets integrated through CC3 and capacitor C_1. The resulting integrated voltage is again converted to current, employing CC2, CC4 and resistor R_2 in such a way that it flows in terminals A and B, simulating FI. Resistor R_3 and capacitor C_2 are added to complete the second-order section for which current–voltage relations are:

$$V_{x1} = V_{o1} = V_{y1}, \ V_{x2} = V_{o2} = V_{y2}, \ V_{x4} = V_{y4} = 0 \tag{16.65a}$$

$$(V_{in} - V_{o1})\, G_3 = -V_{o2}\, G_2 \tag{16.65b}$$

$$I_{z4} = V_{o3}\, sC_2, \ I_{x4} = -V_{o2}\, G_2 \rightarrow V_{o3} = -V_{o2}\, G_2/sC_2 \tag{16.65c}$$

$$I_{x3} = (V_{o3} - V_{o1})\, G_1, \ I_{z3} = V_{o2}\, sC_1 \tag{16.65d}$$

From the relations in equation (16.65), the obtained transfer functions for notch, BP and LP are:

$$V_{o1} = V_{in} G_3 (s^2 C_1 C_2 + G_1 G_2)/D(s) \tag{16.66}$$

$$V_{o2} = -V_{in} \left\{ \frac{sC_2 G_1 G_3}{D(s)} \right\} \tag{16.67}$$

$$V_{o3} = V_{in} G_1 G_2 G_3/D(s) \tag{16.68}$$

$$D(s) = s^2 C_1 C_2 G_3 + s C_1 G_1 G_2 + G_1 G_2 G_3 \tag{16.69}$$

From equation (16.69), expressions of pole frequency and pole-Q are obtained as:

$$\omega_o = \sqrt{\left(\frac{G_1 G_2}{C_1 C_2} \right)}, \ Q_o = \frac{1}{R_3} \sqrt{\left(\frac{C_2 R_1 R_2}{C_1} \right)} \tag{16.70}$$

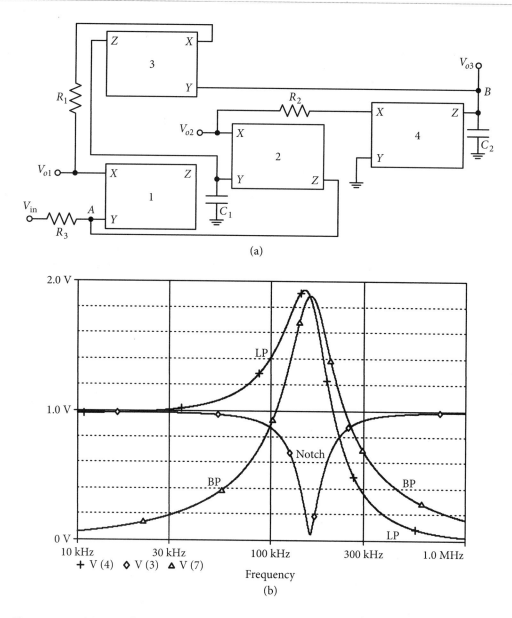

(a)

(b)

Figure 16.16 (a) One of the Chang's biquad topology using four CCs and grounded capacitors {With permission from Springer Nature} (b) its band pass, low pass and symmetrical notch responses.

Example 16.7: Design a BPF for a center frequency of 1000 krad/s and $Q_o = 2$, using the circuit of Figure 16.16(a). Also, show the response of the corresponding LP and notch.

Solution: From equation (16.70), selecting $C_1 = C_2 = 0.1$ nF, $R_1 = R_2$ will be 10 kΩ. For $Q_o = 2$, R_3 needs to be 5 kΩ. With these element values, the circuit is simulated and the responses are shown in Figure 16.16(b). Simulated ω_o = 1004.4 krad/s (159.789 kHz) and with a bandwidth of 83.57 kHz, realized Q_o = 1.91 for the BP. For the LP response, peak gain of 1.93 occurs at 148.39 kHz, which corresponds to ω_{oLP} = 148.39/(1 −1/2 × 1.93^2) = 159.48 kHz. Notch frequency is 159.63 kHz (1003.38 krad/s) with attenuation of 26.3 dBs.

16.10   Inductance Simulation Using Current Conveyors

In most circuits, it is the inductance which has been attended to the most for well-known reasons. Inductance needs to be simulated in the grounded as well as in the floating form. Figure 16.17(a) shows a circuit, wherein direction of currents is shown according to the sign of CCII; expression of the voltage V_{y2} is written as:

$$V_{y2} = V_{x2} = I_{x2}\,R_2 = I_{z2}\,R_2 = I_{in}\,R_2$$

$$V_{y2} = \frac{I_{z1}}{sC} = \frac{I_{x1}}{sC} = \frac{V_{x1}}{sCR_1} = \frac{V_{y1}}{sCR_1} = \frac{V_{in}}{sCR_1}$$

It gives the expression for simulated inductance as:

$$L_{sim} = CR_1\,R_2 \qquad\qquad (16.71)$$

Simulated GI as obtained in equation (16.71) depends on the same idea of voltage to current conversion and current to voltage conversion, which has been used for simulating FI by a number of authors. In this case, I_{in} is converted to V_{y2} using R_1 and C; then V_{x2} drives a current I_{z2} using R_2 and again to voltage V_{x1} using R_1.

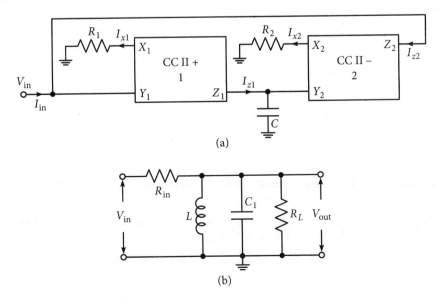

(a)

(b)

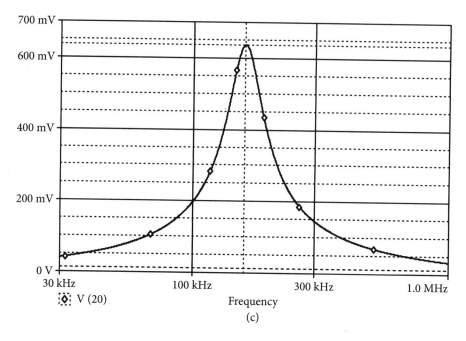

Figure 16.17 (a) Grounded impedance simulator using CCs (b) passive RLC band pass circuit for testing the simulated grounded inductance and (c) its simulated response for Example 16.7.

Example 16.7: Verify the simulation of the grounded inductance of Figure 16.17(a).

Solution: Figure 16.17(b) shows a passive RLC circuit which realizes BP response. Its transfer function is obtained as:

$$\frac{V_{in}}{V_{out}} = \frac{(s/C_1 R_L)}{s^2 + s\left(\frac{R_{in} + R_L}{C_1 R_{in} R_L}\right) + \frac{1}{LC_1}} \qquad (16.72)$$

For the test circuit, center frequency and pole-Q expressions are:

$$\omega_o = \frac{1}{\sqrt{LC_1}} \qquad Q_o = \frac{R_{in} R_L}{R_{in} + R_L}\sqrt{\left(\frac{C_1}{L}\right)} \qquad (16.73)$$

For ω_o = 1000 krad/s, if selected C_1 = 0.4 nF, the required value of the inductance to be simulated will be 2.5 mH. Using equation (16.71), inductance is simulated with $R_1 = R_2 = 5$ kΩ and C = 0.1 nF. Again, using equation (16.73), if values of R_{in} and R_L are 50 kΩ and 100 kΩ, respectively, it gives Q_o = 3. Figure 16.17(c) shows the simulated response with $\omega_o = 2\pi \times$ 159.84 = 1002.8 krad/s. Its bandwidth is 49.9 kHz giving simulated Q_o = 3.2.

Figure 16.18(a) shows a scheme given by Toumazou and Lidgey [16.12] for general floating impedance simulation. Here input voltages are buffered by CC1 and CC2, which also converts voltages to currents using impedance Z_1. Next the currents get converted to voltages using impedances Z_2 and Z_3. Lastly, conveyors CC3 and CC4 convert these voltages to current using impedance Z_4; these conveyors also provide path for the input currents. With CC1 and CC3 as negative and CC2 and CC4 as positive, current–voltage relations are as follows:

$$V_{x1} = V_{y1} = V_1, \ V_{x2} = V_{y2} = V_2$$

$$I_{x1} = -I_{x2} = (V_2 - V_1)Y_1 = I_{z1}, \ I_{z2} = I_{x2} = -(V_2 - V_1)\ Y_1$$

$$V_{y3} = -I_{z1}Z_2 = -(V_2 - V_1)\ Y_1 Z_2 = V_{x3}, \ V_{y4} = -I_{z2}Z_3 = (V_2 - V_1)\ Y_1 Z_3 = V_{x4}$$

$$I_{x3} = I_{x4} = (V_{x3} - V_{x4})Y_4 = -\{(V_2 - V_1)Y_1 Y_4\ (Z_2 + Z_3)\} = I_{z3} = I_{z4} = -I_2 = I_1 \qquad (16.74)$$

Hence, expression of the realized floating impedance from equation (16.74) is given as:

$$Z_1 Z_4/(Z_2 + Z_3) \qquad (16.75a)$$

Choice of elements for these impedances can give inductance, FDNR (frequency dependant negative resistor) or, can function as a capacitance multiplier as well.

Few circuits are available using the idea behind this configuration. Figure 16.18(b) shows one such circuit given by Singh [16.13]. For the circuit shown in Figure 16.18(b), the realized floating inductance expression will be as given below:

$$R_1 R_3 C_2 \qquad (16.75b)$$

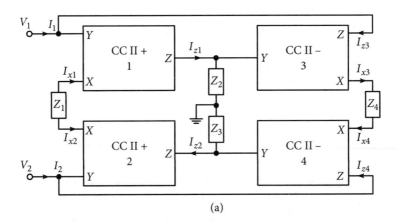

(a)

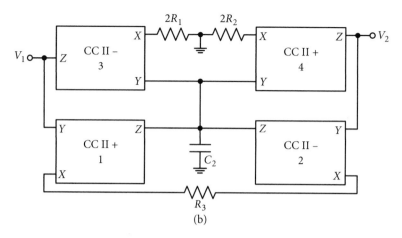

Figure 16.18 (a) A generalized scheme for realizing floating impedance [16.12] {With permission from Springer Nature}, and (b) floating inductance simulators by Singh [16.13] {With permission from Springer Nature}.

Example 16.9: Realize a fifth-order Chebyshev filter having maximum 0.5 dB pass band ripple with pass band edge frequency of 1000 krad/s using the FI simulator of Figure 16.18(b).

Solution: Fifth-order Chebyshev passive filter structure is shown elsewhere as well but repeated in Figure 16.19(a). For the given specifications, frequency scaling factor will be 1000 krad/s and if the impedance scaling factor of 1 kΩ is used, component values will become:

$$R_{in} = R_L = 1 \text{ k}\Omega, C_1 = C_5 = 1.7058 \text{ nF}, C_3 = 2.5408 \text{ nF and } L_2 = L_4 = 1.2296 \text{ mH}$$

FIs are simulated using the circuit in Figure 16.18(b). For both the inductors, $2R_1 = 2.459$ k$\Omega = 2R_2$, $R_3 = 1$ kΩ and $C_2 = 1$ nF; the complete circuit is shown in Figure 16.19(b). The simulated magnitude response of the filter is shown in Figure 16.19(c) with a pass band edge frequency of 157.29 kHz (988.68 krad/s) with maximum ripple width of 0.4917 dB.

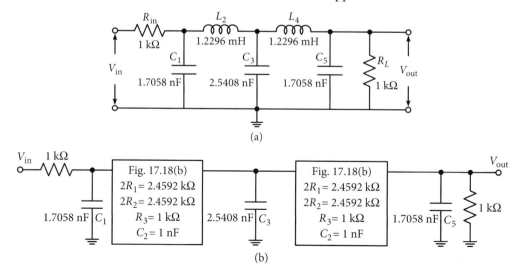

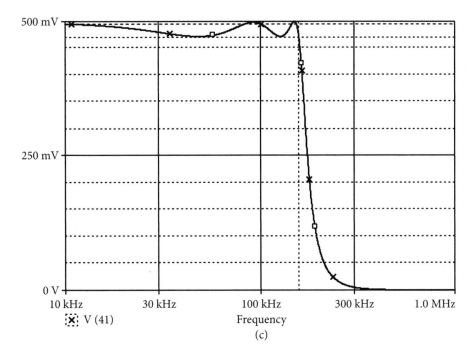

Figure 16.19 (a) Fifth-order Chebyshev passive filter. (b) Chebyshev filter using the floating inductor simulator circuit shown in Figure 16.18(b). (c) its simulated response while using four CC circuit of Figure 16.18(b).

Non-ideal inductance has also been simulated using non-ideal gyrators [16.14]. Figure 16.20(a) and (c) show two such circuits. For the circuit in Figure 16.20(a):

$$V_{y1} = V_{x1} = \left(I_{in} - I_{z2}\right)Z_2, \ V_{y2} = V_{x2} = 0, \ I_{x2} = I_{z2} = \frac{V_{x1}}{Z_3} \rightarrow I_{z2} = \frac{Z_2}{Z_2 + Z_3}I_{in} \quad (16.76)$$

$$V_{in} = V_{y1} + \left(I_{in} - I_{z2}\right)Z_1 \rightarrow \frac{V_{in}}{I_{in}} = Z_3\frac{Z_1 + Z_2}{Z_2 + Z_3} \quad (16.77)$$

For equation (16.77), equivalent of input impedance results in a parallel combination of a few components. For example, for $Z_1 = R_1$, $Z_2 = (1/sC_2)$ and $Z_3 = R_3$, simulated equivalent circuit contains inductance $L = R_1R_3C_2$ as shown in Figure 16.20(b) in parallel with a series combination of R_1 and C_2. It is to be noted that both the conveyors are positive with I_{z2} going into it. Similarly, for the circuit shown in Figure 16.20(c), simulated inductance expression is the same for the same selection of components as that in Figure 16.20(b), and its equivalent circuit is shown in Figure 16.20(d).

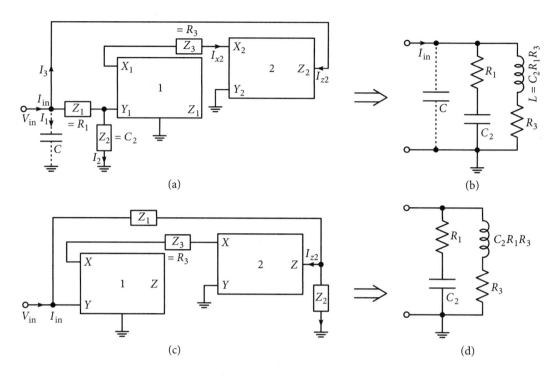

Figure 16.20 (a) and (c) Realizing non-ideal inductor, [16.14]{With permission from Springer Nature} and (b) and (d) are their simulated equivalents.

Example 16.10: Obtain a biquad using the non-ideal inductance simulator of Figure 16.20(a) with a pole frequency of 250 krad/s.

Solution: If a capacitor C is connected in parallel with the non-ideal inductor as shown with dotted lines in Figure 16.20(a), LP, BP and HP transfer functions are obtained. In Figure 16.20(b), the following relations can be easily written:

$$I_{in} = I_1 + I_2 + I_3 \tag{16.78}$$

$$I_1 = V_{in} sC, \, I_2 = V_{in}/\{R_1 + (1/(sC_2))\}, \, I_3 = sC_2 \, V_{in}/(R_3 + sC_2 \, R_1 \, R_3) \tag{16.79}$$

Combining equations (16.78)–(16.79), the following transfer functions are obtained:

$$\frac{I_2}{I_{in}} = \frac{I(R_1)}{I_{in}} = \frac{R_3 C_2 s}{D(s)}, \frac{I_3}{I_{in}} = \frac{I(R_3)}{I_{in}} = \frac{1}{D(s)} \tag{16.80}$$

$$D(s) = R_1 R_3 C_2 C s^2 + R_3 (C_2 + C)s + 1 \tag{16.81}$$

Pole frequency and pole-Q expressions are:

$$\omega_o = \frac{1}{\sqrt{(R_1 R_3 C_2 C)}}, \ Q = \frac{\sqrt{(R_1 R_3 C_2 C)}}{R_3 (C_2 + C)}$$
(16.82)

Selected component for the center frequency of 250 krad/s, values are:

$R_1 = 64 \text{ k}\Omega$, $R_3 = 1.0 \text{ k}\Omega$, $C_2 = 0.2 \text{ nF}$ and $C = 1.25 \text{ nF}$

Theoretical value of pole-Q from equation (16.82) will be 2.758.

Simulated responses are shown in Figure 16.21(a)–(b). Center frequency of the CM BPF is 39.355 kHz (247.37 krad/s) and with a bandwidth of 14.526 kHz, Q = 2.706; mid-band gain is 0.136. For the CM LP response, peak occurs at 37.804 kHz, with a peak gain of 2.748 (almost equal to Q) and CM HP response has a peak gain of 2.872 at a frequency of 40.703 kHz. Deviations in the pole frequency of LP and HP are justified with the value of Q very nearly equal to 2.748 and 2.872, respectively.

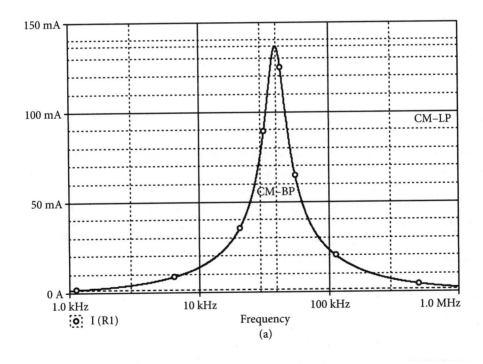

(a)

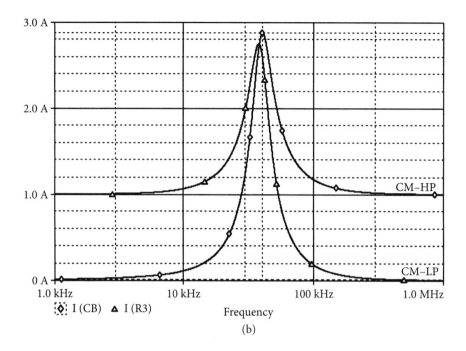

Figure 16.21 (a) CM band pass and (b) high pass and low pass responses from the circuit of Figure 16.20(a).

16.11 Application of CDTA

Recently, current differencing transconductance (CDTA) has been gaining attention for the realization of CM filters. Its behavioral model and symbol for double input double output (DIDO) are shown in block form structure in Figure 16.22(a) and (b) [16.15]. Its transconductance g_m can be controlled by difference current sources and an OTA as shown in Figure 16.22(c) [16.15] with an external load resistance. Though the difference between the input currents I_p and I_n can be multiplied by an electronically controllable gain factor b, which may be more than one, but generally 'b' is equal to unity. Voltage developed across grounded impedance gets converted to current I_x through transconductance. Behavior of the CDTA model can be expressed by the following matrix:

$$\begin{bmatrix} I_x \\ I_z \\ V_p \\ V_n \end{bmatrix} = \begin{bmatrix} 0 & 0 & b & -b \\ \pm g_m & 0 & 0 & 0 \\ 0 & 0 & 0 & 0 \\ 0 & 0 & 0 & 0 \end{bmatrix} \begin{bmatrix} V_z \\ V_x \\ I_p \\ I_n \end{bmatrix}$$

(16.83)

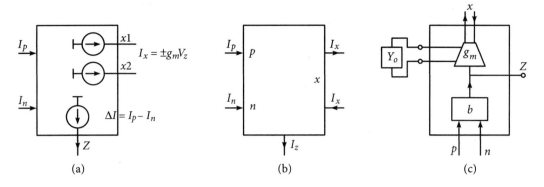

Figure 16.22 (a) Behavioral model of a CDTA and (b) symbol of DIDO type of CDTA. (c) CDTA element with the possibility to choose transconductance by an external admittance Y_o.

A number of options are available for the realization of CDTA. One of the options is through the use of current conveyors and OTAs. Shown in Figure 16.23, two CCII+ are used to implement differential current source and the output current is obtained through the OTA [16.16]. More than one OTA may be used to get as many output currents.

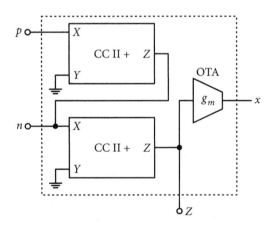

Figure 16.23 Implementation of CDTA using current conveyors and OTA.

Some simple applications of CDTAs are shown in Figure 16.24 [16.15]. Figure 16.24(a) shows a difference current $(I_p - I_n)$ -controlled current source, wherein the current gain K_I equals $(g_m Z)$; Z being an externally connected impedance. Obviously, depending upon the nature of Z, current gain will be linear or frequency dependent. If Z is removed, the CDTA becomes a current operational amplifier (COA). A dual input, duel output COA is shown in Figure 16.24(b). Finite current gain is obtainable from a COA by applying negative feedback, through a current divider as shown in Figure 16.24(c). The current gain is independent of g_m, and depends on the ratio of the elements of the external potential divider.

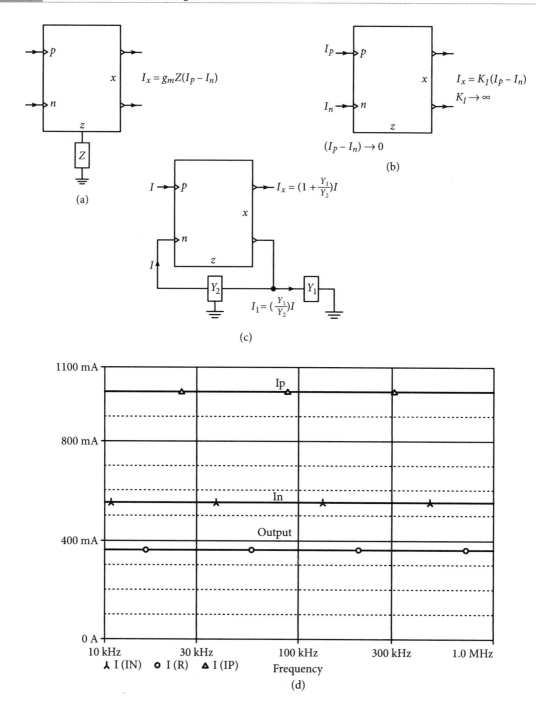

Figure 16.24 CDTA as (a) a difference current controlled current amplifier, (b) COA and (c) finite current gain amplifier using negative feedback. (d) Input and output currents of the CCCA using Figure 16.24(c).

As a simple example, let I_p = 1 mA, I_n = 0.55 mA, g_m = 0.8 mA/V; then, with Z being a resistance of 1.0 kΩ, the theoretical value of output current will be 0.36 mA. Figure 16.24(d) shows PSpice simulated input and output currents confirming the working of the circuit as a current controlled current amplifier (CCCA).

If a grounded capacitor is used as the impedance Z, CDTA becomes an ideal integrator; it becomes a lossy integrator by connecting a resistance in parallel with the capacitor.

16.11.1 Inductance simulation using CDTAs

Inductance simulation technique is one of the most often used techniques in active filter synthesis. To use this approach, grounded as well as the floating form of inductance simulator are needed. Figure 16.25(a) shows a circuit for the simulation of GI. With applied voltage V_{in}, output current from CDTA1 will be $g_{m1} V_{in}$.

It gives $V_{z2} = -(g_{m1} V_{in} b_2/sC_L)$, and with $I_{in} = -b_1 I_{n1}$:

$$I_{in} = g_{m1} V_{in} b_1 g_{m2} b_2/sC_L \rightarrow (V_{in}/I_{in}) = sC_L/g_{m1} b_1 g_{m2} b_2 \tag{16.84}$$

For $b_1 = b_2 = 1$, input impedance is that of an inductor with the inductance expression being $(C_L/g_{m1}g_{m2})$.

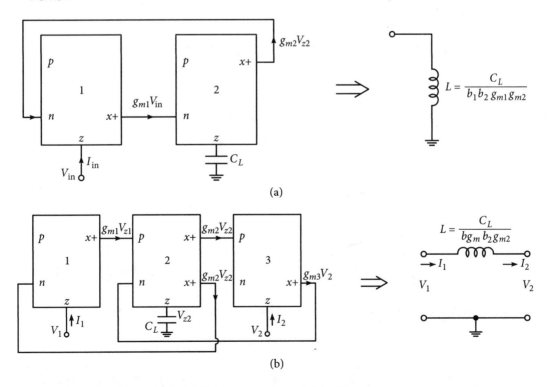

(a)

(b)

Figure 16.25 (a) Grounded inductance simulation using two CDTAs. (b) Floating inductance simulation using three CDTAs.

Figure 16.25(b) shows a circuit simulating floating inductance [16.17]. Here current I_1, I_2 and V_{z2} are obtained as:

$$I_1 = b_1 g_{m2} V_{z2}, I_2 = b_3 g_{m2} V_{z2} \text{ and } V_{z2} = b_2 (g_{m1} V_1 - g_{m3} V_2)/(sC_L) \tag{16.85}$$

For $b_1 = b_3 = b$ and $g_{m1} = g_{m3} = g_m$

$$I_1 = I_2 = \frac{V_1 - V_2}{sL}, \text{ where } L = \frac{C_L}{bg_m b_2 g_{m2}} \tag{16.86}$$

Example 16.11: Use the GI of Figure 16.25(a) in the passive RLC filter of Figure 16.26(a), and test it for a center frequency of 100 krad/s and $Q = 2$.

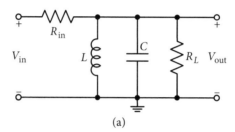

(a)

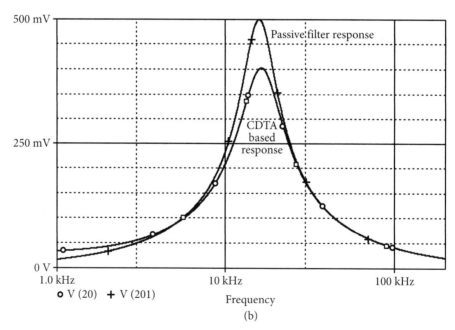

(b)

Figure 16.26 (a) A passive second-order band pass filter. (b) Response of the second-order band pass filter of Figure 16.26(a) and its active version using CDTA based grounded inductance.

Solution: For the passive filter, expressions for the center frequency and pole-Q are:

$$\omega_o = (1/LC)^{0.5}, Q = \frac{R_L \times R_{in}}{R_L + R_{in}} \sqrt{(C/L)} \tag{16.87}$$

For ω_o = 100 krad/s, if C is selected as 10 nF, then the required value of inductance from equation (16.87), L = 10 mH. Inductance is realized with $g_{m1} = g_{m2}$ = 0.1 mA/V, C_L = 0.1 nF and if $R_L = R_{in}$, for Q = 2, $R_L = R_{in}$ = 4 kΩ. With these element values, passive RLC BPF, as well as its active version using CDTA based GI are simulated, and the responses are shown in Figure 16.26(b). Response for the passive filter is near ideal, whereas for the active filter ω_o = 102.86 krad/s and Q = 1.646.

16.11.2 Biquadratic circuits using CDTAs

A number of good performance second-order filter circuits are available in literature. Here, only two circuits and their important relations are given in brief. The circuit shown in Figure 16.27 is given by M. Dehran et al. [16.18], for which the current ratio transfer function are as follows:

$$\frac{I_{LP}}{I_{in}} = \frac{g_m / RC_1 C_2}{D(s)}, \frac{I_{BP}}{I_{in}} = \frac{s / RC_1}{D(s)}, \frac{I_{HP}}{I_{in}} = \frac{s^2}{D(s)}, D(s) = s^2 + (s/RC_1) + g_m / RC_1 C_2 \tag{16.88}$$

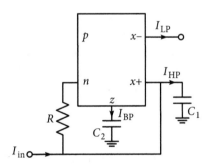

Figure 16.27 A biquad using a CDTA [16.18].

The circuit in Figure 16.28 is given by Bioleck and Biolkava [16.16], which is based on the Tow–Thomas biquad. Important expressions are as follows:

$$\frac{I_{x1}}{I_{in}} = \frac{s(\omega_o / Q)}{D(s)}, \frac{I_{x2}}{I_{in}} = \frac{\omega_o^2}{D(s)}, \frac{I_{c1}}{I_{in}} = \frac{s^2}{D(s)}, \frac{I_{c2}}{I_{in}} = \frac{s(G_{m1}\omega_o R / Q)}{D(s)}, \frac{I_R}{I_{in}} = \frac{s(\omega_o / Q)}{D(s)} \tag{16.89}$$

$$D(s) = s^2 + s(\omega_o / Q) + \omega_o^2, \omega_o^2 = \frac{G_{m1}G_{m2}}{C_1 C_2}, Q = R\sqrt{(G_{m1}G_{m2}C_1 / C_2)} \tag{16.90}$$

Obviously for obtaining BE or AP, either some other circuit is to be used or the well-known techniques of summing two responses can be used.

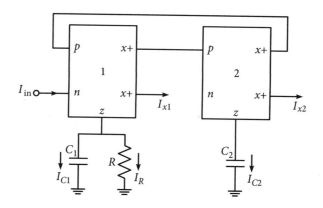

Figure 16.28 Realization of Tow–Thomas biquad using two CDTAs.

References

[16.1]   Mahattanakul, J., and C.Toumazou. 1998. 'Current-Mode Versus Voltage-Mode-Biquad Filters,' *IEEE Transactions on Circuits and Systems-II: Analog and Digital Signal Processing* 45 (2): 173–86.

[16.2]   Smith, K. C., and A Sedra. 1968. 'The Current Conveyor: A New Circuit Building Block,' *Proceedings of IEEE (Letters)* 56: 136801369.

[16.3]   Sedra, A., and K. C. Smith. 1970. 'A Second Generation Current Conveyor and its Applications,' *IEEE Transactions on Circuit Theory* 17 (1): 132–134.

[16.4]   Maheshwari, S. 2008. 'A Canonical Voltage Controlled VM-APS with a Grounded Capacitor,' *Circuits Systems Signal Processing* 27 (123–132): 123–32.

[16.5]   Higashimura, M., and Y. Fukui. 1990. 'Realization of Current-Mode All Pass Networks Using a Current Conveyor,' *IEEE Transactions on Circuits and Systems* 37: 660–661.

[16.6]   Abuelmaatti, M. T. 1993. 'New Current Mode Active Filters Employing Current Conveyors,' *International Journal of Circuit Theory and Applications* 21: 93–9.

[16.7]   Liu, S. J., and H. W. Tsao. 1991. 'Two Single CCII Biquads with High Input Impedance,' *IEEE Transactions on Circuits and Systems* 38: 456–61.

[16.8]   Abuelmaatti, M. T. 1987. 'Two Minimum Component CCII Based RC Oscillators,' *IEEE Transactions on Circuits and Systems* 34: 980–1.

[16.9]   Ananda Mohan, P. V. 1995. 'New Current Mode Biquad Based on Friend-Deliyannis Active RC Biquad,' *IEEE Transactions on Circuits and Systems, Part II* 42: 225–8.

[16.10] Singh, V. K., and R. Senani. 1990. 'New Multifunction Active Filter Configuration Employing Current Conveyors,' *Electronic Letters* 26: 1814–6.

[16.11] Chang, C. 1997. 'Multifunctional Biquadratic Filters using Current Conveyors,' *IEEE Transactions on Circuits and Systems, Part II* 44: 956–8.

[16.12] Toumazou, C., and F. J. Lidgey. 1985. 'Floating Impedance Converters using Current Conveyors,' *Electronics Letters* 21: 640–2.

[16.13] Singh, V. 1981. 'Active RC Single Resistance-Controlled Loss-Less Floating Inductance Simulation Scheme using Single Grounded Capacitor,' *Electronics Letters* 16: 920–1.

[16.14] Fabre, A., and M. Alami. 1992. 'Insensitive Current Mode Band Pass Implementation Based Non-Ideal Gyrator,' *IEEE Transactions on Circuits and Systems* 39: 152–5.

[16.15] Biolek, D. 2004. 'New Circuit Elements for Signal Processing in Current-Mode.' http://www.elektrorevue.cz/clanky/04028/index.html.

[16.16] Biolek, D. 2003. 'CDTA-Building Block for Current-Mode Analog Signal Processing,' *Proceedings of ECCTD 03* III: 397–400.

[16.17] Biolek, D., and V. Biolkova. 2004. 'Tunable CDTA-based Ladder Filters,' *WSEAS Transactions on Circuits* 3 (1): 165–7.

[16.18] Dehran, M., Singh, I. P., Singh, K. and Singh, R. K. 2013. 'Switched Capacitor Biquad Filter using CDTA,' *Circuit and Systems* 4: 438–42.

Practice Problems

16-1 Using CCII, obtain (a) a current source of 10 μA at a frequency of 100 kHz from a voltage source of 1 volt and (b) convert it again into a voltage source of 1.5 volts.

16-2 Verify that the circuit shown in Figure 16.3(a) is a current amplifier. Test it for a current gain of 10, and also show that it is a non-inverting current amplifier.

16-3 Design and test the current summation circuit shown in Figure 16.3(c) with input currents of 2 mA, 3 mA and 4 mA with respective weightage of 0.2, 0.75 and 0.5.

16-4 What will be the magnitude of the output current if direction of the 3 mA current in Problem 16-3 is reversed? Verify the same.

16-5 Design and test VM and CM inverting integrators of Figure 16.4 using CCII with a dc gain of 12 dBs and their gain attenuates by 3 dBs at 100 kHz.

16-6 (a) Design and test a CM LP filter using the general structure shown in Figure 16.5(a) with a dc gain of unity and a 3 dB frequency of 500 krad/s.

(b) Design and test a CM HP filter using the general structure shown in Figure 16.5(a) with a high frequency gain of unity and a 3 dB frequency of 250 krad/s.

(c) Design and test a CM AP filter using the general structure shown in Figure 16.5(a) having a phase shift of 90° at a frequency of 200 krad/s.

16-7 Repeat Problem 16-6(c) using the circuit shown in Figure 16.5(b).

16-8 Design and test a notch filter using the circuit shown in Figure 16.5(a) having a notch at 120 kHz. Find the attenuation at the notch.

16-9 Design and test a VM AP filter using the circuit shown in Figure P16.1(a) with its phase shift becoming –90° at 200 kHz.

16-10 Repeat Problem 16-9 using the alternate circuit shown in Figure P16.1(b).

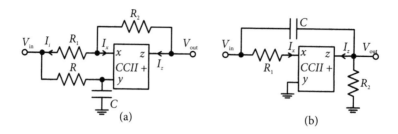

Figure P16.1

16-11 Using the passive structure of Figure 16.17(b), obtain a BP filter for center frequency of 50 kHz and $Q = 5$. Employ non-ideal inductor of Figure 16.20 (a).

Use the voltage following property in CCs for Problems 16-12 to 16-14.

16-12 Design and test the CC based second-order VM LP and a CM HP filter through conversion of an OA-RC single amplifier circuit, having cut-off frequency of 100 kHz and pole-$Q = 2.5$.

16-13 Convert a single amplifier circuit to a CC based second-order VM BP filter having center frequency of 100 krad/s and $Q = 5$. What kind of other CM response can also become available from the circuit? Verify the responses.

16-14 Repeat Problem 16-13 to realize a VM HP filter with a cut-off frequency of 150 krad/s and $Q = 1.5$.

16-15 Realize a BP filter using the multi-functional biquadratic circuit shown in Figure 16.14(a) for $\omega_o = 500$ krad/s and $Q = 5$. What is the peak gain and frequency at the peak for the corresponding LP response? Also realize AP and notch responses.

16-16 Modify the circuit in Problem 16-15 to get a notch when the lowest output voltage is not equal to the input voltage.

16-17 Design and test a maximally flat second-order HP filter section having 3 dB frequency of 50 kHz using the inductor circuit shown in Figure 16.18(b).

16-18 An LP filter has the following specifications:

$\omega_1 = 100$ krps, $\omega_2 = 600$ krps, $A_{max} = 1$ dB and $A_{min} = 50$ dB

Design and test a CC based filter to satisfy the specifications employing the inductor shown in Figure 16.17(a).

16-19 Use the general topology shown in Figure 16.12(a) and the feed-forward concept of Figure 16.12(b) to realize a BP and a LP response with $Q = 2.5$ and center frequency of 100 kHz. Find the mid-band gain realized for the BP and the peak frequency of the LP filter. Convert the structure to realize an asymmetrical notch and an AP filter; test the designed circuit.

16-20 Show that the circuit shown in Figure P16.2 works as a current -controlled current source.

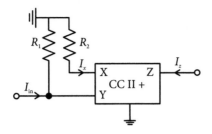

Figure P16.2

16-21 Utilize the circuit in Figure 16.18(b) to realize a second-order BR filter having notch frequency of 60 kHz.

16-22 Show that the circuit shown in Figure P16.3 works as a voltage-controlled current source.

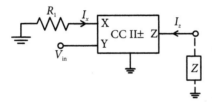

Figure P16.3

16-23 Using the passive structure shown in Figure 16.17(b), obtain a BP filter for center frequency of 60 kHz and $Q = 6$. Employ the non-ideal inductor shown in Figure 16.20(c).

Chapter 17

Active R and Active C Filters

17.1 Introduction

Conventionally, the design of OA-RC (operational amplifiers with resistors and capacitors) circuits assumes that OAs are ideal, having a large frequency independent gain. However, a commercially integrated OA is a non-ideal device, whose gain is frequency dependent and exhibits a LP (low pass) characteristic. A simple model for OAs, given in Chapter 1, is repeated here depicting its frequency dependent gain nature.

$$A(s) = \frac{A_o \omega_a}{s + \omega_a} \cong \frac{B}{s} \tag{17.1}$$

In equation (17.1), ω_a is its first pole, A_o is the gain at dc and $B = A_o \omega_a$ is gain bandwidth product. This effect of the gain bandwidth product, B, being finite, its effect on the magnitude and phase responses of the filter became important criterion for comparison of the performance of filters. Finite B restricts the frequency range of the operation of OA-RC filters to mostly the audio frequency range, particularly when commercially economical OAs, such as OA 741, are used. In fact, in some topologies, product of pole-Q and center frequency ω_o has to be much less than B to avoid undesirable enhancement in Q, and likely instability. Consequently, suitable compensating schemes were developed to design filters with lesser dependence on OA gain characteristics, like those shown for integrators in Chapter 2; in other cases, specially designed but costly OAs were used.

In an alternative approach, instead of considering the frequency dependent gain of the OA as undesirable, it has been exploited by using it directly in the design itself. This approach increases the frequency range of operation, and also reduces or completely eliminates the requirement of external capacitors. In this context, networks which use only OAs and resistors, with no external capacitors in their implementation, and derive their response from the internal dynamics of the OAs, are known as *active R circuits* [17.1], [17.2].

Similar to active R, *active C networks* have also been designed. They utilize the frequency dependent model of the OAs, but use only capacitances, mostly in ratio form. It results in an extended frequency range; the use of small value capacitors in ratio form provides related advantages.

17.2 Basic Techniques in Active R Synthesis

Active R synthesis is an off-shoot of the active RC synthesis, where external capacitors were eliminated. Hence, the techniques used in the active RC synthesis are only tailored to suit active R realizations; these are briefly discussed in this section.

The active R realization approaches can be broadly classified as (i) the *direct form* and (ii) the *cascade form*. As in the OA-RC case, a resistively terminated LC ladder is very commonly used as the basic starting structure in the active R and active C direct form approach. The direct form approach can be further classified into (a) the simulated immittance (inductance and capacitance using OAs and resistors) and (b) the frequency dependent negative resistance (FDNR) approach. For the immittance simulation approach, GI (grounded inductance) and FI (floating inductance) simulation is discussed in Section 17.4 using OAs and resistances (ratios). OA R simulation of FDNR in grounded and floating form is show in Section 17.5. In active R synthesis, as capacitors are also not used, simulation of capacitors in grounded and floating form is discussed in Section 17.6. All the simulated elements are then employed to describe the immittance simulation and the FDNR approaches in Section 17.7. In the cascade synthesis process, active R building blocks (first-order and second-order filter sections) are connected in a non-interactive cascade form to realize higher-order filters; they are discussed in Sections 17.8–10.

Certain limitations of the OA R/OA C circuits are described at the end of the chapter along with some of the methods to overcome these limitations.

17.3 Direct Form of Active R Synthesis

In network realizations using the direct form of active R synthesis approach, the problem effectively reduces to the simulation of inductors and capacitors (and FDNRs) using OAs and resistors. While designing these elements, it is desirable that the simulators possess the same qualities as required in the OA-RC case, such as: (i) low sensitivities, (ii) stability, (iii) reliable high frequency performance, (iv) use of a smaller number of active and passive components, (v) suitability to IC (integrated circuits) implementation, (vi) high dynamic range and (vii) non-critical component or gain matching. It is rare for a circuit to possess all the aforementioned attractive properties simultaneously. In this chapter, only a few active R grounded and floating component simulators will be discussed; a large number of component simulators are available in literature.

17.4   Inductance Simulation

Grounded Inductance, GI-1: A GI simulator is shown in Figure 17.1(a)[17.3]. Using the approximated first-pole roll-off model of the OA of equation (17.1), input impedance is found as:

$$Z_{in}(s) = k\frac{s\tau_1 + 1}{s\tau_2 + 1} \qquad (17.2)$$

$$k = (a_2/(b_2),\ \tau_1 = (a_1/BR_1\ R_3),\ \text{and}\ \tau_2 = (R_1/(BR_3)) \qquad (17.3a)$$

$$a_1 = R_1(R_2 + R_3 + R_4) + R_2\ (R_3 + R_4),\ a_2 = a_1 + A_o\ R_1\ R_3 \qquad (17.3b)$$

$$b_1 = (R_2 + R_3 + R_4),\ b_2 = b_1 + A_o\ R_3 \qquad (17.3c)$$

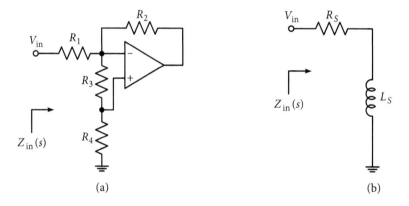

(a)   (b)

Figure 17.1   (a) Grounded inductance simulator G I-I and (b) equivalent circuit of the non-ideal inductor.

Input impedance in equation (17.2) satisfies the condition of a bilinear RL function when $k > 0$ and $\tau_1 > \tau_2 \geq 0$. For $s = j\omega$, we can express $Z_{in}(j\omega)$ as:

$$Z_{in}(j\omega) = R_s + j\omega L_s,\ Q = (\omega L_s/R_s) \qquad (17.4)$$

Series form of the equivalent circuit of the simulated GI is shown in Figure 17.1(b), and the parameters L_s and Q are obtained from equations (17.2) and (17.3), which are frequency dependent.

$$L_s = \frac{R_2 \left(1 + R_4 / R_3\right)}{B \left\{ 1 + \left(1 + \dfrac{R_2}{R_3} + \dfrac{R_4}{R_3}\right)^2 \left(\dfrac{\omega}{B}\right)^2 \right\}} \cong \frac{R_2 \left(1 + R_4 / R_3\right)}{B} \tag{17.5}$$

$$Q = \left(\frac{\omega}{B}\right) \frac{\left(R_2 / R_1\right)\left(1 + R_4 / R_3\right)}{1 + \left(R_2 / R_1\right)\left(1 + R_4 / R_3\right)\left(1 + \dfrac{R_2}{R_3} + \dfrac{R_4}{R_3}\right)\left(\dfrac{\omega}{B}\right)^2} \tag{17.6}$$

It is observed from equation (17.5) that the inductor is realizable even when $R_1 = 0$; however, with the incorporation of R_1, independent tuning of the quality factor Q is possible. While designing the GI simulator, the ratios (R_2/R_3) and (R_4/R_3) are to be pre-selected. The value of L_s is tuned with R_2 and can be further increased by having larger values for the ratio (R_4/R_3). However, the ratio (R_4/R_3) affects Q at high working frequency. Generally, a convenient pre-selection in the design is $(R_4/R_3) = (R_2/R_3) = 1$.

Simulation of Floating Inductance FI-1: FI-1 is realized using GI-1 by un-grounding resistor R_4 in Figure 17.1(a), and connecting it back-to-back with another GI-1, forming a two-port network as shown in Figure 17.2(a). Analysis of the circuit, using an approximate model of the OA, $A(s) \cong \dfrac{B}{s}$, gives the short circuit admittance parameters matrix as:

$$\left[y_{FI-1}\right] = \frac{1}{2k} \frac{s\tau_1 + 1}{s\tau_2 + 1} \begin{bmatrix} 1 & -1 \\ -1 & 1 \end{bmatrix} \tag{17.7}$$

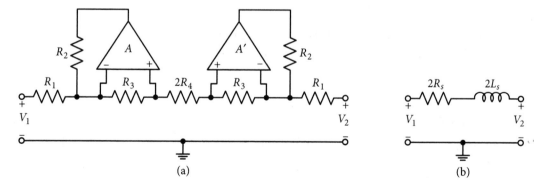

(a)   (b)

Figure 17.2 (a) Simulation of floating inductance FI-1 by ungrounding resistor R_4 of Figure 17.1(a) and connecting it back-to-back; (b) series form equivalent circuit of the non-ideal floating inductor.

where k, τ_1 and τ_2 are given by equation (17.2) and (17.3); equation (17.7) is obtained after simplification of the y parameters. Under the satisfying condition of equation (17.8), parameters of the equivalent floating inductor FI-1 are as given in equation (17.9), and shown in Figure 17.2(b).

$$(R_2 / R_1) \gg 1 \text{ and } \omega^2 \ll (BR_3 / R_4)^2 \tag{17.8}$$

$$L_{FI} = 2L_s \text{ and } Q_{FI} = Q \tag{17.9}$$

Here L_s and Q are given by equations (17.5) and (17.6). The useful frequency range of the FI-1 is given by equation (17.8). For the 741 type of OAs, the circuit gives reliable performance over a range of about 200 kHz; considerably more than the conventional OA-RC case.

It was assumed that back-to-back connected blocks were identical, that is, components were matched. However, in practice, the two blocks may have some mismatch. Simple analysis shows that the circuit still simulates floating inductance provided:

$$y'_{11}y_{22} = y_{11}y'_{22} \tag{17.10}$$

where primed parameters belong to the non-identical block.

Example 17.1: Figure 17.3(a) shows a passive HPF (high pass filter) circuit. Obtain the characteristic of its active version using GI-1. Value of the cut-off frequency needs to be 200 krad/s.

Solution: Transfer function of the passive circuit is derived as:

$$\frac{V_{out}}{V_{in}} = \frac{(1/CR)s^2 + (R_s/L)s}{s^2 + (R_s/L)s + 1/LC} \tag{17.11}$$

From equation (17.11), expression of the parameters of the HPF are:

$$\omega_o^2 = \frac{1}{LC}, \quad Q = \frac{1}{R_s}\sqrt{(L/C)} \tag{17.12}$$

For a selected value of $C = 10$ nF, equation (17.12) gives $L = 2.5$ mH.

From equation 17.5(b), with $B = 2\pi \times 10^6$ rad/s, and a pre-selected value of $R_2 = 10$ kΩ.

$$\frac{R_4}{R_3} = \frac{BL_s}{R_s} - 1 = \frac{4}{7}$$

Hence, if R_4 is selected as 4 kΩ, $R_3 = 7$ kΩ. Resistance R_1 is selected as 0.1 kΩ to keep the series resistance with the inductor small.

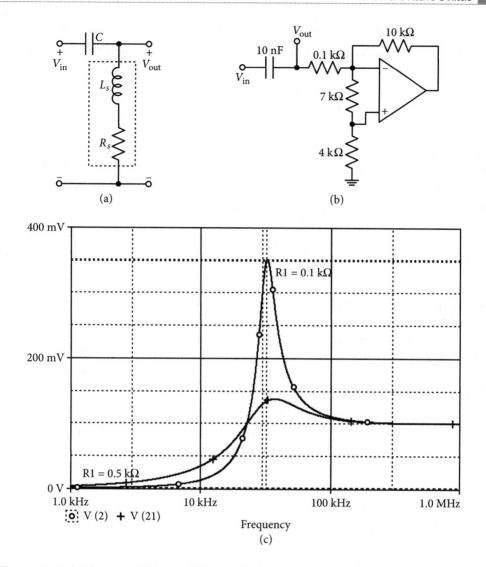

Figure 17.3 (a) A passive high pass filter and (b) its active R version using grounded inductor FI-1.
(c) PSpice simulation of the high pass filter (Example 17.1) using GI-1 with resistance R_1 = 0.1 kΩ and 0.5 kΩ.

Simulated response of the HPF is shown in Figure 17.3(c), in which peak occurs at f_{peak} = 32.219 kHz. Gain at the peak is 3.511, so the realized value of $Q \cong 3.511$, and application of equation (3.46) gives the resonance frequency as:

$$f_o = f_{peak} / \sqrt{\left(1 - \frac{1}{2Q^2}\right)} = 32.219\left\{1 - \frac{1}{2(3.511)^2}\right\}^{-\frac{1}{2}} = 31.56 \text{ kHz}$$

Simulated peak frequency is 198.77 krad/s against the design value of 200 krad/s.

Effective series resistance of the simulated inductor R_s is found using equation (17.12) as:

$$R_s = \frac{1}{3.382}\left(\frac{2.5\times10^{-3}}{10^{-8}}\right)^{1/2} = 147.8\,\Omega$$

To exhibit the control of the quality of the simulated inductance by R_1, it is now selected as 0.5 kΩ. With all other components remaining the same, response of the modified HPF is also shown in Figure 17.3(c). Now f_{peak} is 36.94 kHz, and peak gain is 1.3808 (or $Q \cong 1.3808$).

From equation (17.10), effective series resistance R_s in this case is:

$$R_s = \frac{1}{1.3808}\left(\frac{2.5\times10^{-3}}{10^{-8}}\right)^{1/2} = 362.1\,\Omega$$

Low Component Floating Inductance FI-2: The floating inductor FI-1 has a number of attractive features, such as stability, low sensitivity, and reliable high frequency performance. Moreover, critical component matching is not required. However, the circuit uses seven resistances with sufficiently high resistance spread. Floating inductance FI-2, which we will discuss now, retains most of the desirable qualities of the FI-1. In addition, it uses a minimum number of components; only two OAs and two equal value resistances[17.2] [17.4].

An inductance simulator FI-2 is shown in Figure 17.4(a). For working frequency $\omega \gg \omega_a$, and $A \cong B/s$ for the OAs, analysis gives its short-circuit admittance matrix as:

$$[y] = \begin{bmatrix} \dfrac{s+B}{sR} & -\dfrac{B}{sR} \\[2ex] -\dfrac{B'}{sR'} & \dfrac{s+B'}{sR'} \end{bmatrix} \tag{17.14}$$

For nominal element values, $R = R'$ and $B = B'$, equation (17.14) simplifies to:

$$[y] = \frac{1}{sR}\begin{bmatrix} (s+B) & -B \\ -B & (s+B) \end{bmatrix} \tag{17.15}$$

In the sinusoidal case, with $\omega \ll B$ (say $\leq 0.1B$), deviations in the magnitude of elements in equation (17.15) are less than 1%, and the equation (17.15) modifies to:

$$[y(j\omega)] = \left(\frac{1}{R} + \frac{B}{j\omega R}\right)\begin{bmatrix} 1 & -1 \\ -1 & 1 \end{bmatrix} \tag{17.16}$$

Equation (17.16) represents a lossy FI, and its equivalent in series form is shown in Figure 17.4(c); the parameters are:

$$L_s = \frac{BR}{\left(\omega^2 + B^2\right)}, \quad R_s = \frac{\omega^2 R}{\left(\omega^2 + B^2\right)}, \quad Q = \frac{B}{\omega} \tag{17.17}$$

The circuit is suitable for realizing very high Q values at comparatively lower frequencies. For example, for an OA with $B = 2\pi \times 10^6$ rad/s, $Q > 100$ is realizable at frequencies below 10 kHz. Thus, in the working frequency range of available active RC FIs ($f < 5$ kHz with the same type of OAs), FI-2 realizes very high Q and behaves as a nearly lossless inductor (except at very low frequencies close to ω_a, where the approximated model of OA does not remain valid).

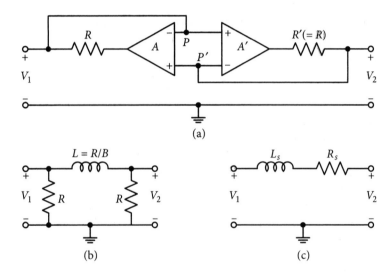

(a)

(b)   (c)

Figure 17.4 (a) Low component floating inductance simulator FI-2, (b) its passive equivalent without mismatch ($R' = R$), and (c) series equivalent of FI-2.

Low Component Grounded Inductance GI-2: FI-2 has the versatility of conversion to GI. Two identical GIs may be obtained from the FI-2 by pulling out the non-inverting terminal P and P' of the OAs in Figure 17.4(a) and connecting to the ground. The resulting GI-2 simulator is shown in Figure 17.5(a) [17.2]. Its analysis gives the RL-driving point admittance:

$$Y_{in}(s) = \frac{1}{R} + \frac{B}{sR} \tag{17.18}$$

Its passive equivalent is shown in Figure 17.5(b).

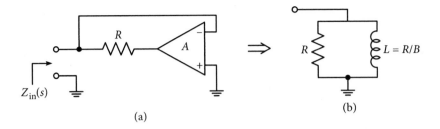

(a)   (b)

Figure 17.5 (a) GI-2 obtained from FI-2 of Figure 17.4(a) and (b) its parallel form passive equivalent.

Example 17.2: Obtain a fifth-order LP Chebyshev filter using FI-2, having a pass band edge frequency of 100 krad/s and ripple width of 0.5 dB.

Solution: Minimum inductance fifth-order ladder for realizing Chebyshev response with 0.5 dB ripple is taken from Table 3.5(b) and shown in Figure 17.6(a). Element values for the normalized ladder are:

$$R_{in} = R_L = 1\ \Omega,\ C_1 = C_5 = 1.7058\ \text{F},\ C_3 = 2.5408\ \text{F},\ L_2 = L_4 = 1.2296\ \text{H} \qquad (17.19)$$

Elements are frequency scaled by a factor of 100 krad/s, and further, impedance scaled by 1 kΩ, which results in the following de-normalized element values:

$$R_{in} = R_L = 1\ \text{k}\Omega,\ C_1 = C_5 = 17.058\ \text{nF},\ C_3 = 25.408\ \text{nF},\ L_2 = L_4 = 12.29\ \text{mH} \qquad (17.20)$$

With de-normalized element values, the fifth-order active R filter realized through using two active R FIs, is shown in Figure 17.6(b). Using equation (17.17), the FIs will employ resistances of 77.36 kΩ.

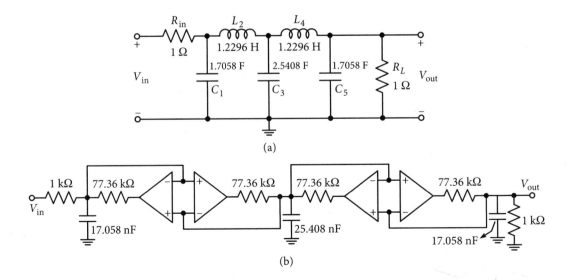

(a)

(b)

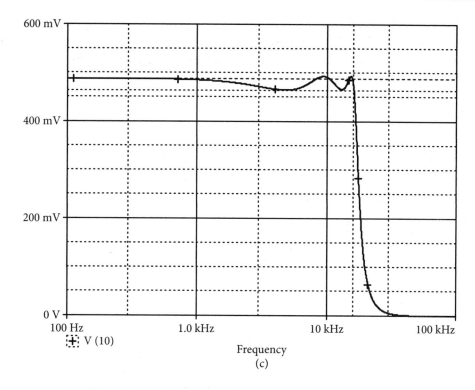

Figure 17.6 (a) Fifth-order Chebyshev filter ladder structure, and (b) its active R realization employing FI-2. (c) Fifth-order Chebyshev filter response employing FI-2 in Example 17.2.

The PSpice simulated response is shown in Figure 17.6(c). The simulated pass band edge frequency is 15.532 (97.63 krad/s) and ripple width is 0.471 dB. Important parameters are very close to the design; however, low frequency gain is 0.487, instead of 0.5, most likely due to parasitic resistance associated with the simulated FI-2, which was shown in Figures 17.4(b)–(c).

17.5 Active R Frequency Dependent Negative Resistance (FDNR)

The frequency dependent negative resistance (FDNR) approach to active RC synthesis, given by Bruton [9.8], is particularly useful when the filter contains a large number of FIs, which get converted to resistors. A number of reliable active RC circuits are available for grounded and floating FDNRs. However, conventional limitations of active RC circuits, mainly the low frequency range of operation, are present. Moreover, the large component count, particularly the use of two external capacitors for grounded FDNR, is not an attractive feature for integration. It has been observed that active R circuits for FDNR simulation can obviate the mentioned drawbacks to a large extent.

Grounded FDNR Simulation: An active R circuit for the simulation of a grounded FDNR is shown in Figure 17.7(a) [17.5]. Straightforward analysis of the circuit, using the approximated model of the OA of equation (17.1), gives the driving point impedance as:

$$Z_{in} = R + (D/s^2) \tag{17.21}$$

where, $R = R_D + R_o$, $D = B_1\,B_2\,R_D$ $\tag{17.22}$

Here, R_o is the output resistance of OA2 and B_1, B_2 are the respective gain bandwidth of OA1 and OA2. Equation (17.21) represents a non-ideal FDNR having a series positive resistance R. For sinusoidal excitation, the frequency dependent impedance of the circuit shall be:

$$Z_{in}(j\omega) = (R_D + R_o) - (B_1\,B_2\,R_D/\omega^2) \tag{17.23}$$

It may be noted that the circuit will exhibit zero impedance at a certain frequency ω_s as:

$$\omega_s = \left\{ B_1 B_2 R_D \,/\, \left(R_D + R_o \right) \right\}^{\frac{1}{2}} \tag{17.24}$$

Similar to the quality factor of inductors, a figure of merit to evaluate the quality of the FDNR is given as:

$$F_D = -\frac{\text{Negatve resistance at frequency } \omega}{\text{Series resistance}} = -\frac{B_1 B_2}{\omega^2}\left(\frac{R_D}{R_D + R_o} \right) \tag{17.25}$$

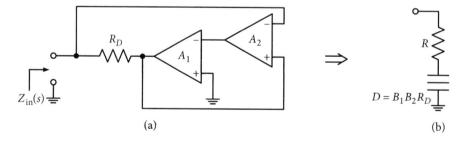

(a)   (b)

Figure 17.7 (a) Active R grounded FDNR simulator and (b) its equivalent representation.

Floating FDNR Simulation: As in the case of active RC, the grounded FDNR circuit of Figure 17.7(a) may be used to form a floating FDNR circuit as shown in Figure 17.8(a) [17.5]. Assuming similar blocks on both sides of the dotted line, analysis of the circuit using $A \cong B/s$, gives the parameters of the floating FDNR, same as that for the grounded FDNR.

$$R = R_D + R_o,\ D = B_1 B_2 R_D \tag{17.25}$$

The circuit is absolutely stable, enjoys low active and passive sensitivities, and is useful for a large frequency range.

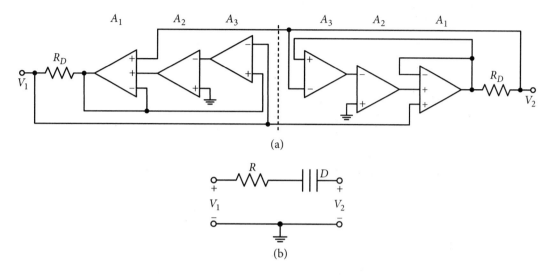

(a)

(b)

Figure 17.8 (a) Active R floating FDNR simulator and (b) its equivalent representation.

Example 17.3: Redesign the filter of Example 17.2 with a pass band edge frequency of 200 krad/s, using grounded FDNRs.

Solution: To realize the filter circuit using grounded FDNR, the passive structure in Figure 17.6(a) is scaled by s. The resulting circuit is shown in Figure 17.9(a), with the normalized circuit elements as:

$$C_{in} = C_L = 1 \text{ F}, R_2 = R_4 = 1.2296 \ \Omega, D_1 = D_5 = 0.5862, D_3 = 0.39357 \qquad (17.25)$$

(a)

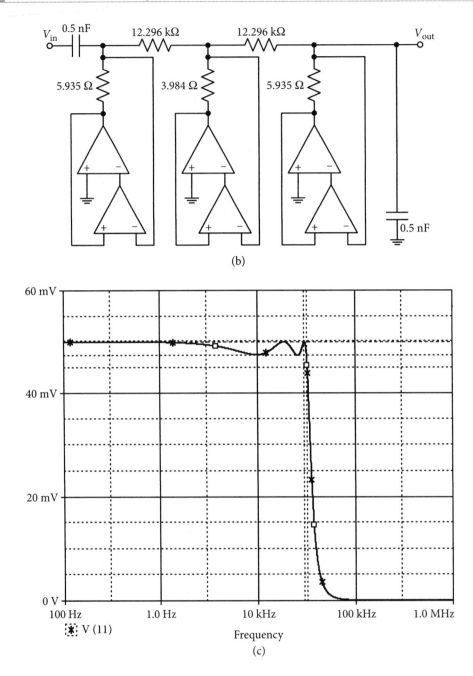

(b)

(c)

Figure 17.9 (a) Fifth-order Chebyshev filter ladder obtained after frequency transformation from Figure 17.6(a). (b) De-normalized version with FDNRs using active R circuit. (c) Simulated magnitude response of the fifth-order Chebyshev filter using grounded FDNRs: example 17.3.

The active R version of the filter is shown in Figure 17.9(b), which employs grounded FDNR of Figure 17.7(a). Its resistances are calculated using equation (17.22):

$$R_{D1,5} = \frac{0.5862}{\left(2\pi \times 10^6\right)^2} \text{ and } R_{D3} = \frac{0.39357}{\left(2\pi \times 10^6\right)^2}$$

Applying impedance scaling of 10^4 and frequency scaling of 2×10^5, element values are

$$C_{in} = C_L = 0.5 \text{ nF}, R_2 = R_4 = 12.296 \text{ k}\Omega, R_{D1,5} = 5.935 \text{ } \Omega, R_{D3} = 3.984 \text{ } \Omega.$$

Simulated magnitude response of the active R filter is shown in Figure 17.9(c). Simulated pass band edge frequency is 31.996 (201.1 krad/s) and the ripple width is 2.523 mV for maximum output of 49.9 mV, which corresponds to 0.45 dB.

17.6 Active R Capacitors

Micro miniaturized capacitors occupy large chip area as compared to active devices. For economic reasons, it is therefore, not only desirable to reduce the number of capacitors, but also the total capacitance on the chip. The concept of eliminating external capacitors, without compromising any of the desirable qualities with them is very attractive. Hence, stable C simulators for grounded and floating capacitors (FCs) will be discussed in this chapter.

Simulation of Grounded Capacitor GC-1: A simple circuit, shown in Figure 17.10(a) is given for the simulation of grounded capacitor (GC) using only one OA and one resistor. With the OA characterized by its first-pole roll-off model, input impedance is obtained as:

$$Z_{in}(s) = R_c \left(1 + \frac{B}{s + \omega_a}\right) \tag{17.26}$$

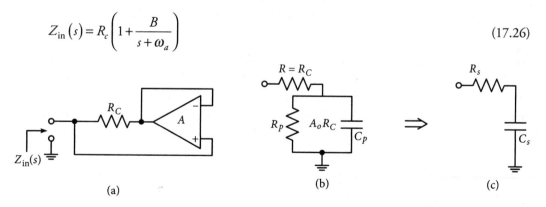

Figure 17.10 (a) Grounded capacitance simulator GC-I, using a single OA; (b) its parallel form equivalent circuit and (c) series form equivalent circuit.

Equation (17.26) represents a lossy capacitor, and its equivalent is shown in Figure 17.10(b). The parameters are:

$$C_p = \frac{1}{BR_c}, \; R_p = A_o R_c \text{ and } R = R_c \tag{17.27}$$

Its series form equivalent impedance as shown in Figure 17.10(c) is given as:

$$Z_{in}(j\omega) = R_s + (1/j\omega C_s) \tag{17.28}$$

where, $C_s = \dfrac{\omega^2 + \omega_a^2}{B\omega^2 R_c} \cong \dfrac{1}{BR_c}$ (17.29)

$$R_s = R_c \frac{\omega^2 + \omega_a^2 + B\omega_a}{\omega^2 + \omega_a^2} \cong R_c \left(1 + \frac{B\omega_a}{\omega^2}\right) \tag{17.30}$$

This gives the quality factor of the simulated capacitor as:

$$Q = \frac{1}{\omega C_s R_s} = \frac{B\omega}{\omega^2 + \omega_a^2 + B\omega_a} \cong \frac{B\omega}{\omega^2 + B\omega_a} \tag{17.31}$$

For $B = 2\pi \times 10^6$ rad/s and $\omega_a = 2\pi \times 10$ rad/s, equation (17.31) shows that the quality factor has a maximum value of nearly 158 at 3.16 kHz and decreases on both sides of the peak value frequency.

It is easily observable that the GC simulator GC-1 uses a minimum number of active and passive components. However, for some applications, where capacitor quality requirements are stringent, it may not be recommended. Hence, a high-quality GC simulator using two OAs will be studied next.

Simulation of High-Quality Grounded Capacitor GC-2: The GC-2 simulator for the simulation of high-Q values is shown in Figure 17.11 [17.6]. It can be analyzed for different values of gain bandwidth products for the two OAs used. However, without any loss of generality, assuming identical OAs ($B_2 = B_1$) and ($\omega_{a1} = \omega_{a2}$), analysis gives input impedance as:

$$Z_{in}(s) = R_c \left\{ 1 + \frac{B^2}{(s + \omega_a)(s + \omega_a + B)} \right\} \tag{17.32}$$

For sinusoidal excitation, equation (17.32) simplifies as:

$$Z_{in}(j\omega) = R_s + 1/j\omega C_s \tag{17.33}$$

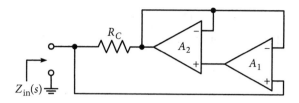

Figure 17.11 High-quality grounded capacitance simulator, GC-2.

Expressions for the equivalent series resistance R_s and simulated capacitance C_s are:

$$R_s = R_c \frac{\omega^4 + B^3 \omega_a}{\omega^2 (\omega^2 + B^2)} \tag{17.34}$$

$$C_s = \frac{(\omega^2 + B^2)}{B^3 R_c} \tag{17.35}$$

Hence, quality of the capacitor is obtained as:

$$Q = \frac{\omega B^3}{\omega^4 + B^3 \omega_a} \tag{17.36}$$

From equation (17.36), finite maximum Q, and the frequency at which it occurs, are:

$$Q_m = 0.25 \left(3B / \omega_a \right)^{\frac{3}{4}} \tag{17.37}$$

$$\omega_m = \left(B^3 \omega_a / 3 \right)^{\frac{1}{4}} \tag{17.38}$$

For nominal parameters of the 741/747 type of OAs, Q_m and ω_m are, respectively 3200 and 43 kHz, which reflects a sufficiently good quality; even better than many practical realizations of capacitors over a large frequency range.

Simulation of Floating Capacitor FC: Similar to the cases of FI or floating FDNR, the circuit for floating capacitor (FC) employs a larger number of elements and requires some form of matching of elements. Figure 17.12 shows a circuit for the simulation of FC [17.2]. Employing the first-pole roll-off model for the OAs, and assuming matched elements and OAs on the two halves of the circuit (primed and un-primed being the same), short circuit admittance parameters are found as:

$$y_{11} = y_{22} = \frac{\left\{ (1 + r_1)(s + \omega_{a2}) + r_1 B_2 \right\} (s + \omega_{a1})}{R_c \left\{ (1 + r_1)(s + \omega_{a1})(s + \omega_{a2}) + r_1 B_2 (s + \omega_{a1}) + B_1 B_2 \right\}} \tag{17.39}$$

$$-y_{12} = -y_{21} = \frac{\left\{(1+r_1)/(1+r_2)\right\}r_2 B_2 \left(s+\omega_{a1}\right)}{R_c\left\{(1+r_1)\left(s+\omega_{a1}\right)\left(s+\omega_{a2}\right)+r_1 B_2 \left(s+\omega_{a1}\right)+B_1 B_2\right\}} \tag{17.40}$$

where, $r_1 = R_2/R_1$ and $r_2 = R_4/R_3$.

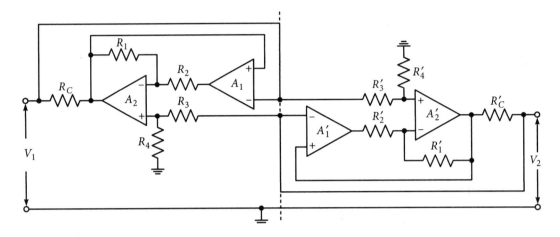

Figure 17.12 Active R floating capacitance simulator for high-frequency operation.

To ensure that the circuit simulates the capacitor, the y parameters must satisfy the following conditions:

i. $y_{11} = -y_{12} = -y_{21} = y_{22}$ (17.41a)

ii. y_{11} yields an RC admittance function (17.41b)

Conditions in equation (17.41) are satisfied (while primed and un-primed elements are the same) when the following relation is satisfied:

$$(1 + r_1)(s + \omega_{a2}) + r_1 B_2 = \{(1 + r_1)/(1 + r_1)\} r_2 B_2 \tag{17.42}$$

If, $r_1 = r_2 = r$, equation (17.42) simplifies as:

$$(1 + r) (s + \omega_{a2}) + rB_2 = rB_2 \tag{17.43}$$

The condition in equation (17.43) is satisfied in the following frequency range, which represents the useful frequency range of the operation of the FC.

$$\omega_a^2 \ll \omega^2 \ll \left\{rB_2 /(1+r)\right\}^2 \tag{17.44}$$

For sinusoidal operation:

$$Z_{11}(s) = \frac{1}{y_{11}(s)}\Bigg|_{s=j\omega} = R_s + \frac{1}{j\omega C_s} \tag{17.45}$$

where R_s and C_s are the effective series resistance and capacitance, which are derived from equation (17.39) while applying the condition of equation (17.43) as:

$$R_s = R_c \frac{(1+r)^2 \omega^4 + B^3 \omega_a r + (r^2 - r - 1)(\omega B)^2}{(1+r)^2 \omega^4 + (\omega^2 + \omega_a^2)(Br)^2} \tag{17.46}$$

$$C_s = \frac{1}{R_c} \frac{(1+r)^2 \omega^4 + (\omega^2 + \omega_a^2)(Br)^2}{rB^3 \omega^2} \tag{17.47}$$

Quality factor of the FC is obtained as:

$$Q = \frac{\omega B^3}{(1+r)^2 \omega^4 + B^3 \omega_a r + (r^2 - r - 1)(\omega B)^2} \tag{17.48}$$

Equation (17.47) can be approximated for the frequency range, $\omega^2 \gg \omega_a^2$ as:

$$C_s \cong r/BR_c = 1/BR_c \text{ for } r = 1 \tag{17.49}$$

17.7 Application of Direct Form Active R Synthesis Technique

Once the basic active R component simulators for inductors, capacitors and FDNRs in grounded as well as in floating modes are available, OA-R filters can be realized in element substitution form.

While applying the technique, selection of proper circuit simulators for a particular application is important as a large number of such circuits are available in the literature. In general, there are three aspects which need consideration while selecting component simulators: (i) complexity and component count in the final realization, (ii) sensitivity of parameters and (iii) stability.

Based on the circuit simulators discussed in the chapter so far, it can be said that all grounded elements, inductor, capacitor and FDNR, and FI use low active and passive component count, and do not require critical component matching. The FC and FDNR not only require a large number of components but also require semi-critical component matching. Therefore, it may be inferred that the active R immittance simulation technique is attractive when the original passive RLC network (generally ladders) either do not have FCs, or their number is small. It can be noted that, in contrast to active RC synthesis, the realization of FIs is not a problem in the OA-R case. Similarly, the FDNR approach is preferred when the original passive RLC network has a minimum of FCs and resistors, unless some better simulators for FC and floating FDNR are available.

Example 17.4: Obtain a fully OA-R filter circuit and its response for the specifications given in Example 17.2.

Solution: In Example 17.2, FIs were replaced by an OA-R simulator, but the resulting circuit was still OA-RC, though it could be used at high frequencies. The same passive ladder was realized in Example 17.3 using the FDNR technique, which is still an OA-RC one. If we want a complete OA-R circuit, both of these circuits could be used. However, in the circuit of Figure 17.9(a), we need to simulate one GC and one FC; whereas for the circuit of Figure 17.6(a), we need to simulate three GCs, in addition to two FIs. Opting for the second choice, the high-quality GC-2 of Figure 17.11(a) is employed, for which resistances are obtained using equation (17.35) with $B = 2\pi \times 10^6$ rad/s as:

$$R_{c1,5} = \frac{1 + \left(2 \times 10^5 / 2\pi \times 10^6\right)^2}{2\pi \times 10^6 \times 8.529 \times 10^{-9}} = 18.67\,\Omega$$

$$R_{c3} = \frac{1 + \left(2 \times 10^5 / 2\pi \times 10^6\right)^2}{2\pi \times 10^6 \times 12.709 \times 10^{-9}} = 12.53\,\Omega$$

For the two FIs having values of 6.148 mH, the required resistances will be:

$$R_{L21} = R_{L22} = R_{L41} = R_{L42} = 38.68\ \text{k}\Omega$$

The final OA-R circuit is shown in Figure 17.13(a) and the simulated response is shown in Figure 17.13(b). Pass band edge frequency is 31.88 kHz (200.4 krad/s). Its dc gain is 0.495, maximum gain is 0.501 and minimum gain in the pass band is 0.466, which corresponds to a ripple width of 0.624 dB. The obtained response is close to the design; small difference in gain value near cut-off frequencies is likely due to the non-idealness of the simulated FIs and GCs.

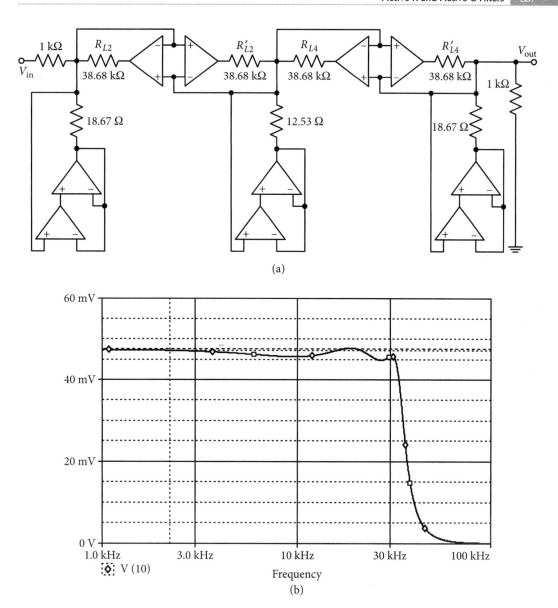

Figure 17.13 (a) Active R structure for fifth-order Chebyshev filter using immittance simulation for Example 17.4. (b) Simulated magnitude response.

17.8   Cascade Form Active R Synthesis

As mentioned in Chapter 10 and other chapters as well, in cascade form of synthesis, the nth order transfer function is realized as a non-interacting cascade of second-order section for even n, and a first/ third-order section is also included for odd n. Hence, it is necessary to develop and study the realization of multifunctional second-order sections, as well as first-order active R sections.

First-order Active R Circuits: A bilinear circuit for realizing active R first-order response is given in Figure 17.14(a) and the LP and HP responses are obtained from it by simply opening and/or short circuiting some resistances [17.7]. Analysis of the circuit using the approximated model of equation (17.1) gives the transfer function:

$$\frac{V_{out}}{V_{in}} = \frac{R_3\left(R_4 + R_5\right)s + B\left(R_1 R_5 - R_2 R_4\right)}{\left(R_1 + R_2 + R_3\right)\left(R_4 + R_5\right)s + BR_1\left(R_4 + R_5\right)} \tag{17.51}$$

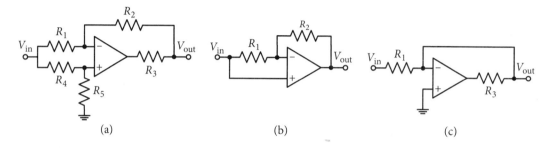

Figure 17.14 (a) A general first-order active R section; (b) low pass section obtained from (a) and (c) high pass section obtained from (a).

The circuit gives LP response if $R_3 = R_4 = 0$ and $R_5 = \infty$, as shown in Figure 17.14(b). Its transfer function from equation (17.51) is:

$$\frac{V_{out}}{V_{in}} = \frac{BR_1 / (R_1 + R_2)}{s + BR_1 / (R_1 + R_2)} \tag{17.52}$$

HP response becomes available when $R_2 = R_5 = 0$ and $R_4 = \infty$, as shown in Figure 17.14(c). Its transfer function from equation (17.51) is:

$$\frac{V_{out}}{V_{in}} = \frac{R_3}{R_1 + R_3}\left\{\frac{s}{s + R_3 / (R_1 + R_3)}\right\} \tag{17.53}$$

17.9 Biquadratic Section Using Two OAs

A number of multifunctional biquadratic active R filtering sections have been made available using two or three OAs [17.8][17.9]. We will first discuss an approach for the realization of a general active R biquadratic filter; this approach uses only two OAs and resistances. By a proper adjustment of resistances, the filter is seen to realize important second-order responses. By using dual OAs on the same chip like 747, the inherently matched characteristics are utilized to advantage.

The basic scheme is shown in Figure 17.15 [17.9]. The voltages at the inverting terminals of the OAs are expressed as a linear combination of terminal voltages V_1, V_2 and V_3 as:

$$V_4 = a_1 V_1 + a_2 V_2 + a_3 V_3 \tag{17.54a}$$

$$V_5 = b_1 V_1 + b_2 V_2 + b_3 V_3 \tag{17.54b}$$

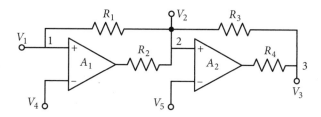

Figure 17.15 Block schematic for active R biquad realization.

The real and positive coefficients a_i' and b_i' are obtained through resistance ratios. Complete filter realization, using the scheme of Figure 17.15, is shown in Figure 17.16, where the coefficients a_i' and b_i', $i = 1$ to 3 are given by:

$$a_1 = \frac{R_6 R_7 R_8}{(\Delta R)}, a_2 = \frac{R_5 R_7 R_8}{(\Delta R)}, a_3 = \frac{R_5 R_6 R_8}{(\Delta R)} \tag{17.55}$$

With $\Delta R = R_6 R_7 R_8 + R_5 R_6 R_7 + R_5 R_7 R_8 + R_5 R_6 R_8$ (17.56)

$$b_1 = \frac{R_{10} R_{11} R_{12}}{(\Delta R)'}, b_2 = \frac{R_9 R_{11} R_{12}}{(\Delta R)'}, b_3 = \frac{R_9 R_{10} R_{12}}{(\Delta R)'} \tag{17.57}$$

With $(\Delta R)' = R_9 R_{10} R_{11} + R_{10} R_{11} R_{12} + R_9 R_{10} R_{12} + R_9 R_{11} R_{12}$ (17.58)

A little manipulation of equations (17.55) and (17.57) gives the resistance ratios:

$$\frac{R_5}{R_6} = \frac{a_2}{a_1}, \frac{R_5}{R_7} = \frac{a_3}{a_1}, \frac{R_6}{R_7} = \frac{a_3}{a_2} \tag{17.59}$$

$$\frac{R_9}{R_{10}} = \frac{b_2}{b_1}, \frac{R_9}{R_{11}} = \frac{b_3}{b_1}, \frac{R_{10}}{R_{11}} = \frac{b_3}{b_2} \tag{17.60}$$

Since the coefficients a_i' and b_i' are obtained through passive circuits, the following condition has to be satisfied.

$$a_1 + a_2 + a_3 \leq 1 \tag{17.61a}$$

$$b_1 + b_2 + b_3 \leq 1 \tag{17.61b}$$

With OAs represented by approximated first-pole roll-off model, analysis of Figure 17.16 gives the following relations for pole frequency and pole-Q as:

$$\omega_o = B \left[\frac{\{a_2 b_3 + a_3 (1 - b_2)\} G_2 G_4}{(G_1 + G_2)(G_3 + G_4) + G_3 G_4} \right]^{\frac{1}{2}} \tag{17.62}$$

$$Q = \frac{\left[\{a_2 b_3 + a_3 (1 - b_2)\} G_2 G_4 \{(G_1 + G_2)(G_3 + G_4) + G_3 G_4\} \right]^{\frac{1}{2}}}{b_3 G_4 (G_1 + G_2 + G_3) + a_2 G_2 (G_3 + G_4) + a_3 G_2 G_3 - (1 - b_2) G_3 G_4} \tag{17.63}$$

A large number of filter realizations are possible from Figure 17.16 by taking output at terminal 2 or 3. Out of this large number of possibilities, two cases are discussed here; the other cases are left as exercises.

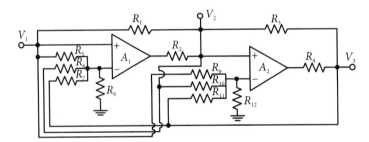

Figure 17.16 General active R biquad from Figure 17.15 with multiple feedbacks.

Case 1: In addition to other responses, BP (band pass) response becomes available when the output is taken at terminal 2. For its realization, based on the choice of feedbacks, the obtained resulting transfer function, filter parameters and design equations are given in Table 17.1.

Table 17.1 BP realization when output is taken at terminal 2 in Figure 17.16

Required conditions	Resulting transfer function	Filter parameters	Design equations
$G_1 = G_3 = 0$ $G_2 = G_4 = G$ $b_1 = b_2 = b_3 = 0$	$\dfrac{B(1 - a_1)s}{s^2 + a_2 Bs + a_3 B^2}$	$\omega_o = B(a_3)^{\frac{1}{2}}$ $Q_o = (a_3)^{\frac{1}{2}} / a_2$ $H_o = (1 - a)/a_2$	$a_1 = 1 - \omega_{on} H_o / Q_o$ $a_2 = \omega_{on} / Q_o$ $a_3 = \omega_{on}^2$

Sensitivity figures of the general biquadratic section can be obtained using the incremental sensitivity measure discussed in Chapter 6. Sensitivities can also be found for specific filters, while assuming gain bandwidth products as B_1 and B_2. Evaluation of sensitivities of ω_o and Q_o for the BP filter gives the following values.

$$S_{G_1,G_2,G_3,G_4}^{\omega_o} = 0,\ S_{a_1,a_2}^{\omega_o} = 0,\ S_{a_3}^{\omega_o} = \frac{1}{2},\ S_{b_1,b_2,b_3}^{\omega_o} = 0,\ S_{B_1,B_2}^{\omega_o} = \frac{1}{2}$$

$$S_{G_1,G_2,G_3,G_4}^{Q_o} = 0,\ S_{a_1}^{Q_o} = 0,\ S_{a_2}^{\omega_o} = -1,\ S_{a_3}^{\omega_o} = \frac{1}{2},\ S_{b_1,b_2,b_3}^{Q_o} = 0,\ S_{B_1,B_2}^{\omega_o} = -\frac{1}{2}$$

Example 17.5: Using OAs with $B = 2\pi \times 10^6$ rad/s, design a BP filter with center frequency of 100 kHz, pole-Q = 5 and mid-band gain of 10, when the output is taken at terminal 2 in Figure 17.16.

Solution: For the second-order BPF obtained from Figure 17.16 and shown in Figure 17.17 (a), $\omega_{on} = \omega_o/B = 0.1$. Using the design equation from Table 17.1, we get:

$$a_1 = 0.8,\ a_2 = 0.02,\ a_3 = 0.01,\ \text{and}\ b_1 = b_2 = b_3 = 0$$

From equation (17.59), the resistance ratios are:

$$\frac{R_5}{R_6} = \frac{1}{40},\ \frac{R_5}{R_7} = \frac{1}{80}$$

Selecting R_5 = 3 kΩ, we get R_6 = 120 kΩ and R_7 = 240 kΩ.

From the expression of a_1 in equation (17.55), R_8 = 16.1 kΩ. The choice of $R_2 = R_4$ is arbitrary; hence, it is chosen as 3 kΩ equal to R_5.

Simulated magnitude response of the BP filter is shown in Figure 17.17(b). Center frequency is 100.23 kHz, mid-band gain is 10.122 and with a bandwidth of 19.55 kHz, the obtained value of pole-Q = 5.12; excellent results.

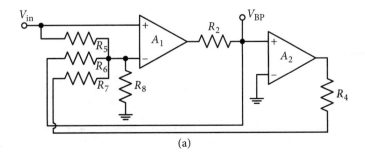

(a)

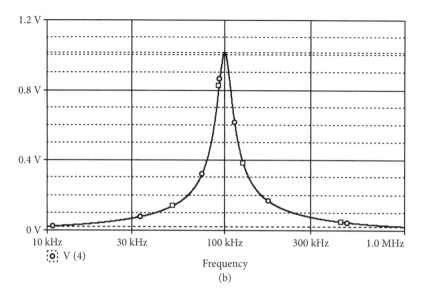

Figure 17.17 (a) Band pass section from Figure 17.16, with output taken at terminal 2, and (b) its simulated magnitude response.

Case 2: When the output is taken at terminal 3, different responses are also available. A BR (band reject) response is available with a proper choice of resistance, which is shown in Table 17.2, along with the filter parameters and design equations.

Table 17.2 Band reject realization when the output is taken at terminal 3 in Figure 17.16

Required condition	Resulting transfer function	Filter parameters	Design Equations
$G_1 = G_2 = G$ $G_3 = G_4 = G$ $a_2 = 0$ $b_1 = \dfrac{(2-a_1-b_2)}{3}$ $b_3 = (1-b_2)/3$	$\dfrac{1}{5}\dfrac{s^2+(1-a_1)(1-b_2)B^2}{s^2+\dfrac{a_3B}{5}s+\dfrac{a_3(1-b_2)B^2}{5}}$	$\omega_o = B\left\{\dfrac{a_3(1-b_2)}{5}\right\}^{1/2}$ $Q_o = \left\{\dfrac{5(1-b_2)}{a_3}\right\}^{1/2}$ $\omega_1 = B\{(1-a_1)(1-b_2)\}^{1/2}$	$a_1 = 1-\dfrac{\omega_{1n}^2}{\omega_{on}Q_o}$ $a_3 = 5\omega_{on}/Q_o$ $b_2 = 1-\omega_{on}Q_o$

Sensitivity figures of the BR filter under consideration were evaluated, which are as follows:

$$S_{G_1}^{\omega_o} = -\frac{1}{5},\ S_{G_2}^{\omega_o} = \frac{1}{10},\ S_{G_3}^{\omega_o} = -\frac{3}{10},\ S_{G_1}^{\omega_o} = -\frac{1}{5},\ S_{G_4}^{\omega_o} = \frac{1}{5},\ S_{a_1,a_2}^{\omega_o} = 0,\ S_{G_3}^{\omega_o} = \frac{1}{2},\ S_{b_1,b_2,b_3}^{\omega_o} = 0,\ S_{B_1,B_2}^{\omega_o} = \frac{1}{2}$$

$$S_{G_1}^{Q_o} = -\frac{1}{5}\left(1-\frac{Q_o^2}{3}\right),\ S_{G_2}^{Q_o} = -\frac{3}{10}\left(1-\frac{Q_o^2}{9}\right),\ S_{G_3}^{Q_o} = -\frac{7}{10}\left(1-\frac{4Q_o^2}{21}\right),\ S_{G_4}^{Q_o} = -\frac{4}{5},$$

$$S_{a_1,a_2}^{Q_o} = 0,\ S_{a_3}^{Q_o} = -\frac{1}{2},\ S_{b_1,b_2}^{Q_o} = 0,\ S_{b_3}^{Q_o} = -\frac{Q_o^2}{15},\ S_{B_1}^{Q_o} = -\frac{1}{2},\ S_{B_2}^{Q_o} = \frac{1}{2}$$

17.10 General Biquad Using Three OAs

Some filter sections using two OAs were discussed in the last section; many more are available in the literature. Though most of the responses were possible, all types of responses are not available. The ones which are available, are with restricted independent control of the response parameters like pole-frequency, pole-Q and gain at the low, high or mid frequencies. Sometimes, design also results in high resistance spread. Moreover, when the response is taken from a terminal other than the output terminal of the OA, its performance becomes load-sensitive for cascading purpose. In such a case, an additional buffer is required, which effectively makes it a three OAs filter. It is therefore, suggested to use the well-known techniques of adding the input signal with two other responses from a two OAs filter, to obtain the general biquadratic function, as shown in Chapter 8.

17.11 Basic Techniques for Active C Synthesis

As the active C network synthesis is also an off-shoot of the active RC synthesis, like the active R case, the basic techniques remain the same.

While employing direct form synthesis in the active R or active RC filters, the starting point is an RLC prototype network; most often the doubly terminated ladder. Same approach is used in the case of direct form of active C filters; the final network would contain only OAs and capacitors.

Discussion on the cascade form of the active C synthesis begins in Section 17.14. Its basic steps are the same, that is, cascading non-interactive second-order sections (Section 17.15) and a first- or third-order section, if needed. Hence, efforts will be made to get active C sections which have low component sensitivities, wide frequency range tunability, and high functional versatility. Low and medium Q filters are realized and examples of OA-C cascade design are shown in Section 17.16. Both active R and active C filters have practical limitations on account of the dependence of their parameters on the gain bandwidth product of the OAs employed. These limitations and some methods to minimize these limitations are briefly discussed in Section 17.17.

17.12 Active OA-C Simulation of Immittance Functions

This section is primarily concerned with the OA-C simulation of important types of immittance functions. The basic object of these simulations is to use them in the active C direct form synthesis for the realization in MOS (metal–oxide semiconductor)-compatible circuits.

Realization of Active C Inductors: Inductors can be realized in the ideal form, as well as in non-ideal form. Generally, ideal form realization requires a greater number of components and need some matching of components but may give better performance when employed in a filter circuit; though real comparison can be done only after their usage. Figure 17.18(a) shows the circuit realization of an ideal GI. Employing OA model of equation (17.1), analysis of the circuit in Figure 17.18(a) gives the driving point impedance as:

$$Z(s) = \frac{1}{sC} \frac{(B_2 - B_3)s^4 - (B_2 - B_3)B_4 B_5 s^2 + B_2 B_3 B_4 s^2}{(B_2 - B_3)s^4 - (B_2 - B_3)B_4 B_5 s^2 - (B_1 - B_4)B_2 B_3 s^2 + B_1 B_2 B_3 B_4 B_5}$$ (17.64)

In equation (17.64), if $B_1 = B_4$ and $B_2 = B_3$, then the expression of impedance simplifies as:

$$Z(s) = s/(CB_1 B_5) = sL_e$$ (17.65)

Expression in equation (17.65) represents an ideal inductance, shown in Figure 17.18(b).

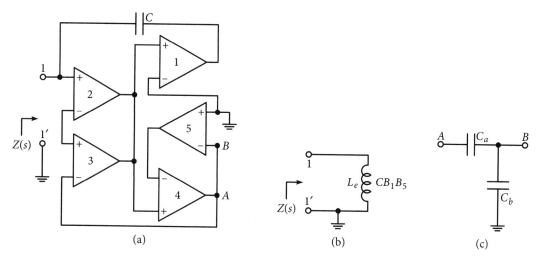

Figure 17.18 (a) Ideal OA-C inductance simulator; (b) simulated ideal inductor when $B_1 = B_4$ and $B_2 = B_3$, and (c) a capacitive attenuator.

Realization of the ideal inductor requires matching of two pairs of matched OAs in Figure 17.18 (a), which is not difficult to achieve when OAs are fabricated on the same chip like the LM 747, or while integrating the entire circuit on a chip and placed close by.

Value of the inductance can be tuned with C and B_5, without affecting the condition of the required matching of $B_1 = B_4$ and $B_2 = B_3$.

Realized value of the inductance can be enhanced by a factor $k = (1 + C_b/C_a)$, by inserting the capacitive attenuator of Figure 17.18(c) between points A and B of the inductor circuit in Figure 17.18(a). The attenuator affects the effective value of B_5 and realizes inductance value as:

$$L'_e = \frac{1}{CB_1(B_5 / k)} = kL_e$$ (17.66)

A large number of non-ideal inductors have been realized [17.10] in which parasitic resistance and capacitance appear in series or parallel form. One such circuit is shown in Figure 17.19(a) for which the following driving point admittance is obtained.

$$Y(s) = B_1C_1 + s(C_1 + C_2) + (B_1B_2C_2 / s) \rightarrow \frac{1}{R_e} + sC_e + \frac{1}{sL_e}$$ (17.68)

Equation (17.68) represents a parallel combination of resistance and capacitance with the realized GI, with the following expressions as shown in Figure 17.19(b):

$$R_e = 1/(B_1\ C_1),\ C_e = (C_1 + C_2)\ \text{and}\ L_e = 1/(B_1B_2C_1)$$ (17.69)

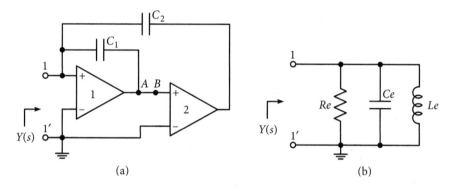

(a)   (b)

Figure 17.19 (a) An RLC simulator and (b) its equivalent representation.

Though the circuit simulates a parallel combination of R, L and C, it employs only two OAs and two capacitors compared to the five OAs and one capacitor needed in the realization of an ideal GI. In addition, the circuit does not require any matching constraint and presents a fairly good flexibility in the design of R, L and C parameters. The sensitivity figures of the circuit are less than or equal to unity in magnitude as shown here.

$$S^{R_e}_{B_2,C_2} = 0,\ S^{R_e}_{B_1,C_1} = -1,\ S^{C_e}_{B_1,B_2} = 0,\ S^{C_e}_{C_1,C_2} = \frac{C_1}{C_1+C_2},\ S^{L_e}_{C_1} = 0,\ S^{L_e}_{B_1,B_2,C_2} = -1$$

Realization of Active C FDNRs: Many circuits have been made available for the realization of ideal and non-ideal FDNR in reference [17.10]. One such non-ideal FDNR simulators is shown in Figure 17.20(a). Its admittance function is found as:

$$Y(s) = sC + 1/\left(\frac{1}{CB_1} + \frac{B_2}{s^2C}\right) = sC_e + 1/\left(R + \frac{1}{s^2D}\right)$$ (17.70)

Equation (17.70) shows that the circuit realizes a non-ideal FDNR, having parasitic resistance and capacitance, which are shown in Figure 17.20(b). Expressions of the realized grounded

FDNR, its parasitic elements and the enhanced value FDNR after introducing capacitive attenuators are:

$$D = C/B_2, \ C_e = C, \ R = 1/CB_1 \text{ and } D' = kD \tag{17.71}$$

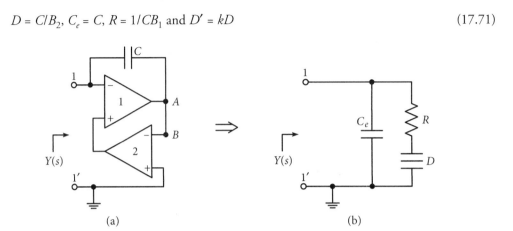

(a)　　　　　　　　　　　(b)

Figure 17.20 (a) Parallel C: (R, D) simulator and (b) its equivalent circuit.

These expressions for C, R and D parameters indicate a convenient and flexible design over a fairly wide range, and the realization of the circuit does not require any matching of bandwidth of OAs. Again, enhancement in the D value and its easy passive and active tuning is possible by connecting a capacitive attenuator of Figure 17.18(c) between terminals A and B in Figure 17.20 (a); incremental sensitivity values are also low, being less than or equal to unity.

Realization of Active C FDNCs: Frequency dependent negative capacitance (FDNC) is an unconventional and relatively less used element [17.11]. The s^2 transformation on a capacitor C, as shown in Figure 17.21(a), yields the following admittance function:

$$Y(s) = s^3 \ F \tag{17.72}$$

In equation (17.72), F is a real positive constant, and for $s = j\omega$, we get:

$$Y(j\omega) = -j\omega^3 \ F \tag{17.73}$$

Hence, the circuit element shows the characteristics of a frequency dependent negative capacitor.

Figure 17.21 Frequency transformation applied to obtain FDNC from a capacitor.

A non-ideal FDNC-based circuit realization is shown in Figure 17.22, along with its equivalent circuit. Its analysis gives the driving point impedance function as:

$$Z(s) = \frac{C_{1-2}}{sC_1C_2} + \cfrac{1}{\cfrac{s^2C_1C_3}{B_1C_4} + \cfrac{s^3C_1C_{3-4}}{B_1B_2C_4}} = \frac{1}{sC_e} + \frac{1}{s^2D + s^3F} \quad (17.74)$$

In equation (17.74), $C_{i-j} = C_i + C_j$ (the symbol will be used later as well), and elements of the equivalent circuit are expressed as:

$$C_e = \frac{C_1C_2}{C_{1-2}}, \; D = \frac{C_1C_3}{B_1C_4} \text{ and } F = \frac{C_1C_{3-4}}{B_1B_2C_4} \quad (17.75)$$

The non-ideal FDNC is economical from the point of element use and does not require any component matching. As expressed in equation (17.75) circuit in Figure 17.22 simulates a series combination of a capacitor C_e with a parallel D: F impedance function. The D and F parameters are controllable with B_1 and B_2.

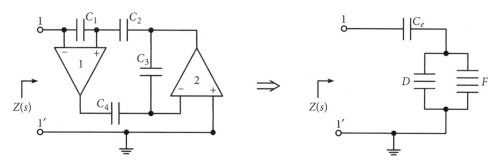

Figure 17.22 Series C: (D||F) simulator and its equivalent representation.

Realization of Active C Resistors: In addition to the requirements of simulating inductance, FDNR and FDNC, sometimes resistances are also to be simulated in terms of OAs and capacitors; two such circuits are discussed in brief.

In addition to the simulation of an ideal resistor, some other useful immittances have been obtained. For example, circuit analysis for Figure 17.23 gives the driving point admittance as:

$$Y(s) = sC\frac{s^2(B_3 - B_2) + sB_2B_3}{s^2(B_3 - B_2) - B_1B_2B_3}$$

$$= -s^2(C/B_1) = -s^2D \text{ with } B_3 = B_2 \quad (17.76)$$

Equation (17.76) implies that under matching conditions, it realizes a frequency dependent positive resistance (FDPR), as for $s = j\omega$, $Y(j\omega) = \omega^2D$.

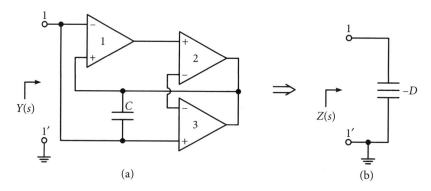

Figure 17.23 Ideal OA-C FDPR and its equivalent circuit when $B_2 = B_3$.

Another useful circuit shown in Figure 17.24 uses only one OA and a capacitor and realizes a parallel combination of a resistance and a capacitor as obtainable from the following driving point admittance function:

$$Y(s) = CB + sC \rightarrow (1/R) + sC_e \qquad (17.77)$$

The circuit needs no matching; hence, sensitivity is always low. The circuit also realizes a parallel combination of same value components, but the resistance becomes negative, if the input terminals of the OA are interchanged.

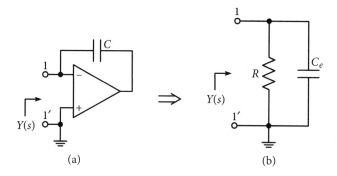

Figure 17.24 Parallel R:C simulator and its equivalent circuit.

Simulation of Floating Components: A floating parallel LC simulator is described in this section using the back-to-back parallel technique [17.2]. The BBB (basic building block) for the realization is shown in Figure 17.25(a). The non-inverting terminal of the OA1 is ungrounded, and the floating form is realized by inter-connecting the two identical BBBs in a back-to-back manner, as shown in Figure 17.25(b). Routine analysis of the circuit yields the admittance matrix:

$$Y(s) = \begin{bmatrix} sC(1+A_1A_2) & -sCA_1A_2 \\ -sC'A_1'A_2' & sC'(1+A_1'A_2') \end{bmatrix} \qquad (17.78)$$

When BBBs are identical with: $C = C'$, $A_1 = A_1'$, $A_2 = A_2'$ and the OAs are represented by the approximate model, equation (17.78) will modify in the frequency range $\omega^2 \ll B_1 B_2$ as:

$$Y(s) = \left(sC_e + \frac{1}{sL_e} \right) \begin{bmatrix} 1 & -1 \\ -1 & 1 \end{bmatrix}$$

(17.79)

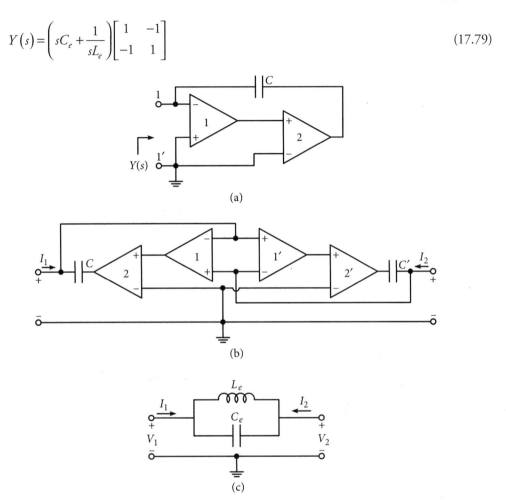

(a)

(b)

(c)

Figure 17.25 (a) Grounded parallel L:C simulator, (b) parallel L:C floating immittance simulator with back-to-back parallel technique and (c) equivalent circuit of the FI simulator.

Equation (17.79) satisfies the condition for the floating components of the parallel LC simulator, viz., $y_{11} = -y_{12} = -y_{21} = y_{22}$ [17.12], which is shown in Figure 17.25(c). Parameters of the realized equivalent circuit are given as:

$$C_e = C \text{ and } L_e = 1/(B_1 B_2 C)$$

(17.80)

The parameters are identical to those for the corresponding grounded simulator from which the circuit has been derived.

17.13 OA-C Filters: Direct Form Synthesis

In the present case, direct form synthesis approach implies realization of first-, second-, and higher-order OA-C filters. The synthesis technique aims to realize the final network in a form suitable for micro miniaturization in MOS technology [17.13].

The following three techniques are proposed [17.14]: Inductance-based approach; FDNR-based approach; and FDNCAP-based approach. As in earlier cases, the direct form synthesis starts with the prototype passive network whose elements are to be simulated so that the final realization contains OAs and capacitors. The choice of a particular technique will largely depend on the configuration of the prototype.

FDNR Based OA-C Filters: The FDNR approach has been used in OA-RC and active R cases and does not need any further explanation. The designer needs simulators for FDNR and resistors. However, there are two choices for the selection of simulators. Circuits simulating ideal FDNRs with ideal resistor simulator can be used. Otherwise, a clever combination of non-ideal FDNR simulators can be used. It has been observed that the latter choice results in a smaller component count and requires lesser component matching. Hence, using a non-ideal FDNR simulator, the synthesis approach is illustrated with the following examples.

A prototype second-order BE (band elimination) filter is shown in Figure 17.26(a) and its transformed FDNR version is shown in Figure 17.26(b) [17.14]. To realize the circuit in OA-C form, the non-ideal FDNR simulator of Figure 17.20(a) is used in the direct simulation of the entire shunt branch (shown within the dotted line) of the transformed circuit. It is to be noted that a capacitive potential divider in the form of C2 and C3 is connected between terminal A and B of Figure 17.20(a). Final realization of the circuit in active C form is shown in Figure 17.26(c). Modeling OA by equation (17.1), gives the transfer function as:

$$H_{BE}(s) = \frac{V_{BE}}{V_{in}} = h_{BE}\frac{s^2 + \omega_n^2}{s^2 + (\omega_o/Q)s + \omega_o^2} = \frac{N(s)}{D(s)} \tag{17.81}$$

where, $h_{BE} = \frac{C_0}{C_{0-1}}$, $\omega_o = \omega_n = \left(B_1B_2\frac{C_2}{C_{2-3}}\right)^{0.5}$ and $Q = \left(1+\frac{C_0}{C_1}\right)\left(\frac{B_2}{B_1}\frac{C_2}{C_{2-3}}\right)^{0.5}$ (17.82)

It is to be noted that all the parameters of the second-order filter are dependent on capacitor ratios. Incidentally, the circuit also provides BP and LP responses at the output of OA1 and OA2, respectively, and transfer functions of these responses are:

$$H_{BP}(s) = \frac{V_{BP}}{V_{in}} = h_{BP}\frac{(\omega_o/Q)s}{s^2 + (\omega_o/Q)s + \omega_o^2} \tag{17.83}$$

$$H_{LP}(s) = \frac{V_{LP}}{V_{in}} = h_{LP}\frac{\omega_o^2}{s^2 + (\omega_o/Q)s + \omega_o^2} \tag{17.84}$$

where, $h_{BP} = -\frac{C_0}{C_1}$, and $h_{LP} = \frac{C_0}{C_{0-1}}$ (17.85)

The circuit possess independent bias-voltage tuning of pole frequency $\omega_o = \omega_n$. Moreover, independent passive tuning of pole-Q is possible with C_o and /or C_1. Sensitivity figures of the filter parameters are found to be less than unity in magnitude.

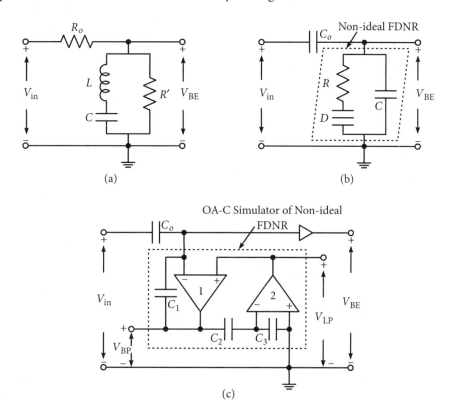

Figure 17.26 (a) Prototype-RLC band elimination filter, (b) its frequency scaled version and (c) OA-C version using non-ideal FDNR simulator.

Example 17.6: Design and test a BE filter using the circuit of Figure 17.26(c), having a notch frequency of 50 kHz and $Q = 5$. Also, find other responses available from the circuit.

Solution: In a practical circuit, values of the bandwidth of the individual OAs have to be found out or estimated according to the fabrication process used. However, in this chapter, a fixed value of $B = 2\pi 10^6$ rad/s is used for the 741 types of OAs. Using equation (17.82), ratio $(C_2/C_{2\text{-}3})$ will be 1/400 that will yield the ratio C_o/C_1 as 99; hence, the selected values of the capacitances are:

$$C_o = 99 \text{ pF}, \ C_1 = 1 \text{ pF}, \ C_2 = 20 \text{ pF and } C_3 = 7.98 \text{ nF} \tag{17.86}$$

To enable the flow of bias current in all OAs, 10 MΩ resistance were to be connected across C_o and C_2 in Figure 17.26(c). Simulated response of the BE and BP is shown in Figure 17.27(a).

Simulated notch frequency, as well as center frequency of the BP is 49.776 kHz. Bandwidth of the BP response is 7.853 kHz, resulting in $Q = 6.337$ and mid-band gain is 125 (theoretical value is 99), whereas the BE filter gain at low frequencies is 0.99 and unity at high frequency. LP response is also shown in Figure 17.27(b) having unity gain at dc and peak gain of 6.28 at 49.57 kHz. Slight tuning in terms of changing C to 1.21 pF from 1 pF, increases bandwidth to 9.918 kHz, resulting in the value of Q as 5.016, mid-band gain becomes 99 for the BPF, and the peak gain of the LP response is the near theoretical value of 5 as shown in Figure 17.27(b).

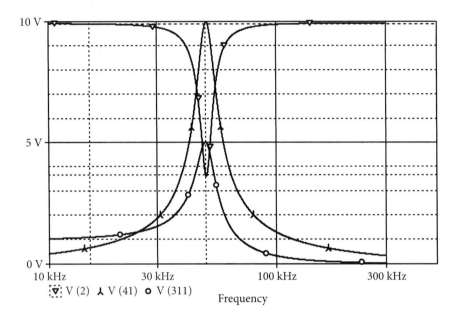

$\mathbf{V}$ V (2) λ V (41) $\mathbf{o}$ V (311)

Frequency

Figure 17.27 (a) Band elimination, band pass and low pass second-order responses of the circuit in Figure 17.26(c).

Inductance Simulation Based OA-C Filters: In the inductance-based approach, resistances and inductances of the prototype filter are simulated in the OA-C form and substituted in it. A passive prototype second-order HP section is shown in Figure 17.28(a). For obtaining its OA-C version, the RLC circuit is directly replaced by a non-ideal GI simulator. The simulator is shown inside the dotted line in Figure 17.28(b), which yields a parallel combination of a resistor, a capacitor and an inductor which is shown in the dotted line in Figure 17.28(a). Analysis of the circuit in Figure 17.28(b) gives the transfer function as:

$$H_{\mathrm{HP}}\left(s\right) = \frac{V_{\mathrm{HP}}}{V_{\mathrm{in}}} = \frac{b_{\mathrm{HP}}s^2}{s^2 + \left(\omega_o / Q\right)s + \omega_o^2} \qquad (17.87)$$

$$\text{where, } b_{\mathrm{HP}} = \frac{C_0}{C_{0-1}}, \ \omega_o^2 = B_1 B_2 \frac{C_2}{C_{0-1-2}} \frac{C_3}{C_{3-4}}, \ Q = \frac{C_2}{C_1}\left(\frac{B_1}{B_2} \frac{C_{0-1-2}}{C_2} \frac{C_3}{C_{3-4}}\right)^{0.5} \qquad (17.88)$$

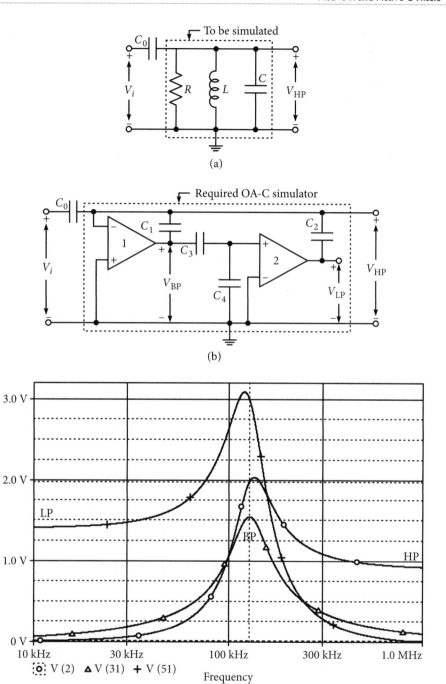

Figure 17.28 (a) Prototype passive high pass filter and (b) its OA-C version employing a non-ideal grounded inductance simulator. (c) High pass along with band pass and low pass response for Example 17.7.

In addition to the LP response, the circuit also gives BP and LP responses as shown in Figure 17.28(b). Expressions for gain of the BP and the LP, the frequency ω_o and Q remaining the same, are:

$$h_{BP} = -(C_0/C_1); \; h_{LP} = -(C_0/C_2) \tag{17.89}$$

It is observed that filter parameters are in terms of capacitor ratios. Sensitivity figures of the filter parameters, with respect to active and passive elements, are equal or less than unity in magnitude. To provide biasing, high valued resistances are required across the capacitors C_2 and C_3.

Example 17.7: Design a second-order HP filter having 3 dB frequency of 125 kHz and $Q = 2$ using inductance simulation approach.

Solution: Use of gain bandwidth value for 741 type OAs, $B = 2\pi (10)^6$ rad/s, equation (17.88) gives the following values for the required capacitors.

$$C_0 = 14 \text{ nF}, \; C_1 = C_2 = 1 \text{ nF}, \; C_3 = 1 \text{ nF}, \text{ and } C_4 = 3 \text{ nF}$$

Simulated responses of the HP along with the BP and LP are shown in Figure 17.28(c) (however, with modified value of $C_1 = 1.45$ nF for adjusting the maximum gain of the HP filter as 2). For the designed 3 dB frequency of 125 kHz, peak gain of 2.02 for the HP response occurs at 134.45 kHz, which corresponds to the simulated 3 dB frequency of 125.76 kHz. Center frequency of the BP is obtained at 127.05 kHz; bandwidth of 59.45 kHz gives $Q = 2.013$. In the case of LP, peak gain of 3.04 occurs at 120.78 kHz, resulting in the 3 dB frequency of 124.28 kHz.

FDNC Based OA-C Filters: In this approach, the passive RLC prototype network is transformed by scaling its each admittance by s^2 as shown in Figure 17.29. The resulting network N' is obtained by converting a resistor to an FDNR, an inductor to a capacitor and a capacitor to an FDNC. The FDNC has impedance of the form $(1/s^3)$ F, where F is the parameter of the element. FDNR and FDNC are then replaced by OA-C simulators. The technique is seen to be very useful in the realization of OA-C versions of those structures which have excessive number of inductances. In this approach as well, realizations using non-ideal FDNC simulators are preferred over ideal simulators. Realization of a second-order LP filter will illustrate the use of FDNC.

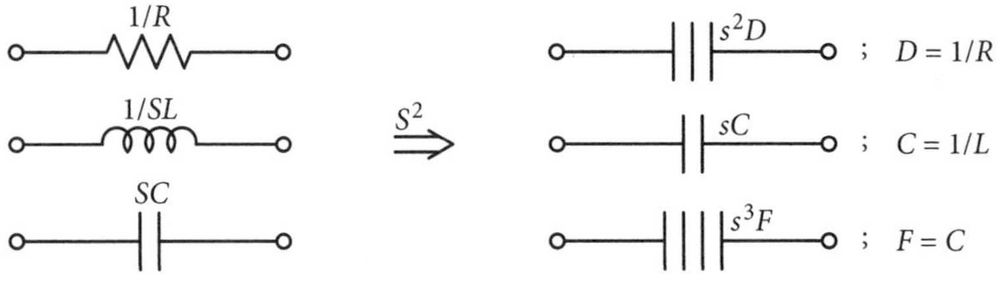

Figure 17.29 Transformation of RLC elements to DCF version by s^2 scaling on each admittance.

Figure 17.30(a) shows a prototype second-order LP filter, and its s^2 scaled version is shown in Figure 17.30(b). It can be observed that the transformed network may be completely simulated by the FDNC based network of Figure 17.22 by taking the responses at the output of OA2 as shown in Figure 17.30(c) where capacitor C is split as parallel combination of capacitors C_1 and C_2. Analysis of the circuit yields the transfer function as:

$$H_{LP}(s) = \frac{V_{LP}}{V_{in}} = \frac{h_{LP}\omega_o^2}{s^2 + (\omega_o/Q)s + \omega_o^2} \qquad (17.90a)$$

where, $h_{LP} = 1$, $\omega_o^2 = B_1 B_2 \dfrac{C_2}{C_{1-2}} \dfrac{C_4}{C_{3-4}}$, and $Q = \dfrac{C_2}{C_1} \dfrac{C_{3-4}}{C3} \left(\dfrac{B_1}{B_2} \dfrac{C_2}{C_{1-2}} \dfrac{C_4}{C_3} \right)^{0.5}$ (17.90b)

Parameters of the filter are in terms of capacitor ratios, and their sensitivity figures are also low. In order to provide dc biasing, high value resistors are required across C_2 and C_3.

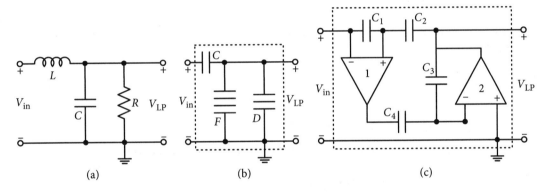

(a)   (b)   (c)

Figure 17.30 (a) Prototype RLC low pass filter, (b) its FDNC transformation, and (c) OA-C version.

17.14 OA-C Filters: Cascade Approach

The basic realization approach remaining the same, it is the second-order section, which forms the basic function to be realized, in addition, to the first-order section. When realized in OA-C form, the performance criteria remain the same which need not be repeated here.

First-order OA-C Sections: Three types of first-order circuits are studied here; LP, HP, and AP [17.15]. Figure 17.31 shows the two first-order LP sections; part (a) realizes the non-inverting section and part (b) realizes the inverting section. The transfer function for the circuit in Figure 17.31(a) will be:

$$H_1(s) = \frac{\alpha B}{s + \alpha B} \qquad (17.91)$$

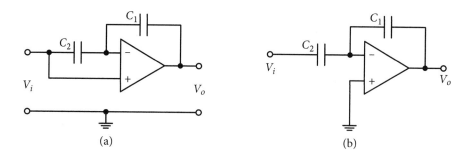

Figure 17.31 (a) Non-inverting and (b) inverting first-order low pass OA-C sections.

In equation (17.91), $\alpha = C_1/C_{1-2}$ and dc gain of the section is unity. For the inverting section shown in Figure 17.31(b), dc gain is (C_1/C_2) and the transfer function is as follows:

$$H_{11}(s) = -\frac{(1-\alpha)B}{s+\alpha B} \tag{17.92}$$

The circuit shown in Figure 17.32(a), realizes an HP section, for which transfer function is obtained as:

$$H_{12}(s) = \frac{(1-\alpha)s}{s+\alpha B} \tag{17.93}$$

Its high frequency gain will be $(1 - \alpha)$.

For the circuit shown in Figure 17.32(b), the obtained transfer function is:

$$H(s) = \frac{1}{1+\alpha_2}\frac{s-(\alpha_1\alpha_2-1)\{B/(1+\alpha_1)\}}{s+\{B/(1+\alpha_1)\}} \tag{17.94}$$

where, $\alpha_1 = C_1/C_2$, and $\alpha_2 = C_3/C_4$. $\tag{17.95}$

For $\alpha_1\alpha_2 - 1 = 1$ or $\alpha_1\alpha_2 = 2$, the circuit realizes the following AP function:

$$H_{AP}(s) = \frac{1}{(1+\alpha_2)}\frac{s-\{B/(1+\alpha_1)\}}{s+\{B/(1+\alpha_1)\}} \tag{17.96}$$

When the circuit is used as a phase shifter, its magnitude and phase shift are given as:

$$|H_{AP}(j\omega)| = \frac{1}{(1+\alpha_2)} \text{ and } \varphi = \pi - 2\tan^{-1}\left\{\frac{(1+\alpha_1)\omega}{B}\right\} \tag{17.97}$$

All the parameters of the first-order sections are in terms of capacitor ratios; sensitivity figures of all the parameters are also small being equal to or less than unity in magnitude. To provide the bias current, a high value resistance is to be connected across capacitor C_1 in HP and LP cases.

Example 17.8: Design a first-order LP and HP filter in OA-C form having a 3 dB frequency of 50 kHz.

Solution: For the LP filter of Figure 17.31(a) for the 741 type OA, with $(\omega_o/B) = (50/1000)$, $\alpha = C_1/C_{1-2} = 1/20$. Hence, for a selected value of $C_1 = 0.1$ nF, $C_2 = 1.9$ nF. Similarly, for the HP filter of Figure 17.32(a), for the same cut-off frequency $C_1 = 0.1$ nF and $C_2 = 1.9$ nF. Simulated responses of the LP and HP are shown in Figure 17.32(c). Cut-off frequency for the LP response is 51.2 kHz and dc gain is unity. For the HP case, cut-off frequency is 50.7 kHz and high frequency gain is 0.973 against 0.95.

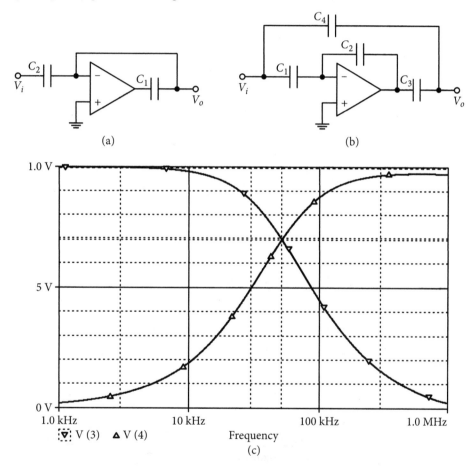

(a)   (b)

Figure 17.32 First-order (a) high pass and (b) an all pass OA-C sections. (c) Responses of the first-order low pass and high pass filters of Figure 17.31 and 17.32(a).

17.15 Second-order OA-C Filters

In this section, two circuits for the realization of low to medium Q filters, and two circuits for medium to high Q filters are discussed [17.16]. Only OAs and capacitor ratios are employed in all the configurations. These circuits simultaneously realize LP and BP responses at the output of the OAs. These may therefore, constitute the non-interactive basic building blocks for the realization of higher-order filters.

Low to Medium Q Filters Two multifunctional OA-C filters for the realization of LP and BP responses are shown in Figure 17.33(a) and (b). The expressions for the voltage transfer functions are given in Table 17.3, where $\alpha_1 = C_1/C_{1-2}$, and $\alpha_2 = C_3/C_{3-4}$.

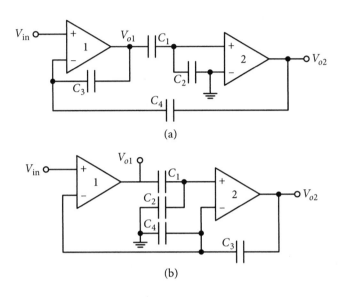

(a)

(b)

Figure 17.33 Low to medium-Q low pass and band pass second-order filters.

Table 17.3 Voltage ratio transfer functions of the filters of Figure 17.33(a) and (b)

Fig. No.	H_{BP}		H_{LP}
17.33(a)	$\dfrac{V_{o1}}{V_{in}} = \dfrac{sB_1}{s^2 + sB_1\alpha_2 + (1-\alpha_2)B_1B_2}$	$= \dfrac{sB_1}{D_1(s)}$	$\dfrac{V_{o2}}{V_{in}} = \dfrac{\alpha_1 B_1 B_2}{D_1(s)}$
17.33(b)	$\dfrac{V_{o1}}{V_{in}} = \dfrac{B_1(s + \alpha_2 B_2)}{s^2 + sB_2\alpha_2 + \alpha_1\alpha_2 B_1 B_2}$	$= \dfrac{B_1(s + \alpha_2 B_2)}{D_2(s)}$	$\dfrac{V_{o2}}{V_{in}} = \dfrac{\alpha_1 B_1 B_2}{D_2(s)}$
	$\cong \dfrac{sB_1}{D_2(s)}$ if $\omega \gg \alpha_2 B_2$		

The circuits are basically suitable for low to medium Q values, for which it is required that by design:

$$C_1 > C_2 \text{ and } C_4 > C_3 \tag{17.98}$$

Parameters can be tuned electronically, but passive tuning is interactive. An attractive feature is the low value of active and passive sensitivities of the filter parameters.

Medium to High Q Filters: Figure 17.34(a) and (b) are two representative OA-C circuits for the simultaneous realization of LP and BP responses. Analysis based on the approximated frequency dependent model of OAs gives the transfer function included in Table 17.4. Parameters of interest for these filters are given in Table 17.5.

Table 17.4 Voltage ratio transfer function of the filters of Figure 17.34(a) and (b)

Fig. No.	$H_{BP}(s) = V_{o1}/V_{in}$	$H_{LP}(s) = V_{o2}/V_{in}$
17.34(a)	$\dfrac{B_1\left(s+B_2C_3/C_{3-4}\right)}{s^2+s(B_2C_3/C_{3-4})+B_1B_2C_1/C_{1-2}} = \dfrac{B_1\left(s+B_2C_3/C_{3-4}\right)}{D_3(s)}$ $= \dfrac{sB_1}{D_3(s)}$ if $\omega \gg B_2C_3/C_{3-4}$	$\dfrac{B_1B_2C_1/C_{1-2}}{D_3(s)}$
17.34(b)	$\dfrac{sB_1C_1/C_{1-2}}{s^2+s(B_1C_3/C_{3-4})+B_1B_2C_2C_5/(C_{1-2}C_{5-6})} = \dfrac{sB_1C_1/C_{1-2}}{D_4(s)}$	$\dfrac{-B_1B_2C_2C_5/(C_{1-2}C_{5-6})}{D_4(s)}$

Table 17.5 Parameters of the filters of Figure 17.34(a) and (b)

Fig. No.	h_{LP}	h_{BP}	ω_o	Q
17.34(a)	1	$\dfrac{B_1}{B_2}\dfrac{C_{3-4}}{C_3}$	$\left(B_1B_2\dfrac{C_1}{C_{1-2}}\right)^{1/2}$	$\left(1+\dfrac{C_4}{C_3}\right)\left(B_1B_2\dfrac{C_1}{C_{1-2}}\right)^{1/2}$
17.34(b)	$-\dfrac{C_1}{C_2}$	$\dfrac{C_1}{C_{1-2}}\dfrac{C_{3-4}}{C_3}$	$\left(B_1B_2\dfrac{C_2}{C_{1-2}}\dfrac{C_5}{C_{5-6}}\right)^{1/2}$	$\left(1+\dfrac{C_4}{C_3}\right)\left(\dfrac{B_2}{B_1}\dfrac{C_2}{C_{1-2}}\dfrac{C_5}{C_{5-6}}\right)^{1/2}$

The filter of Figure 17.34(a) realizes the non-inverting LP response at the output of OA2 and gives mixed BP response at the output of OA1, which becomes near ideal BP response for $\omega \gg B_2C_3/C_{3-4}$. Only four capacitors are used and the circuit parameters are in terms of capacitor ratios only. It has independent passive tunability of Q with C_3 and/or C_4. Circuits are suitable for medium Q filters; in the high Q case, component spread between C_3 and C_4 will become high.

The multifunctional filter of Figure 17.34(b) employs an additional pair of capacitors and provides ideal non-inverting BP and inverting LP characteristics at the output terminals of OA1 and OA2, respectively. Other aspects regarding tunability, component spread, etc., are similar to the previous circuit. Incremental sensitivities of both the filters are small.

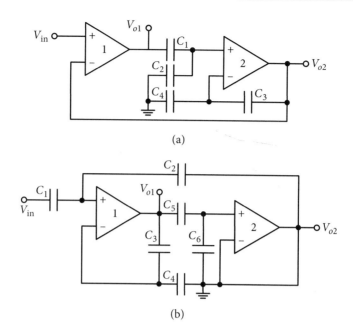

(a)

(b)

Figure 17.34 Medium to high-Q OA-C second-order low pass and band pass filter sections.

It is well-known in literature that a biquadratic building block can be realized from a second-order filter having two standard responses simultaneously. Techniques have been used in previous chapters and need not be repeated here. It can be successfully employed in the OA-C case, using circuits shown in Figures 17.33 and 17.34 or similar circuits.

Example 17.9: Design the second-order filter of Figure 17.34(b) which is to be used to realize BP and LP responses for a center frequency of 62.5 kHz having Q = 2, 5 and 10.

Solution: With B_1 and B_2 taken as $2\pi\,10^6$ rad/s for 741 type OAs, to get center frequency of 62.5 kHz, the ratio of capacitors C_2/C_{1-2} and C_5/C_{5-6} each, are taken as 1/16 from Table 17.5. With these ratios of capacitors, ratio of C_4/C_3 becomes 31, 79 and 159 for Q = 2, 5, and 10, respectively. The selected value of the capacitors is as follows:

$$C_1 = 75 \text{ pF, } C_2 = 5 \text{ pF, } C_3 = 2 \text{ pF, } C_5 = 5 \text{ pF and } C_6 = 75 \text{ pF}$$

Values of C_4 = 62 pF, 158 pF, and 318 pF, respectively, are for increasing values of Q. For completing the path for dc biasing, high value resistances were connected across C_2, C_3 and C_5. It was observed that with these values of C_4, the value of Q was enhanced and trimming of either C_4 or C_3 was required. The values used for C_4 was 61 pF, 116 pF and 167 pF with the other capacitors remaining the same. Simulated responses of the BP and LP responses for the three values of Q, but with same value of center frequency, are shown in Figure 17.35(a) and (b). For the BP case in Figure 17.35(a), simulated center frequency and Q for the three cases are 64.3 kHz, 62.99 kHz and 62.99 kHz, and 2.16, 5.24, and 10.01, respectively. For the LP case of Figure 17.35(b), dc gain for the three cases is 15, and peak gain of 33.15, 77.65 and 149.3 occurs at 60.11 kHz, 62.33 kHz and 62.45 kHz; which corresponds to the cut-off frequency of 63.44 kHz, 62.92 kHz and 62.6 kHz, respectively.

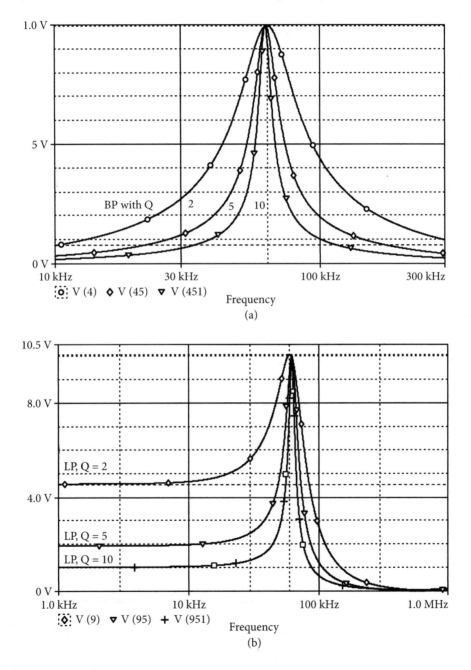

Figure 17.35 (a) Second-order band pass responses for the circuit in Figure 17.34(b) and (b) its low pass responses.

17.16 Cascade Design

Cascade design has been discussed before in detail. In this section, a design example shall be taken to illustrate the technique for the OA-C case [17.16]. A fourth-order LP Butterworth filter is realized by cascading second-order LP sections of the circuit shown in Figure 17.34(a). The cascaded filter is shown in Figure 17.36(a), for which overall transfer function is given as:

$$H(s) = \frac{B_1 B_2 B_3 B_4 \left(\dfrac{C_1}{C_{1-2}} \dfrac{C_5}{C_{5-6}} \right)}{\left(s^2 + sB_2 \dfrac{C_3}{C_{3-4}} + B_1 B_2 \dfrac{C_1}{C_{1-2}} \right)\left(s^2 + sB_2 \dfrac{C_7}{C_{7-8}} + B_3 B_4 \dfrac{C_5}{C_{5-6}} \right)} \tag{17.99}$$

From equation (17.99), the pole frequency and pole-Q of the individual second-order section are, respectively, given as:

$$\omega_{o1} = \left(B_1 B_2 \frac{C_1}{C_{1-2}} \right)^{1/2}, \quad Q_1 = \left(1 + \frac{C_4}{C_3} \right)\left(\frac{B_1}{B_2} \frac{C_1}{C_{1-2}} \right)^{1/2} \tag{17.100}$$

$$\omega_{o2} = \left(B_3 B_4 \frac{C_5}{C_{5-6}} \right)^{1/2}, \quad Q_2 = \left(1 + \frac{C_8}{C_7} \right)\left(\frac{B_3}{B_4} \frac{C_5}{C_{5-6}} \right)^{1/2} \tag{17.101}$$

Then, the two sections are designed to realize fourth-order Butterworth response for cutoff frequency of 25 kHz. For the standard Butterworth denominator polynomial, the corresponding Q-values are available, which are: $Q_1 = 1.31$ and $Q_2 = 0.54$.

For 741 type OAs, $\omega_o/B = 1/40$, it gives, from equations (17.100) and (17.101):

$$(C_1/C_{1-2}) = (C_5/C_{5-6}) = (1/1600) \tag{17.102}$$

Application of relation in equation (17.102) in equation (17.100) and equation (17.101), gives:

$$(C_4/C_3) = 51.4 \text{ and } (C_8/C_7) = 20.6 \tag{17.103}$$

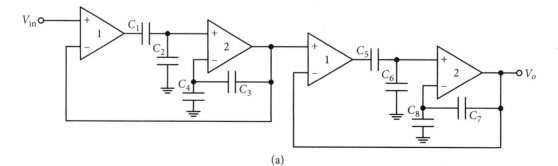

(a)

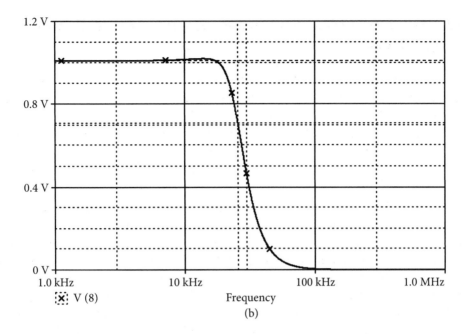

Figure 17.36 (a) Cascaded fourth-order OA-C low pass filter for equation (17.99) and (b) its simulated response.

The following values of capacitors were used for the relations in equations (17.102) and (17.103).

$$C_1 = 100 \text{ pF}, \ C_2 = 159.9 \text{ nF}, \ C_3 = 200 \text{ pF}, \ C_4 = 10.28 \text{ nF} \qquad (17.104a)$$

$$C_5 = 100 \text{ pF}, \ C_6 = 159.9 \text{ nF}, \ C_7 = 200 \text{ pF}, \ C_8 = 4.12 \text{ nF} \qquad (17.104b)$$

The realized filter with capacitor values was simulated, while 10 megΩ resistances were connected across C_1, C_4, C_5 and C_8. Simulated response is shown in Figure 17.36(b), having unity dc gain and cut-off frequency of 25.7 kHz.

Practical Considerations: There are many advantages in utilizing OA-R/ C circuits using the first-pole roll-off model of OAs. Circuit parameters are in terms of capacitor (or resistance) ratios and the frequency range of operation is appreciably extended. However, it is necessary to know and account for the various factors which affect the idealized behavior of the OA-R/ C filters.

In the practical fabrication of MOS-based filters, parasitic capacitors are always present. Their effect has to be reduced considerably through a proper layout of the system. In many designs, effort is made to assimilate parasitic capacitances with physically used capacitances.

It is necessary to provide complete dc path in active C circuits for input bias currents and dc stability. As has been shown in examples, a high value resistance is connected in parallel with the capacitors blocking the dc path.

17.17 Effect of OA's Non-idealness on Active R and Active C Filters

Degradation in the performance of active RC networks using 741 type of OAs, especially beyond audio frequencies, is due to a number of reasons, including the limitations and non-idealness of OAs. The main reason for performance deviation at higher frequencies has now been shown to be significantly reduced in active R and active C circuits. As a result, the useful frequency range has been shown to be increased in examples. However, there are some other imperfections in OAs, which can affect filter performance. Obviously, the amount of non-idealness should be known, so that corrective measures may be taken.

Effect of Temperature and Bias Voltage Drifts: Dimensional considerations show the dependence of active R/C circuit parameters on the gain bandwidth of the OAs used in their realization. For example, in two OA second-order filters, the pole frequency ω_o, and quality factor Q are, respectively, proportional to $(B_1 B_2)^{1/2}$ and (B_1/B_2). Similarly, in the simulation of inductance, FDNR and capacitance or resistance, the realized values and parameters are strongly dependent on bandwidths of the OAs. The bandwidth of a commercially available OA in ICs has large tolerance; it can be somewhere between 20% and 50%. In addition, it is sensitive to variations with respect to temperature and bias voltage drifts. In the temperature range $0°C \leq T \leq 70°C$, the variations in B may be as large as 20%; such changes in B are undesirable [17.17]. It is found that B is inversely proportional to temperature T but varies almost linearly with supply voltage. A combination of large tolerance and value drifts due to temperature and supply voltage variations in B makes active R and active C circuits unsuitable for practical applications unless corrective measures are taken.

Some of the important compensation/corrective methods for reducing the aforementioned problems are: (a) schemes using variations in capacitors or resistors to match the variations in B, (b) schemes using phase-lock ω_o stabilization and (c) schemes using temperature compensated OAs with regulated power supply. A combination of the two schemes can also be used for more stringent requirements.

Rao and Srinivasan [17.18] gave a simple yet effective method for the B-stabilization against temperature and bias voltage drifts. In another effective method for the stabilization of the bandwidth B, basically, the dependence of B on the bias conditions in an OA is used to overcome temperature drifts, either in (i) open-loop, or (ii) closed-loop. The closed-loop systems are superior and derive the actual monitored variation of $B(V, T)$, via ω_o by using phase-lock ω_o stabilization technique. The schemes are known as ω_o stabilization techniques and are given by a number of researchers [17.8].

The phase-lock loop ω_o stabilization techniques are a bit more complicated than the other schemes, and are clearly suited for systems of several filters, where they can be implemented in the inexpensive integrated form. They, however, provide excellent stabilization not only against temperature variations, but also against drifts in bias voltages and aging.

A simple and convenient method of B stabilization is the use of temperature compensated OAs. Such OAs, like LM 324 are commercially available and have about 3.5% variation in B over a range of 0-70° C. For bias voltage drifts, use of regulated supply is recommended.

References

[17.1]   Allen, P. E., and J. A. Means. 1972. 'Inductor Simulation Derived from an Amplifier Roll-Off Characteristic,' *IEEE Transactions* CT-19: 395–7.

[17.2]   Siddiqi, M. A. 1979. 'Network Synthesis Using Internal Dynamics of Operational Amplifiers.' PhD. Thesis, Aligarh Muslim University, India.

[17.3]   Siddiqi, M. A., and M. T. Ahmed. 1978. 'Active R Simulation of Lossy Inductor for High Frequency Applications,' *Proceedings of. IEEE IACAS (USA)* 924–6.

[17.4]   Soliman, A. M. 1978. 'A Novel Inductor Simulation Using the Pole of the Operational Amplifier,' *Frequenz* 1 (32): 239–40.

[17.5]   Siddiqi, M. A., and M. T. Ahmed. 1991. 'Direct Form Active R Synthesis and their Critical Assessment,' *International Journal of Electronics* 71: 621–35.

[17.6]   Siddiqi, M. A., and M. T. Ahmed. 1979. 'Realization of Grounded Capacitor with Operational Amplifier and Resistance,' *Electronic Letters* 14: 633–4.

[17.7]   Siddiqi, M. A., M. T. Ahmed., and I. A. Khan. 1980. 'Realization of a First-order Active R Network,' *23rd Midwest Symposium on Circuit and Systems*, Toledo.

[17.8]   Brand, J., and R. Schaumann. 1978. 'Active R Filters: Review of Theory and Practice,' *IEE Journal of Electrical Circuits and Systems* 2: 89–101.

[17.9]   Ahmed, M. T., and M. A. Siddiqi. 1978. 'Realization of Active R Biquadratic Circuit,' *12th Asilomar Conference of C.S. and Computers*, Montery (USA).

[17.10] Khan, I. A., and M. T. Ahmed. 1983. 'An Active C Resonator and its Applications in Realizing Monolithic Filters and Oscillators,' *Microelectronic Journal of England* 14 (1): 61–66.

[17.11] Ishida, M., T. Fukui, and Ebistuam. 1984. 'Novel Active R Synthesis of Driving Point Impedance,' *International Journal of Electronics* 56 (1): 151–8.

[17.12] Mitra, S. K. 1969. *Analysis and Synthesis of Linear Active Networks*. US: John Wiley.

[17.13] Schaumann, R., M. A. Soderstrand, and K. R. Laker. 1981. *Modern Active Filter Design*. New York: IEEE Press.

[17.14] Khan, I. A. 1987. 'Realization and Study of MOS-compatible Active C High Frequency Filters and Oscillators.' PhD Thesis, Aligarh Muslim University, India.

[17.15] Khan, I. A., and M. T. Ahmed. 1981. 'Realization of MOS Compatible First-order Active C Networks,' *Journal of IETE India* 27 (6): 204–6.

[17.16] Khan, I. A., and M. T. Ahmed. 1986. "Realization of a MOS Compatible Multifunctional Active C Biquadratic Filter for High Frequency Applications". *Microelectronic Journal of England* 17(4): 233–7.

[17.17] Fairchild Semiconductors. 1973. 'Linear Integrated Circuit Catalog.'

[17.18] Rao, K. R., and S. Srinivasan. 1974. 'A High Q Temperature Insensitive Band Pass Filter Using the Operational Amplifier Pole,' *Proceedings of IEEE* 2: 1713–4.

Practice Problems

Note: OAs will be considered non-ideal, represented by its approximate model of equation (17.1), $(A \cong B/s)$ with bandwidth $B = 2\pi \times 10^6$ rad/s.

17-1   (a) Figure P17-1 shows a passive BPF circuit. Design and test the filter circuit simulating the inductance employing GI-1 of Figure 17.1(a). Center frequency of the BPF is to be 5×10^5 rad/s and $Q_o = 5$.

   (b) Explain the effect of any parasitic resistance in parallel with the simulated inductance L.

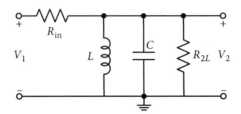

Figure P17-1

17-2   Show that the circuit shown in Figure P17-2 simulates a non-ideal GI. Obtain its equivalent circuit and the expression of the simulated GI and associated parasitic. Assume $R_1 = R_4$ and $(R_2/R_3) = k < 1$.

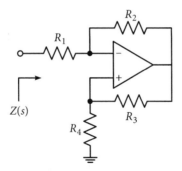

Figure P17-2

17-3   Employ the GI of Problem 17-2 to get a first-order HPF of Figure P17-3, having cut-off frequency of 4×10^5 rad/s.

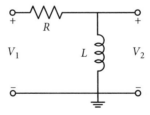

Figure P17-3

17-4   A 5 mH GI is to be simulated using the circuit of GI-1 shown in Figure 17.1(a). Plot its quality factor in the frequency range of $0.01 \leq (\omega/B) \leq 0.5$; assume $R_1 = 0$ and $(R_4/R_3) = (i)$ 1, (ii) 2 and (iii) 5. Explain the effect of these values of the resistance ratio (R_4/R_3) on the variation of the quality factor of the simulated inductance.

17-5   Low component FI-2 of Figure 17.4(a) is modified as shown in Figure P17-4. Show that it still realizes inductance L_p in parallel with a parasitic resistance R_p with the following expressions for these parameters:

$$L_p = \frac{kR}{B}, R_p = \frac{k}{1+k}R \text{ and } Q = \frac{B}{\omega(1+k)}$$

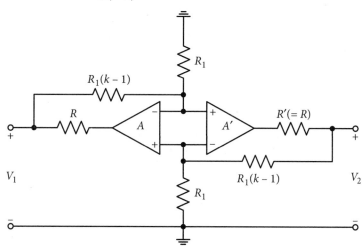

Figure P17-4

17-6   Design and test a sixth-order LP Butterworth filter using FI-2 of Figure 17.4(a) having pass band edge frequency of 100 krad/s and 0 dB gain at dc.

17-7   Design and test a sixth-order LP Chebyshev filter using FI-2 of Figure 17.4(a) having a pass band edge frequency of 120 krad/s and ripple width of 0.1 dB. What is the attenuation at 150 krad/s?

17-8   Redesign the sixth-order LPF for Problem 17-7 using the grounded FDNR of Figure 17.7(a).

17-9   Show that the circuit in Figure P17-5 realizes a second-order notch filter. Design the filter using floating capacitor of Figure 17.12 and floating inductor of Figure 17.4. Frequency of the notch is to be 100 kHz. Assume a suitable value for the load resistance and find the attenuation at the notch frequency.

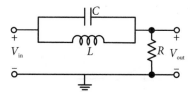

Figure P17-5

17-10 Figure P17-6 shows a second-order passive prototype LPF. Design it having Butterworth characteristics with a cut-off frequency of 10^5 rad/s. Replace the inductor and capacitor to obtain a fully active R version of the circuit and verify it using PSpice. The high-quality GC simulator of Figure 17.11 may be used.

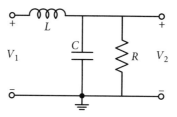

Figure P17-6

Note: Problems 17-11 to 17-13 are based on the realization of second-order filters using the basic scheme shown in Figure 17.15.

17-11 If the output is taken at terminal 2, show that the realized filter is a mixed HP, for which transfer function is given here, provided the following selection is used: elements $G_1 = G_3 = G$, $G_2 = G_4 = (k_1 G_1)$ and the parameters in equations (17.55–17.58) as $b_1 = b_3 = 0$, and $b_2 = (1 - a_3)$.

$$H_{HP}(s) = k_2 \frac{s^2 + k_1(1-a_1)Bs}{s^2 + k_1 k_2 a_2 Bs + \left(k_1^2 a_3^2 B^2 / 1 + 3k_1 + k_1^2\right)}$$

where, $k_2 = \dfrac{1+k_1}{1+3k_1+k_1^2}$

17-12 If the output is taken at terminal 2, show that the realized filter is a BEF with $H_{BE}(s)$ given here, provided the following selection is used: elements $G_1 = G_2 = G_3 = G_4 = G$ and the parameters in equations (17.55–17.58) as $a_2 = 0$, $b_1 = 2(1 - a_1) + b_3$ and $b_3 = (1/3)(1 - b_2)$.

$$H_{BE}(s) = \frac{2}{5} \frac{\left[s^2 + (1-a_1)a_3 B^2 + \left\{(1-b_2)(1+a_3 - a_1)/6\right\}\right]}{s^2 + \dfrac{a_3 B^2}{5}s + \dfrac{a_3(1-b_2)B^2}{5}}$$

17-13 Using OA with $B = 2\pi \times 10^6$ rad/s, design a BR filter with notch frequency of 50 kHz and $Q = 5$, when the output is taken at terminal 3 in Figure 17.15.

17-14 Derive the driving point function for the circuit in Figure P17-7 and find the expression for the elements in the obtained equivalent circuit. Utilize it in the formation of a second-order BP filter with center frequency of 50 kHz and $Q = 2.5$.

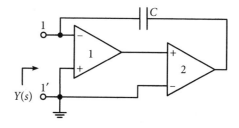

Figure P17-7

17-15 Show that the circuit in Figure P17-8 simulates a parallel combination of a resistance, inductance and a capacitance.

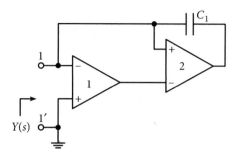

Figure P17-8

17-16 Derive the driving point function for the circuit in Figure P17-9. Use it to design a second-order BP filter for a center frequency of 50 kHz and $Q = 5$. Connect a capacitor attenuator between terminals A and B such that center frequency of the filter enhances to 100 kHz.

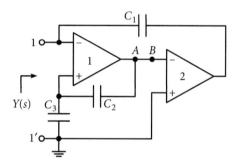

Figure P17-9

17-17 Show that the circuit of Figure P17-10 simulates a grounded FDNR. Find the expression for the FDNR and parasitic elements.

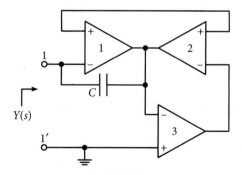

Figure P17-10

17-18 Show that the circuit of Figure P 17-11 simulates a grounded FDNR. Find the expression for the FDNR and parasitic elements.

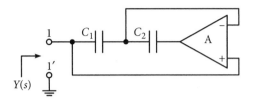

Figure P17-11

17-19 Derive the input admittance function for the circuit in Figure P17-12 when $B_1 = B_2$.

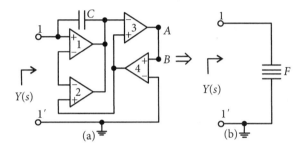

Figure P17-12

17-20 Design and test a second-order BE filter using the circuit shown in Figure 17.26(c) for having notch frequency of 62.5 kHz and $Q = 4$. Also find the maximum value of voltage gain and the frequency at which it occurs for a simultaneously obtained BP response from the circuit.

17-21 Design and test a second-order OA-C APF using the circuit shown in Figure P17.13 having phase shift of $90°$ at a frequency of 100 kHz.

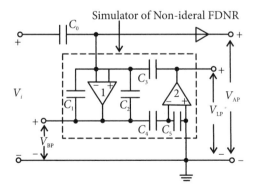

Figure P17-13

17-22 Design and test a second-order Butterworth OA-C HPF using the circuit shown in Figure 17.28(b) having a cut-off frequency of 50 kHz.

17-23 Design and test a second-order LPF using the circuit shown in Figure 17.30(c) having 3 dB frequency of 125 kHz and $Q = 2.5$.

17-24 Design and test a first-order OA-C AP section of Figure 17.32(b), such that its phase becomes $-270°$ at $2\pi \times 100$ krad/s.

17-25 Design and test a third-order Butterworth OA-C filter using cascade approach having cut-off frequency of 100 krad/s.

Index